Powers

of

Perceptual

Control

VOLUME I

AN INQUIRY INTO LANGUAGE, CULTURE, POWER, AND POLITICS

M. Martin Taylor

Living Control Systems Publishing, Menlo Park, CA

Publishers Cataloging in Publication:

Taylor, Maurice Martin, 1935-2023
    Powers of Perceptual Control:
    An Inquiry into Language, Culture, Power, and Politics
    xxxii, 459 p. : ill. ; 28 cm.
    978-1-938090-14-1 (paperback)
    1. Perceptual Contrrol Theory. 2 Control theory. 3. Feedback systems.
    4. Culture. 5. Language. 6. Politics. 7. Economics. I. Title.

This is the first volume of a four part series:

Volume I: Introduction to Perceptual Control Theory
        ISBN 978-1-938090-14-1 (paperback)

Volume II   Creativity, Consciousness, Communication
        ISBN978-1-938090-15-8 (paperback)

Volume III: Collectives, Truth, Trade, and an Invisible Rabbit
        ISBN978-1-938090-16-5 (paperback)

Volume IV: Stabilities and Instabilities, Truth and Lies
        ISBN978-1-938090-17-2 (paperback)

This book

Powers of Perceptual Control
An Inquiry into Language, Culture, Power, and Politics
Volume I: *Introduction to Perceptual Control Theory*

can be downloaded from www.livingcontrolsystems.com, and www.archive.org/details/perceptual-control.

File name: PPC_I_Taylor2024.pdf

Before you print, check the modest price from your favorite Internet bookstore.

For more, search *Perceptual Control Theory*

To my long-suffering wife, who put up with my hours at the keyboard and even encouraged me when we might have been doing things together.

We thought we understood everything
but then we got more data
and saw how naïve we were.

— Malcolm Collins

Every sentence I utter must be understood
not as an affirmation, but as a question.

— Niels Bohr

The first principle is that you must not fool yourself.
And you are the easiest person to fool.

— Richard Feynman

VOLUME I

INTRODUCTION TO PERCEPTUAL CONTROL THEORY

We build not on bare ground.
What we build below
allows what we can build above.
Can we make the foundations solid?

Table of Contents

Abbreviations and Acronyms
(Used in Volume I: Powers of Perceptual Control)

AC	Artificial Cerebellum
BLC Model	The Bilateral Cooperative Model of Reading (Taylor & Taylor 1983)
B:CP	*Behavior: The Control of Perception*
BI	Behavioural Illusion
CCEV	Collective Corresponding Environmental Variable
CEV	Controlled Environmental Variable; Corresponding Environmental Variable; Complex Environmental Variable
CV	Controlled Variable
CSGnet	Mailing list of the Control Systems Group (CSG)
Df	Degrees of freedom
ECS	Elementary Control System
ECU	Elementary Control Unit
EEG	Electroencephalogram
EMG	Electromyography
GPG	General Protocol Grammar
GVC	Giant Virtual Controller
HaH	Hebbian-anti-Hebbian learning
HPCT	Hierarchical Perceptual Control Theory; Hierarchical PCT
LCS	Living Control Systems
LPT	Layered Protocol Model
MOL	Method of Levels
OOP	Object-Oriented-Programming
PIF	Perceptual Input Function
QoC	Quality of Control
PR	Perceived Reality; Perceptual Reality
RIF	Reference Input Function
RR	Real Reality
RREV	Real Reality Environmental Variable
TCV	Test for the Controlled Variable

Author Biography

My paternal forebears, at least as far back as my great grandfather (a well-loved Professor at the Royal College of Music), have been skilled musicians. Both my grandmothers were skilled visual artists, especially in the art of miniature painting. My father and his father were prominent engineers and inventors. Grandfather was in the Royal Army, where he became the General in command of the Royal Engineers along with being appointed a Knight Commander of the Order of the Bath and awarded for battlefield bravery the Belgian Order of Leopold and the British Distinguished Service Order. In his little spare time, especially between the two World Wars, he composed a lot of music that has received considerable commendation from professional musicians, and some of which has been performed in public concerts.

My father was an important contributor to Anglo-US electronic development during the Second World War. Among his 75 patents were several that as of a couple of decades ago were still official secrets. After the war we were often visited by a pleasant Russian Colonel, who I now suppose must have been a KGB agent with a mission to glean secrets from my father. My father was also an accomplished pianist who performed jazz music, sometimes with a friend and my younger brother forming a trio which played weekly at the Toronto Cricket, Skating and Curling Club of which he had been a founding Director. So I have a lot to live up to in my four-fold family background as an engineer, inventor, painter, and musician.

While I was still in high school in Toronto, my father introduced me into computing, first by taking me to see the computer that was being built in the Electrical Engineering Department of the University of Toronto, and then to watch the University's first commercial computer being brought in many sections through a second floor window into the Physics building.[1] A couple of years later, while I served a summer as an assistant maintenance technician on that machine, I was to learn programming from Cristopher Strachey, one of the early leaders in computing, on that computer (in a code of five-bit teletype letters that symbolised instructions).

Academically, I was educated in a private English primary school, with public high school in Scotland and Canada (Toronto), followed by an Engineering Physics Bachelor's degree (1956) at the University of Toronto. My later degrees in Operations Research and Experimental Psychology demonstrated the validity of my father's assessment that from Engineering Physics one might take a wide variety of paths. In my free time in high school and as an undergraduate Engineering Physics student, I played a lot of cricket and wound up playing for my Province (Ontario) and for Canada.

---

1 "It was originally destined for the Atomic Energy Research Establishment in Britain, but before the computer could be delivered there was a change of government that resulted in the cancellation of any contracts over £100,000. Thus the computer became available to the University of Toronto at a greatly reduced price." <historyofinformation.com/detail.php?id=703>

While I was in my last year of Engineering Physics, my father took part in some kind of Royal Commission on education, on which one of the other Commissioners was a Dr. R. B. Bromiley, the then President of the Canadian Psychological Association. My father asked Dr. Bromiley to read my Bachelor's 'Thesis', which contained a chapter on some applications of the then rather new 'information theory'. On reading my Bachelor's 'Thesis', he said he thought I should become a psychologist, and he wrote on my behalf to four heads of reputable psychology departments that he knew personally.

Dr. Bromiley thought, however, since I had not been really aware of what 'Psychology' and psychological research entailed, I should begin with Operations Research, and his letter to the four psychology heads asked them to refer me to the heads of Operations Research. That request led to three dismissive responses and one useful response, from Dr. W. R. Garner at The Johns Hopkins University, which resulted in my admission to the Operations Research division there.

After getting a M.S.E. in Operations Research in 1958, I transferred to the Psychology Department headed by Dr. Garner, where I took courses to learn what research psychology was about and did a study on anchoring and immediate visual memory that became my thesis. That study was based on a well-known party game in which a whispered sentence is passed from one player around to the next until it returns to the originator with the meaning much changed. The classic English example was an original "Send reinforcements, we are going to advance" which returned as "Send three-and-fourpence [English money at the time], we are going to a dance."

I graduated in 1960 with a Ph.D. (and a psycholinguist fiancée who had financed her own way from Korea to Johns Hopkins), after which I went on to a nearly 50-year research career with the Canadian Department of National Defence, with Dr. Bromiley as my first Division head. I had three previous summers as an intern there before joining full time with a new Ph.D. at $4000/year (which was raised to $5000 just before I joined).

I worked at the Defence and Civil Institute of Environmental Medicine (DCIEM) officially for 35 years and unofficially for another ten or fifteen. For most of that time, I split my time about 50-50 between computer hardware and software development and research into human perception in vision, hearing, and touch, before the two threads came together around 1980 in the form of multi-modal human-computer interaction research. I retired officially in 1995 but continued as a volunteer researcher for another ten or fifteen years.

During that time, I published some hundred refereed papers, chapters, and books (on which my wife was the first author) in ten topic categories, as well as several unrefereed reports on innovative computer hardware and software design. In the hardware area, I was responsible in conjunction with the PDP-9 series Chief Engineer for the specific variant design of the first true (memory-mapped) time-sharing computer built by DEC, the PDP-9T, which was delivered in 1967 and used at DCIEM to run several unrelated experiments at the same time, in which the programmers could (but never did) use individual microprogramming, and with pre-bookable guaranteed response timing for interrupts for subject responses and the like. (These software novelties taking advantage of my hardware novelties were designed and implemented by two young Harvard students who had met as finalists in a US national science competition.) Unfortunately, DEC sold only two PDP-9Ts, the other one to the Harvard Psychology Department, though DEC did use some of its technical innovations in their later rather popular PDP-11 series.

After work in visual and auditory perception research, I worked with a summer-student interested in touch (Susan Lederman, who subsequently became a prominent researcher in that field). One study in the early 1970s was on the reason why active touch gave the impression of feeling an object, whereas the same sensations imposed passively gave the sensation of being touched, without the perception of an object. We used what turned out to be a three-level perceptual control model to account for the effect. Later I used a multi-level control model in the 1980s to produce my Layered Protocol Model (LPT) as a theoretical framework for the design of multimodal (voice, text, and pointing as input media, with visual output) human-machine and then human-human interfaces.

In retrospect, I, and I suspect many others, had every opportunity to discover Perceptual Control Theory (PCT) as a natural outgrowth of these ideas but did not have the genius of Bill Powers to see through the complexities into the underlying simplicity and generality of PCT to all living things.

Around 1990 I discovered that my Layered Protocol Model was actually a special case of PCT, which had been promulgating little heeded since before the time I first went to Johns Hopkins. As I say early in this book, it takes a genius such as Bill Powers to dream up a theory so simple that one says to oneself "Why didn't I think of that", and yet so powerful in its precision where tested by experiment and so wide in its range of application. Since then, I have been in awe of the power of PCT, and as I learned more about it in the process of writing this book, that awe has not diminished. My awe suggested the book title, with its deliberate pun on the name of Powers.

Preface

A theory is the more impressive the greater the simplicity of its premises, the more different kinds of things it relates, and the more extended its area of applicability.

— Einstein

What PCT offers

It takes a certain genius for a person to create something about which other people say to themselves "How obvious. Why didn't I think of that?" Once you understand it, you cannot easily go back to your previous way of seeing the world. Perceptual Control Theory (PCT) is a creation of that kind. However, simple exposure is not by itself sufficient for one to 'see it', as my own experience attests. You have to explore it for yourself, and you probably will not do that unless you have some reason to believe the exercise will be worth the effort. Why did I think it might be worth the effort? Should you?

PCT as explained by W.T. (Bill) Powers appeared in my mental universe as a first ray of sunlight from a rising sun, illuminating many things of which I found I had only a superficial understanding after an academic lifetime of experimental psychology. The simple point is that the core of all observed behaviour is nothing but control of things one perceives. As many have said: *"If you see that something is not the way you want to see it, do something about it."* That is the essence of perceptual control. I had been feeling my way to local theories of many diverse phenomena from touch and hearing to language and economics, thinking I was beginning to understand them. Only the rising of the PCT sun over my mental horizon illuminated the basic relationships that are the foundations of them all.

There is a tendency to think that if you have a hammer, everything looks like a nail, and all problems look like planks that must be nailed together. Is PCT like that in my mind? Do I intrude PCT where it has no right to be? I, of course, cannot be sure that I don't, and of course I don't believe that I do. From my Engineering Physics background I argue that fundamentally PCT is a concept that is almost required by non-equilibrium thermodynamics in a through energy flow. From my background as a researcher in various fields of psychology I see that, even without elaboration such as I offer in this book, PCT accounts for much that had been otherwise puzzling. In Einstein's terms, PCT seems both simple in its premises, relates a very wide range of apparently different things, and has a very extended area of applicability, only a very small sample of which is investigated in this long book. PCT is, in Einstein's word, 'impressive'.[2]

---

2   The epigraph is from Lieb and Yngvason (2000). A mathematical description of the interactions among Einstein's criteria, together with a potential measure of how 'impressive' a theory might be, are presented in my 1972 Working Paper "To Sharpen Ockham's Razor", reproduced as Working Paper 1 at the end of volume IV.

Over the time I have been writing this book, my image of the sun of PCT has continued to rise, and my ability to perceive the scope of the landscape it increasingly illuminates has grown with it. Most writings on PCT largely ignore its relationships with other sciences and the insights they can give into the use and functioning of perceptual control by living things, individually or in groups. The better my understanding of PCT grows, the more 'impressive', in Einstein's word, it seems to be.

My current understanding is that what PCT gains from the so-called 'hard sciences', it returns manyfold in its contributions to the 'soft sciences', some of which I try to suggest in this book. This book is a tour through what I have learned about PCT by writing about it.

The Scope of PCT

Most important insights in science and everyday life are created by combining in new ways ideas that have been current for some time, and PCT is no different. The stroke of genius that became Perceptual Control Theory was the melding of these and other ideas, such as those based in cybernetics (e.g. Wiener, 1948/1961, 1950) into a coherent structure that was in many cases open to precise testing. It has proven valid when so tested on a small scale, and this book argues that it remains valid even on the largest scale of society, international relations.

W.T. Powers titled his seminal book *Behavior: The Control of Perception*. At first, this may seem to be an unjustifiably large claim, but once you have a basic understanding of perceptual control, it seems to become a tautology. According to PCT, every intentional action is for control of some internal variable called a 'perception'. Unintentional actions, such as muscular spasms, interfere with such control. Though it might sometimes be hard for an external observer to tell the difference between intentional and unintentional actions, the actor has no such problem. However, one must distinguish between an unintended action and an unintended (side-effect) consequence of an intended action.

The word 'perception' has a specialised meaning in PCT, and explicitly does not necessarily refer to a perception of which one is conscious. What often remains obscure, however, even to oneself as 'the actor', is what variable, what 'perception', is being controlled by some action. How often does one say to oneself "*Why did I do that?*".

Writing the book is itself an entire individual action, just as is the depression of a key on the keyboard of my computer. A major difference is the duration of the action, years in the case of the book, portions of a second for pressing keys. Book writing is a compound action, which can be decomposed into many other actions, such as doing background research to improve my understanding of many of the concepts I want to deal with. All of these have the same kind of reason why they are being done, the 'why' of the action, which is to advance the book toward completion.

I begin by describing in Volume I the nature of perceptual control (the control of one's own perception, no-one else's) before I turn to Volumes II to IV, in which I address many topics that are in themselves often 'siloed' research specialties with little in common. I am a 'professional expert' in none of these, but careful consideration of the implications of PCT has led me to make suggestions in each, with the hope that some of my suggestions are not already everyday 'truths' or 'idiocies' to the relevant scientific communities and that they may offer novel avenues for research useful in some of them.

Despite an apparently mechanistic foundation more suited to the building of machines, PCT has been used as the theoretical basis for a successful psychotherapeutic approach called 'The Method of Levels' (e.g. Carey 2006, 2008; Mansell, Carey, &Tai 2013; Carey, Mansell, & Tai 2015), for designing human-computer interaction (e.g. Marken 1999; Farrell et al. 1999; Engel and Haakma 1993) and for architectural design (e.g.,Wise 1988). So far as I know, it has not hitherto been applied to behavioural economics, but such a use seems very natural, and is discussed in Volume III of this book.

I consider the 'why' of large-scale social phenomena such as those discussed in archaeology, rather than simply describing the phenomena themselves. For example, I speculate in Volume IV how a change in the Indian Ocean monsoon track some 5,000-6,000 years ago could be responsible for the observed political fact that even now women are much more likely to have important political positions in northern than in southern Europe.

Sometimes PCT investigations suggest that commonly believed 'good things' may not be so good at all. One example is the commonly held belief that national governments should strive to attain and maintain a balanced budget. A PCT analysis (following the lead of an IEEE conference paper by Samuel Bagno (Bagno 1955), suggests that if an economy is not to stagnate and run down, with increasingly tall and narrow islands of extreme wealth among oceans of poverty, governments should instead aim for an average annual deficit of perhaps 2% to 3% of GDP with a similar rate of inflation. A PCT analysis of why these 'islands of wealth' matter for social stability is developed in Chapter IV.7.

The increasing tide of autocratic populism in the developed world may perhaps be largely attributed to the fact that Bagno's analysis has been totally ignored by those economists to whom politicians listen. Economic advisors to government leaders continue to claim, I think falsely, that balanced budgets, or even government surpluses, are targets to be aimed at on average. Both Bagno's and the PCT analysis argue that such economists are dangerously wrong. The very survival of Democracy worldwide may depend on politicians ceasing to listen to them. We might, perhaps, relate the political rise of Donald Trump to the much approved budgetary surpluses produced by his two-decade-earlier predecessor as US President, Bill Clinton.

Although Powers did not use PCT much in the social realm, yet it is a measure of the genius of his insight that his simply described construct of hierarchic perceptual control applies very naturally, with powerful results, in such disparate areas quite outside his own primary domain of interest. For example, Williams (1989, 1990) used PCT to explain the 'Giffen Effect', according to which more of a particular good will be purchased if its price rises (see Wikipedia 'Giffen Good').

Most people never notice the infrastructure of a city until part of it fails. Then "all hell breaks loose" as citizens complain about poor maintenance, the sins of the current municipal government, or whatever is most convenient to complain about. But since most psychological theories and applications are constructed as isolated entities or as descriptions of observations like "if you do this, that often happens", they simply notice the city's phenomena, rather than explaining or predicting it from more fundamental sciences.

PCT is not like that. It explains and predicts, making the extreme claim that just as the interplay among atoms gives rise to all the material structures we observe, so the interplay of control loops in control systems accounts for everything that *all* living things do, even octopi, bacteria, trees, and slime moulds, to bring their biochemical variables into desired states. Perceptual control simply is the link between that inner world and the physical environment. Because perceptual control uses, but does not directly influence, the biomechanical properties of the physical living body, PCT has been called a science of psychology. It may be that, but it is a lot more as well.

PCT and History

An important theory provides both questions and answers outside the topic area within which it was first developed. Perceptual Control Theory is such a theory. Elaborated in many publications over the years by Powers and others, it was (and is) basically a theory of individual psychology, though a few researchers use it in wider domains such as sociology and language. PCT is, however, ultimately based on fundamental principles of classical physics such as the laws of motion and of thermodynamics. The same principles and 'Natural Laws' are the basis of engineering control theory, though PCT applies them slightly differently.

The powers of PCT are not easy to appreciate at first, any more than was the power of the Atomic Theory of Matter, proposed as early as the fourth century B.C.E., by Democritus and the ancient Greek Atomists, or of the steam turbine demonstrated by Hero of Alexandria around two millennia ago, but not used effectively until the 20th century. I discuss such 'before their time' insights as an expected consequence of 'creative autocatalytic networks' of perceptual control, starting in Volume II.

But just as we now understand everything material, in all its complexity, to consist of atoms combined in different ways, so according to PCT all of our individual and social behaviour is the result of perceptions controlled in different ways relative to each other. The same may be true of all living things singly or in groups, as Philip Runkel (2003) suggested by entitling his book on PCT *People as Living Things*. Or it may not be true at all, just as the construction of molecules from atoms may not be true at all. We simply recognise that so far we have no evidence to suggest it is not true, and much to suggest that it is.

When one first comes across PCT as a basic underpinning for all of psychology, having previously been exposed to different schools of psychological thought, it is easy to fall into the trap of thinking that PCT is 'nothing but' something one has heard of or studied before, because many of these schools have incorporated one or two of the ideas or Natural Laws from which PCT is built.

Many precursor approaches with considerable surface similarity to PCT have been described, all the way from Sun Tzu (2,500 or so years ago as applied to land warfare and the politics of opposing regimes) to Aristotle, and through John Dewey to the people who analysed tracking behaviour as control in the 1940s and 1950s, and the followers of Karl Friston's 'Free Energy' or 'Predictive Coding' approach today, which I will argue collaborates with PCT in human functioning, as it probably does in several other species as well.

Many have come very close to developing PCT on their own, as did I. Perhaps the closest might be Kenneth Craik, whose early death in 1945 preceded the publication by his friends and colleagues of many of his writings. Nor was the idea of levels of different kinds and complexities of perception novel, having been proposed as early as the mid-19th century (Donders, 1862), and having found its way even into a children's encyclopaedia given me around the age of nine or ten.

PCT and the Building Codes of Nature

Who is this book written for? I have tried to pitch the book at a level suited for a person who has a serious interest and is prepared to think a little for themselves, if only to judge the likelihood that what I say might actually be false, given the evidence that I claim supports it. I sometimes think that this whole book is about discovering the 'Building Codes' of Nature for the behaviour of living systems, ranging in scale from bacteria to the global ecology of life. So this book is for anyone who might be interested in the level of Nature's 'Building Codes', akin to the legal codes for regulating the structural integrity of physical constructions in which we live, work, and play, and on which we travel. PCT explains how living entities maintain their structural integrity individually or in groups.

Nature does not publish a book of official Building Codes, or even a catalogue of them, whereas human-constructed Building Codes are published and builders are supposed to consult and obey them. Scientific enquiry tries to 'reverse-engineer' Nature's unpublished Building Codes to find out what those codes are from observing living things going about their business, and tickling them in certain ways to see what they do.

If we try to violate Nature's Building Codes, we soon find ourselves in trouble, with even more assurance of that fate than is likely to befall the violator of published Building Codes trying to save money on building a structure that violates them. Many, if not most, of our experiments test the limits of those codes by trying to violate what the experimenter imagines might be one of them.

Published Building Codes describe the constraints of what will survive legal challenges if they are used in constructing a building or its electrical or water circuitry and something fails. Nature allows us to discover her building codes by showing us only what has not yet been found to fail. Her laws are very strict, and her judicial verdicts, up to and including a death sentence, admit no appeals. So the book is for anyone interested in what PCT might have to say about those of Nature's building codes that relate to life forms, at a level from casual personal interest to careful scientific research. Throughout, I try to appeal to both kinds of reader in different ways.

Just as an architect or building contractor may try using innovative materials or techniques not mentioned in published Building Codes, and those techniques may survive challenges on the grounds that they achieve more effectively the objectives of the relevant parts of the codes, so does Nature allow for innovation in the structure of individuals, allowing and integrating into the Building Codes of Life those that work and discarding those that don't. Trees, fish, dinosaurs, bacteria, camels, and we humans are Nature's trial 'buildings', and their interactions in the form of ecologies are her evolving designs. None are perfect, but all that we see are 'OK so far' (though the Code violations we humans make daily that lead to climate change may soon alter that assessment).

The domains of building codes range in scale from the specifications of materials in pipes and wall-paint or the shape and materials of a household electrical socket, to the structural integrity of residential and commercial buildings and the planning and zoning of entire cities. Building Codes involve the reuse in many different situations of the same form and material of smaller components. In the same way, what I call 'motifs' of PCT have emergent properties that allow for their reuse in appropriate larger configurations of perceptual control, just as subroutines are reusable chunks of code that help in writing software.

I hope both casual readers and serious researchers appreciate the roles and importances of these repeating motifs, because once you understand their emergent properties — what they do that produces something different, and why they might be useful — much else often falls into place about life itself.

On top of the infrastructure of the city, its water and electricity conduits, its sewer system, and much else unseen, we see dwelling houses, office towers, banks, shops, and a myriad of different buildings used for different purposes, all apparently independent. They are the superstructure of the city. But they are independent only in the ways they are used by their occupants. All of them depend on the same interconnected infrastructure and the same Building Codes that, if followed, ensure the safety of the building.

In our metaphor, each building or each building type, each branch or twig of the 'Tree of PCT life' is analogous to an 'independent' line of research about the way living things, including people, behave and interact. They seem independent at first, each working more or less well as its independent specialised theory says it will if the research was properly done, but with little or no coherence among the theories represented by the architecture of the different buildings. The uses of the buildings affect how their neighbour buildings may be effectively used, but these effects were little considered when each building was independently designed.

Why This Kind of a Book About PCT?

Why should there be a common underpinning between, say, archaeology and economics, the learning of language by a baby and the ancient, modern, and worldwide century-scale rises and falls of populist autocracies? PCT doesn't answer why there should or should not be such a common underpinning; it just accepts as data that there is and suggests why the confluence of many people controlling their own perceptions leads to it happening.

PCT explains many of the processes and interactions involved, the 'why' of what is. As a result, I see this book as a homage to Bill Powers, building on and elaborating his insights both written and spoken, his generosity, and his unfailingly helpful critiques of my misunderstandings of his ideas. The main title *Powers of Perceptual Control* is a deliberate pun on his name.

The book, however, is far from being a rewrite of what Bill said so much better in so many ways. Rather, by suggesting research directions that others might choose to take, it tries to shorten the timing implied by his belief that it will take a long time, perhaps centuries, to develop PCT to its full potential. By trying to show how so much of what we see people do every day is a direct consequence of their perceptual controlling, I hope to develop a foundation of PCT-literate members of the general public who might thereby be able to avoid political extremism and to counter it when it arises.

To continue the 'tree' metaphor, the book builds on the roots Powers planted and nurtured for so many years, developing in directions he often suggested but seldom explored as well as in some directions of which he did not approve. It speculates and offers tentative implications of PCT in many different domains of ordinarily 'siloed' research. It offers an organic structure with branches that will certainly not be as strong as the rooted trunk provided by Bill Powers. Nevertheless, I hope that its rambling branches may perhaps show some of the power of Bill's vision.

Although most, if not all, of the branches and twigs are amenable to mathematical analysis and might be the stronger for it, I have tried throughout to avoid mathematical explanation as much as I could. Instead, I have tried to concentrate on the conceptual relationships that are often cryptically hidden in published mathematical derivations. I use verbal analysis by choice, despite my largely mathematical engineering physics training, in the hope that the results will be more easily understood by readers who might be more interested in the topics than the analytic details.

The branches of this book-tree necessarily poke their twigs into many areas of research with which I am professionally unfamiliar, from microbiology and the origin of life to sociology, from archaeology to linguistics and communication, from culture to technological and political revolution. In all these domains of application, I fear that the second half of the description *"Jack of all trades, and Master of none"* probably applies better than the first. Nevertheless, these topics and more are where the study of PCT has led me, and may with luck lead the reader who comes to the topic with an open mind, no matter whether she or he is a specialist in one of the subject areas or a person simply interested in science. I hope that readers versed in current theories in these specialties do not find too apt a parallel aphorism *"Fools rush in where angels fear to tread"!*

Why Did I Want to Write This Book (Or Any Book)?

Why have I wanted to advance the book toward completion through these years of writing in spite of my general ignorance of the great body of research into the topics I address, and despite assuming that I will make many mistakes of detail in the process? That question leads to another. Why do I want to write the book at all? As we shall see, the question in PCT is "What perception(s) am I controlling by the actions involved in writing it?" Overtly, I perceive myself to be controlling for making public material that some may find interesting or useful, but is this a sufficient answer? Is it even true? The third epigraph on the title page is an important statement by Richard Feynman: *"The first principle is that you must not fool yourself, And you are the easiest person to fool."*

This last question, about what perceptions I might be controlling in writing the book, is one I found hard to answer when I asked it of myself half-way through the writing. When a book is finished and published, it is of no value to the author, other than perhaps the royalties that it may generate and that it may give pleasure to see one's name printed on its cover or cited by other authors in connection with their own work. Most authors, including me, hope that other people will read, enjoy, and use the finished product. Why? Some other variable must be being controlled, but what? One possibility is my self-image as a capable person. But do I want myself to perceive that to be true, do I want other people to perceive me as capable, both, or neither?

There are two kinds of self-image, one I perceive directly and another that I perceive only through the eyes of others. The two kinds of self-image are not the same thing, as Robert Burns pointed out when he said (roughly translated from the original Lallans by me): [3] "*Would some spirit allow us/to see ourselves as others see us.*" I call these two different self-images the 'self-self-image' and the 'other-self-image'. The self-self-image is the way we *see ourselves* (or want to), while the other-self-image is the way '*others see us*' (or as we would like them to do). In different circumstances, 'other' might be a particular person or group, or an identifiable class of living control systems such as our pets or our neighbourhood birds.

This book should be read with a *highly* critical attitude throughout. One epigraph on the title page quotes the physicist Niels Bohr saying "*Every statement I make should be treated as a question,*" and his comment applies to just about everything in this whole book. As you read it, you should always be thinking about in what ways its statements might be wrong, and what evidence might be sought by direct experiment or from other research domains to falsify any of the claims, hypotheses, and proposals scattered throughout. The further we go through the book, the more of what I state as though it is established fact is really unsupported by experiment or mathematical analysis, but seems (to me) to follow from the better supported infrastructure material earlier in the book. The biologist Malcolm Collins suggests that to do otherwise is naïve ("We thought we understood everything, but then we got more data and see how naïve we were"). [4]

Throughout, I have tried to claim only what seems to follow from what can be supported, but over the thousands of years in the history of science and 'natural philosophy' similar conjectural extrapolations from what seemed firm at the time have many times been proven wildly in error. A very important case that affected the whole history of the 20th century was Einstein's assertion of the equivalence of mass and energy, an equivalence never previously suspected except possibly by widely ignored mavericks.

I hope this is not the case in too many places in this book, but I do warn the reader to guard carefully against the possibility, by thinking for herself or himself about the issues discussed. As I write this branching tree of a book, I go out on many limbs. It is up to the reader to determine how far out on any of these limbs to follow me without a safety harness.

---

3   Lallans is a Scottish dialect of English spoken widely in the Scottish Lowlands before that age of broadcast radio and television. I don't know whether communities of Lallans speakers still exist.

4   Quoted by A. Curry (Curry 2018:626).

The Straits of Magellan [5]

In the early 1500's the 'Spice Islands' (an archipelago near Indonesia and Malaysia) were a part of the world reached from Europe by a long and arduous passage past the Cape of Good Hope and India. Because their spices fetched huge sums of money in Europe, European maritime powers contested to colonise them and monopolise their particular products. Columbus had hoped to reach them by travelling west, but had been blocked by the Americas. Within fifteen years of Columbus's first voyage of discovery, Amerigo Vespucci had shown South America to be a large land mass blocking further westward travel, and Henry VIII of England had sent John Cabot to try to find a way west around North America.

In 1520, less than thirty years after Columbus, Ferdinand Magellan found a passage a long way south along the South American coast, and that passage opened onto a vast ocean, an ocean that was not unknown, but that until then could be reached from Europe only by the eastward route or overland across Mexico or Central America. In early 1521, Magellan reached the Philippines having discovered a couple of Pacific islands on the way, but was killed there. Eventually a few of the original crew arrived home, having completed the first circumnavigation of the Earth.

What the small remaining crew brought home was the news that a way existed to get to the Spice Islands westward by sea. There would be a map of the narrow and difficult strait, and maps of a few islands new to European knowledge, including the Philippines. To European eyes, the Pacific world now contained more than the Spice Islands. The way was open to them and the rest of the broad Pacific from Europe by sea.

But was the way open? Even now, with modern technology and power, it is not easy to use the Strait of Magellan, especially against the prevailing Westerly winds. It took a long time to map the Pacific. Even 200 years after Magellan Baja California was an island, Australia was not known to be an island and it took another 50 years before any European saw its east coast. Magellan's maps would have been far from charting all the details even of the passage that is now named after him.

What has this to do with Powers? I think there are several analogies worth thinking about. Let's think about the Spice Islands, a rich region that grew spices that fetched huge sums of money back in Europe. It was a region much coveted and fought over by European colonial powers. I think of this and the rest of the riches of the Pacific as analogous to Psychology, much fought over by different schools that are all based on the same underlying concept, the 'Eastward' or 'unidirectional' concept, to which we now oppose the "'Westward' or 'negative feedback' concept.

---

5   Taylor posted this analogical essay about progress in science to the CSGnet listserv on 15 September 2016. It can be found in the CSGnet archive on the IAPCT discussion forum discourse.iapct.com with the identifier [Martin Taylor 2016.09.14.15.10].

Just as Magellan opened an entirely new way to approach those islands, Powers opened up a new way to approach Psychology. Just as Magellan mapped the Strait in gross detail, so Powers mapped his entry-way in gross detail. Just as Magellan's maps did not list every rock and shoal in his strait, so Powers acknowledged that there were many uncertainties yet to be explored within the gross structure of his control hierarchy. Just as Magellan found a few islands in the Pacific unknown to European commerce, so Powers found a few aspects of psychology not known to those who approached it from the other direction. And just as the maps Magellan made were guides for later explorers, so the guidance Powers offered to those who would follow his footsteps helped and continues to help later explorers.

There's another parallel, as well, which is that although Magellan's maps showed the way for ships to sail from the Atlantic into the Pacific, the route was never easy for wooden sailing ships; it is not so easy even for modern powered ships with GPS, radar, and other technologies. Likewise, Powers's map of the possibilities of control hierarchies is not easy for others to follow, few researchers having all the necessary skills and understanding. To follow Powers and extend our understanding of how his system actually works requires expertise in experimentation, simulation, mathematics and physiology. Possibly nobody has all those skills, so, just as with modern exploration, most real advances depend on the work of teams or taking advantage of what other disciplines can offer. Even Powers often said that he was often surprised by the way the hierarchy worked. And like Magellan's maps, Powers's maps always remain subject to revision as later explorers learn more about the terrain.

Not every European ship-borne expedition that explored the west coast of America started by using the Straits of Magellan; several Spanish expeditions launched from Mexico or elsewhere along west coast of the Americas. Again we have a parallel, there being other negative feedback theories of psychology such as 'ecological psychology', but as with the coastal explorations starting from west coast harbours, they seem to have an ad-hoc feel to them, bits and pieces having situation specific components, in contrast to the 'all-by-sea' purity of the control hierarchy route pioneered by Powers.

I offer the Magellan analogy as a salute to Powers, not as a man who explored the whole world of Psychology, but as one who through the control hierarchy opened that wide world to coherent exploration from a new direction, a world in which well known phenomena can be seen as belonging to a whole rather than being colonised by specialists in different areas, in the way the fighting colonial powers colonised the different Spice Islands, each island separate and distinct. As with Magellan, the world he opened will probably not be fully explored for a very long time, but all future explorers should acknowledge a debt to W. T. Powers.

Acknowledgements

Finally, I must try to give at least a little credit where much more is due. I cannot overemphasise the importance of Bill Powers in helping me to understand his insights, and his gentle persistence in correcting what he (but not always I) saw as my errors. Many times, when I thought I had made a new insight, I found that Bill had preceded me but had subtly, almost imperceptibly, guided me to discovering it for myself—the best kind of teaching. I'm fairly sure that were he still alive and able to do so, Bill would be quietly heckling me, trying to get me to substantiate or reconsider much of what is in this book. Anyone else I mention here pales in comparison to Powers in their influence on my thought, but they are all still important, as is my total academic heritage.

Among the others, let me mention first someone who probably never heard of PCT or Powers, Wendell R. (Tex) Garner, chairman of the Psychology Department at the Johns Hopkins University, who in 1956 invited me to try graduate school at Hopkins in a discipline apparently far removed from Engineering Physics and from his domain of Psychology in Operations Research. This I did for two years before transferring into Garner's care as a budding experimental psychologist. Apart from his academic influence on me, which remains important in many places in this book, had Tex Garner not invited me, I would not have met my wife of over 60 years, then Kim Insup (or, in North America, Insup Kim).

Among the followers of Powers, I must first mention Rick Marken, who has usually disagreed with my ideas if they were not backed up by experimental data, and whose criticisms always made me think very carefully about what ideas I continued to espouse. Much of what is in Chapters I.4 to I.7 of Volume I of this book has survived the fire of his criticism in my mind, if not in his.

In a quite different vein, Kent McClelland has influenced me strongly as a PCT-oriented sociologist. Many of my ideas are explicitly based on his work, but more, I am sure, have been influenced by his careful criticism and frequent collaboration, not to mention his unfailing encouragement. Over the years of my fitful development of this entire book, Kent has helped me by extensive discussions both on PCT and on sociology. Over many years Bruce Nevin, likewise, has been an important influence on my thinking in many ways, especially about language.

It is very hard to do justice to the many others, most of whom were on the now superseded mailing list CSGnet, but I should be remiss if I did not mention at least Warren Mansell, for his long-time encouragement and the impetus to begin this book, which started as one half of an overlong draft version of a chapter he solicited for The Interdisciplinary Handbook of Perceptual Control Theory: Living Control Systems IV (Elsevier, 2020).

Among the followers of Powers, I thank John Kirkland for years of encouragement and much copy-editing, and Eetu Pikkarainen, Bruce Nevin, and Eva de Hulu for their unstintingly provided expertise in their own fields and for their questions and suggestions that have made me think and often rethink. I also thank too many others to name, who have, from time to time, engaged in discussions on various mailing lists and fora on topics that in my mind involved PCT.

One other author to whom I should give honorary credit, someone who knew nothing about PCT because he died 400 years ago, but who my mentor Wendell Garner said knew more psychology than did any academic psychologist, was William Shakespear. I use his writings in several places in the four volumes of this book. Indeed, I devote a whole section at the end of Volume I to his sonnet: "Shall I compare thee to a summer's day".

Finally I should mention my wife and occasional co-author of 60 years, Insup (Ina) Taylor, a prominent psycholinguist author of several books that concentrate on literacy questions, especially among the unrelated East Asian languages of Korean (her native language), Japanese (which she was forced to use in school), and Chinese, and between those languages and ones that use alphabetic scripts, such as English. Writing and reading in these different scripts and languages have much in common, but also much that differs. I have benefited greatly by learning from her, as well as for her emotional support.

Over the years Ina gave me co-authorship on several psycholinguistics books that were mainly written by her. She urged me to write this book on a topic outside her field of interest, and supported me well in the years it took me to write it.

Martin Taylor

Martin Taylor (14 June 1935 - 17 March 2023) dedicated his last years to this survey of the foundations and ramifications of Perceptual Control Theory. Some new developments in the mathematics of 'rattling' and 'crumpling' evoked rethinking of many parts of it. The consequent reorganization process was reaching equilibrium when illness required him to release it to us. Being unable to consult with him is a challenge for the editorial team, to be sure, but Kent McClelland and I have enjoyed more than three decades of correspondence and collaboration with Martin on topics featured here, and Warren Mansell almost as long. Martin's intentions are everywhere clearly discernible, and it has been our privilege to help them come across clearly. Our content editing has mostly tidied repetitions and lacunae, and sought to make some hurried passages more clear and felicitious. Richard Pfau did the meticulous work of copy-editing and indexing, with assistance by John Kirkland. Martin approved the design and layout of this book in collaboration with Dag Forssell.

For the editorial team,
Bruce Nevin, Managing Editor
Living Control Systems Publishing

Part 1: Overview of this Book

This book was started because the editor of The *Interdisciplinary Handbook of Perceptual Control Theory* asked me to halve the length of a chapter on communication, and I wanted to use the excised half as the core of a publication. The work to complete that other half chapter as an academic paper grew into something completely different — an attempt to suggest a variety of lines of research in different domains, taking PCT as a fundamental premise. The book delves into areas that William T. (Bill) Powers knew were important but knowingly avoided, such as synaptic developmental processes, the strands that Powers wove into his concept of 'neural current', and lateral inhibition, quite apart from the application of PCT to the social topics in the book's subtitle — language, culture, power, and politics. The title, *Powers of Perceptual Control*, is a deliberate pun on Bill's surname.

In writing this book, I hope to suggest to interested parties how consideration of PCT as an underlying process might clarify a lot of apparently independent topics, in the same kind of way as some aspects of chemistry are clarified by even an incomplete understanding of the underlying physics, for example the concept of valence electrons in the outer shell of an atom. I try to avoid mathematics as much as possible, but there are occasions where some mathematical concepts seem unavoidable. I hope these are few, far between, and intelligible. Part 1 (Chapters I.1 through I.3) presents a kind of abstract of the whole book. Part 2 starts with a tutorial on PCT that is a bit different from most, and then begins to elaborate the theory, its applications, and the philosophically challenging relationship of the theory to our experience of the perceptual universe. Part 3 dives back into more technical properties of interacting control systems within a hierarchy, affecting their capacities and limitations. Part 4 brings in information, reorganisation and the perception of structure, and a deeper look at how all this relates to subjective experience.

Much of the book is intended to make at least some sense if you simply drop in and read a chapter or two at random. Certain concepts and constructs that recur are explained in earlier chapters simply and without detail, and later more precisely with back references to the earlier statements. This may occasionally give the impression of contradiction, as the later and more careful account supersedes the earlier description.

As used in PCT, 'control' is a neutral term that does *not* mean getting others to do what you want. It is merely what all living things do to stay alive and healthy. Control allows you and your microbes, your pets, your friends and your trees, to avoid dangers, to take advantage of opportunities, and to get where and what they want. To do so they must be in (or create) an environment that offers the necessary means, and they must have the necessary skills or capacities to employ those means for their purposes.

PCT deals with the functioning of active systems, living or not. In Volume IV, for example, we spend a little time wondering about hive minds and the consciousness and social interactions with and of robots. PCT does not care whether the functions it describes are performed by chemicals and/or neurons in a brain, hydraulic circuits, steel and steam, swarms of ants, fungal mats, simulation programs in computers, mixtures of those, or whatever fantasy support circuits might be devised, so long as their functions are appropriately connected into what are called 'control loops'.

If some physical (including biological) support can be shown to perform any of the required functions in practice, so much the better. For example, early in Volume II, I show how chemical concentrations, reactions, and catalysis can form control loops. If the underlying mechanism is not clear, the functional, often mathematical analysis still is worth pursuing (though I limit the mathematics in this book).

PCT investigation of living things has generally ignored biochemical processes, except for the stable workings of the biochemistry that keep us alive and healthy. Powers lumps these into a single category of 'intrinsic variables' that need to be controlled but are inaccessible individually to our perceptions. We can feel hunger when our blood sugar is too low, but we do not perceive the actual level of sugar in our blood stream, nor the differences among the various molecules that taste sweet but come from different food sources, such as sucrose (table sugar), fructose, and maltose, among others.

We do not perceive our gut microbiota directly, but we do perceive discomfort when their condition causes us to perceive, say, nausea. Powers argues that this kind of problem leads us to change what we do in our environment — what perceptions we control — that has an effect on the 'intrinsic variables'. We will learn fairly early in the book about 'reorganisation' induced by our displeasure with how we feel, and likewise about how we do not change what we do when we feel ourselves to be comfortable with its results.

Nevertheless, little in this book deals directly with biochemical physiology. I lack training in it, but I can take for granted that the biochemical work which explains the effects of the 'intrinsic variables' on the organisation of perceptual control systems has been properly done.

I can say the same about 'cognitive psychology'. Conscious cognition presumably does influence perceptual control, and I try to address that relationship and its importance in Volume II. Powers considered conscious perceptual control as a continuation of the same structure as non-conscious perceptual control. I do not, and I hope that the reasons will become apparent before the distinction becomes critical midway through Volume II.

The crucial difference between non-conscious and conscious control is speed, because it is unnecessary to use computational resources in consciousness if the necessary computations have already been done and 'compiled' into the non-conscious processes. Powers himself frequently commented on these differences in the ability of a human to counter rapidly changing disturbances. I hope this will be evident if you dip into the middle of Volume II without having the basic understanding of the perceptual control hierarchy provided in Volume I.

I aim to present a level of detail appropriate to the topic at hand. As with quantum physics, it is possible to go as deep into the microscopic details as you want, but to analyse, say, a complete protein molecule starting with its electrons and the quarks and gluons inside its protons and neutrons is rather too complicated to be worth doing because it affords no insight into what the protein does in its normal biochemical context. Often, the external shape of the protein is enough to determine its possible interactions with other molecules, and even that is usually much too detailed when one is interested in the functional interactions of hormones and enzymes, which in this book I often lump together in words such as 'biomolecules'. At the level of understanding that concerns us here, what they do matters, but how they do it does not. What matters is their functional possibilities. Discussion of complex control systems and their interactions has similar characteristics. I go into detail in many cases, but I try to base novel constructs on what has gone before, rather than referring all the way back to the individual control loops.

Powers was interested in *Iseds*, which encompass everything we would unequivocally call 'alive', as well as organisations of living things such as crowds, at scales from a bacterium or possibly even a virus to trees and forests, fish and algae, ants and ant swarms, to people and politics. He concentrated on what could be definitively demonstrated by experiment and simulation, but I take the opposite tack, and ask what would be likely to develop if what Powers demonstrated can be extrapolated to domains in which he expected his Perceptual Control Theory eventually to be applied. Others, I hope, will subject my extrapolations to experimental test.

Those domains have in most cases been subjected to a great deal of careful research of a sort that I call 'descriptive' rather than 'explanatory' because it describes what happens when some factor is changed but doesn't give any underlying reason why it should be expected to happen. Seldom (but not never) do I expect PCT to predict that things should be other than the specialised researchers have found them to be. What PCT does in such cases is to explain why they are as they have been found to be, and in addition PCT may suggest other aspects of that topic which might be fit subjects for research.

When we investigate the behaviour of a feedback loop, what we observe is usually what should be expected, but not always. Often observations of what actually happens is the basic material of specialised research. As we progress through this book, frequently we will find why the 'often' is not 'always', and what to expect when it isn't.

Powers believed that when we understood what perceptions are controlled in such situations we would be able to say with high precision what would *always* happen in these uncertain situations. Some students of PCT argue that the main, if not sole, purpose of PCT is to find out what perceptions are controlled. PCT doesn't ask about correlational analyses, though they can be useful as guides to places to seek mechanisms. Instead, PCT asks about mechanisms and about when and why correlations might be observed.

In this book, we try to suggest where specialised research based on PCT might yield useful and often practical results not available to correlational observational research. The first three chapters introduce some different fields of enquiry that we study in the four volumes of the book. In Chapter I.1, we look at Perceptual Control and compare it with other approaches or theories which claim to cover much of the same ground, but which all appear to address only parts of the range of PCT. Chapter I.2 concentrates on how perceptual control systems interact with the environment in which living organisms live, again comparing other theories that purport to cover the same ground. We ask what is 'real' about what we perceive, and why evolution and life-time experience often seem to combine to bring about a match between properties of the real world outside the physical boundaries of the living organism and the perceptions created by its limited sensory systems.

In Chapter I.3 we outline some problems of language and culture, such as the problem of how linguistic and cultural stabilities over millennia can be reconciled with the rapid changes of youthful slang, a form of language that dates one as ancient if one uses words and short forms that were novel as recently as last year. Although such changes are transient, the processes that cause them to be transient are as enduring as the underlying stabilities that allow language histories to be traced back over time. In the same way, the details of cultural norms and forms of Government come and go year by year, while the perceptual control processes involved and the larger cultural forms that they shape, such as the treatment of women as inferior to men in Abrahamic cultures, remain consistent over millennia or longer.

Complicated as is the internal organisation of every living control system, the organisation of groups of interacting individuals is much more complicated. Just as has always been done for individuals, the study of social structures must be greatly simplified if we are to make any sense of it at all. The situation for the researcher is analogous to the situation for the student of material structures. Disciplines are segregated by levels. A civil engineer is concerned with the strength of pillars and girders and wires and reinforced concrete beams, and so on, but cannot ignore the chemistry that over time may rust the metal components or turn a rock-solid moulded material into sandy dust.

If a civil engineer worried about the interactions among the various flavours of the quarks in the atoms of her girders and wires, she might in a lifetime achieve a minuscule degree of increased precision, but she is more likely to want to know now how much steel and concrete to use if her bridge is not to fall down. External events she cannot estimate might have far larger effects on her bridges than would be gained by her improved understanding of molecular dynamics. The engineer builds in a 'safety margin' instead.

Likewise the engineer constructing a non-living control system or servomechanism would build in tolerance levels that compensate for unknown disturbance rhythms and unavoidable time lags. Time matters, and acceptance of unavoidable variations that can be quickly corrected requires simplified visions of what in leisure could be studied in unlimited detail. A living thing must tolerate and use these time limitations if it is to survive and thrive in an ever-changing and sometimes intentionally malevolent environment. A prey animal that moves too slowly is likely to get eaten, which is not good for the propagation of the species, and a predator that moves too slowly will likely starve to death.

When we talk about 'rattling' in connection with organisations, as we increasingly do beginning in Volume II and more so in the last two volumes, we do not treat as unique each individual in the organisation and their precise interactions with all the others. We ask about a rattling measure for the entire organisation or some part of it. We do likewise for structures such as the perceptual control hierarchy and the biochemistry within the individual organism. We recognise the foundational importance of the interactions of our hormones and enzymes to our health, but seldom do we refer to any specific physiological mechanism other than to use it as a specific example of a general process.

In Volume II we also begin to introduce conscious thought processes, in contrast to the perceptual control processes that are often performed entirely without conscious thought, such as the angles of the joints when walking or picking something up. Conscious thought about how to achieve some end is what we do before we have learned an ordinarily effective way to achieve it. Once we have, and have reached what is sometimes called an 'overlearned' state where, as the phrase goes, we could "do it in our sleep", the perception and the action tend to become non-conscious. Just about any controlled perception, however, can become consciously available for thought, and the interaction between conscious and non-conscious control is a salient topic of Volume II.

As I said in the Preface, what we are doing in this book amounts to the discovery of Building Codes and motifs used by Nature during evolution. They are publicly available, but refer to components, often themselves motifs, whose properties we do not yet know. Finding those components and their properties is the objective of the scientific enterprise, whatever the specialty. This book is about the consequences of believing in the motif embedded in the title of Powers' seminal book *Behavior: The Control of Perception* (Powers, 1973/2005). Accordingly I call several of the structures built by interacting control loops, 'motifs' of control, taking the control loop, itself a motif of a physical level of structure, as a unit to be used where it fits.

'Motifs' are regular, repeated forms of structure that have consistent effects in their local interactions. Many motifs seem to be built using simpler motifs as their basic components. The 'Trade Motif' (Chapter III.9), for example, incorporates among others two 'Conflict' motifs (Sections I.5 & I.6). The use of a motif is analogous to that of a subroutine by a classical computer programmer. It simplifies understanding by the psychologist and may simplify the developmental processes of the maturing mind.

In perceptual control, motifs exhibit emergent properties that are dependent on the precise way in which the component units are related, analogously to how a subroutine in a computer program depends on its array of possible arguments. Two structures composed of these same components, but with different interrelationships, exhibit different emergent properties. A useful motif is useful both from the point of view of the developing mental system and from the point of view of the psychologist-analyst.

Like a software subroutine, a motif is available to be used, where appropriate, in many different contexts. This is what the civil engineer does when in one design and then in a different design he uses materials with known properties such as good weathering and strength in compression and tension, deploying them in arrangements similar to how he and others have used them successfully. Such motifs tend eventually to be incorporated into Building Codes that must be followed by a designer, reducing the opportunity for creativity and inventiveness, but saving much time and effort that would be required to create a totally novel design and verify its trustworthiness.

The rest of this book is elaboration and explanation of the three introductory chapters. Volume I is largely explanation of the infrastructure, Volume II elaborates some of the things that might be built on the infrastructure, Volume III deals with the interactions of small groups such as dyads and families, using some of the motifs discovered in Volume II. Volume IV extends the ideas in Volume III into larger formal and informal organisations up to global scale, such as multinational corporations. Most of the book after Volume I (and to a large extent also in Volume I) could be considered as offering pointers to places where topic specialists might usefully do research based on PCT that might help clarify their field of interest.

Before we start, one very important foundational thing must be understood: the meaning of '*feedback*', positive and negative, which we use in their technical sense as do engineers and mathematicians, and as did Powers. In popular language, 'positive feedback' is encouragement, and 'negative feedback' is discouragement or perhaps even punishment. In contrast, this book is almost entirely about feedback *processes*.

This technical engineering-style use of 'feedback' has very little to do with encouragement to or correction of one person by another. A 'feedback process' exists when the effects of change in something or other has effects that eventually return to influence the same thing. Feedback occurs in a loop, and only in a loop. If the loop feedback is positive, the returning effect enhances the initial change; if it is negative, the returning effect opposes the original change. Positive feedback often makes things unstable, and can lead to runaway effects that are usually unwanted, whereas negative feedback tends to maintain stabilities that are usually desirable.

All 'control loops', about which you will read much in this book, are negative feedback loops, but not all negative feedback loops are control loops. Likewise, not all positive feedback loops cause unwelcome instability. But we leave these niceties until the appropriate points in the book, after the basic control loops have been described and some of their uses previewed. As indicated before, we are concerned instead with the 'building codes' for living control systems, ranging in size and extent from bacteria to the vegetable ecology of the entire world.

The 'building codes' we seek are the components that can be linked into motifs that can serve as components of other motifs at ever increasing scales. The codes, when properly interpreted, do not limit what could be built. Just as maverick builders may use innovative materials and techniques to build a new bridge or house, so may Nature invent from time to time new structural motifs that build on the old in new ways. And so, the book can be construed as an effort to trace ecologies of perceptual control motifs through the fractal scales of life, where we often find the same or related motifs at very small and very large scales. We begin by examining the nature of perceptual control itself, which we do in the early part of Volume I, because from that seed grows everything else. From there we branch both down the root structure of the tree to ever smaller scales, before branching upward among the tree's branches to ever larger social scales. At whatever scale we look, we will find many of the same motifs, but as we increase the scale, the conceptual space available for motifs widens, and so there are motifs found at large scales that are simply unavailable at smaller scales. The 'Trade' motif, for example, cannot occur without at least two independent parties to the trade (Chapter III.9).

The first three chapters lay out the landscape, but only as a crude map, which, as Korzybski (1933) famously said, is not the territory it represents. So let us begin to unfold the map. But first…

Ways to Use This Book

One could read this book straight through, if one has the energy to do what I would consider a Herculean task. As with the labours of Hercules, in many places my intent is to give my ideas to the reader while providing sufficient material to allow the reader to break off and develop their own ideas, experiments, and tests, because (as is usually true in science), even if my ideas happen to be correct, the task of 'sweeping that stable' can never be completed. There is always more to find out, more to understand, and more to correct of what has been written. So a linear read-through, while I suppose possible, may not be the best way for everybody to take what advantage they may from my ideas.

A quite different approach is to treat the fifty-odd chapters in the four volumes as a buffet, taking samples from here and there, where the chapter titles or section headings sound as though they might be interesting. In several places, two or three chapters form a natural group and are probably best read together.

Longer groups are collected into 'Parts', of which there are ten. In my mind, though perhaps not yours, the chapters within a 'Part' have a certain coherence that they do not necessarily share with chapters elsewhere in the book.

Yet another way might be to use the first two, three, or four chapters of Part 2 as a basic tutorial and then start the buffet wherever you please in the book. Many of the chapters in Volumes III and IV might be easier to follow if the idea of autocatalytic creativity as described in Chapter II.2 is fully understood. The same is true of 'Rattling' described in Chapter II.5. Autocatalytic creativity and 'Rattling' have a lot to say individually and together about social issues.

As you may have gathered, I am just 'blue-skying' some of the ways a resourceful person interested in some of the potential applications of PCT might use the book. I am sure there are many more, just as effective and efficient for you as those few I picked out of thin air. What I do hope, whatever way you choose to use the book, is that from it you will get sufficient understanding to be able to tell the long-gone author (me) "*That's nonsense and here's the evidence to prove it.*" I also hope that there are not too many places in the book where you will feel like saying this, but are able to say instead "That all makes good sense."

Chapter I.1. Why Perceptual Control?

"There's glory for you!", [said Humpty Dumpty].

"I don't know what you mean by 'glory'", Alice said.

Humpty Dumpty smiled contemptuously. "Of course you don't — until I tell you. I meant 'there's a nice knock-down argument for you!'"

"But 'glory' doesn't mean 'a nice knock down argument',"Alice objected.

"When I use a word," Humpty Dumpty said in a rather scornful tone, "it means just what I choose it to mean — neither more nor less."

"The question is," said Alice, "whether you CAN make words mean so many different things."

"The question is," said Humpty Dumpty, " which is to be master — that's all."

— Lewis Carroll, Through the Looking Glass

What do we mean by 'controlling'? Is it Humpty-Dumpty's claimed ability to be master of the meanings of his words? Carroll (1871) used this passage not to show Humpty-Dumpty as a master controller, but to show him as a silly egomaniac. Nobody can control the meanings of their words. Those meanings, if they reside anywhere, reside in the effects they produce on the hearer (or reader). Maybe Humpty-Dumpty is producing on Alice exactly the effects that he wants to produce, such as her confusion. If so, he has been successful as master of his meanings, though Alice does not see it that way. But would an onlooker accept that "glory" and "a nice knock down argument" mean the same thing? For me, as for Alice, my answer is "No".

I cannot say what 'control' means to you. I can only explain what it means to me, which I will start to do here. In a way, this whole book is dedicated to explaining what 'control' means to me. I hope that by the time you finish reading, either 'control' will mean to you something like what it means to me or you will have a clear understanding of why it does not.

Suppose I want to visit Aunt Maude, but I am at home, two blocks away. My location is not where I wish it to be, so I change it by walking over to Aunt Maude's. Now my location is much closer to where I want to be, but it still isn't quite right. I move so that I stand on her doorstep in front of a closed door. That is not where I want to be, which is inside, conversing with her. I want the door to be open so that I can complete my change of location to where I want to be, but I can't open the locked door, so what do I do? I want Aunt Maude (or someone) to open that door from inside, but how can I act so that they do what I want?

I expect Aunt Maude will act to open the door if she hears the doorbell ringing, but right now I cannot hear the doorbell ringing. I have experienced that doorbells sometimes ring when a button is pushed and I perceive that Aunt Maude has such a button beside the door. I want to perceive the button being pushed, but right now it isn't, so I act to change that situation and use my muscles to push it. Now I hear that the doorbell is ringing, as I wanted, and soon the door opens and Aunt Maude lets me in, as I wanted.

I acted. But how did I act? I performed a sequence of actions, part of which is the sequence of walking joint movements (correcting the 'error' or discrepancy between what I wanted and what I perceived to be my current location) and getting the door to open (correcting the error that I perceived it to be closed). But in order to do either of those things, I had other errors to fix. To walk, I had to change my leg positions many times. To get the door open I had to get Aunt Maude to open it, which meant I had to get the doorbell to ring, which meant I had to get the button pushed, which meant I had to perform certain muscular actions, which would be different in detail every time I visited Aunt Maude. Every time I acted, I was correcting something about what I perceived that was not what I wished it to be — I 'controlled' some perceptions.

Figure I.1.1 shows two simple components of 'Visiting Aunt Maude', the processes involved in arranging to hear the doorbell ringing. Everything happens in loops in which actions are performed in order to reduce the difference between something I perceive and what I would like that perception to be (the 'error' in my perception).

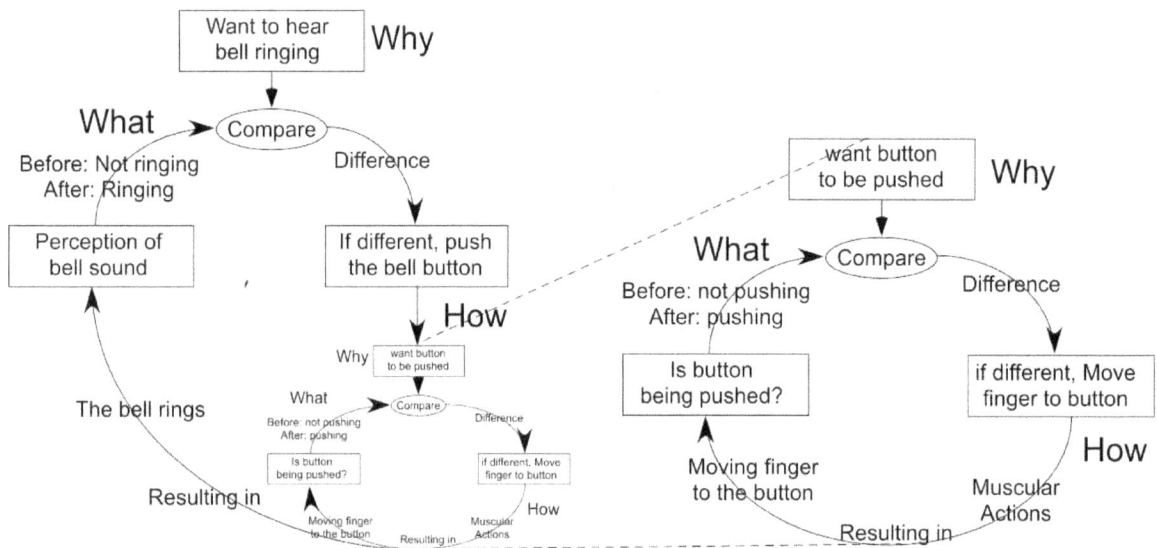

Figure I.1.1 A two-loop control structure. A person wants to perceive the sound of a doorbell ringing, and to bring about this perceptual value, wants to push the appropriate button, so acts to move a finger to push the button, which causes the doorbell to ring.

These 'What-Why-How' loops are control loops. Above each is the reason 'Why' for that loop, what is wanted — what state the actions of the loop are to bring about. 'How' to achieve it specifies the actions that will achieve it if the current state is 'What' it now appears to be. In Figure I.1.1 the highest depicted 'Why' is that I wanted to hear the bell ringing (and above that would be a loop for which the 'Why' is that I wanted someone to open the door). The 'How' is to have the button pushed, because the 'What' is that the bell is not now ringing.

At the next level, the 'What' is that the button is not being pushed, the 'Why' is that I want it to be pushed, and the 'How' is to tell various muscle groups to produce the effect of the button being pushed. Below the 'button pushed' loop are many more un-shown 'What-Why-How' loops concerned with muscle tensioning and relaxing. But these two may be enough to illustrate the general idea. You will see this Figure I.1.1 again as Figure I.5.6b. These are *negative* feedback loops because they *negate the pre-existing error*, changing a state you didn't want to the one you want.

I didn't control Aunt Maude, who could have chosen not to open the door, or she might have been lying unconscious on her floor. But I did act so that she actually came to open it. I controlled my perception of her actions. If she had not come to open the door, I probably would have tried something else to get her to come, perhaps banging loudly on the door, perhaps calling her name so that she would know it was me. No matter what I did, I never would have been controlling Aunt Maude. I have no ability to do that, but nevertheless, I can often get her to do what I want. I act to control my perception of the world, and that's all I can do. The world in this example includes the open-closed state of Aunt Maude's door, which I can see, and Aunt Maude's location, which I can only imagine.

Very crudely, 'control' means to me the ability to see that something is not as I wish it to be, and by acting, to change it to a state I prefer. If the state is just as I want it, then I don't need to do anything other than what I am already doing unless some external influence (which in PCT is called a 'disturbance') acts to change it. I have been continuing to control the state while it was as I wanted, but my control did not involve changing my actions at all until something happens to make the state less like what I want. This is the essence of negative feedback, which leads toward a desired outcome — an emotionally positive state. It may sound paradoxical that it takes negative feedback to approach an emotionally positive state, and that positive feedback could lead to a disaster, but it is true as this book illustrates in many ways.

The important statement that *negative* feedback acts to change things in a desirable (*positive*) direction can be taken at any level of complexity. Maybe I don't like the direction the government is taking, and I may act to change it for the better by voting, by communicating with my representative, or perhaps by more violent means. Maybe I don't like the wording I just wrote, and I can edit it to make it better — but only until it has been published in print form. Maybe I would prefer that the cup of tea in front of me was actually at my lips and slanted to allow me to taste the tea, and I act to pick the cup up off the table. All of these are 'control', but some control is more successful than others. I will probably succeed in tasting the tea; I may find wording that I really like by carefully editing what I wrote; but I am not very likely to find that my control actions change the direction of the Government by very much. Control need not be effective or efficient in order for it to be 'control'.

As we progress through Volume I and beyond, we will find that a simple perceptual control loop may not be so simple after all. It may have several stages at which 'things may happen', and we start to limit the term 'control loop' to a particular kind of loop among many kinds of loops and networks that can be simplified for the purposes of controlling one perception into the kind of control loop discussed in Volume I. A critical function of perceptual control is the use of a through energy flow to reduce the entropy (uncertainty) of relations between 'insides' and 'outsides' — the internal structure and variable values of an entity (organism or machine) and that of the environment outside the entity. In everyday language, control is to act in ways that make the words look more as we would like them to look.

In the end, we will see that 'explaining what control means to me' can be translated as 'feedback all the way down', a phrase that probably makes no sense to you right now. I hope it will soon. Maybe a look at a stage in 'Visiting Aunt Martha', when I am standing on her doorstep wanting the door to be open so I can enter, may give a tiny hint.

Perceptual Control Theory, very simply, is about control loops that cause actions that make some aspect of the world look more as one wants, what complex structures of control loops can do, how they interact with each other, and how understanding such control loops may be useful for research in a wide range of fields of study that involve living organisms.

I.1.1 Perceptual Control Theory

Perceptual Control Theory (PCT) has many possible forms, but all of them are founded on the same basic concept, which has been understood for millennia. Your body needs to be fed, to be not too hot nor too cold, to be not too damaged by outside events, to be able to produce descendants, and so forth. If you are hungry, you try to find food and eat it; if you are too hot in the sun, you move into the shade; if you see a rock flying at your head, you dodge.

According to PCT, all these examples are based on one single idea: if you perceive that something is not as you want it, you act and try to fix the discrepancy (the 'error'). More precisely, if you perceive (the 'What' of the loop) that something is not as you would like it to be (the 'Why' of the loop), you act (the 'How' of the loop) in ways that bring your perception of it closer to how you want it. That, put simply, is 'control of perception'.

This is by no means a new idea. Powers liked to cite Aristotle and John Dewey as his intellectual ancestors in noticing that everything you do, you do to serve your own purposes. No matter whether others see your actions as selfish or altruistic, they are all done to make the world as you see it become more as you would like to see it. Powers's main new realisation was that all the mathematical and engineering tools for analysis of servo-mechanical control systems applied equally to biological control systems of any complexity. Importantly he demonstrated that stable control of great complexity could be achieved by layering control systems, one level supporting the next, as in the two loops of Figure I.1.1. Over half a century, he argued and demonstrated that these same tools could be applied to the foundation and framework of Psychology, and in this book I argue that it is true for Sociology as well.

Powers had a second new realisation, perhaps a more important one. Much of modern control theory is concerned with the output of the controller and the discovery of algorithms that allow complicated sequences of movements and forces to get a robot to do something as apparently simple as moving a hand at the end of a jointed arm in a straight line. Powers's idea is much simpler. If the robot is to 'want' to move the hand in a straight line, it watches to see how the hand is moving and corrects deviations at any level while it moves. It doesn't have to calculate anything more complicated than determining whether the corrections are in the right direction. *"If you perceive that something is not as it should be, you act to make it nearer to what you want."*

According to Powers, *control is of input* (how you perceive something about the world), not of output (how to act to produce a result). The important byproduct — or maybe it is the main product — of seeing control as being of *input* rather than of output is that the controller need not know anything of the sources of external influences that might disturb the environmental state being kept under control. If such information is available or if the immediate future of the disturbance is predictable, the controller can use it, but it will work well with no such knowledge, something an algorithmic controller of output cannot easily do.

A third important idea produced by Powers was that perceptual (input) control of a complex process *needs few computing resources*. Roboticists often complain about not having enough computing power to do the necessary computations in real time. The resources needed for perceptual control are much, much less, because the components can be treated individually and independently as control operations, rather than as problems in which the partial solutions of complicated equations based on knowledge of the current environment interact in non-linear ways.

Non-linearity is irrelevant if the control is of one perception rather than of how to fix the state of a complex world. As the example of going to visit Aunt Maude shows, each complication is represented by control of simpler perceptions in simple ways. The simpler perceptions (inputs) hierarchically build the complex ones whose control actions (outputs) are so difficult to compute. This devolution of responsibility is at the heart of Powers's hierarchic control structure, the structure on which we build throughout this book.

HPCT (Hierarchical Perceptual Control Theory) was developed by Powers in many publications, initially in Powers, McFarland and Clark (1957, 1960a, 1960b), but most clearly in Powers (1973, 2005), known here as B:CP, short for *Behavior: The Control of Perception* and in three collections of his writings that I refer to as LCS I (Powers 1989) and LCS II (Powers 1992), and a stand-alone book, LCS III (Powers 2008).[6] A posthumous Festschrift for Powers by several independent authors (including myself) has the title *The Interdisciplinary Handbook of Perceptual Control: Living Control Systems IV* (Mansell et al. 2020) and will be called LCS IV in much of this book.

Forms of PCT other than pure HPCT may structure the relationships among the units differently, but all of them subscribe to the mantra *"All (intentional) behaviour is the control of perception"*, the primary statement of Perceptual Control Theory. That word 'all' is significant, because it includes behaviour directed at other people, including the linguistic and cultural behaviour that is the meat of Volume II and Volume III.

In this work, only Powers's HPCT version of PCT is used, with the H dropped from the acronym to conform to the normal usage in current discussions of Perceptual Control Theory. Such discussions usually assume that the Powers version is the only one there could be. I do not make this assumption, but I regard HPCT as so successful by itself in generating ideas and problem solutions that I take it as a skeleton on which a complete theory will be the flesh, the muscles, organs, and skin. Powers himself considered it only as a start to a comprehensive theory that might be developed over the coming decades or centuries of careful research.[7] Where we deviate from or augment HPCT in this book, and there are indeed places where we do so in important ways, the change or addition is usually noted.

---

6   To which I frequently refer either with or without reference to an original source. LCS is an acronym for 'Living Control Systems'.

7   Where I refer without attribution to what Powers thought or claimed, I mostly rely on my memories of much person to person interaction with him face-to-face, electronically one-to-one, or in the mailing list known as CSGnet, now archived at http://discourse.iapct.org.

Since PCT is the study of control of perceptions only *within an individual*, it may seem strange to use it to study a 'language' or a 'culture' as an artefact outside any individual, let alone to call such an artefact 'malleable', as we will do. Nevertheless, if we follow the process of '*reorganisation*' (a technical term in PCT, discussed in several places in all three volumes), we find that such a designation is both reasonable and natural.

I.1.2 Control Loops

Quite often, in discussions of PCT or in everyday speech, you will hear talk of something in the environment being controlled. You may, for example, say you control the temperature of your oven or the selection of clothes you put on in the morning. In PCT discussions, this is just a shorthand way of talking, which can mislead the general reader. Your perception of your environment is your only contact with the world, which implies that you can control nothing but your perception. Nobody knows what is actually in your environment, you least of all. At this moment you know what you perceive right now, and what you perceive is all you can really know, whether or not it has any relationship with the world in which you actually live.

A control loop is an example of a negative feedback system. It is the example we will use most in this work, but we nevertheless should mention that it is not the only kind of negative feedback loop. Appendix 1 (at the end of Volume IV) describes a few others. The examples of longer loops in Appendix 1 illustrate the fact that even when feedback stabilises a variable through the action of a long loop with many stages, nevertheless the influence of that variable and disturbances to it may be quite closely localised.

Protocols, which become important in Volume II of this book, are based on a longer negative feedback loop that involves two people, as described in my chapter in the *Handbook of Perceptual Control Theory* (Mansell et al. 2020 = LCS IV). Also, a physiological homeostatic loop that maintains mainly biochemical concentrations such as of hormones and enzymes (Chapter II.3) may incorporate perceptual control loops (Chapter II.8, esp. Figure II.8.1a).

Coming down to earth to consider only the properties of a single control loop, we consider the canonical control loop of Figure I.1.2a. This diagram shows the Controlled Environmental Variable (CEV) as being affected in several ways by the action of the control loop and the CEV's influence on sensors in many ways. The reason for this is that this generic control loop might control a complicated perceptual variable high in a control hierarchy, so the Perceptual Function in the diagram might be a complex that includes many lower-level perceptual processes, each of which produces a simpler perception that might be controlled by its own component action from its own Output Function. If that sounds complicated, I hope it will not remain so for very long.

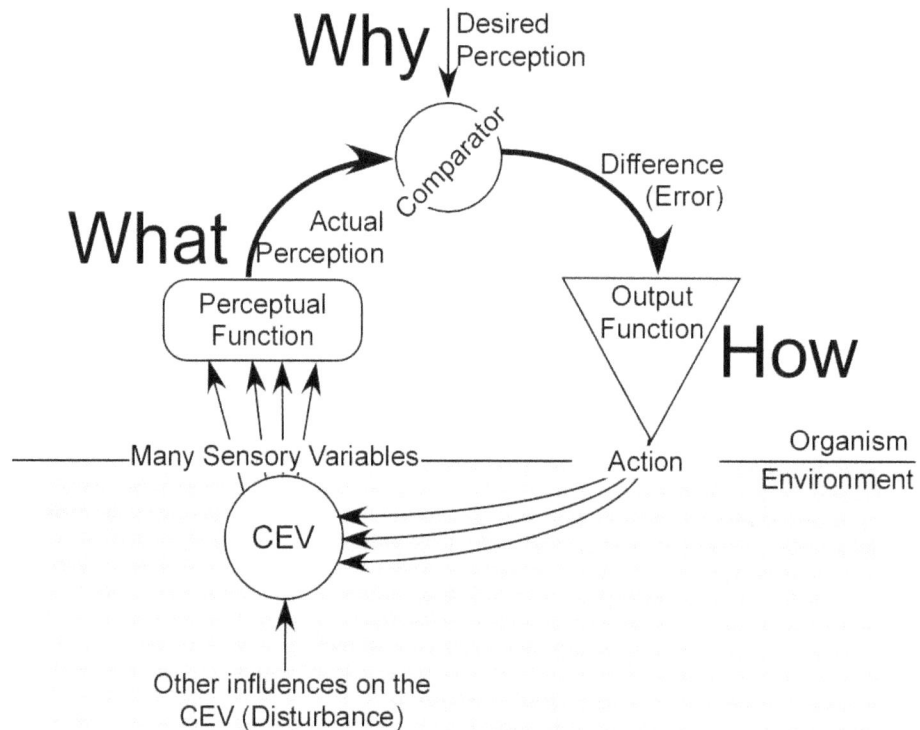

Figure I.1.2a. The canonical control loop. If the actual perception is not what is wanted, the difference, known as "the error" is processed by the output function to produce action, which through the Environmental Feedback Path influences the CEV (Corresponding Environmental Variable), which may be subject to other influences collectively called the Disturbance. The sensory input is processed by the Perceptual Function to produce the actual perception. The form of the Perceptual Function defines the CEV.

Control in PCT is the same as engineering control in every respect, though the names applied to different parts of the loop are not. In any Perceptual Control Loop the part of the loop within the organism, specifically the composite consisting of Perceptual Function, Comparator, and Output function, is called an 'Elementary Control Unit' or ECU. The Engineering control loop does not name the corresponding components.

All the analytical techniques used in engineered control systems could be used in thinking about and analysing biological perceptual control systems. But there is a difference in approach, and therefore of naming, as suggested in Figure I.1.2b. The two systems are analytically identical, but interpretively far apart.

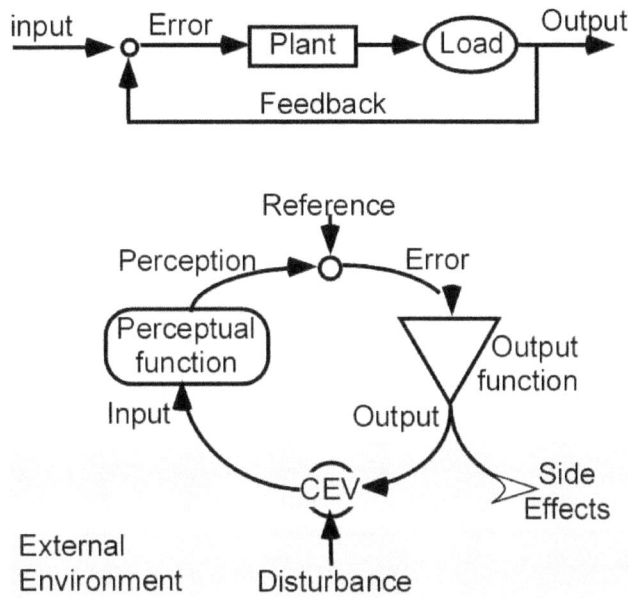

Figure I.1.2b Comparing an engineered control unit (top) with an Elementary Control Unit as conceived in PCT (bottom). "Disturbance" and "Side Effects" are explicit in the Perceptual Control Loop diagram, but are ignored in the engineering diagram (though they exist in practice). Otherwise the elements and signal paths in the two diagrams map onto each other one-to-one.

The components of the two loops map onto each other, as in Table I.1.2.

Table I.1.2 Comparing engineered and biological control (PCT) loops

Engineering	PCT
Input	Reference: ("Why" in Figure I.1.2a)
Error	Error
Plant	Output Function: ("How" in the Figure)
Load	Disturbance
Output	CEV
Feedback	Perception ("What" in the Figure)

In the engineering world, the 'input' is a value obtained from somewhere — another process or perhaps a manual setting — and the 'output' is a desired result, perhaps revolutions per minute of a spinning 'Load' or the temperature of a furnace in which glass is melted. The engineer wants to produce a particular output, and enters that value into the input.

If you want to get a particular result from a biological system other than yourself, you can't directly enter a reference value to make it happen. In an engineered system, the reference value (engineering 'input') is obtained from outside the controller. Inside a biological body, it comes from somewhere within the organism, not from the external environment. In HPCT, the reference value comes from the output of one or more 'higher-level' ECUs, except at the highest level, where the reference value is supposedly a fixed quantity.[8] In Volume II of this book (Chapter II.14) we discuss 'protocols' as a way Nature has found to get around this problem and to influence reference values in other control hierarchies in both human and non-human societies.

The biological controller knows what is happening in the Real Reality environment only from changes that occur in a perceptual value. This is equally true of the engineered controller, but that fact is easily obscured by the fact that the person using the engineered controller can see both its input and its output, and therefore can make a separate estimate of the effectiveness of the controller.

The perception is the only signal in the PCT loop that corresponds with anything in the environment accessible to another person (it corresponds to the CEV), whereas all the signals in the engineered loop are accessible to the person who designed or who maintains the system. In both systems, what is controlled is the value of the signal that is compared with the reference (the engineering 'input') but what matters is the effect on the Real Reality that most closely corresponds to the CEV (the engineering 'output'). We may propose, but Nature disposes.

In the CD that is included with LCS III (Powers 2008), a demo associated with his Chapter 10 contains a powerful demonstration that we do not control output (at least we do not do so consciously) when we control our perceptions. A square is presented on the screen and the subject is asked to use a mouse or joystick to drive a 'cursor' closely around its edge. The subject can do this quite well, which is no surprise to the subject — until Powers reveals the path the mouse took during the trace around the square. It is a perfect, if noisy, circle. According to Powers, nobody with whom he tried this out had any idea that their mouse track was not a square, let alone that it was a circle.

We must distinguish between the word 'perception' as used in PCT and 'perception' as it is used in everyday conversation. In everyday parlance the word 'perception' refers generally to things of which we are aware, whether from the outside world or from our memory and imagination. In PCT, the meaning is related, but different. A PCT 'perception' is a variable such as a neural firing rate in an organism, that ultimately depends on data from the senses or from memory and imagination. It has no necessary relation to awareness or consciousness. Most PCT analysis deals with perceptions we have 'reorganised' to control non-consciously, within the perceptual control hierarchy described by Powers.

---

8   Except possibly for effects of changing the chemical environment of the neurons active in that highest level loop. Such 'fixed quantities' are called 'intrinsic variables' by Powers.

Typically, perceptions that are well controlled during interactions with the outer world are not conscious at all, and making them conscious may disrupt control. If you are a skilled car driver, are you usually aware of the angle of the steering wheel? Probably not, but if you set it wrongly, you may die. There's an old mantra that goes *"Don't think; just do it"* that expresses this possibility. Whenever you read anything based in PCT, it is important to keep clear this distinction between the two meanings of 'perception': on the one hand the internal variable perceptions that are or might be controlled, on the other the everyday language version that means the contents of conscious awareness.

What we know, especially what we know how to do, is not necessarily available to consciousness, nor may we be able to express it overtly. Much of it is embodied not in things we can speak about, but in the ways we act to control our perceptions, ways that were developed through a process of learning (called *reorganisation* by Powers) of the behaviours used to control them.

The ability to ride a bicycle is a popular example. Without special training in teaching, a skilled cyclist (or golfer, tennis player, or even orator) may be unable to describe explicitly how they do what they do so well. This learned ability to do something is sometimes called 'procedural memory' or 'muscle memory', in contrast with other forms of memory such as semantic memory, working memory, or memory for facts and events. According to PCT, procedural memory is one effect of effective reorganisation on the complex inter-relationships among the control loops within an individual.

After Volume I, this book largely concerns those elements of procedural memory that constitute our ability to use language and act effectively within a culture whose common procedures and rituals we use in controlling our perceptions. For example, we may control a perception of hunger by eating, but to be able to do that, most of us have to use a raft of cultural and linguistic protocols and rituals that we refer to collectively as 'shopping'.

'Shopping' is done differently in a North American supermarket, a Turkish bazaar, or a Chinese laneway, but each has a consistent package of protocols. If you fail to use the protocols in the expected way, your shopping may not turn out as you want. Haggling over price is a proper protocol in a bazaar, but in a supermarket it might get you escorted from the shop. We argue that packages of such protocols and rituals can properly be called 'artefacts', or even 'things'. Some of them we will call 'motifs' of control.

The point of this quasi-philosophical introduction is to point out that though we cannot know just what is 'out there', nevertheless 'out there' is where things important to us happen. Our muscles affect what is 'out there' and what is 'out there' influences our sensors. The CEV is a construction that we try to control. Only if it corresponds well to some portion of Real Reality will we control it well. Sometimes we call what is 'out there' Real Reality, to distinguish it from the 'Perceptual Reality' content of our perceptions. But mostly we ignore the difference and temporarily assume that the two Realities are effectively the same. Usually, that works for us in everyday life.

What is important is that our internal structure continues to function reasonably well, despite the inevitable 'slings and arrows of outrageous fortune', more prosaically known as 'entropic decay'. We must either shield our interior from external events that might damage it, or we must counter them by action. This is a truth of thermodynamics and a central truth of life itself.

To 'counter by action' potentially damaging effects from the outer world is the province of perceptual control. We cannot counter what we cannot perceive, and we cannot effectively counter dynamically varying perceived effects other than by negative feedback control. Only if our perceptions correspond fairly well with things that matter in the environment, so that controlling our perceptions implies controlling against real-world dangers, will perceptual control help in our survival. And only if our actions in controlling our perceptions also affect our internal physiological states will perceptual control be useful at all. 'Control' means acting to maintain a perception close to a desired value (its 'reference value') by influencing relevant properties of Real Reality.

I.1.3 Neural Bundles and Neural Current

As a functional theory, PCT is agnostic as to the mechanisms that serve the individual processes that together form a complete control loop. Nevertheless, much of this entire book contains a hidden assumption, that many of the processes are performed by the firings of individual nerves within or outside the brain, such as the nerves that cause muscle contractions throughout the body. To base a theory with measurable consequences on the entire neural connection network, with its trillions of synaptic connections in the brain alone, would be totally unwieldy and humanly impossible to comprehend.

Even just the timings underlying nerve firings, let alone synaptic variation of firing likelihoods, are also too much for an analytic human theorist to encompass usefully. Accordingly, as a theorist, Powers resorted to statistical measures in order to develop an intelligible theory. One of these measures was the 'neural current', underlying which was another, the loosely defined 'neural bundle'.

Powers did not define the neural bundle within *Behavior: The Control of Perception* (Powers, 1973/2005, which I will frequently refer to as B:CP), other than in his introduction to the idea of a 'neural current': "As the basic measure of nervous system activity, therefore, I choose to use *neural current*, defined as *the number of impulses passing through a cross section of all parallel redundant fibers in a given bundle per unit time*" (Italics are by Powers 2005, p. 24). Powers initially thought that such a definition might lead to predictions of control within 10% of experimental results, but in practice the predictions are usually better than that.

The Powers definition of a neural bundle depends on redundancy among the firing patterns of nerve fibres. His concept of 'redundancy' is, however, unclear. From the definition cited above, it seems he thought there was a clear division between fibres that were redundant and fibres that were not, and that a bundle could be precisely defined. But although 'redundant' technically implies that a fibre could be omitted, the omission would not be without loss. A redundant fibre could be omitted because what it would convey could be approximately computed from the signals on the other fibres in the bundle, and the precision of the approximation is an indication of the degree of its redundancy with the other fibres.

Accordingly, there is no clear discrimination between nerve fibres that are members of a particular bundle and fibres that are not (Figure I.1.3). Neither is there initially a clear core membership that defines what fibres are redundant with any particular fibre, at least in the *tabula rasa* assumption of a totally naïve newborn with no genetically defined inter-neuron connection structure (obviously untrue in practice). The only way bundles could be defined for a (fictional) *tabula rasa* baby is the correlation pattern of its sensory input. (See Section I.11.4 for what, following Norbert Wiener (1950) we will call White Boxes, a functional representation of neural bundles.)

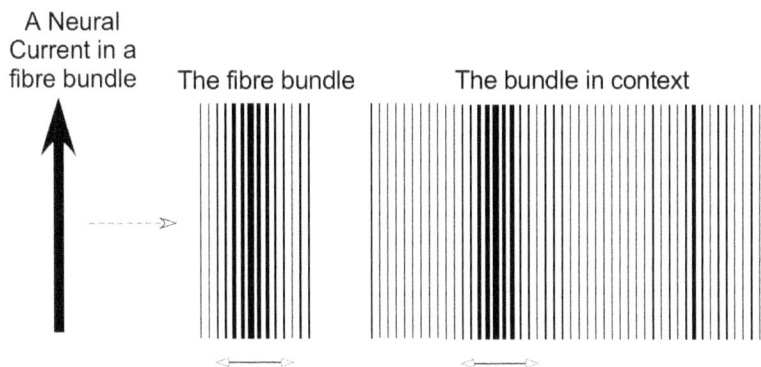

Figure I.1.3 Schematic of a fibre bundle responding to a specific pattern of input in isolation and in the context of other neural fibres, some of which also respond to the same pattern. The fibres in the bundle also respond similarly to other input patterns, whereas the fibres not in the bundle are unlikely to respond much to other input patterns. Line thickness indicated the firing rate of that fibre, and a neural current is determined by the firings per second summed over the entire bundle.

Put shortly, neurons whose correlation patterns match each other over some patterns of input will form a neural bundle for Powers's purposes. Some will correlate very closely, some less well, but any neuron will contribute something to the neural current ascribed to the bundle if its correlation patterns with the highly correlated bundle core are above random noise.

Why do I spend such introductory detail on the nature of a neural current and a neural bundle when the statistical end result is the same? Because later (Chapter I.10 to Chapter I.12), these distributions of bundle membership around a core will turn out to be central to our perceptions of belief and certainty, concepts that are only clumsily addressed, if at all, by the basic neural current analyses of PCT. How strongly we believe that man in the hazy distance is the person we came to meet is something we experience, but it is not incorporated in the Powers perceptual control hierarchy. It is, however, covered in the same hierarchy if each perception is automatically covered by the kind of diffuse neural bundle introduced here (Chapter I.12).

Belief and uncertainty are very important in our social relations, and quite often define social groups, as we will see in Volume III, so if PCT is to fulfil its promise of addressing all intentional behaviour of every living organism, belief and uncertainty must be properly incorporated in the theory, as I have begun to do in this section.

I.1.4 Measurement and Perceptual Control

Now we take a different look at what we mean when we say 'control' or 'perceptual control'. What follows is a hypothetical situation to which we will return from time to time. It illustrates a control loop in which all the components are open to public view (which is not true of control loops partly inside people, with which we will be largely concerned). The example also illustrates the close link between control and measurement, a link that is not always appreciated.

Oliver wants to see how heavy a rock is that he has picked up. To do that, he simply puts the rock on the left pan of a pair of scales and adds weights to or takes weights from the right pan until the scale ceases to tilt one way or the other or until he has no smaller weights available. At that moment, the weight of the rock in the sample pan is less than the sum of the weights in the scale pan that tips the pointer one way, and more than the sum of weights that tips it the other way.[9] The weights are labelled, so Oliver can add them up to find that the weight of the rock is somewhere between the two sums. To make his job easy we give Oliver weights of 2kg, 1kg, 500g, 250g, 125g, and so forth, down to some minimum that determines how closely Oliver will be able to determine the weight of the rock, which he hopes will be less than 4kg, since he cannot measure anything as heavy as that.

---

9   The mathematically inclined may see this example as a mechanisation of a 'Dedekind cut' (Dedekind, 1901) that defines a real number by dividing the number line into rational numbers that are larger and those that are smaller than or equal to the number in question. In the control example, the weights used are analogous to the rational numbers in the Dedekind cut.

We can diagram Oliver's weighing process as in Figure I.1.4a.

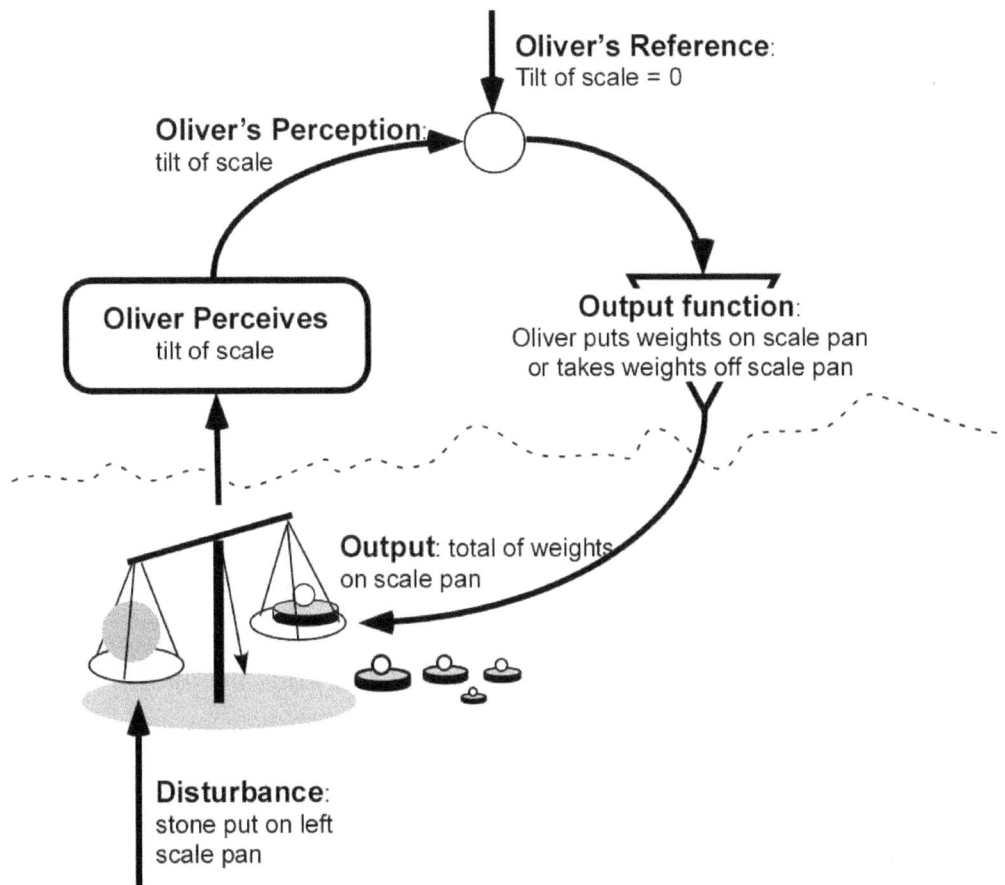

Figure I.1.4a The perceptual control loop that describes Oliver weighing a rock. This is a prototype for every instance of measurement in which the result is compared to some reference scale or value. It takes longer to make a fine measurement than to make a coarse one on the same thing. This is equally true of artificial scales and of perception by an organism.

Oliver can see whether the scale pointer is left or right of vertical and can act by putting weights in the right pan or taking weights off. He naturally has many other perceptions, but for now we are interested in his control of only one, the scale pointer position, for which he has a reference value of vertical. We assume that no matter what the scale pointer shows, it takes Oliver a fixed amount of time, say one second, between weight changes.

When Oliver puts a weight onto the pan, the scale will tilt either to the rock side or to the weight side. Oliver wants to perceive the scale pointer stopping at vertical, but unless he is incredibly lucky or the scale is a sticky bargain-basement one, the scale pointer will never be exactly vertical. His '*perception*' is whether the scale pointer is vertical or left or right of vertical. If it isn't where he wants to perceive it (vertical), the difference is called '*error*'.

If the error is 'rightward', the rock being heavier than the weight in the right pan, Oliver acts to correct the error by adding the next lighter weight as in Figure I.1.4b; if the error is 'leftward', Oliver acts to correct it by taking the last weight off the pan and adding the next lighter one in its place. These weight changes influence which side of vertical the scale pointer points, completing a Perceptual Control Loop when Oliver perceives the result.

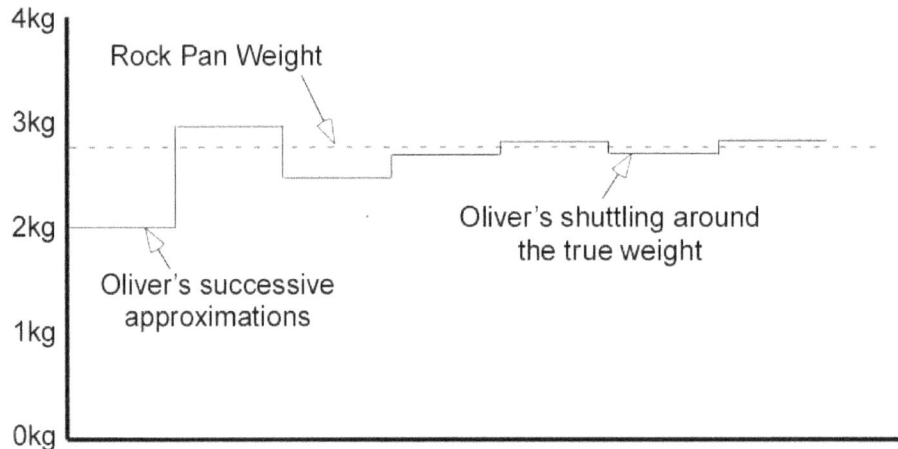

Figure I.1.4b The total weight in the scale pan as Oliver places and removes weights to balance the weight in the pan containing the rock.

The set of weights in the pan, if the scale remains centred, can be read as the weight of the rock in kg represented in binary notation, a 1 representing a weight that remains in the pan, a zero a discarded weight. If what remains in the pan is, say, the 1, 1/8, 1/16, 1/64, ...kg weights, the rock weighs (in binary) 1.001101... kg. It is up to some other perceiving system, perhaps also in Oliver, perhaps in someone else, to actually count the weights to determine its measured 1.203125...kg weight.

Physically, the scale will never be exactly centred, but if including Oliver's smallest weight makes the right pan too heavy, and taking the smallest off makes the pan too light, Oliver knows that the true weight of the rock is between the two values so obtained, and that he can't do any better than that. He has run up against a problem faced by every measuring instrument, limited resolution. One's eye has a certain blur, and can't distinguish two dots from one if they are closer than that; one's ear cannot discriminate between two pitches if they are too similar, and so forth. Oliver's scale is a perceiving aid that allows him to judge the weight of the rock more finely than he could by simply hefting it, just as a microscope or telescope is a perceiving aid that allows us to discriminate things that are too similar for the eye to discriminate.[10]

---

10 We aren't, at this point, interested in the scale as a 'weight-microscope', so much as in its use to demonstrate a control process, but the 'weight-microscope' concept should nevertheless be kept in mind.

What, at some higher level, is Oliver asking, really? In our loop of Figure I.1.2a, the 'What' is the angle of the scale pointer and the 'How' is the manipulation of the scale-pan weights, but how about the 'Why' that comes from a higher level?

At the higher level, he isn't at all interested in the pointer. He is interested in the weight of the rock. The pointer only tells him whether the weights in the right pan total more or less than the weight of the rock, a *relationship*. Oliver wants to perceive — has a reference value for — the relationship to have the value 'equality', and he keeps changing the relationship from 'too heavy' to 'too light' and back again by adding and taking weights on and off the right pan. At the lower level he doesn't perceive the relationship, but he can perceive whether the pointer is on one side or the other of vertical. What Oliver perceives at the higher level is simply the count of the weights when they total the same as the weight of the rock within one unit of the lightest weight.

Oliver's control of the relationship is by a 'higher-level' control loop that uses the scale operation as a 'lower-level' supporting control loop in a hierarchy of which we have noted two levels. Oliver doesn't actually have to move the weights himself. He could have a machine or an assistant move them, telling the supporter only 'too heavy' or 'too light' and letting the supporter translate that into the appropriate action with the weights.

Nor does Oliver need to look at the scale pointer. Another assistant might look at it and tell Oliver whether he should say 'too heavy' or 'too light' to the weight-manipulator. Oliver would know only what he was told about the pointer. As for the actions that happen when he says 'too heavy' or 'too light', all he knows is that when he tells his supporter one or the other, what he gets told by his assistant is likely to change. The assistants take the place of lower-level control processes, or (at the lowest level) of action and perception processes.

More crucially, Oliver's higher level control unit knows nothing of how the assistants do their jobs. It is computationally isolated from the details of perceiving the scale pointer and changing the weights. The assistants doing their jobs are links in the environmental feedback path of the higher level controller. We call such links '*atenfels*' (atomic environmental feedback links), which we explain further in the next chapter (Section I.2.4).

Figure I.1.4a contains an arrow for which the rock weight is labelled 'Disturbance'. This is almost the final piece in the description of a canonical control loop. Disturbances can change without the controller knowing why or by how much, but a perceptual controller can deal with them nevertheless if the changes are not too rapid or too violent.

Imagine that some prankster keeps randomly adding or taking away small pebbles or sand grains to or from the left-hand pan containing the rock, while Oliver wants always to keep the relationship between the weights in the two pans at 'equality', and therefore the pointer maintained at vertical. The prankster is the source of changes to the disturbance, an influence that would change the value of the pointer angle perception if it were not countered by adding and removing weights to compensate. In a control loop, the 'output' continuously opposes the effect of the disturbance.

The prankster is not the disturbance. Nor is the rock. The disturbance is the weight in the pan, which changes with the addition or removal of the sand or pebbles. The prankster is simply the source of the changes. Weight is just a property of the rock and of the pebbles.

Oliver does not know anything about the prankster, or even that the disturbance is varying. All he knows at any moment is that the scale pointer now shows 'too heavy' or 'too light', and he (or his assistant) must add and subtract weights on the right pan to keep the pointer from staying on just one side of vertical.

He does this exactly as indicated by Figure I.1.4b, but now he cannot guarantee that the weight is less than the last time he added a weight and found the result 'too heavy' or greater than the last time he removed a weight and found the result 'too light'. Now he should test whether the weight he is measuring is still within his most recently determined upper and lower bounds. We will not suggest how Oliver should choose when and by what method he should add or remove how much weight, as this is a tricky problem ill suited to this introduction to control loops.

Oliver consciously changes the weights, one at a time after seeing which way the scale pointer moves. The control described by Perceptual Control Theory may be conscious, and operate by discrete moves, but it is primarily non-conscious, continuously acting and perceiving changes in its perception caused by its actions and by an ever-changing disturbance, all at the same time. Whereas Oliver and the example of visiting Aunt Maude, with which I started, are both examples of conscious perceptual control, we will treat non-conscious, highly overlearned skilled control as the base phenomenon, finally linking non-conscious and conscious control in Chapter II.10.

Reference Value "r"

Comparator

Perceptual Value "p" (=v)

Error Value "e" (=r-p)

Output Function

Gain "G" (free parameter)

Perceptual Gain Assumed to be 1.0

Perceptual Function

Interior

Environment

Input Value "v" (=d+o)

Output Value "o" (=G*e)

CEV

Disturbance Value "d"

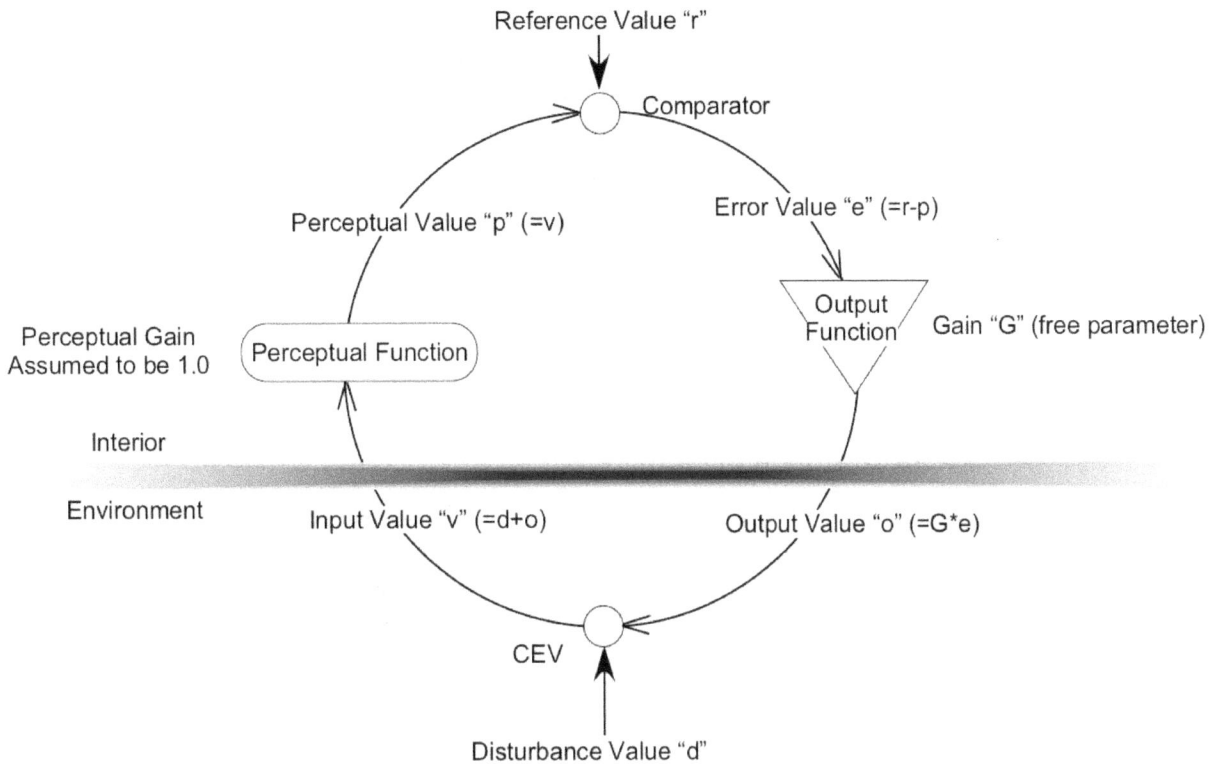

Figure I.1.4c A simplified view of a basic control loop. The expressions for the variable in the different legs of the loop are asymptotic, valid for a state that would never occur in practice, an infinite time since the last change of disturbance and reference values. The processes in the loop are taken to be linear, so that everyday arithmetic is allowed. This assumption also is unlikely to be justified in a control loop in a living being.

Figure I.1.4c shows the skeleton of a basic control loop. A perceptual value p is compared with a reference value *r* obtained from somewhere — it is what the controller would like the perceptual value to be, and we will discuss much later where that reference value (desire) might come from. Suffice it to say here that it may have deep evolutionary roots that are connected with the biochemistry of the controller's body. In this basic skeletal loop, we ignore all of that, and continuing around the loop, say that the difference between reference and perception produced by the 'comparator' is passed along to the Output Function as a value called the 'error' (often labelled 'the perceptual error' in the text of this book).

Note: As with any other physical science, the properties of the model have a precise mathematical specification. The remainder of this section requires some fluency with algebra. If you don't have that background, feel free to skip to the next section.

For a loop that has reached an invariant asymptotic state, the error value is simply multiplied by a Gain value to produce an output value o, which is added to the disturbance value d to produce the Input Value v, and since we assume the Perceptual Gain is precisely 1.0, the perceptual value p is exactly v.

Simple algebra allows us to compute the sensitivity of p to changes in the disturbance value d and the reference value r, as follows:

$$p = v = d+o = d+G^*e = d+G^*(r-p)$$

Collecting terms in p gives

$$p + G^*p = d + G^*r$$

from which

$$p = d/(1+G) + r^*(G/(1+G))$$

If G >> 1.0, then the effect of variations in d will be negligible, and p will be close to r, both of which are required for control to be effective. Note the importance of the negative sign associated with p in computing the error value. If the error value were (p-r), the loop would exhibit positive feedback, and in a dynamic situation a linear loop would produce a p value that asymptotically approaches infinity.

As we talk about dynamic situations, we must ask how this asymptotic analysis might apply to a control loop with this basic structure. As soon as we incorporate time, we have to think about the effect that it takes a finite time for a change to propagate through any process and to reach the end of a channel that is not infinitely quick. There is a 'loop transport delay' or 'loop transport lag' u between the time *t* when the effect of a disturbance value change first affects the perceptual value and the time *t* when the effects of that change on the output return to influence the perceptual value.

The implication of the loop transport lag is that the output function cannot be a simple multiplier, but must have some time-binding property. In most simulation studies by Powers and others, the output function is assumed to be a leaky integrator with gain rate g and leak rate k per time unit. Such an output function has an asymptotic value G = g/k. When such a control loop serves as a model for human tracking behaviour with the three parameters g, k, and u optimised, it is usually found to fit very closely to what the human actually does.

Actual control loops are clearly not linear, if only because the maximum output a living system could produce is nowhere near enough to resist the effects of disturbances that Nature can produce. Nature can overwhelm the best efforts of any life form to control its own destiny! We will treat a slightly less basic but general form of control loop in Chapter I.4, and elaborate it in later chapters, mostly in Chapter I.5.

I.1.5 Evolution, Perception, and Real Reality

This section is largely quoted from a message I sent to the mailing list of a group interested in exploring complex adaptive control systems (Martin Taylor 2018.01.27.14.38). It was addressed to that group, but I think it is directly relevant to the general theme of this book, and serves as an introduction to ideas that are developed in several places later in the book. The message also repeats and extends some of the ideas discussed earlier. I have made minor edits to eliminate references to the mailing list.

> *Does PCT offer an approach to the relationship of perceived reality (PR) to the unobservable 'real reality' (RR), as Bill Powers called the environment in which we live, however well or badly our perceptions reflect it? To approach this problem, I imagine a simplified world in which organisms live and evolve, as a metaphor for the world in which our first ancestors did the same.*
>
> *Let me begin at the beginning: I assume that there really is a 'Real Reality' and in it everything that has ever existed, whether living or not, has been a part. Every organism that has ever existed has survived as long as it did by taking advantage of what RR has to offer, and avoiding the effects of RR that would have otherwise killed it. To survive, an organism doesn't have to know anything about RR or even to act in any specific way. Maybe it has been just lucky, but eventually its luck will run out and it will die. Or maybe it has known and understood enough of RR to be able to act in ways that help it avoid some dangers and take advantage of some opportunities as well as being lucky. Even so, eventually its luck will run out and it will die.*
>
> *We have to assume that some of the very early organisms in this hypothetical world can make copies of themselves, and that some manage to do so before they die. Maybe initially all of them have the same structure, but that's not necessary.[11] All we need to know is that occasionally a copy is not an exact duplicate of its creator, and that at least one of the original structures makes enough copies that there will on average be more than one lucky enough to survive long enough to make another generation of*

---

11 The concept of 'structure' is used extensively throughout the book, so perhaps it should be explained. A 'structure' is a network of relationships among elements we might call 'nodes'. These relationships stay coherent for some period of time that depends on the structure being described, from picoseconds and shorter in high-energy physics to millennia or hundreds of millions of years in archaeology or geology. A structure need not be between directly sensed variables such as the locations or colours of two objects. It could occur between the perceived desirability of an object and its perceived size, for example. In many structures, the nodes are themselves structures. For example, a family is a structure, within which the important constructs are the relationships among the family members, each of whom is an immensely complicated structure. The family is itself a node in a structure of social relationships in a neighbourhood, and each person might be a node in an internet structure of 'friends'.

copies. *If there is not, then life will die out entirely and must be restarted from scratch, until a structure comes into being that does produce copies fast enough to make the average number of survivors that live long enough to make more than one survivable copy per generation. Such a structure will exist in numbers that increase exponentially until the resources to make it are depleted. The structure survives across generations, while its embodiments in individual entities do not.*

The first structure that produces on average more than one copy that survives to make copies is not just luckier than the ones that die out. It is luckier because in some way it acts so that RR is less likely to kill it in any short time interval. Maybe it has a slightly harder shell than most, or it happens to live in a gentle part of RR, or maybe something about its structural resilience renders it less susceptible than others to the effects of its interactions with RR. Maybe the structure includes a passive membrane that tends to hold its bits and pieces together, for example, or it has a stretchy interior that rebounds from a blow. Maybe it lives in little holes in rock, so that it is less exposed to RR influences. Who knows? And it doesn't matter.

What we know is that if there are more of structure X generation by generation, eventually there will be millions or trillions of X's copying themselves (and most of them dying by the million or trillion before making further copies). For a given entity X, RR contains not only what it did for the first X, but also a lot of other Xs. Of course, none of the Xs know anything of this, since at this primitive stage they have no sensors with which to create perceptions of their environment.

Not all copies are exact. If they were, we wouldn't be here. So the initial structure X will be accompanied by some slight variants X1, X2, and so on. Most of these Xn will probably be unluckier than X, and will not produce on average more than one copy per generation. Those structures will die out, but at some point an Xy will be produced that 'fits' RR better than X did, in the sense that if X produced on average $1+\partial$ copies before dying, Xy will produce $1+\Delta$, where $\Delta>\partial$. Soon there will be more of Xy than of X, but X doesn't die out unless something about Xy changes the average number of lucky copies made by an X so much that a X on average would make less than one copy of itself. With the advent of the Xy structures, RR changes for a member of the X type. For an X, RR now contains a bunch of Xy, as well as a bunch of X. Again, neither an X nor an Xy knows anything of this.

So far in the story, X, Xy, and Xz and all the other structures that on average make more than one copy that survives to make its own copies are just lucky enough to survive by making many, many copies, almost all of which die before making more copies, but some of which live to produce another generation. They need not know anything about their RR environment, and they may not even act on it, but all of them use the surrounding energy flow to build their copies (and they may use it to maintain their structure against entropic decay).

But suppose one of the copying mistakes creates a kind of structure Y that can actively move. Maybe it is a structure with some internal tension that is released so that it changes configuration (flips) if touched in some particular place. A flip has a chance of moving the entity with that structure away from what might have been a damaging force, very slightly enhancing its probability of long survival. Maybe the structure in tension was part of the copying mechanism in its ancestors, but the energy was stored rather than being released into the environment in the form of a copy of itself.

Is this Y-series 'flipping structure' a control system? Perhaps it is. Something about the structure allowed it to extract energy from its RR surroundings to store that energy in its internal tension. It may be said to sense the touch that causes the flip, and though this is a pure stimulus-response effect, it does have a 'meaning', at least to the outside observer. The 'meaning' is that the entity reduced its probability of imminent death. Of course, the entity knows nothing of that 'meaning'. It knows nothing of anything. It just is. But the Y-series produces more progeny on average than do the X-series of structures, which do not actively get out of the way of that particular kind of threat.

The Y-series of copies proliferate faster than their immediate X-series ancestor, but not necessarily faster than other members of the X-series. All of the ones that do proliferate do so because on average their 'lifestyles' fit RR better than did the ones that have died out long since. Each of them has the others as part of its RR environment, which is therefore a more complex environment than the RR encountered by entities that had the early X-series structure. As the Y-series has been described, the X entities present in its real environment do not affect its probability of survival to copy, but in a more complicated organism, the presence of different kinds of neighbours might make a difference.

To create a copy of itself out of material in an unstructured environment, an entity requires energy. The X-series may use energy only for copying, but the Y-series uses it also to build the tension required for the life-saving flip. A Z-series of structures descended from a different X-type copying error might use energy for something quite different, such as to enhance the probability that a Z-series entity will find itself in an energy-rich environment. How might it do this? One possibility is that something about an energy-poor environment irritates it in the sense that it moves more the lower the energy flow in its RR environment.[12] Again, would this be perceptual control? Yes, it would, because its actions depend on an effect of RR on it that results in its movement to a place in which it is more likely to survive and produce copies. No it would not,

---

12 In Chapter I.5 we will introduce the concept of 'rattling', a measure akin to variance, which has a relatively high value in this example. In Volume II and more generally in Volume III on social structures, 'rattling' will become a very useful measure.

because it is not changing the 'irritation' to match any prior reference value, built-in or provided as a variable from somewhere else. Either way you think of it, a Z-series structure better fits the RR environment in which it lives than would its X-series ancestor.

Y-series and Z series structures both are better fitted to the RR environment than are their X series ancestors, but they differ in why they are. Y-series structures escape a few possibly lethal interactions with RR, whereas Z-series structures use characteristics of RR to their advantage. In both cases. there is no question of whether they correctly perceive their environment, or whether what they perceive is 'really there'. By doing what they do, they increase their probability of surviving to produce many copies of themselves, some of which are inexact. In Darwinian terms, they are 'fitter' than the X-series without being in any kind of competition with X-type entities. They 'fit' RR better than do X-type.

Remember how 'intrinsic variables' are described in HPCT. They are variables that are not controlled as perceptions are, but keeping them near genetically determined reference values enhances the survival probability of the organism. This happens as a side-effect of controlling perceptual variables that are not intrinsic variables. Our Y and Z series structures enhance their probabilities of survival through the side-effects of their actions. The immediate effects of their actions are to remove the effect of a touch in Y-series 'flippers' and to reduce irritation in the Z-series 'swimmers'. It just so happens that these immediate effects change the way RR influences the internal workings of the structure, because it is RR, not some intermediary representation of it such as perceptual reality (PR), which determines survival. If what induced the flip of a Y or the increased motion of a Z was not caused by something in RR, the actions would not affect the entity's survival probability.

Fast forward a few more generations with very occasional copying errors, some of which enhance survival probability. By now there may be dozens or hundreds of different varieties of descendants of X, Y, and Z series entities, all of which are fitter to survive in the RR of their local environment than were their direct ancestors. They are likely also to be more complex, perhaps having duplicate copies for portions of their structure, perhaps having developed from Y-series ancestors with touch sensors in different parts of their surface that induce flips directed away from the touched surface. We can call them YY-series entities. Some descendants of Z series might have developed irritation sensors at opposite ends of an elongated structure, so that they move in the direction of decreasing irritation, and therefore move more directly than the Z series toward a high energy flow region in the manner of e-coli. Let's call those ZZs.

*By now, we can call the Xs, Y,s, Z's and their descendants 'species'
rather than 'series'. Each has all the others in its RR environment, a Real
Reality that is therefore more complex than the Real Reality in which the
original X progenitor lived. There will be interactions among members
of the same species and among members of different species. Some of
these will be inimical to one or other of an interacting pair, most will be
neutral, and some will be beneficial to one or other. The terms 'inimical'
and 'beneficial' should be understood purely in terms of probability of
survival long enough to produce a new generation of copies.*

*Two probably rare types of mutually beneficial interaction are important
here. One is a pairwise interaction, while the other is the interaction of
one entity with a host of others simultaneously, in other words the so-called
quorum effect. The sprouting bodies of slime moulds are examples of the
latter, as are the huddling behaviour of Antarctic penguins that enhances
the survival of all by conserving the body heat of those in the huddle. The
'behaviour' is the continual flow of penguins between the middle and the
cold periphery of the huddle. Species of YY and ZZ descendants that act in
beneficial ways with respect to others of their kind are likely to have higher
survival-to-copy probabilities than those that do not.*

*At the moment, I am not interested in the interactions (though they
do suggest that 'altruism' is a very basic property of life), except to suggest
how quickly RR can increase in complexity in the presence of life, even
when the only life is as simple as the X Y and Z-series entities. What I am
interested in is the relation between perceptual reality and Real Reality. Is
the mechanical effect of the flip-inducing touch on the tensed structure of
the original Y entity a perception in the PCT sense? Is the internal effect I
have called 'irritation' of a Z? It would be easy to say Yes, to both questions,
but would it be in the spirit of a PCT 'perception'? I think in the case of Z
it surely would be, and in the case of Y it probably would be.*

*But what now of YY and ZZ species, for which the actions change
depending on how the sensing surfaces of the Y and Z series relate to each
other and to the acting parts of their structures? Both YY and ZZ act
as though their actions depend on the direction from which an effect of
RR comes. Do they perceive 'direction' despite having only touch (YY) or
irritation (ZZ) sensors? There is no way of knowing by observing their
actions, since as I described YY and ZZ types, the effects are due to the
activations of individual specialised surfaces we might now call 'sensors'. In
the case of YY, just one 'sensor' is touched, and the entity flips away from
the direction of touch. In the case of ZZ the part of the structure near an
irritated surface moves faster the greater the irritation and if irritation
sensors at the two ends of the structure are differentially irritated, the whole
structure will move in the direction of less irritation -- greater energy flow
in the RR environment. The actions suggest that direction is perceived and
acted on, but since we know the mechanism, we know that it isn't.*

Nowhere internal to either YY or ZZ is there a quantity that relates to the differential activation of sensors. But there could be, and in the course of generation by generation copying errors, the common sensors might become linked, not an unlikely thing to happen when you already have multiple copies of something in what we might as well call the genome. Maybe two YY touch-sensitive places might get linked to the same place, which wouldn't initially be much use, though multiple touches might both induce a bigger flip and (to an outside observer) occur in times of greater danger to survival. Cross-linked versions of YY might survive better and produce more surviving copies on average than plain YYs, and the same might be true of ZZs.

In both cases, however, cross links that produce differences rather than sums offer the possibility of something new, a new type of perception rather than just more of the same. For example, a differentially cross-linked ZZ would have an enhanced ability to approach a high-energy region of its RR environment. The differential might suppress the movement at the less irritated end of its structure and enhance it at the more irritable end, increasing what a PCT-inclined observer might think of as the Gain of a direction-control loop, or perhaps better, a gradient-control loop. A ZZZ structure created by some copy error might replicate the irritable parts of the surface on the sides of the entity, and these might alter the lateral symmetry of the movement, making the entity turn to point up the steepest gradient.

And so it goes. Each time a copy error does something that enhances the likelihood that the resulting entity will generate more than one copy of itself that survives in RR long enough to make copies, the entity becomes a better fit to its local RR environment, both in the colloquial sense of the pieces of a jigsaw puzzle fitting together, and in the sense of 'survival of the fittest' (which means being 'survived by' offspring that can have offspring).

But what else is happening? The more different kinds of sensors a species has and the more actions it can produce, the less its survival depends on blind luck and the more it depends on the side-effects of control of something that has the quality of a non-blind PCT perception. The more surviving copies it produces, on average, the fitter it is in the Darwinian sense. It tends to 'do the right thing' more than did its pre-copy-error ancestors. It perceives (in the PCT sense) more of the real world in which it lives. We know this not by comparing the perception to our omniscient knowledge of RR, but from the entity's genetic survival, which depends only on its interactions with RR.[13]

---

13   In an old message (to which this was and is a posthumous reply) Powers said, These complex systems not only do not 'care' about what is actually going on in the 'real' environment, they cannot even know what is going on 'out there.' They perform the sole function of bringing their feedback signals, the only reality they can perceive, to some reference-level, the only goal they know.

To recap, we have three basic kinds of species so far in our imaginary tale of early evolution, the X-Series that survives by the luck of the draw alone, the Y-series that acts to reduce the probability of potentially lethal interactions with RR, and the Z-series that takes advantage of opportunities offered by RR. The Y and Z series both use something akin to perceptual control, though they are both reactive, reducing the effect of what, in a more conventional PCT context, we would call disturbances. We do not have a proactive species yet, one that would act because a reference value changed rather than because something impinged on it from outside.

Nor do we have the kind of hybrid of Y and Z that could both avoid danger and acquire resources. Nevertheless, YY and ZZ species can combine their 'sensor' data into what might be called a higher-level perception. The effect of this higher level perception is to enhance the probability that they will survive to reproduce. The addition of multiple copies of the same kind of higher-level perception in the ZZZ species enhances that probability even further. All of these survival probabilities depend on the Real Reality encountered by the individual entities, of which there are myriads of copies, most of which die before reproducing. If they did not, the resources needed for building copies would be very quickly depleted. Even with the very low probability that any specific entity lives to reproduce, an average reproduction rate greater than unity eventually produces a population greater than any pre-specified number, leading to resource depletion.

Resource depletion is the means by which RR imposes the Malthusian limit to growth. But this is not determined solely by the population of one species (say YY5). When different species require for their reproduction some of the same resources (as is likely), the depletion of a given resource is determined by the sum of the requirements of all the species which use that resource. The competition for finite resources is what Darwin and Wallace saw as the basis for the relative fitness of species, and that competition is constrained by the facts of Real Reality.

Suddenly we are in the realm of potential chaos in the mathematical sense, exhibited for a single resource in the Lotka-Volterra ('predator-prey') equations.[14] The actual survival probability of a species in RR is no longer a simple number, but is a dynamic variable, changing over time, and with different parameters in different local RR environments, whether the mathematics is a complete description or a mere sketch of a tiny part of RR. Of course, no equation can represent all the nuances of a real world situation, but equations don't lie about the consequences of the assumptions that go into them, so I think we can be confident

---

14   A pair of first-order, non-linear, differential equations frequently used to describe the dynamics of biological systems in which two species interact, one as a predator and the other as prey.

that when several species use a common limited resource, the real-world consequence can be dynamically volatile and not necessarily periodic. A consequence of this is that a species may suddenly die out despite having been steadily increasing in numbers until shortly before the collapse. If that happens, it is a feature of Real Reality, not of mathematics.

Species interact in ways other than by being in competition (as seen from outside) for limited resources. If, for example, a YY and a ZZ merged, or if the separate copy-errors that led to them becoming separate lineages had occurred one after the other in a single lineage, the resulting YZ type could both reduce RR risk and take advantage of RR opportunities. Such a type would be more likely than either YY or ZZ to survive to produce descendants.

Let's think of the energies involved rather than of the functional relationships involved in control. The Y and the Z have different energy requirements. Y needs to store just enough energy as potential to allow it to flip when its surface is lightly touched, and Z needs to use energy continuously without much storage. If the concept of 'energy' applies in RR (always an assumption), then it is quite probable that the Y type and the Z type access it differently and convert it differently to movement, and that their waste products differ (again dependent on an assumption that the concept of 'entropy' applies in RR). There is a possibility that the waste product of one might enhance the survivability of the other, as the oil spilled from a tanker might ease the control of a small boat in a gale.

Always we are talking about the survivability to produce copies as the major effect of RR on an organism. If RR does not allow a particular organism to produce copies, then no instances of that structure will exist at a later time. It doesn't matter at all how the organism senses its environment or what it does with the effects produced internally by its sensors. What matters is that whatever it does enhances the probability of keeping the intrinsic variables in a condition that allows the organism to survive in RR long enough to make on average more than one surviving copy per individual per generation. The requirement then is that whatever the organism does must be consistent. The high probability way of ensuring the necessary consistency is to influence the local RR environment so that the sensors generate internal effects which keep the intrinsic variables in good condition on other occasions. This will happen only if the sensors consistently report to the internal structure about the RR effects on them, and if the actions influence the local RR more consistently than randomly.

I am not talking about reorganisation (yet), but about the survival benefit of control, and its emergence through copy errors through the generations. Control of something that depends on some state not closely related to RR is unlikely to produce effects that are consistent within RR. Effects that are not consistent within RR are unlikely to have consistent results on survivability, so species that expend energetic resources on controlling such effects are less likely to produce an average of one surviving copy than are species that expend the same amount of energy controlling something that is truly related to an aspect of RR.

Our X series of self-copying entities depended on pure luck for survival, but their luck depended on where and how they lived. For example, those that had shapes that would fit in small pores in rocks would be less subject to mechanical shocks, but would also have less access to energy supplies than those that floated freely in their environment. The Y series and the Z series depended slightly less on luck than did the X series, which enhanced the probability that at least one of their millions of copies would survive to reproduce. The YY and ZZ copies and then the YYY, YZ and ZZZ series found other ways to enhance their luck, simply by virtue of their ancestors' having had many copies, one of which turned out to be felicitous.

All these improvements in luck were consequences of controlling some internal quantity based on input from sensors, 'touch' in the Y series descendants, 'irritation' in the Z series descendants (quotes because these are labels applied from outside, not 'meanings' to the organism; the 'meaning' to the organism, if there is one, is in its control action, the flip or the change in wiggle movement).

RR has more ways of damaging one of these structures than by touching and then squashing it, just as it can provide energy by more means than by locating the entity in an energy flow. These ways are as unknown to us now as they would be to these primitive XYZ structures. We do not sense radioactivity, but it can kill us. The YY, ZZZ, and YZ structures do not sense chemical variations in RR. We do, to some extent, but we may not sense some chemicals in our environment that could kill us, or that could make life better for us. The Y-series structures always flip when a sensitive part of their surface is touched, but perhaps there is a quality of touch they do not sense which would determine whether some benefit is available rather than a danger.

The improvements during the evolution of the Y to YY and YYY, or Z to ZZ and ZZZ series of structures are not in the development of new sensor types to detect different aspects of RR. The YY and ZZ series only developed more of the same sensor type, through copy errors that duplicated part of the genetic design of the structure. That allowed them to detect the sensed property of RR at different parts of their surface. The YYY and ZZZ series did develop something quite new, but not new sensor types. They developed linkages among the sensors already in place in their structure. In PCT, those linkages might be seen as higher-level perceptual signal channels.

Some kinds of linkage allowed the newly complex structures to detect the direction of variation in RR, and if these linkages were coupled with actions that had a directional preference, the entities that related directional action to directional 'perception' would be more likely to survive to make further copies than would ones that did not, and the more precise the relationship between sensed direction and action direction, the more their luck would be enhanced — a primitive version of "Fortune favours the prepared mind".

The one aspect of PCT which the Y-based and Z-based species lack is a variable reference value for the perceptions they control.[15] Without that, they are indistinguishable from reactive 'S-R' (stimulus-response) machines. Their perceptual values are simply the values of the stimulus and their actions are built into their structure. Those actions do bring the corresponding perceptions nearer to some implicit reference value (no touch and minimum irritation) more often than not. They do so in ways that produce side-effects that enhance the probability of short-term survival and hence the likelihood that the entity will produce more copies of itself. In other words, even this rudimentary version of perceptual control serves to maintain in good condition whatever intrinsic variables the organism may use internally.

In the Powers version of PCT called HPCT (Hierarchic PCT), all reference values come from higher-level perceptual control systems, and the top-level controlled perceptions have only fixed reference values with no input from higher levels. Our Y and Z-based controllers started as Y and Z series structures that control individual sensor values (badly). They evolved into YY and ZZ species that controlled multiple individual sensor values independently, possibly producing conflicting actions if, say, a Y experienced two simultaneous touches. The YYY and ZZZ species produced a higher-level perception, a sum or a difference of the sensor values in opposite locations on the surface of their structure, avoiding these possible conflicts if that higher-level perception is controlled.

What is this 'higher-level perception' as a concept in the mind of the analyst (you or me)? In the 'mind' of a YYY or a ZZZ, what the analyst identifies as a high-level perception is only a signal value on a connecting 'wire', which might be implemented as a hormone concentration flow, a mechanical linkage, an electrical connection or some Rube Goldberg apparatus. But in the mind of the analyst, the higher-level perceptions are the overall intensity (sum) and gradient direction (difference) of the environmental states that produce sensor output. The analyst's concepts of 'meaning' include perceptions of the inside and the environment of the organism. The analyst's concept is a relationship which is embodied in but not sensed by the organism.

---

15   This is not relevant to X because it doesn't perceive anything.

Now let us imagine the action that a YYY or ZZZ entity might perform if its sum perception is large but its difference perception is zero. In a YYY, the analyst sees that there are equal and opposite touches on the surface, whereas in a ZZZ both ends (or sides) are equally irritated. A YYY would presumably try to perform two equal and oppositely directed flips simultaneously, which would not result in its escaping from the possible danger of being crushed. But this opposition is in just one direction, which we can call the x-direction. If at the same time there is any asymmetry in the y- or z-directions, the opposed flips could have the same effect as a strong push on the ends of a stiff but slightly bent drinking straw — a rapid movement away from the line between the two x-direction opposed flips, a strong escape from being squashed by the opposed x-direction forces that caused the touch perceptions.

In the case of a ZZZ entity caught in a low energy zone, high and equal irritation perception at both ends of the structure would result in strong movement at both ends, but with no tendency for that motion to progress in either longitudinal direction. The effective movement would be a random walk, which would take the entity into a different part of the environment where there might be a gradient that would guide it toward a source of energy.

Let's imagine a ZZZ entity in a region of high energy, where it will 'eat' all the energy it can get, and both ends of the structure perceive equally low irritation. Physically, this would be a dangerous situation, since the entity would be acquiring energy but not dissipating it to the environment. Eventually, it would burn up or explode. The intrinsic variables for its health and its very survival would then not be in a condition conducive to its production of copies of itself. For the entity to survive, we have to include another assumption, but one that is intrinsic to the very idea of intrinsic variables. We must assume that there is some kind of effect of the deviation of the intrinsic variable from its optimum value. That effect must be to tend to bring the intrinsic variable nearer its reference value. The intrinsic variable must be controlled as a perception that is not part of our developing hierarchy.

The mode of action of this intrinsic variable control loop must be to influence the perceptual hierarchy in some way. I use 'must be' advisedly, because by definition its value does not derive from the values produced by the different sensors. Powers called the action effects of intrinsic variable control 'reorganisation', influencing the connections and parameter values of the connections within the perceptual control hierarchy. With our YYY and ZZZ (and YZ) species, there isn't much to reorganise. Cross-links among touch sensors and among irritation sensors already exist, and cross-links in YZ species among touch and irritation sensors are unlikely to have much effect on the intrinsic variables. What can vary is the strength of the connections.

In the mind of the analyst, the strength of the connections is represented by the concept of loop gain. Considering the stored energy available to cause movement, an intrinsic variable (a Z-type variable in both Z and YZ types of species), a deviation in either direction suggests that the entity would be better off moving to a different part of RR. The movement of the organism is caused by its wiggle movements, and its direction of movement by the difference of the output of its opposed irritation sensors. Error in the intrinsic variable either 'too much' or 'not enough' thus suggests that an increase in loop gain of all the control loops might be useful. We can imagine that the error in the intrinsic variable induces a flood of some generic sensitiser throughout the control hierarchy that in effect says 'Do whatever you are doing more strongly' to the different perceptual control loops. Conversely, near zero intrinsic error says 'We are happy, so don't change.'

Here we are back to the issue of how perception relates to Real Reality. Why is the reference value for the intrinsic variable what it is, and how can there be any assurance that the value compared to that reference value fairly represents the actual value of the intrinsic variable? The answer to both questions is the same. It is RR that determines whether the entity lives to make copies or dies before making them. The better suited to this end is the relation between the reference value and the value (of the intrinsic variable) to which it is compared, the more copies of the structure will exist in the next generation.

The reference value and the perceived value of the intrinsic variable are some kind of transformation of the intrinsic variable in RR. The reference value and the perceived value may be neural impulses, mechanical stresses, or whatever, while the actual intrinsic variable is a chemical concentration, but there is a consistent relation between the RR value of the intrinsic variable and the values of the reference and the measure to which it is compared.

There are also collective consequences.[16] If there is any competition for resources, the entities that work together effectively in RR, in effect optimising their collective use of scarce resources, will be the ones most likely to continue to exist.

Returning to the perceptual control hierarchy of the YY and ZZ ancestors, the linkages across the multiple sensors were described as being only across sensors on opposite parts of their bodies to produce YYY from YY or ZZZ from ZZ. That was a convenience which I, as their creator, imposed. But we can assume that the lucky or luckless entities that do

---

16   We take up collective control in Volume III.

or do not produce on average more than one copy of themselves have no such initial bias. If they have links at all, their connections will be haphazard, and will form a network among the sensors and the action elements. In the case of the YY species, the action elements are directed flips, while in the case of the ZZ, one pair affects longitudinal motion while the others create turning motions that change the direction of longitudinal movement.

If RR happened to be unstructured apart from the effects of the actions on the sensors of an entity,[17] the situation would be just what Powers arranged for his Arm 2 demo of the reorganisation of a control hierarchy with 14 degrees of freedom. His intrinsic variable was control quality, but ours translates to the probability of surviving long enough to create copies, whatever the mechanism might be.

Furthermore, we do not want to presume, as Powers did, that RR is unstructured. Perhaps the risk of death to a YY following a touch is really (and I mean 'really') only that of a pair of pincers, so that touches are dangerous only when two touch sensors on opposite sides of the structure provide output together. In the case of a ZZ, equal movement at each end does not change the entity's location, so a ZZ descendant that disallowed equal and opposite end to end action might survive longer than one that allowed it.

In millions or billions of descendants of YY or ZZ types, some will have cross-links among sensors and some will have cross links among action producing regions. Of these, some few will have cross-links that lead to better survival-to-copy probabilities than the rest. For example, one variety of ZZZ might have produced mutually inhibitory links between the action outputs for movement at the two ends of the structure, creating either an enhancement of the difference between the actions invoked by their respective sensors or even a flip-flop relation between them, either of which would enhance the speed of longitudinal motion toward an energy source. Another might have produced a ring of links among the action outputs of the side-sensors with similar mutually inhibitory connections, enhancing the ability of this kind of ZZZ to turn accurately up-gradient toward the energy source.

Such mutually inhibitory connections are not unlikely if link creation is itself probable and if RR permits an inhibitory connection with reasonable probability. Purely for descriptive convenience, we can divide the ZZ sensors into two classes by their location on the entity's body. Those on the end quarter we can call longitudinal, and those in the middle we can call lateral. The same applies to the action units, though we assume that Z class species already linked one sensor to a corresponding action

---

17 'Unstructured' can be thought of as 'empty', or as being without relationships among entities (apart from the relationships of the actuators to the sensors of an entity).

unit and that this connection persisted into their ZZ descendants. So we presume that in ZZ types, longitudinal sensors will already be linked to longitudinal movement actions, whereas lateral sensors are already linked to turning movement actions. The evolutionary possibility we are considering includes both adding new links (sensor to sensor, sensor to action, action to action, or action to sensor) and eliminating existing links.

This is not the place to do simulations involving millions of possibilities over several generations in an artificially designed pseudo-real world. The point is only to suggest that what links survive depends largely if not entirely on what structures may exist in RR, of which the shape of the organism's structure is one. If the effects of RR on survival are related to the effects of a constricting circle that might cut a YY entity in two if four lateral touch sensors simultaneously experienced touch, then some kind of a ring or star linkage among them might be expected to survive, with an action mode that involved a flip in the longitudinal direction.

The YYY and ZZZ structures with differencing links across their sensors and/or action outputs do perform like a two-level hierarchy, in that their difference perceptual values depend on multiple lower-level inputs, and the outputs of those connections send commands (not yet reference values) to lower-level action units. They are commands rather than reference values because the action units (effectors) are simply transformers of internal values to energetic influences on RR, just as the sensors are transformers of influences of RR into internal values. However, it is easy to imagine another stage of evolution that allows for sensors attached to the action units (like muscle tension sensors in mammals), in which case the action outputs of the second-level systems would produce reference values for the action units rather than simply commands. Such action units would act more consistently than would the commanded action units of their ancestors.

What does 'act more consistently' mean? An action is an influence on RR, not a sensation in an internal part of the entity. The consistency of an action can only be determined by its effect on RR, but that is not directly accessible to the entity. What is accessible to the entity is the influence of RR on its sensors. 'Consistency' can be assessed only in relation to those influences. How consistent is the relation between the reference value sent to the action output and the value reported by the sensor? An action is more consistent only insofar as its effect on some sensor or set of sensors is more consistent. That effect is mediated by RR, and only by RR at this stage of the evolution of the XYZ family of entities. Much later in the evolutionary process, it will be mediated also by the history of the individual entity and by internal processes we might call 'imagination' or 'thinking', which may not behave in the same way as RR.

If the effect of a commanded action on the sensors is reliable and if the action also enhances the probability that the individual survives to make copies of itself, then, from the analyst's viewpoint, it is likely that the sensor, or the pattern of sensor outputs 'perceived' at higher levels, is producing output dependent on some reasonably isolatable property of RR. It is not an illusion whose changes are divorced from changes in RR. The perception may be an illusion, but the effects of its actions on RR rather than on its perceptual value improve the condition of the intrinsic variables of the entity. What matters always is that controlling this particular perception enhances the probability that the entity survives to make copies. If this particular perception were unrelated to RR, the survival enhancement would not happen, and instead reduction in the survival probability would be more likely.

I have been slowly building the case that any perceptions that can be controlled well by an entity and that enhance the state of the intrinsic variables when they are controlled are highly likely to correspond to static and dynamic states in RR. The organism in which they reside can function without perceiving that they are. However, the organism controls its perceptions, survives, and perhaps thrives because it perceives properties of RR or something closely related to RR. Control of perceptions unrelated to RR would tend to be eliminated in simple organisms by evolution, in more complex ones and in ones that can survive a variety of environments, by the faster process of individual Perceptual Reality reorganisation.

But there is no reason why an organism might not also perceive the contingencies we are discussing. If they have been correctly described, the contingencies are themselves aspects of RR. Such perceptions are likely to be conscious, at least initially.

One of the correspondents to the mailing list had said:

Where I start is that a person (or organism) navigates the world. The world they navigate in ordinary living (as opposed to post-hoc analysis as I do now) is an interpretation (including filtered details*) of all that is 'there' but not a representation removed or isolated from what is 'there'. Knowing that 'there is a puddle' or a 'child' in the way is crucial to navigating the pavement. If the pavement and the puddle is not the 'real environment' for the purposes of the conversation you will let me know.[18]

To which I had replied:

The 'filtered details' in the Y, YY, and YYY, and in the Z, ZZ, and ZZZ entities are what PCT calls 'perceptions'. I don't know what your 'knowing' is, here, but the PCT approach to avoiding a puddle does not

18 Angus Jenkinson 2018-01-27.

require knowing that it is a 'puddle'. It requires a perhaps non-conscious perception of a real-world property that an outside observer might label 'to-be-avoided'. So would the child. That's all the navigating organism needs. There may, of course, be no puddle. It might be a sheet of glare ice or a glossy picture of a puddle, but the perception that is relevant for navigation is the property of being good or not being good to step on. For that, I would not use the word 'knowing', any more than I would for the fact that in walking we control perceptions of the muscle tensions all around the body. I tend to use the word 'knowing' for a much higher level of perception that involves at least labelling a category — what I sometimes call 'seeing something as …'.

The person of whom you speak may perceive puddles where there are none, but most often when a puddle is perceived, it will be a puddle (or something with a lot of the properties of a puddle) in RR. The child probably learned those properties by splashing through every puddle in sight, seldom trying to splash through a sheet of ice or a photograph. As they say, "If it looks like a puddle, and splashed like a puddle, it's a puddle." And if it isn't, maybe RR will let your intrinsic variables know as you fall into a deep water-filled pit.

This long message, which at the time I called an "essay", explains some thinking that underlies much of this book. A quick summary of it is that what we perceive creates a 'Mirror World', a Perceptual Reality that we perceive to be our real environment. In Figure I.1.5, the Mirror-World's 'corresponding CEVs' are shown in an undifferentiated grey, because they are creations of the perceptual functions, with no independent reality, much like Plato's 'shadows on the wall of the cave'. Those shadows must have some relationship to the Real Reality outside the cave, and it is on Real Reality that the perceptual functions that become stable ultimately depend.

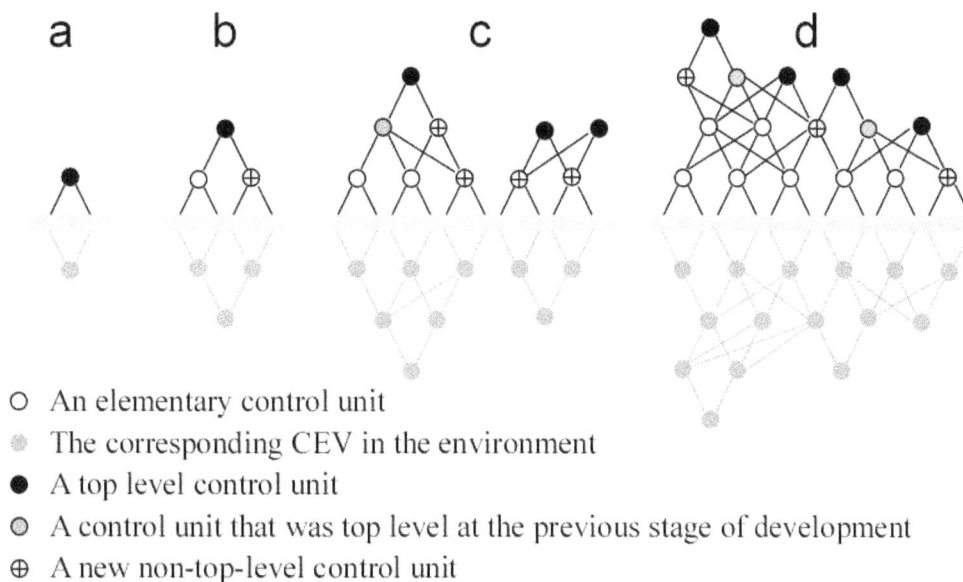

○　An elementary control unit
◦　The corresponding CEV in the environment
●　A top level control unit
◉　A control unit that was top level at the previous stage of development
⊕　A new non-top-level control unit

Figure I.1.5 (Repeated later as Figure I.11.2) A developing control hierarchy builds control of ever more complex perceptions (with correspondingly complex environmental variables) onto previously reorganised control units. A "top-level" unit is one that receives no reference input from any higher-level unit. For example, in Panel d, there is a top-level unit at level 2, two top-level units at level 3, and one top-level unit at level 4. The grey mirrored structures below the line are the "mirror world" created in the perceptual world by the developing hierarchy. The mirror world is what we see as our environment, but the perceptual world is but a distorted mirror of a small part of real reality.

Real Reality (RR) determines the inputs to our sensors and the effects of our outputs on environmental variables as we perceive them to be in the Mirror World. RR also determines how well our control of these particular perceptions helps us to stay alive and healthy. The better they do that, the more likely it is that we and our descendants will continue to control those perceptions, as opposed to other complexes of our sensory inputs. In this way, RR moulds our perceptual control hierarchy into a mirror of a very small part of itself. The environment we perceive as real is a distorted mirror view of a distorted mirror view, with the distortions being progressively reduced by ongoing evolution and reorganisation.

To conclude, it is my hope that this section's 'essay' allows the reader of this book to place my understanding of PCT in the context of whatever their own view happens to be about the relationship between what we perceive and what might be 'really true.' That theme becomes more central, the further you go in this book. Later, we will discuss hallucinations and illusions, perceptions that do not correspond to the 'real world' of other people.

I.1.6 'Deep Learning', 'Predictive Coding', and 'Enactivism'

Perceptual Control theory lives in the same conceptual world as do three popular approaches to the issues of living in a complicated world, *Deep Learning, Predictive Coding,* and *Enactivism*. Moreover, PCT captures the essential attributes of each of these approaches, though they are quite dissimilar.

Deep Learning.

This is an approach to Artificial Intelligence that builds on neural network concepts developed in the 1950s. Oliver Selfridge's 'Pandemonium' may be the best known early example (Selfridge, 1959). The basic idea of most artificial neural networks is that there are multiple levels of pattern recognition units — pseudo-neurons, 'demons', etc. — with varying forms of feedback connecting the layers. 'Sensory' data is entered at the bottom level, and a 'correct answer' or an objective is usually provided at the top level. Different schemes have different kinds of inter-level feedback connections, but the objective is generally to have the network report in some human-intelligible form what is in the scene that is presented to it.

'Deep Learning' has Learning as part of its title. What it learns relates to the content of the data, usually images, that the neural network is shown, both while it is being trained and when it is tested. During training, the network is usually not entirely disconnected from the world outside its picture catalogue. It may be given feedback by being instructed that "This picture contains a dog" or "… an office building" or "… a cedar tree". In that use, Deep Learning involves much feedback, but that feedback is almost entirely within the learning structure itself, and involves its environment in a feedback loop only insofar as the trainer observes how well it performs and modifies its input as a consequence, if that happens at all.

Until relatively recently, Deep Learning has not been connected to the world in the sense that its perceptions at the different levels have effects that change the outer world. Its use in game playing, where Deep Learning Systems can now beat GrandMasters in a variety of games, does involve it in dynamic variation of its environment when its environment is 'social', consisting of other game players that it learns to outperform. In both varieties, pattern recognition and game-playing, what changes within the neural network structure is usually difficult for a human to disentangle in any meaningful way. To quote Steve Jobs on the original Macintosh computer, "It just works."

The perceptual side of the Powers PCT hierarchy is a similar kind of multi-level neural network, as is the output side of the hierarchy. The actual levels are the result of Powers' speculations based on his personal introspection about the nature of his own perceptions. He surmised eleven levels in this perceptual network, and treated them as a basis for experiment and as suggestions for future research. Despite this, in much writing on PCT, these eleven levels are taken as cast in stone. In this book, they are ignored almost completely.

On the other hand, Powers does specify a learning procedure ('reorganisation'), which we address with growing levels of sophistication from Chapter I.11 onwards through Chapter IV.4. Controlling low-level patterns by acting on the perceived environment allows effective and *useful* perceptions of the environment to be built, perceptions that later form the components of higher-level perceptions. 'Useful' here means contributing to the well-being and survival probability of the organism. This we will retain, and see where it leads us.

In 'Deep Learning' and similar schemes, the structures that develop over time and training depend largely or entirely on relations among components of the set of input patterns presented. In PCT, learning may use these relations, but depends for its stability on how each of the internal 'nodes' of the structure individually relates to something in the outer world that can be altered by actions on that world. A criticism often levelled at Deep Learning is that the resulting structure is impenetrable, so that nobody can guess why a particular input results in the output it does. The corresponding PCT structure is 'transparent' in that each node has an identifiable place in higher level perceptions, based on how different actions influence the output of that node (the perceptual signal value).

It is true that when Deep Learning is used in systems that beat human masters at games such as chess and Go, the game positions evolve over time as a consequence of decisions made by the learning network, but they do so discretely. The position after a knight is moved in chess is a different position from the one before the move, and the system or its opponent caused the difference, but there is no continuity of change during the move of the knight, or of the transition between a square on a Go board being empty and being occupied. PCT accommodates such transitions, but is founded on perceptions of the continuous dynamic change in process.

When one Deep Learning structure competes with another similar structure over a game board in order to learn, then its developing perceptions of the complexities of the game situation do involve feedback from players back to themselves through the effects of their 'plays' on the game-board environment. Game players trained this way seem to be both simpler and more powerful than those that are built by being shown millions of exemplars.

A PCT version of this training technique would begin by building a perceptual structure of the game board, such as perceiving the relationship among the squares, hexagons, or whatever, on which game pieces might legally be placed, then on that base to perceive legitimate moves, following which the player would learn the objective of the game — in other words, learn the rules of the game before playing it competitively. The rules of the game would form the basis for the PCT game-player, just as the Laws of Nature determine the rules of the Game of Life. Much of this book discusses how a living control system, and then a society of living control systems, works to win as much and as long as possible in a competitive Game of Life while acting within the Laws of Nature.

Within the PCT structure, but not the Deep Learning structure, high-level patterns such as words can be learned and stabilised without being constructed from as yet unlearned lower level perceptions of individual letters. The same seems to happen in life (e.g. 'three-stage learning' in Taylor and Taylor 1983), and we shall look into it in connection with perception of categories in Chapter II.6 on 'Crumpling'. The layers are not pre-specified in PCT, though commonalities of the environment in which organisms of the same species grow up may well lead to different individuals developing similar perceptual types.

It is the process of control that fixes the perceptual functions, though. As with 'Deep Learning', it is the correlational patterns of the data incoming from the sensors through lower levels of the network that form the early approximations to the stable perceptual functions which are then refined by the reorganisation process. This is further developed in the discussion of 'Black Boxes' and 'White Boxes' in Chapter I.11. This process tends toward a state in which control is both effective and useful for the health and well-being of the organism, be it human, or anything else in the living world from trees to dinosaurs to bacteria.

Predictive Coding.

This name is applied to a popular range of systems, some of which also call themselves 'control'. 'Prediction' in this sense is the computation of what actions will create desired effects in the environment. To make these predictions requires considerable computation, even to work out what commands to send to the shoulder, elbow, and wrist muscles in order to get one's hand to a glass one wants to drink from. We will concentrate on one of the kinds of Predictive Control that has gained a considerable following that developed with the leadership of Karl Friston. We concentrate on that one because it seems to mesh very well with PCT, filling in with conscious control where PCT cannot suffice with non-conscious control, and providing a mathematical background that applies equally to both.

For Predictive Control, the basic question is "Now that I know *this* situation, how best can I get *that* to happen?" The question deals with conscious control, and PCT never asks such a question …

> *What, never?*
> *Well, hardly ever.*
> (Gilbert and Sullivan, HMS Pinafore)

It is possible to merge Predictive Coding with PCT, despite their fundamental differences in approach, and we will begin to do so in Chapter II.9. Predictive Coding seems to be required for control in situations that are sufficiently different from those previously encountered for the living control system (perhaps but not necessarily human) to have developed a non-conscious means of control. If, as Seth and Friston (2016) suggest, Predictive Coding control is hierarchically organised like the PCT hierarchy, we can ask whether one way of creating new perceptual functions as part of the reorganisation of the PCT hierarchy might be by solving control puzzles that recur in a Predictive Coding hierarchy operating in parallel to the behavioral hierarchy. Powers considered 'consciousness' as the core of reorganisation, and Predictive Coding offers a mechanism to support his intuition.

In a version of Predictive Coding based on 'free-energy', Friston and colleagues (e.g Seth and Friston, 2016) construct a structure that can be mapped directly onto the hierarchic perceptual control structure proposed by Powers (e.g. 1973/2005), with a change of names. Their 'prediction' of what to do to produce an effect maps onto the Powers 'reference value' for the result an action should produce, and their 'surprise', the difference between a predicted and observed value, is Powers's 'error', the difference between a desired and observed value. The key difference is in the interpretation, rather than in the operation of the structure. However, this difference in interpretation does result in different proposals as to how 'surprise' and 'error' are reduced.

PCT incorporates Predictive Coding predictions into the same multi-level network learning process as was mentioned above in connection with Deep Learning. Perceptual control and the Seth and Friston version of Predictive Coding both occur at many hierarchically organised levels in which lower level control units support ones at higher levels of the hierarchy. PCT reference values perform the function that 'prediction' does in Predictive Coding theory so closely that the inter-level connection circuitry proposed by Seth and Friston is equally useful and perhaps more useful than the conventionally accepted connection circuitry proposed by Powers in the PCT hierarchy.

Above the lowest levels, a perception controlled in the PCT hierarchy can be made conscious and controlled in the predictive hierarchy, slowly and with much logical computation. Conversely, a perception which is repeatedly controlled by similar actions through the Predictive Coding hierarchy is likely to be reorganised into the non-conscious PCT hierarchy, where it becomes what we call a skilled ability.

A newborn baby soon finds that the perception of tensing and relaxing *these* muscles allows it to perceive its arm moving, and that the perception of tensing and relaxing *those* does the same for its leg, while tensing and relaxing *these other* muscles allows it to perceive changes in the noises it hears itself making. When the older child wants to pick up a glass, no computation is necessary, because the lower level systems 'know' how to do their parts already. This suggests that as we go down levels from the more complex and context-specific to the simpler and more widely useful, no matter whether the high-level control starts within the PCT hierarchy or the Predictive Coding hierarchy, at low enough levels it is likely to be entirely PCT.

The organism has reorganised over time so that the actions at the lowerlevels usually ensure that the high-level error essentially vanishes, because as control improves at one level of the hierarchy, less of the random variation caused by the disturbance remains to be passed up to higher levels. Powers demonstrated in *Living Control Systems* III (2008) the reorganisation of a simulated arm with 14 degrees of freedom that accomplished this smooth integration of the different arm components, starting with a random set of relationships among them.[19]

Mathematically, Perceptual Control Theory implies the information-theoretic free energy principle on which Friston and colleagues base their many publications. However, since Friston and colleagues seem to argue that correct output requires prior explicit computation of the actions needed to produce predictable results, Friston's work does not as directly imply PCT.

Enactivism

In this popular approach, the basic premise is that our perceptions and actions are tuned to the facts of the world in which we live. PCT is technically an enactivist theory. The only things we can perceive are those that excite existing perceptual functions, so those functions create and limit our view of the environment. But the perceptual functions are created because *these* patterns appear in the world consistently, and *those* patterns (the overwhelming majority of possible patterns, as we shall see in Chapter I.10) do not. Furthermore, our actions can consistently influence many of the patterns that do recur, which helps to distinguish what we call 'reality' from a mirage or an illusion. The email 'essay' quoted above discusses this, and we go a bit further further in Section I.2.1 and later in Chapter I.12.

It is not unusual for students of one or another of these three related approaches to think that PCT is "just a version of what we are working on." This is seldom true, even if 'what we are working on' is a control theory. PCT is not just Deep Learning, Predictive Control, or Enactivism. Each of those could be seen as a way of thinking about one aspect of PCT, and Predictive Coding can be coupled with PCT, but the reverse does not hold.

PCT is an overarching conceptual structure that includes them all.

---

19   Also at http://www.livingcontrolsystems.com/demos/tutor_pct.html, 'Arm with 14 degrees of freedom'.

I.1.7 Sun Tzu and *The Art of War*

As with most important ideas and theories, many precursor theorists have had parts of the theory but did not put them all together in one powerful and fundamental structure of thought. Powers himself included Dewey and Aristotle among his predecessors. Earlier than both, however, was Sun Tzu, a Chinese General and Theoretician about 550 B.C.E., whose writings influence military theorists even today (e.g. Yuen 2014), despite his military successes having been achieved some 2500 years ago. So, why should a book on Perceptual Control Theory such as this be concerned with an army general, let alone one who lived so long ago?

The answer to why military affairs should be of any concern here is that we will be enquiring as to how people get what they want in a world full of other people also wanting things that might not be compatible. If a person has control of a military, whether it be a street gang, a scattered group of terrorists, or a modern technically developed army, navy, air force, and cyber force, that person has a powerful tool at hand to impose what they want on people who might want something quite different.

Why should we be interested in Sun Tzu in particular? If the interpretation by Yuen (2008, 2014) of Sun Tzu's treatise *The Art of War* is anywhere near correct, Sun used many of the techniques that we will be describing in the latter half of this book, techniques that can be derived directly from PCT in the context of social interactions. Sun used these techniques to defeat an enemy, but the underlying thought can be inverted to describe how to create good relationships in place of enmity.

According to Sun Tzu, the successful general understands patterns of the enemy's thought, while working to ensure that the enemy cannot learn his own patterns. To do this, the general uses unorthodox manoeuvres, but not to the exclusion of orthodox actions, because to be continually unorthodox would be as much a pattern as would rigid orthodoxy. Even if apparent retreat followed by an enveloping manoeuvre is often an effective tactic, the general who uses it once too often may find his army defeated by an enemy prepared for it. In PCT terms, what Sun is advocating is that the good general does not provide the opponent the opportunity to reorganise to produce higher-level perceptual functions defined by the general's prior actions. The enemy will not be able to determine which of his actions will produce what effect. In PCT terms, the enemy will not succeed with a 'Test for the Controlled Variable' (Section I.2.5).

Sun Tzu obviously knew nothing of the technology of servo-mechanisms or of negative feedback loops, but his monograph makes clear that his advice is based on a shrewd understanding of what we now are able to explain using PCT. Why was he successful, and how can we use either his precepts or the theoretical underpinning provided by PCT to make life more nearly as we would like it to be? According to Yuen (2008), Sun was clear that the objective was not simply to win a battle, or even a war, but to win them in ways that would reduce the necessity of participating in future battles or wars. Indeed, Yuen emphasises one of Sun's tenets, that the best victory is achieved by getting the enemy to want to do what you want him to do, avoiding battlefield conflict entirely. The best victory is the one you do not fight, because the enemy has become a friend.

I.1.8 Ockham's Razor and the Powers of a Theory

William of Ockham [20] (c. 1287-1347) [21] is reputed to have been the author of a principle called 'Ockham's Razor'. Simply put, of various explanations that purport to account for observed facts, the simplest is the most likely to be true. How William used this principle in his life as a monk, I do not know, but on the surface the simplest explanation of any observed fact seems to be "*God willed it to be so.*" Indeed, thousands of years ago, ancient people seem to have attributed much of what they observed to the whims of Gods, Goddesses, and the spirits inherent in both living and non-living things. Even now, many people pray to their God to ask him or her to do them some favour. This behaviour is very rational if one accepts 'the simplest explanation is likely to be the true one' version of Okham's razor.

Rational it may be, but it is hardly useful for the purposes of Science, which I take to be the discovery of the ways the world works, whether by the wilful manipulations of omnipotent powers or by 'Natural Laws' of as yet unknown provenance. "God willed it" sounds very simple, but on analysis turns out to be not so simple. Science progresses in untangling the thicket of things observed by asserting forcefully that "God did it" is not a sufficient explanation, because of the past, present, and future lack of evidence for the truth or falsity of this assertion. Science wants to find how things we perceive connect with each other, whether or not there is a God or Gods who arrange the décor in our little living space.

Science wants to pull together various apparently unrelated phenomena, and show that rather than treating them separately as though each observation were independent of other phenomena, as one might if one said "*the cause of browning skin is that the sun has tanning properties*",[22] it would be more appropriate if instead we studied the sun and the properties of skin and learned that the sun emits ultraviolet (UV) radiation (which we cannot see) and that UV stimulates melanophores (or however the skin does tan), and said "*to understand the cause of browning skin, you must know some other things that have been discovered about*

---

20 Or Occam, Ockam, Ogham, and various other spellings.

21 Wikipedia article https://en.wikipedia.org/wiki/Occam%27s_razor, retrieved 2019.02.19.

22 Powers would have called this a 'dormitive principle' along the lines of "we feel sleepy because a dormitive principle in our bloodstream increases while we are awake." Bateson (1976:xx) popularised the term as a critique of specious claims in science.

the sun and about the interactions of processes in the skin. If you already know these other things, then I can tell you that they are how solar UV causes tanning."

The latter formula is much longer than "the *sun* has tanning properties," but it explains a lot more. It asks about your prior knowledge, tells you that you should learn some things not superficially related to tanning if you don't know them already, and then finishes with a flourish that can be paraphrased as "*It works like that.*" Skin tanning is just one among a lot of things that relate to the sun's radiation spectrum and to processes in the skin that affect our well-being.

Ockham's Razor says that for the person whose background does not include an understanding of various background facts and processes, "The cause of skin browning is that the sun has tanning properties" is simpler, but for the person who knows a lot of the background, "It works like that" is much simpler. Both are correct, but "It works like that" is not an explanation for someone without the necessary prior knowledge.

To evaluate theories using Ockham's razor as a guide in comparing hypotheses or theories, one must ask what facts require description by the competing hypotheses or theories. One must ask how precisely the theory or hypothesis describes a range of facts, and how wide that range is. Some theories encompass a wide range of facts, but describe them loosely, whereas others describe a narrowly defined set precisely, making no claims at all about the wider range. If it requires parameters to have particular values in order to make numerical predictions, those values must be included in the description of the theory. And finally, as pointed out above, 'simplicity' depends on what *you* already know, because all else that is necessary for application of the theory must be included in its description.

To put it succinctly, the questions about the relative value of one theory rather than an equally plausible other theory are about the range of observations that each claims to say something about, how precisely each says it, and how simply each says it, given that each must include what you already know. Einstein's famous equation e=mc^2 is very simple, but doesn't tell most people very much at all. To understand what it means, one must take a course in relativity theory, all of which is entailed in understanding that simple equation.

How do we put together these three criteria for simplicity? In 1972, I circulated privately a working paper in which I discussed this question mathematically. It is reproduced at the end of this book's volumes as Working Paper W1, but here is the non-mathematical gist.

The theory "God willed it to be so" is very simple, but predicts nothing about any future observation. In the 'simpler is better' version of Ockham's razor, this is about as simple as one can get, but if you add "the more precise the explanation" and "the wider the range of explanation", it ceases to be the preferred theory to account for any observation at all.

Taking the "God did it" hypothesis as an example, how succinctly does it actually describe a particular set of facts? Not very succinctly at all, since to describe the facts in question, one must describe them one at a time. According to the hypothesis, the facts conform to one definition of randomness, so to add the description of the theory just makes things worse. The theory may well be correct, but it is scientifically useless. At the other extreme, if you want to describe the observation of this number sequence: '0, 1, 1, 2, 3, 5, 8, 13, 21, ...' to someone who understands the language, you need only say "It's a Fibonacci sequence starting at zero", and that person will (with the aid of a computer) be able to tell you the eleventh or hundred and eleventh number in the list, assuming always that the theory that it is a Fibonacci sequence continues to make correct predictions.

In order to explain anything, one needs a language. A language is made of a sequence of some kind of symbols, whether the language is written, spoken, literary or mathematical. Using this language, which has, shall we say, 4 different symbols (as does DNA), using 21 of these symbols one can distinguish as many as 4^{21} (about 44 quadrillion) things. Each extra symbol in a description reduces its simplicity, because there are more possibilities for the reader or listener to choose one from.

Each hypothesis can be described in numerous ways. Of these, one is the shortest and therefore simplest. For example, F=ma and 'Force is equal to mass times acceleration' say the same thing to someone who knows the conventional meanings of the symbols. F=ma is simpler, but not to someone who does not know what the symbols mean. Such a person must be told something like F means force, = means equals, m means mass, a means acceleration, 'putting two symbols together' means multiply their values, and F=ma. That is longer than 'Force equals mass times acceleration', but it has wider application, as those same symbols can be re-used in quite different formulae.

The simplicity of a theory is 'in the mind of the beholder'. It depends on what the person is using the theory for and what they knew beforehand. To someone who knows the theory, an explanation of one of its implications may be very simple. To someone who does not, the explanation of the specific implication may be either long and complex because it must include the underlying theory, or it must be specialised, depending on what the person already understands about other theories that apply to this smaller part of the Universe.

In this book, I am exploring the implications of a theory that I somewhat understand, but the reader may not. I explore its implications in many different specialised domains, each of which has different explanations for the same set of observations, depending on who is offering the explanation and who is listening to it. Simplicity is neither simple nor sufficient in deciding between such competing theories.

The second thing one must ask about when evaluating a theory is what are the facts to be explained. Given a set of facts, each competing theory explains them to a certain degree of precision. If some observation says that X is one nanometre, and a theory says that X should be around one kilometre, that theory fails to explain the fact. A theory that says X must be smaller than a millimetre does explain it, though not very precisely. And so, the greater the range of facts or the more precisely the facts are explained by equally long descriptions, the better the theory.

These last are mathematical information-theoretic issues. Without the theory and not having seen the fact to be described, you have a certain level of uncertainty about it from your background knowledge. With the theory, you are less uncertain about it, but you probably have not removed all your uncertainty. In the God case, you have not removed any uncertainty; in the Fibonacci example you have removed it all. Most theories lie somewhere in between. It is good, for example, for the theory to say that X, Y, and Z will all change in the same direction when V increases, but better to say which direction that is, and better still if it describes how much they will vary relative to each other.

Range and precision trade off against one another — a greater range of facts about which the theory makes some claim compensating for another theory's greater precision over a smaller range of facts. The way these trade off is described in working paper W1 of Volume IV. The 'range of facts' extends into the future. It is better to describe accurately a fact that has not yet been observed than to find that an already observed fact fits accurately into its place in the theory. There is no technical difference between these cases, but if the fact has been observed, the possibility exists that it might have been used somehow in formulation of the theory.

Into all these considerations, where does PCT fit? What is its range of claim, and how well does it explain facts over that range, compared to other theories that have claims only over more specialised areas? PCT is often claimed to produce very precise descriptions of experimental data, but these experiments have been in very restricted domains, such as tracking and demonstrations of particular phenomena, such as that a computer can be programmed to discover with high accuracy which of several objects moving on its screen is under the control of a human using the mouse, even when an outside observer would say they were all moving randomly.

In this book, I consider not the precision with which PCT predicts a narrow range of data, but the range of data about which PCT is able to say something useful, especially when it is combined with already 'known' theories from the so-called 'hard sciences' — basically, sciences dealing with non-living matter. PCT is about life, but life exists in an environment that includes and is composed of non-living matter. The extreme claim for PCT is that it applies to all living things and everything they do, and that it can explain their behaviour, at least functionally. In this book I enquire how PCT explains or predicts phenomena in many domains in which many organisms interact, from language to culture, money and power politics, to lies, laws and morals, governments and revolutions.

The powers of theories depend on the ranges of facts explained, at least in principle, and the precision with which these facts are described by those theories. I claim that in domains related to the behaviours of single organisms and of large and small collaborative or competing organised or unorganised groups of individuals, the powers of PCT are very considerable indeed. This book is an attempt to justify this claim.

To justify my claim, I must first explain how I understand PCT. My understanding is founded entirely on the insights of Bill Powers, but it builds upon them in ways that Powers did not propose, to my knowledge, though he mentioned some of these possibilities in his various writings, published and otherwise. It explores applications of PCT which Powers did not pursue. And it situates PCT in the context of principles which apply to all physical systems, including control systems. Because of this wider scope, my understanding of PCT differs in some respects from that which many readers take from *Behavior: The Control of Perception* (Powers 1973, 2005).

Where does PCT fit into the world of life? That is what this book is about, starting with how I understand PCT as based on the work of Bill Powers.

Chapter I.2. The Environment of Control

We live in a complicated world full of obstacles, opportunities and dangers. Most demonstrations and discussions of PCT ignore these complexities. To do so is helpful when one is introducing PCT to someone as yet unsure of how control works — and perhaps more importantly how it does not work. We do the same, at least in this introductory part of the book, but as we progress, characteristics of the environments of control become increasingly important.

I.2.1 'Real Reality' and Consistency

Powers frequently referred to one's perceptions as being the only sure facts that one can have. What is truly in the environment is a great mystery, the central problem of scientific research. Physicists and astronomers discover new kinds of things all the time. Was the stone in your distant ancestor's hand solid, or was it mostly empty space? Yes it was solid, and yes it was mostly empty space. Or at least so the physicists of today tell us — but 'solid' and 'empty space' mean different things to a physicist of today than they meant to your ancestor or to most people living today. Even the physicist ignores the 'mostly empty space' problem when she goes shopping for groceries.

Does it matter that atomic nuclei are very tiny compared to the spaces between them? In everyday life, no it does not. You can use your mostly-empty-space stone to hold open a mostly-empty-space door just as if both were quite solid. Your ancestor made his kill and brought home the bacon using the stone he felt was solid, sharp, and the right weight, and that's what mattered to him (and to the prey he killed). He perceived solidity, sharpness, and weight, and used those perceptible properties to satisfy his hunger for food. But if you are developing nuclear power systems, the emptiness matters a lot. A neutron can pass through an atom very easily without hitting anything at all. Only when it hits an atomic nucleus do 'interesting things happen'.

Your ancestor had flaked the stone into the shape he wanted it to be for the purpose of making the kill, because when he first picked it up, his perceptions of its weight and shape were not what he wanted to perceive. The nuclear engineer of today cannot do that. The important details of what the engineer is working on are not directly available to her or his senses. To perceive them, the person must use physical and analytical tools created earlier by other living control systems, other people.

All we can know of the real world is based on the evolutionarily developed capabilities we have been born with, supplemented by what we can gather through our senses. To augment our senses, over time various people have developed a wide array of devices to indicate to our senses things that seem to be in the real world but are, for example, objects too small to be seen, electromagnetic wavelengths to which we are not sensitive, vibrations too fast, too slow, or too gentle to be felt, and so forth. These devices (microscopes, radio receivers, seismometers, Geiger counters and the like) depend on conceptual models that we make for ourselves of the worlds beyond our senses.

People have developed other kinds of tools, 'conceptual devices' such as mathematical theorems and manipulations, that allow us to perceive other things that seem to be in the real world, such as atoms, force fields, energy, and warped spaces. Our neurological tools — our perceptual functions — show things that we can perceive with our senses. Our theories and models, some of which may be 'baked into' our neurophysiology, link what we see, hear, or feel with what might 'really' be 'out there'.

If we have skills to use physical and mathematical devices, whether they be tangible or conceptual, we can use them to perceive a world. The world we so perceive may not be any 'real' world, but so long as we are effectively able to control our perceptions, we can never know that the world is not as we perceive it. If, on the other hand, we fail to control well, the reason might be that our senses and devices have led us to perceive a world that differs from the 'real' world in which our actions have their effects. We may be trying to control a mirage or an illusion. Or, perhaps we perceive very well, but are imperfectly able to 'bend the world to our will'.

While he is weighing a rock, Oliver (Section I.1.4) may perhaps see the scale pans, their contents, and the pointer all together as a configuration, but why should he be justified in assuming that in some Real Reality there is a real pan with real weights, and that his perception of the 'heavy-or-light' location of the scale pointer represents a property of a real scale pointer that indicates something about a property of another scale pan and its contents? Is he so justified?

There are two opposed facile answers to this. One is "You have access only to your perceptions, and can know nothing about the world that appears to be 'out there' since your perceptions might be created by something entirely different, such as a manipulative super-intelligence." The other is "The world is whatever it happens to be, and what you perceive is what it is for you, but perhaps for nobody else." Neither answer is really helpful, though either may be true. Let us contemplate a different kind of answer, perhaps no more true, but perhaps more useful. This is based in two assumptions about Reality.

Although we could never prove it, we must act as though our perceptions are not entirely self-referential (solipsistic) or created by some super-being just when we need them, and we must act as though we are not merely inhabitants of some grand super-software simulation project. All of the foregoing are possibilities, but for the sake of doing science, they are not very useful. That's why, even though they may possibly be true, we should assume that they are false.

We must also assume that there is a distinction between what we perceive and what there is to be perceived — that there is a 'real world' of which we are a distinct part separate from the rest of the world. We assume that what happens in the real world sometimes alters the tiny part of it that affects our perceptual apparatus, and that how we act sometimes influences a little of what happens in our local environment, which is only a tiny part of the enormous Universe that is outside of us. If these assumptions are wrong, we cannot know that they are wrong, so there is no point in discussing that possibility. We must act as though they are valid.

Given these basic reality assumptions, we can say a surprising amount about what is in our real-world environment. For example, Oliver can say that no matter what the 'weights' and 'scale pointer' really are, when he perceives himself to be adding or subtracting 'weights', the 'scale pointer' changes its angle consistently from one side of vertical to the other depending on whether he is adding or subtracting weights. It rarely if ever changes from 'too heavy' to 'too light' when he adds a weight, or the reverse when he subtracts a weight.

More generally, we often find consistencies between what we do and what changes in what we perceive. Few of these consistencies are always observed, and some happen in contexts we rarely encounter, but many happen almost always in frequently encountered circumstances. We perceive something being put onto our outstretched hand and feel weight on that hand. A stage magician might be able to arrange conditions in which this didn't happen, but usually it does. We throw something into the air and perceive it to rise and then fall. We put it on a table and perceive that it stays there, while the perception of weight in the hand is reduced. Assuming that there is a real world out there, something about the interactions between it and our senses creates these normally observed consistencies. When they don't happen, we seek explanations for why they don't rather than thinking we were wrong to expect them to happen.

Perceptual Control Theory (PCT) is based in the philosophically obvious fact that if there is a 'real world out there', we can know of it only what we obtain through our senses. As we said before, we assume that the 'real action' is in the 'real world' outside our bodies, and thereafter simply treat it as a given that the real world has effects on our sensory systems and is acted on by our muscular and chemical (and for some species electrical) outputs. What happens in the real world, whether we can perceive it or not, determines whether we live or die, are sickened by radiation poisoning, bruised by hits from hard objects, are well nourished by our food, enjoy good social relations, and so forth.

Whatever is in the real world, our interactions with it have allowed us to stay alive long enough to be able to perceive these little consistencies and allowed our ancestors to stay alive long enough to propagate their genes into their descendants, including us. We can perform certain actions and expect certain things to happen, which would not be the case if our perceptions were entirely divorced from what is 'really' out there. In particular, if by acting on the environment we can control a perception such as the angle of Oliver's scale pointer, that in itself is evidence (under our basic assumptions) that our perception corresponds to something in the real world. We do what in a laboratory would be called 'experimenting', influencing the environment in different ways and seeing what happens.

I.2.2 The Taste of Lemonade

We put a glass to our lips and perceive the taste of the liquid as 'lemonade'. Something about the way that liquid affects our perceptual apparatus creates the perception of 'lemonade'. But unless we act to influence the environment, we cannot know whether the 'lemonade' taste is the consequence of drinking liquid, or is dependent on some 'real' properties of the liquid. So we try putting a liquid from a different source into the glass, and find that it does not taste like 'lemonade'. This would not be the case if 'lemonade' were purely a construction of our perceptual apparatus that happens when it is exposed to liquid. Consciously we perceive 'liquid in the mouth' but not 'lemonade'. There is something special about the 'lemonade'-tasting liquid that is not found in the other liquid.

To pursue the external reality of the 'lemonade' taste, we might try various actions and see whether they affect the 'lemonade' perception. We might try adding different substances, such as salt or turmeric, and we would find that the liquid tastes less of 'lemonade', but if we squeeze a lemon into it it might taste more of 'lemonade'. We might try extracting substances from the liquid by filtering or distilling it. In other words, we do what scientists are supposed to do. We experiment. Eventually, we may be in a position to say something about how functions of physical and chemical variables in the liquid make us perceive 'lemonade'. When a liquid has those physical and chemical properties, we can say that, for us though perhaps not for anyone else, the liquid *has* the taste of lemonade. It may not *be* lemonade, but we perceive that taste.[23]

The taste of lemonade is not a perception that one can compare with anyone else's perception of that taste, other than to have them drink some of the liquid and say something like: "That tastes like lemonade to me; what does it taste like to you?" The taste seems ethereal, not really 'out there'. But compare it with a perception such as the position of a glass on a table. One can act to move the glass to some other place, and if the action succeeds in changing one's perception of the location of the glass, the glass has some 'out there' reality, and so does its location.

One cannot determine whether someone else sees the glass on the table as one does other than by a related test. If someone else acts to move the glass back to its original position, one may assume that for the other person, the glass, the table, and their relationship also have some 'out there' reality independent of you both. In other words, you are doing a crude Test for the Controlled Variable (Section I.2.5).

Can one do a similar Test with the taste of lemonade, and thereby determine whether the taste has some 'out there' reality that other people can perceive in much the same way as you do? Yes, one can. The analogy to moving the glass is to change the ingredients of the liquid tasted by the other, as we did to determine

---

23 I can personally attest to the inconsistency of taste, though not necessarily of lemonade, since a few years ago, something happened in my brain that resulted in sweet things like sugar tasting bitter. They still do, and I tend to avoid sweet desserts, though if I have something such as coffee that normally would taste bitter along with something nominally sweet, I can often taste the sweetness.

whether for us ourselves a particular organisation of what appeared to be 'out there' corresponded with the taste of lemonade. When we change the ingredients, we ask the other person whether the resulting liquid tastes like lemonade. If it turns out that the same physicochemical mix that produces 'lemonade' for us also produces 'lemonade' for the other, while mixtures that do not taste of 'lemonade' for us, also do not for the other, we can say that this organisation of ingredients corresponds to the perception of 'lemonade' taste as a property of that mixture. The production of 'lemonade taste' in other people 'out there' is indeed a property of the liquid.[24]

These contrived examples by themselves suggest very little about the 'realness' of the outer world, but when one multiplies the number of perceptions that can be consistently varied by our actions, and that seem to have correspondences in the perceptions of other people, judging by their actions if we disturb variables that would alter our own perception, it seems (under our assumption that there exists a 'real world' to be perceived) that our apprehension of the nature of that reality is strongly constrained by our ability to control our perceptions and to disturb those of other people.

Oliver's measurement of the weight of the rock is a procedure that could be repeated by another person who would perceive the result to be just as Oliver would perceive it. The weight is not in Oliver or the other person; it is 'out there' in the real world, a property of the rock.[25] But the same cannot be said of perceptions that we cannot influence by our actions, such as the honesty of another person. Only by their interactions with perceptions that we can control do such perceptions effectively belong to the 'real world', and, even then, that is something of which no individual can be sure.

Another person, an Observer watching Oliver's actions, might not see the scale pointer, and might perceive Oliver to be playing with the weights so as to enjoy watching his pan move up and down or, perhaps because Oliver is a scientist, examining the dynamics of pan movement. Different people observing different aspects of the same part of the universe may perceive different things, as the old allegory of the 'Blind Men and the Elephant' suggests.[26]

---

24   This example has been used to say that only the sources of input to sensory receptor cells are really in the environment, and that higher-level perceptions constructed from them are not. I do not assert that a single thing 'the taste of lemonade' exists in the environment, but something that we perceive as sweet, something that we perceive as sour, something that we perceive with this or that aroma, etc. exist together in a complex, and that is why these perceptions are concurrent, which is prerequisite for constructing the higher-level "taste of lemonade" perception.

25   More properly, of the rock at that location on this earth; 'mass' is a property of the rock, weight is not.

26   Source: https://en.wikisource.org/wiki/The_poems_of_John_Godfrey_Saxe/The_Blind_Men_and_the_Elephant, retrieved 2011.10.26. According to Wikipedia, this allegory was first (1872) presented in English in the following rhyme by John Godfrey Saxe, though the tale is much older. I apologise to anyone of the Hindu faith who is offended by his use of the word 'Hindoo' in his version of the rhyme. Clearly the blind men in the older picture were not Hindu. Nor were they in the non-poetic version I heard as a child.

Figure I.2.2. Blind men examining an elephant. **Hanabusa Itchō** *(1652 –1724). https://upload.wikimedia.org/wikipedia/commons/thumb/4/45/Blind_ monks_examining_an_elephant.jpg/1280px-Blind_monks_examining_an_ elephant.jpg, retrieved 2011.10.26. Image is in the public domain.*

THE BLIND MEN AND THE ELEPHANT. A HINDOO FABLE.

I.

IT was six men of Indostan
To learning much inclined,
Who went to see the Elephant
(Though all of them were blind),
That each by observation
Might satisfy his mind.

II.

The First approached the Elephant,
And happening to fall
Against his broad and sturdy side,
At once began to bawl:
"God bless me! — but the Elephant
Is very like a wall!"

III.

The Second, feeling of the tusk,
Cried: "Ho! — what have we here
So very round and smooth and sharp?
To me 't is mighty clear
This wonder of an Elephant
Is very like a spear!"

IV.

The Third approached the animal,
And happening to take
The squirming trunk within his hands,
Thus boldly up and spake:
"I see," quoth he, "the Elephant
Is very like a snake!"

V.

The Fourth reached out his eager hand,
And felt about the knee.
"What most this wondrous beast is like
Is mighty plain," quoth he;
" 'T is clear enough the Elephant
Is very like a tree!"

VI.

The Fifth, who chanced to touch the ear,
Said: "E'en the blindest man
Can tell what this resembles most;
Deny the fact who can,
This marvel of an Elephant
Is very like a fan!"

VII.

The Sixth no sooner had begun
About the beast to grope,
Than, seizing on the swinging tail
That fell within his scope,
"I see," quoth he, "the Elephant
Is very like a rope!"

VIII.

And so these men of Indostan
Disputed loud and long,
Each in his own opinion
Exceeding stiff and strong,
Though each was partly in the right,
And all were in the wrong!

MORAL.

So, oft in theologic wars
The disputants, I ween,
Rail on in utter ignorance
Of what each other mean,
And prate about an Elephant
Not one of them has seen!

Werner Heisenberg's elephant joke inverts the story in the following way:[27]

Six blind elephants were discussing what men were like. After arguing they decided to find one and determine what it was like by direct experience. The first blind elephant felt the man and declared, "Men are flat." After the other blind elephants felt the man, they agreed.

Moral:

"We have to remember that what we observe is not nature in itself, but nature exposed to our method of questioning."

But what was our method of questioning before the invention of language? Perhaps one of the blind 'Hindoo' might have stuck a pin in the metaphorical elephant. The elephant might then have ensured that the Hindoo became one of Heisenberg's flat men, after which the other 'Hindoo' might have been a little more careful with their pins, having learned something about what the 'elephant' was capable of doing.

---

27   https://en.wikipedia.org/wiki/Blind_men_and_an_elephant#John_Godfrey_Saxe, retrieved 2011.10.26.

Before changing the metaphor, we can wring a little more from it, or rather from the version in the poem. Each 'Hindoo' observed some property of the elephant, and proclaimed that the elephant was 'very like' something else with which the others would have been familiar, a wall, a spear, a snake, a tree, a fan, or a rope, each of which had among its many properties the property that man had observed the elephant to have.

Our 'elephant' is, of course, a metaphor for some part of the real world, or what Powers often called Real Reality (RR) to distinguish it from Perceived Reality (PR). The 'very like' entities such as the wall, the snake, or the fan were the PR experienced by the different Hindoo observers, not the RR truth, 'part of an elephant'.

At the end of the poem, the Hindoo dispute among themselves because they do not recognise that not only the elephant but also the entities chosen as similes have coherent bundles of properties, not just the single property each observed and likened to a property of the elephant. Indeed, the point of the poem might be that the elephant and the wall, spear, and so forth are, so far as we can perceive, nothing more than bundles of properties that are encountered together often enough to be perceived as an instance of 'that kind of object', which is not 'the other kind of object'.

We perceive only some of the properties of the coherent bundle; we never perceive the nature of the processing that might underlie the bundle — the bare entity in the real world, if such a thing exists. The 'taste of lemonade' is a property of all liquids that have a particular mixture of ingredients, whether or not any specific person is able to perceive that taste. But is it in Real Reality? Are the ingredients which go into the mix and that result in many different people perceiving a taste that has the label 'lemonade'?

We discuss the relationship between Real Reality and Perceptual Reality further after dealing with reorganisation in Chapter I.11. Until then, we consider only control of perceptions that are influenced by actions on the real environment of the controller.

I.2.3 Command versus Control

In military operations 'Command and Control' is a common phrase, as though the two words were almost synonyms. When we consider them in the light of PCT, the difference between them is not subtle. It can be well illustrated by this small snippet from Shakespear's *Henry IV Part 1*. Owen Glendower is plotting with Harry Hotspur to overthrow the King.

Glendower: *I can call spirits from the vasty deep.*

Hotspur: *Why, so can I, or so can any man;*
            But will they come when you do call for them?

Glendower says he can Command the spirits, though he intends Hotspur to understand that he can Control them. Hotspur does not fall for it, and points out that Command is easy, but not everyone can Control.

If there is such a stark difference between Command and Control as Hotspur points out, why are the words so conjoined that many people use them almost interchangeably? A possible answer is that if the environment of a control system is stable and protected from most disturbances, then Command is likely to have the desired effect, and to bring about the desired result, just as though Control had been used to correct the perceptual 'error' that induced the Command.

You turn a key to lock a door. Perhaps you then perceive that the door is locked without testing to see whether it 'really' is. Maybe on this occasion the lock failed to work, leaving the door unlocked, but you do not perceive that; instead you perceive what ordinarily would have been the result of turning the key. You commanded the door to be locked, but did not control it. Your set of perceptions of the current state of the World now includes a perception of a locked door, whether or not the door is actually locked.

If a military commander tells a subordinate to do something that is in the normal range of the subordinate's duties, the commander may well 'predict' that it will be done, and go about his business with a World Model (Section I.7.7) which includes the subordinate's success. That would be Command. Or the commander might tell the subordinate "Report back to me when you have done it", in which case, the commander Controls. But what actually does the Commander control? It is not necessarily his perception of the result of the subordinate's having followed (or not followed) orders.

Perceiving the subordinate's reporting back that the job was satisfactorily completed is not the same as perceiving that the subordinate correctly and effectively achieved the result the commander intended. However, if the subordinate has proved reliable in the past, the effect on the commander's perception of 'the way the world is' could be almost the same as it would have been had the commander actually observed the effects of the subordinate's successful actions.

When a reference value is established for a perception, the Elementary Control Unit (ECU) controls the perception to match that reference value. That controlled perception contributes an input to the perceptual functions of several higher-level ECUs, among which is likely to be one whose output contributed to the reference value in question. If a controlled perception is stable, then so is its contribution to a higher-level perceptual input.

Lower-level stability is one advantage, though not the only advantage, to multi-level control. However, this stability does not by itself allow Control to be superseded by Command. Other circumstances determine whether the effect of the action output can be trusted to have the desired effect. If other people have door keys and you can't keep watching the lock, you cannot know whether the door is locked even if the lock works perfectly, and if you have a reference value that it be locked, you have to try the door from time to time.

Why would one ever use Command without Control, given all the circumstances that have to be just so if command is to work? As Rabbie Burns wrote: "*The best laid plans of mice and men gang aft agley*", and when they do, you need control. On the face of it, one might assume that it would be necessary to control all the perceptions for which one has reference values, all the time. But is it? The example of the commander and the subordinate shows one situation in which it is not. We will come across others later in this book.

Provided that the subordinate correctly interprets the commander's order and the commander trusts that the subordinate is willing and capable and taking orders from nobody else, the issuance of the Command is all that the commander requires. The commander's World Model — the current state of all the perceptions of '*the way the world* is' — will include a perception that the order had the intended result. The commander's actions to set the subordinate's relevant reference values will almost always win any conflict the subordinate has about achieving the result the commander wants, unless he is an agent of the enemy.

One swallow does not a summer make, and one example does not prove a complex point. Critical properties of the example, however, may do the job. In the example, the commander commands a subordinate to do something the commander probably could do well, and would have been asked to do before being promoted. This being so, why command and not control? The answer is that having been promoted, the commander has many extra things to do that were someone else's job in his earlier life. He does not have time to do the things he previously did. So conflict is one of the important properties of the example — conflict in the commander between the things Commanded and other perceptions that the commander might now need to control.

A person has only a few degrees of freedom for output, but a group of commanded subordinates has many times more. A series of commands involves far less conflict at the physical output level than would an attempt by the commander to perform all the commanded tasks, and at the subordinate's level the conflict possibilities are reduced by the fact that each subordinate has a separate musculature and they can all be in different places doing different jobs at the same time. Even though a subordinate may not perform a task exactly in the manner the commander had envisioned, the result is likely to be close to what the commander intended.

I.2.4 Atenfels, Contingencies

To influence something in the environment and thereby control the corresponding perception requires some connection or 'Environmental Feedback Path' between the output of an ECU to its CEV, and continuing onward to its perceptual input. The importance of an accurate and rapid environmental feedback path is encapsulated in Kenneth H. Craik's (1970) comment on ancient and modern architectural design:

> *Once upon a time the architectural design process was unselfconscious. There was a right way to make buildings and a wrong way. Building skills were learned through imitation; design decisions were referred to custom, and the same form was erected over and over again. Learning form-making meant learning to repeat a single familiar physical pattern. Building practices were supported by a wealth of myth and legend, assuring the stability of the architectural tradition. Because men in these cultures built the shelters they inhabited, they were alert to shortcomings of the physical form. Thus, when change was compelled, the recognition of misfits was immediate and correction precise. The process of building, use, feedback, and alteration was continuous and resulted in well-fitting forms.*
>
> *In cultures with a self-conscious architectural design process, master-craftsmen control the form-making activities. Traditional design is not invincible, and wilful change, purportedly reflecting the inventiveness and individuality of the designer, becomes acceptable. Reaction to misfits is indirect and delayed, if it occurs at all. Social and administrative channels, which allow inhabitants to communicate misfits to designers, are ineffective or nonexistent, undermining the process of feedback and corrective action.*

A CEV is never a whole physical object, but it may well be a property of a physical object. The design of a house is not a physical object (though the house is). Whether it is physical or abstract, the feedback links in the environmental feedback path through which we may influence it may use the properties of concrete objects when the CEV is abstract, and vice-versa. For example, some elements of house design use aspects of the local building codes, while others use the mechanical properties of structural materials, and yet others use predictable properties of the yearly cycle of temperatures, sunshine, and rainfall.

If we want to perceive ourselves to be on the other side of a river, the means of getting there might be provided by a strong log. Properties of such a log can serve as an 'atenfel' ('atomic environmental feedback link') which can be used for crossing the river if the controller using it has adequate control of things like confidence and balance.[28]

---

28 Kent McClelland and I coined the term 'atenfel', from *ATomic ENvironmental FEedback Link*. An atenfel may be a physical or abstract property of a physical object, or it may be a supporting control loop, functioning as a link within an environmental feedback path. In a series of postings to the CSGnet mailing list in the 1990s, I used the word 'effordance' for this. The analogy with J. J. Gibson's (1966) concept of 'affordance' sometimes led readers to confuse the two concepts, and the concept of 'atenfel' is more precise. Gibson's 'affordance', a property of some structure, is a physical environmental component of an atenfel.

An atenfel is simply a means that a perceptual controller might use in control of a specific perception. A control loop in which an environmental component offers the 'atenfel' is called an 'atenex' (atomic environmental nexus), comprising both the external object property and the skill to use it effectively. The items listed in the house design of the last paragraph are all atenfels, so long as the builder has the skill to use them. Some are properties of concrete objects like bricks, some are properties of abstract entities such as building codes and historical meteorological records. All are properties, functional possibilities of the entity relevant to the perception being controlled.

Most objects we think of as physical objects are atenexes, in that they offer many different properties that could be the environmental part of atenfels for controlling different perceptions. Later, we will come close to claiming that all we can know of their possible existence in Real Reality is the total bundle of atenfels that they might embody. In other words, we will claim that the atenex is the reality, and the object is just a possible means of producing the atenfels. In Volume II and in Volume IV, we will use a metaphor based on crumpling paper to clarify a complex relationship between consciousness, category, and object perception, and the relation between language and the analogue hierarchy. But we must understand a lot of preliminary material before we arrive at that nexus where PCT will meet consciousness, independently of the concepts of Predictive Coding Theory.

In Chapter I.12, when we deal with the relation between Real Reality and Perceptual Reality, we will use an analogy that is already appropriate here — the analogy between the perceived entity and an Object in Object-Oriented Programming (OOP). In OOP, an Object is a 'private' chunk of code that is accessed only through a set of inputs and a set of outputs. The specifications of the Object determine what needs to be done to produce specified functional relationships between the inputs and the outputs, but those specifications set no constraints on how the Object might produce the specified relationships.

The programmer of an OOP Object may change its internal programming language in any way at all, so long as the processes that are coded produce the correct relationships in all respects including timing. Each of the functions which the Object specifications define is analogous to an atenfel, so an OOP Object is an atenex (unless it has only one output terminal, in which case it would just be a simple atenfel). A direct equivalent of an OOP Object, and therefore of an everyday object such as a teacup or a street, is what we will call a 'White Box' after Wiener (1948/1961). A white box is a knowable model of an unknowable 'black box'. Known internal processes in a White Box relate input (sensory) terminals to output (action) terminals in a way that emulates what an inscrutable 'Black Box' such as Real Reality does when it is acted upon so as to generate observable effects (Section. I.11.4).

The three constructs of perceived object, OOP Object, and White Box will prove useful when we look at PCT from different starting points, but for now it is necessary only to point out that all of them take the semiotic position that an object is defined by what it does when acted upon, not by any philosophical concept of 'thingness'. As the old saw tells us: "Handsome is as handsome does." To put it another way, an object might be defined entirely by its properties as an atenex. What can it do that might someday be useful to someone to control some perception? Answer that, and you know all there is to know about the object.

A car often provides an atenfel for perceiving oneself at a distant location, but also might provide an atenfel for seeing one's face in a mirror, or for perceiving oneself to have a little more cash on hand (e.g. by selling it, or using it as collateral for a loan). A microscope can provide an atenfel for control of perceptions of very small things or for pressing a tack into the wall. A telephone can provide an atenfel for controlling perceptions of the sounds made by a distant person or a reference value for controlling the colour desired for a new curtain. Most objects can provide many different atenfels, of which usually only one is part of the environmental feedback path in the control of any particular perception.

One must always remember, too, that it is not the object's property alone that can provide the atenfel. For it to be useful in control, the property must be coupled into a control loop together with the skill to use it for the given purpose. A car can indeed provide an atenfel for moving you to a distant location, but it can only do so if you have the skill to drive it or to use as a second atenfel the skill of another person to drive you in it to where you want to go. Although we will usually talk about the atenfels as though they are provided simply by the external environment, the complete atenfel always includes the skill coupled into a control loop together with the environmental property.

In the preceding paragraphs, I used the phrase "*can provide an atenfel for X*" several times. This phrase is a shorthand for "*has a property that could form a link in the environmental path for controlling a perception of X*". To use the full form every time would be cumbersome; in practice we often say that the object or environmental stability "*can provide an atenfel for*" or "*can be used to*", as in "*A car can provide an atenfel for moving to a distant location*" or "*A car can be used to move to a distant location*", eliminating also the phrase "*controlling a perception of*", which is always implied. However the word is used and whatever words are elided, the essential point is that it refers to only a section of the environmental feedback path for a specific controlled perception.

Atenfels can be treated as controlled perceptions of the properties of passive objects or of the actions of other people. As a child matures, it encounters different environmental situations in which it must use different actions to control a perception (such as the acquisition of food to control the perception of hunger). Initially, the baby has only one source of food, attained perhaps by crying, as we discuss further in Chapter II.11. The entire 'crying action' loop is an atenfel for controlling the hunger perception.

As an adult, one must choose what to eat, even in one's own house, but when one ventures into foreign places the range of choices of atenfel for controlling the hunger perception grows enormously. One may learn how to cook, how to find a good restaurant and order a meal, how to plant a vegetable garden, how to identify edible mushrooms in a forest, how to catch and kill a deer, and so forth. In the appropriate environments, all of these are potentially available atenfels for controlling the perception of hunger (and other perceptions as well).

Figure I.2.4 reproduces Figure I.1.2a with the inclusion of any atenfels that might be used by the action output to influence the CEV, or by the sensory input to affect the perception of the CEV. By the end of Chapter 4, the placement of the atenfels totally in the environment in this Figure I.2.4 should be clear, but here is the capsule summary: The 'Organism' part of the figure actually shows only what we will be calling an 'Elementary Control Unit', which is a small part of what is actually involved in control within a organism. The 'Environment' in the figure includes the rest of what happens inside the organism, including the lower-level control loops that implement the skills to use the objects in the exterior environment as atenfels.

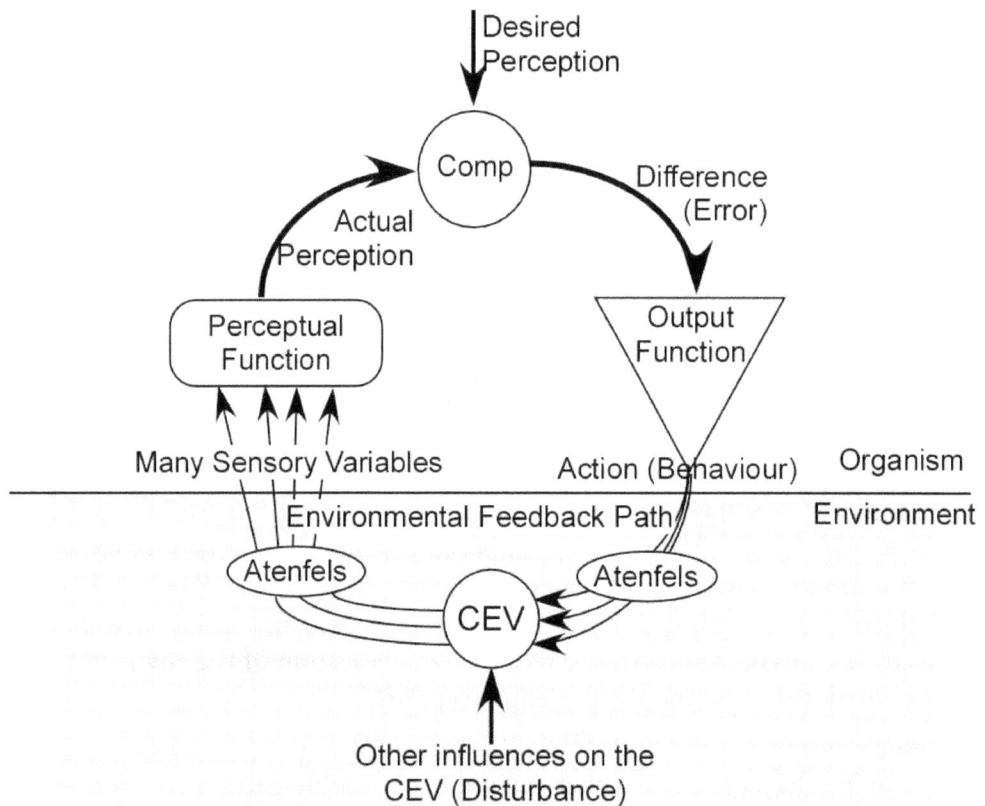

Figure I.2.4. The canonical control loop showing atenfels in the environmental feedback path, where "the environment" consists not only of the portion of the Universe outside the organism's skin, but also parts interior to the organism's skin affected by the signals from the "Output Function" and that contribute to its "Perceptual Function" inputs.. Nothing on the organism side of the loop senses its atenfels directly. They only determine how the output action influences the CEV, or how the CEV is perceived (e.g., spectacles to sharpen vision or a Geiger counter to sense invisible nuclear radiation).

Reorganisation, which we discuss beginning in Chapter 11, allows the developing perceptual control hierarchy to use more and more atenfels, a greater variety, singly or in combination, to control any one perception. It builds a repertoire of skills, which are atenfels if the environmental properties required to use the skills are available. A useful metaphor is the development of a carpenter's toolkit, from the child's first play hammer and saw, to the great complex of tools used by a professional cabinet maker, which would be of no use to the child who had not the skill to take advantage of them. The atenfels to which one has access, including the requisite skills, are one's personal toolkit for all of the perceptions that one controls. We treat access to atenfels at length later, particularly toward the end of Volume IV, when we consider some social implications of PCT, such as power relations.

I.2.5 Viewpoints and the Test for the Controlled Variable (TCV)

We will consider here three viewpoints, the controller, the observer, and the experimenter. The controller controlling a perception has no access to the state of the outer world beyond the current state of its perceptual signal. The controller's view of the outer world is limited to the effect of the changing state of its CEV upon the value of its perceptual signal. The controller has no knowledge of its atenfels or even of how its output influences the CEV or how the CEV relates to whatever is being influenced in the Real Reality to which it corresponds more or less closely. All it 'knows' is its perception, and all an organism 'knows' of the state of the outer world is the sum of all its perceptions.

An outside Observer looking at a controller doing its controlling might be able to perceive the actions induced by the controller's output, together with all the atenfels and her own CEV based on a portion of the same Real Reality as it is influenced by the controller. The Observer hopes that her CEV matches the part of Real Reality being influenced by the controller and that it otherwise corresponds to the controller's CEV. However, she sees output, atenfels, and CEV mixed with the effects of all the other perceptual controls being simultaneously performed at many levels of the subject controller's control hierarchy. In order to disentangle them, the observer must become an experimenter.

The difference between an Observer and an Experimenter is that an Experimenter acts upon her personal perceived world to see what happens, whereas an Observer simply watches what happens. In this case, what is observed or is the subject of experiment is what appears to be some component of the environmental feedback path of a controller. Most studied in the context of PCT is the case known as 'The Test for the Controlled Variable', in which the Experimenter acts directly to disturb a function of environmental variables which she hypothesised to be the CEV of the controller.

The Observer's view consists of anything that happens in the part of the control loop which she can perceive in the loop's external environment. An Experimenter is also an Observer, and the 'external environment' is usually outside the organism's skin, unless one is recording neurological or chemical values inside the body. Just as the Controller cannot see the environmental variable corresponding to the perceptual variable, so an Observer or Experimenter can see only whatever their own perceptual functions make of the sensed properties of the environment, many of which may be available also to the Controller.

The Observer can, however, assert that if the Controller is controlling some perception, then in the controller's environment there must exist a CEV, and the CEV will appear to be controlled in the sense that it will be stabilised against disturbances.[29] An experimenter can use this fact to seek a CEV in the hope that the Controlled Variable (CV) described by Powers (1973/2005 and elsewhere) is based entirely on sensory inputs, and thereby infer what function of environmental variables the Controller is controlling. To do this, the experimenter hypothesises one or more possible CEV functions, and disturbs them in ways that would be resisted if they were the actual CEV. This procedure is called the "Test for the Controlled Variable" (TCV). [30]

Two and a half millennia ago, the Chinese military theorist and successful general Sun Tzu advocated disturbing the enemy to determine his intentions (which we would call the reference values for his controlled perceptions). In a section on discovering the enemy's plans, he says: [31]

> *Rouse him, and learn the principle of his activity or inactivity.*
> *Force him to reveal himself, so as to find out his vulnerable spots.*
> — *Sun Tzu, ca.600 B.C.E., Tr. Giles, 1910; VI.23)*

The translator gives an example from another ancient Chinese commentator on Sun Tzu in which a commander sent the enemy commander an unexpected gift as a disturbance to see whether he would then launch an attack. In many places, Sun Tzu describes different kinds of disturbance that could lead the enemy to reveal his intentions. The principle is exactly the same as that of the formal TCV.

---

29 In the environment of the controller, but not necessarily in the external environment of the person, and therefore not necessarily accessible to the Observer.

30 The TCV can be used properly only if the hypothetical CEV is as accessible to the experimenter's senses as it is to the senses of the controller. However, even if some of the inputs to the perceptual function that defines the CEV come from the subject's imagination, nevertheless the imagined components may be sufficiently stable to allow the Observer/Experimenter to discover a function of environmental variables that defines an external CEV contingent on the stable value of the perceptual component of the actual function.

31 Various versions of this famous work are available on-line. The Giles translation is available in Apple iBooks form from Apple iTunes store, and from Project Gutenberg http://www.gutenberg.org/ebooks/12407, in which the cited item is Chapter VII.12.

Yuen (2008) describes the subtlety of Sun Tzu, and we shall discuss in several places how obviously in tune he was with the PCT of millennia later.

The so-called 'coin game' is a demonstration of the TCV. In it, the Experimenter gives the subject a few coins and asks the subject to lay them out in a pattern that the subject freely chooses and can describe in words (such as 'three in a row with two others on opposite sides of the row'). The Experimenter then moves a coin, and if the subject moves it so that the arrangement resumes satisfying that criterion, the Experimenter deduces that this coin must be part of whatever pattern the subject has chosen as a reference (Figure I.2.5).

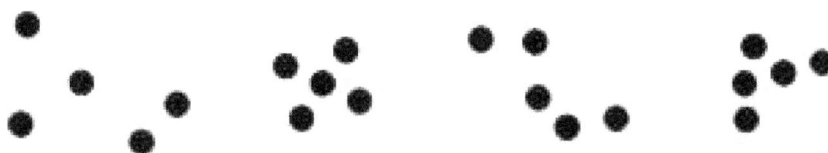

Figure I.2.5. Four arrangements of five coins, three of which would be accepted by the subject as conforming to the (secret) reference pattern (three in a straight line with one on either side of the line).

After a while, the Experimenter may be in a position to predict accurately which moves the subject will or will not resist. At that point the Experimenter is perceiving something related to the CEV, though when they finally compare notes what the Experimenter describes as the CEV may well differ from what the subject describes as the reference pattern. For example, if there were, say ten coins, the Experimenter might deduce that the subject is controlling for the coins to be laid out in the shape of an 'N', whereas the subject was controlling only for 'any zig-zag pattern'.

When the choices of what might be being controlled are few and well-defined, the Experimenter can be highly accurate in discovering the CEV. The key is that the effect of the disturbance on the true CEV almost vanishes except for a short transient (for example, until the subject has had a chance to return the coin to its 'correct' position).

In a continuous system with ongoing variable disturbances, the matter is not so clear-cut, because control is never perfect and some influence of the disturbance will always remain (apart from random moments) whether or not the Experimenter has correctly guessed the CEV.

A second observation available to the tester, even as Observer, is that if change is continuous over time, then the disturbance is less correlated with the CEV of a controlled perception than it would be if the environmental state did not correspond to a controlled perception. However, this will also be true if the hypothesised CEV is not the actual CEV of a controlled perception, but is highly correlated with it over the range of disturbances used in the test.

One thing the Observer/Experimenter can say by using the TCV in a continuously variable domain is that the true CEV is related to the hypothesised one, and that the less the influence of the disturbance, the closer the relationship. The perceptual variable corresponding to the CEV has near zero correlation with the disturbance, whereas the observed output to the environment has a highly negative correlation with the disturbance, i.e. almost completely cancelling its effect..

We will argue later that a General Protocol Grammar (GPG) implements the TCV in interpersonal interaction. Anticipating that, it is important now to emphasise that although the TCV can come close to determining what the Controller is controlling, it can be exact only when the possibilities are from a small, discoverable set of discrete possibilities.

Whereas the Controller can know only the state of its own perception, and the Observer/Experimenter can see only the part of the control loop outside the organism which contains the Controller, there is yet a third viewpoint, that of the Theorist or Analyst. The Analyst can imagine the entire loop, in the same way that someone maintaining a mechanical control system can investigate every component of the loop.

The Analyst studying the loop can analyse what would happen if this or that property of the loop had this or that value, and can assert that if the controlled perception is built by such-and-such a perceptual function, then the CEV in the environment will be influenced thus and so, or if this or that property of the perceptual function's input is obtained from imagination instead of from direct sensory input, then this and that consequence will follow.

The Analyst's viewpoint includes both the Observer/Experimenter view and the Controller's view, so it becomes very easy to mix the viewpoints when discussing the implications of perceptual control. This is especially true in a multi-person situation, in which the Analyst sees all the participants, some of whom are Controllers of some perceptions while being Observers of, or Experimenters on, other participants. No doubt viewpoints will be inadvertently mixed in the sometimes involved discussions later in this book, despite my best efforts to keep them clear. I trust that such occasions will not lead to unnecessary confusion. Most of the time, the Analyst's viewpoint is the default assumption and when other viewpoints are used, the occasions either are mentioned explicitly or should be evident from the context.

I.2.6 Words: Skeletons in the Flesh

I gotta use words when I talk to you.
But if you understand or if you don't,
That's nothing to me and nothing to you.

These wise words are from T. S. Eliot's "Fragment of an Agon", composed in 1927 for an unfinished play in verse called "Sweeney Agonistes".[32] We may question the last line, but the first is true of the text of an interaction when the 'I' and 'you' of the poem are separate. It is not true if you understand 'talk' to include gestures, facial expressions, tonal contour of speech, and other continuously variable effects that the 'I' may do that the 'you' might be able to perceive. Talk accompanied by these continuously variable modulating actions may allow nuances of understanding that would be very difficult to produce using words alone, but those understandings can never be exact.

Eliot's words are a mere skeleton of what Sweeney might have been talking about. They can be seen, but the flesh, the connotations and emotional relationships, is not in the words. The conformable flesh lies elsewhere, not in any words that could be written. Our mind is a closet of perceptions, which we can open only to show the hidden skeletons within.

The 'skeleton in the closet' metaphor refers to something deliberately hidden, probably that the hider would prefer not to be observed. But the skeleton of a vertebrate is normally hidden under flesh, and without it we could not live. The skeleton gives us our visible shape. If we were pure flesh, we would fall into a mushy pile on the floor. But if we were pure skeleton, we would be like any other skeleton in a museum, a static arrangement of bones. We would be unable to act. Our skeleton gives us shape, our flesh gives us life, but only when its muscles work with the skeleton to change the angles between the bones. Muscles pull, bones are unchanging. Muscles change their length as we act, bones have clear and nearly constant shapes and dimensions apart from their growth from infancy to maturity. We will refer to this contrast, and its analogues in control systems and their relationships, when we discuss 'tensegrity' starting in Chapter I.8.

I liken a theory to the bones under the flesh of natural or experimental observations. Perceptions which we consciously experience are the only truth of which we can be sure, and they are the flesh that has some shape, under which we build theories that could be possible truths that define the shape. Our theories are hard-edged and hold their shapes, like the bones of a skeleton. Some theories define the bones, some define their connections, which allow the bones a constrained set of angles that underlie the changing shapes of observations.

A good theory describes observable shapes and collections of relationships that have not previously been described — but how do you 'describe' an observed shape, whether physical or metaphorical? You have to use words (including mathematics, which is a precise shorthand for words in specialised technical

---

32 Eliot (1963).

parts of language). This presents problems when variables which are not discrete, not individually identifiable, are identified with words, which are discrete.

This problem was resolved by Cantor, who proved in a simple way that there are more points along a line than there are integer numbers, though both are infinite. The number of integers is typically a countable infinity called 'Aleph-null', and the infinitely larger number of points along a continuum line, area, volume, etc. is called 'Aleph-one'. All the integers could be placed along a line of indefinite length, but no matter how closely they were spaced, there would always be an infinite number of points on the line between any two of them.

Here we come to the crux of the problem of using words to describe shapes. The number of possible words and word sequences is infinite, but it is a countable infinity, because each is distinguishable from all the others, whereas shapes vary continuously. Using words, one could approximately describe a shape as closely as one wished, but could never in a million lifetimes describe it exactly. And therein lies the truth of T.S.Eliot's words. His words can describe words, but no words can precisely describe what Eliot's Sweeney intended to be understood. Sweeney says that his interlocutor's understanding is not a perception that he controls; neither does a theorist have the means to control to a precise shape anyone else's understanding of their theory, with its connotations and consequences.

Words are labels, labels for things, labels for actions, labels for events, labels for feelings, labels for relationships, and so forth. All labels reference categories — equivalent to stretches along Cantor's continuum line — as befits their individual label identities. As labels, they identify what they label as being something different from what might be identified by any other word. The difference might not be in denotation, the ability to select, but in connotation, things, actions, etc. which often go along with or are associated with that which is selected by the word. Think of the difference in connotation among these words, all of which denote the same action: 'kill', 'murder', 'assassinate', 'rub-out', 'eliminate' (when talking of a person), and so forth. They all denote acting so that a person ceases to be a living thing, but their 'meanings', which includes their connotations, are rather different.

If we accept these connotative variations among words that denote or label 'the same' thing, in this case an action with a specific consequence, the labelled 'thing' is different from other 'things' that have other labels. To 'assassinate' is indeed to 'kill' and to 'murder', but as a label for a category of killing, it is different in having different category boundaries. The others do not suggest that the person is killed for a political purpose.

Words label categories, and categories imply the possibility of 'Yes-No' decisions. A penguin is or is not a member of the category 'bird', a crystal is or is not hard enough to etch glass. There can be no nuances if a decision is needed. To say that the penguin is a non-flying kind of 'bird' will not do if the problem solution requires a bird that flies. The decision is 'Yes it is' or 'No it isn't' (though we will soften this in Chapter I.9). The problem here is that theories expressed in words, as they are in a book like this, are based in hard-edged logic, and any theory developed using classical logic alone must necessarily fail to describe continuously variable observations exactly.

Powers's Perceptual Control Theory is based on a hierarchy of perceptions with a continuous range of variation, based on inputs from our myriad impulse-producing sensors in several sensor systems (e.g. vision, audition, taste, smell, touch, kinaesthesia) in which each sensor has a continuous range of variation in the time between output impulses. The perceptions produced by his hierarchy of perceptual input functions are continuously variable, until we reach what Powers called 'the category level' where ranges of perceptual variation are merged into separably identifiable categories (we offer an alternative approach to category perception in Volume II) that can be individually labelled in words or as 'rational' numbers.[33]

I will refrain from pursuing the link between consciousness and labelling here. It will be an integral part of the complex interaction between the categorical operation of Predictive Coding and the continuous operation of the Powers perceptual control hierarchy. We will discuss this interaction at several points in Volumes II, III, and IV of this book — using words, of course, to describe the skeleton of understanding that Eliot's Sweeney considered inadequate.

I.2.7 Motifs and Emergents

At various places in this book, we will introduce a new 'Motif'. But just what is a motif? As ordinarily used, the term applies to some form of art, typically visual or musical, but also culinary, architectural, or in whatever field some practitioner might be called an 'artist', including the writing of any form of literature. A motif is a structural element which is used by the artist appreciably more often than other possible structures, in such a way that instances of its use within the art object are similar in effect. For example, the 'da-da-da-boom' rhythm is a motif that begins Beethoven's Fifth Symphony and recurs with minor transformations throughout the Symphony, and a major triad collection of pitches such as C-E-G is used in most forms of Western music. Within PCT, the simple control loop is a motif for which the artist is Nature, a motif that is used by all living things to the extent that its use might almost be taken as a definition of life.

As the word 'Motif" (capitalised) is used in this book, it refers to a structural arrangement of components that has some emergent property that cannot be attributed to any other arrangement of its individual components. The control loop Motif has the emergent property of control. It controls the value of some variable. The property is attributable only to the control loop's specific structure,

---

33 A rational number is one that can be expressed exactly as a fraction in which the numerator and denominator are both integers. A number that cannot be so described, such as π ('pi'), is called 'irrational'. We have names for a few irrational numbers that, like π, appear in many mathematical formulae, but the infinite majority of the points on the number continuum are nameless.

not to its components, since a different arrangement of the components would not have that property.

The uncapitalised 'motif' refers to the repeated structure itself, without the new emergent property that the structure might have. For example, our first Motif is the arrangement of functions in a loop (perceptual function, comparator function, output function, environmental feedback function, perceptual function…) which we call a control loop, with the emergent property of control or stabilisation. This property of 'control' is an 'emergent' property which occurs only when the structural components of the Motif are correctly organised. We do not consider a particular structure to be an instance of a distinct Motif unless it has a distinct emergent property not exhibited by any other Motif.

A Motif can be varied and elaborated, provided the components still influence each other in the same way and the emergent property is not compromised. For example, Volume II starts with an extension of the control Motif into the 'homeostatic loop', a longer version of the control loop in which multiple interacting variables are stabilised against independent disturbances. The homeostatic loop might have been considered a Motif of its own, but it is not, because it produces no new emergent property, just more instances of the stabilisation or control that is emergent.

Different arrangements of the same components may occur and form different Motifs, each with its own emergent property. For example, when two control loops are used in conjunction, the different ways they interact can produce different emergent properties, sometimes more than one new emergent. We will discuss some more Motifs and their emergents in later volumes, but in this volume we will meet two emergent properties from an arrangement in which two control loops oppose each other's attempts to change the value of some environmental variable. These two emergents are 'stiffness' and 'conflict', both of which will turn out to be important when we later discuss personality.

The same opposition structure also induces an emergent that is more than a simple property, that we will call a 'virtual control loop'. The emergent virtual control loop structure is treated as a Motif of its own, serving as a component in other structures. When a 'conflict' structure is seen in two or more dimensions, we call the emergent virtual control loops 'Giant Virtual Controllers' (GVCs), which appear to control variables affected by the conflict structure, such as the environmental variable in contention in the conflict. When we deal with social structures, GVCs are often more influential than controllers in individuals, and in a model can often be substituted directly for individual controllers.

As we progress through the different specialised fields for which PCT seems to have useful application, more complex Motifs appear, most of which involve more than one individual actor. Many, if not most, of these complex Motifs have simpler ones among their components. An important one is what I call the 'Protocol' (Chapter II.11), a Motif for which the primary emergent is mutual comprehension or its opposite, deceit. The Protocol Motif, in its turn, is a component of the Trade Motif (Chapter III.9), which uses two of them and also includes four 'conflict' Motifs. These Motifs and others, and their emergent properties, will become prominent when we trace the implications of PCT in different areas of research.

Chapter I.3. Language and Culture

What is Language? What is Culture? What is a language, and what is a culture? These four different questions are seldom asked in the literature of Perceptual Control Theory. (For exceptions in respect of language, see Runkel (2003), Nevin, in LCS IV (Nevin 2020), and in respect of culture McClelland, in LCS IV (McClelland 2020) and other writings referenced below.)[34]

Language is sometimes simplistically thought of as being defined by collections of words whose meanings can be found in a dictionary together with rules for the ways these words can fit together, while *a language* is considered to be a specific selection of words and rules. According to this view, you could, in principle, find out all there is to know about Language and a language from books, exactly as you might if you wanted to find out all about an electric motor. And for a formal language such as FORTRAN, C++, or Python, you can. But we are not talking about computer languages here. We are interested in how language is used between living people, and how it comes to be the way it is.

Language between people is *used* differently from an electric motor or FORTRAN. How you talk to somebody depends on whether the other is a close friend, a colleague, a new acquaintance, or an enemy, whether they seem to be feeling happy or sad, whether the situation is formal or festive — one could extend the list of 'whethers' indefinitely. But like an electric motor, language is a tool, a tool that works not with the inanimate objects that seem to surround us, but with other people, and of those other people, with only those who understand the particular language that we use. Books cannot tell how any language is used in every different situation or by random people pursuing their random purposes. For that we need to consider principles other than selections of words and rules. We have to deal with how people interact, and how they *learn* to interact within a particular culture.

To define a 'language' is not as easy as it might seem. There are the great languages such as English, French, Chinese, Arabic, Swahili, and Russian, but within these there are considerable variations. Chinese comprises at least seven different languages (united by a writing system),[35] and English varies so much from region to region that some 'dialects' might well be called different languages, just as French, Italian, Spanish, and Romanian are considered to be separate languages rather than different dialects of Latin.

Should we call the dominating and submissive gestures of wolves, the varied alarm calls of monkeys, or the communication signals of porpoises or bees 'languages'? For some purposes we may, but for now we will not. Later, we will see that they can sometimes take the same place as human language in the analysis of interactions, and then we may call them 'languages' in that context.

---

34   Also writings of Ted Cloak presented and cited at https://www.tedcloak.com/. — Ed.

35   Hundreds of Chinese languages form seven groups;
        https://en.wikipedia.org/wiki/Languages_of_China. — Ed

Different cultures use the same 'language' differently, using different vocabulary, pronunciation, and even syntax. Within English spoken in England, for example, consider this example. Some time around 1970, I was playing cricket in Somerset (southwest England) and there was a small dog of unknown breed lying just on the wrong side of the 15 cm high fence that marked the boundary of the playing field. I asked a local watching the game "What kind of dog is that?" Transcribed into standard English words from the very different pronunciations he used, his answer was: "*Her be a 'Sooner* Hound', *her be. Her be sooner on that side of the fence than this.*" The sounds were as different from those of the English to which I was accustomed as were the vocabulary and syntax. The final 'this' sounded more like 'dyeeez', but it was intelligible, if barely, and funny. As another example, when I was on Sabbatical in England in the 1960s, one of my local colleagues claimed to be bilingual in 'English' and 'Manchester'. He said he spoke 'Manchester' when he went home some 60 km (about 40 miles) away, and 'English' when talking with academic colleagues.

Not only do major languages sometimes have dramatic differences in their vocabulary and syntax among cultures that use 'the same language', but sometimes languages blend across geographic boundaries. There is a famous sentence: "*'Good butter and good cheese' is good English and good Fries.*" (Friesland, pronounced 'Freezeland', is a northern region of the Netherlands). The Fries language is often hard for a person from Amsterdam to understand, but it is said that Fries and East Anglian English fishermen can speak together when they meet at sea.

The same was also said to be true of fishermen from northeast England or southeast Scotland and those from West Jutland (Denmark). In my own experience as an originally English child with four years of schooling in southeast Scotland, on first coming to Canada I was able to follow, without using the subtitles, perhaps half of the dialogue of a Danish movie set in West Jutland.

When that kind of cross-language intelligibility happens across a land boundary, we sometimes say that the people of the intermediate area speak a separate language or an interlingua, such as Piedmontese (between French and Italian) or Catalan (between French and Spanish). But the boundaries of such inter-languages are never clear, and quite often a speaker of one of them will also use a more 'standard' version of one of the major languages in the blend.

Members of groups based on different interests may use 'the same language' differently. Technical groups have their jargon or sublanguage within which some words are not used as they would be in the larger public. For example, "It's all perception" means something quite different in PCT than it does to a politician attempting to divert attention from a difficult situation. Many families use language with each other in ways that are subtly distinct from the ways they speak to outsiders, without even noticing the difference. These groups that use language differently represent different, though possibly overlapping cultures.

I.3.1 Language and Culture as Artefacts

There's that word 'culture'. What is 'Culture' and what is '*a culture*'? The Oxford English Dictionary (OED) offers many meanings for the word, most of them related to farming, agriculture, or microbiology, in which someone encourages the growth of something of which they are in charge. We don't mean anything like that in this book. Here we deal with the patterns of interactions among and within social groups, usually but not always human. It is what we mean when we talk about 'Italian culture' or 'the sporting culture', but not when we talk about '*a bacterial culture*'.

What are those patterns of interaction? In a culture, people use certain *protocols* and may perform certain *rituals* when meeting, when dining, when dealing with strangers, when needing assistance, when competing as a sports team, and so forth. A protocol is a flexible way of interacting, which I discuss in detail both in Volume II and in a chapter in LCS IV (Taylor 2020). Rituals, in contrast, are agreed sequences or patterns of action that have some public effect within a particular culture.

'Culture' refers generically to sets of rituals and of protocols of interaction, whereas 'a culture' refers to the specific protocols and rituals used by a particular defined group, whether it be a family, a sporting club, those professing a particular religion, the citizens of a particular region or country, a secret society, and so forth. A given person might belong to many of these groups, and would exist within a different culture in each of them.

Now we come to the word 'artefact'. We claim that both language and culture are artefacts, having the same kind of status as more concrete artefacts such as houses and ships. They exist in the perceptual environment of the perceiver just as do bicycles and trees. The Oxford English Dictionary has this to say about the word 'artefact': "*(rare) a thing made by art, an artificial product*". The Unabridged Random House Dictionary is more forthcoming, having six definitions for 'artefact':[36]

1   Any object made by human beings, esp. with a view to future use.
2   A handmade object, as a tool, or the remains of one, as a shard of pottery, characteristic of an earlier time or cultural stage, esp. such an object found at an archaeological excavation.
3   Any mass-produced, usually inexpensive object reflecting contemporary society or popular culture: *artefacts of the pop rock generation.*
4   A substance or structure not naturally present in the matter being observed but formed by artificial means, as during the preparation of a microscope slide.
5   A spurious observation or result arising from preparatory or investigative procedures.
6   Any feature that is not naturally present but is the product of an extrinsic agent, method, or the like: *statistical artefacts that make the inflation rate seem greater than it is.*

---

36 The American spelling is of course a cultural difference.

Most of these definitions suggest that an artefact is a tangible, concrete object, but even when an object is a tangible object, the artefact may be abstract. The last definition describes an abstract relationship between abstractions, the true inflation rate and the computed inflation rate, and definitions 4 and 5 could be construed as referring to abstract structures or features.

These dictionary definitions do not seem to help very much when we talk about culture and language as artefacts, so let us try another: *An artefact is perceptible by humans, is susceptible to influence from humans, and exists in its current form only as the result of human perceptual control.* This definition is agnostic as to whether the artefact is tangible, but it does cover the essence of the OED definition and all of the Random House definitions, if we ignore the connotation of 'thing' as necessarily being a tangible object. The definition also suggests why an artefact, perceptible to and influenced by humans, is often 'malleable' (literally 'deformable by hammering').[37] What humans can create, humans may be able to change.

Consider a statue by Michelangelo, quite literally a malleable artefact. The block of marble is a completely natural phenomenon, created millions or billions of years ago by geological processes. The artefact is not the marble, but the shape imposed on the marble by the sculptor, the relationships of planes and curves, the likeness to a human form. The shaped marble is clearly tangible, but is the shape itself? One can touch the shaped marble left behind when Michelangelo finished chipping and polishing, and one can perceive its shape, but one cannot touch the shape. Michelangelo, however, could perceive and have a reference for the shape, and could control for perceiving the marble to take on that shape.

Suppose Michelangelo had started to carve a statue, but left it after he had only roughly cut out the general shape of the upper part and had not touched the native marble of the lower half. Wherein is the artefact? Is it in Michelangelo's reference for the shape it would have become? Michelangelo's intention is not perceptible to another person who looks at the rough-cut block after his death. To define the artefact as a reference value is not useful in a social context. We must consider only what another person can perceive, and so the *social* artefact can be only the actual shape achieved by the sculptor.

An 'artefact', in this context, is not only an artificial object whose properties are the environmental consequences of perceptual control, but also one whose properties might be used in controlling some other perception or perceptions. An artefact can provide atenfels, but atenfels are also available from non-artefacts such as the weather, pebbles on a beach, or the colour of a leaf. I mention the provision of atenfels by artefacts here because the grand artefacts of language and culture are primarily atenfel providers.

Artefacts exist outside any individual, in the public environment. Anyone can pick up a knife and, if sufficiently skilled, could use it to carve a wooden sculpture or slice an apple. Those properties of a knife depend on the knife, not

---

37 Latin *malleus* 'a hammer'.

on the individual using it. Nobody could use a baseball bat to slice an apple with any precision. The skills required in order to use the knife in these different ways are properties of the individual. Only to a person with the skill is a knife useful to carve a delicate sculpture out of a block of wood. But is that true of a language or a culture? Can anyone with the appropriate skill 'pick up' a protocol and start using it as freely as they could pick up a visible knife? I claim that the answer is "yes" if the person with whom they want to use it has complementary skills, such as being able to speak or at least understand the same language.

This book will argue that, along with a culture, a language can be treated as an artefact distinct from the words that are actually written and spoken, and that this artefact is malleable. We will argue that the artefact that is language is created and maintained through a process called 'collective control' that we will explain in Volume II of this book. Just as with any physical artefact, if a language is not maintained, it will erode, decay away or disintegrate, and finally become unrecognisable as a distinct entity except by specialised 'linguistic archaeologists' called paleolinguists.

In all the above, I have used 'person', a human, as the unit of a culture, but many other species have characteristics that seem to allow them to belong to cultures. In most of what follows, I will continue to talk about humans, but it should always be kept in mind that much of what is said that does not depend on linguistic skill will apply in whole or in part to many primates, to elephants, to pack hunters such as wolves, to sea mammals, to many birds, to ants, and to a whole list of other species that live together and may sometimes collaborate in their activities, even, sometimes, bacteria. In some cases, the concept of 'language' may be purely chemical, as with the pheromones emitted by ants (and mammals) or the chemicals that affect the sprouting behaviour of a slime mould.

The artefact that is language is not in its physical manifestation any more than the shape of a Michelangelo sculpture is a property of marble. The artefact is in its effects on the perceptions of the receiver. Nevertheless, in most of what follows, we will treat 'language' as an artefact created, shaped, and used by humans, ordinarily perceived using auditory and visual senses. The extension to non-human communication may be mentioned or can be easily imagined, but that is not our main concern.

Note: The next two sections further develop an understanding of language and culture as artefacts. Discussion explicitly of PCT resumes with Part 2: Simple Perceptual Control and Chapter I.4. Basic Aspects of Control.

I.3.2 Language Drift Over Time

I have said that artefacts are malleable. Human languages change over time. For example, corpus studies (https://cqpweb.lancs.ac.uk/) of spoken English in Britain show that the use of split infinitives like "to actually get", "to really want" and "to just go" has almost tripled over the last three decades; Americanisms such as awesome and cheers have replaced some of the more typically *British* sayings like marvellous and cheerio; and there is rapid expansion in using a noun as a verb, as e.g. to send a text becomes to text.

In my youth, 'impact' was purely a noun describing the effect of a short, sharp blow, but is now both a noun and verb with no sense of suddenness or sharpness, almost entirely supplanting 'influence', 'effect', and 'affect'. Similarly, 'gift', which used to mean a thing or service voluntarily provided without recompense, is now a verb 'to gift', which seems to be in the process of replacing 'to give'. At the same time, the noun 'gift' has become 'free gift', so that whereas one might have said "I gave Susie a gift", now one might even say "I gifted Susie with a free gift", which would have been complete nonsense half a century ago. But personal impressions like this are often deceptive, and an apparent innovation may only be a resurgence in popularity. Evidence of 'impact' as a verb is about two centuries older than that for the noun usage, and **gift** as a verb is at least four centuries old. Resurgence of 'to gift' was probably boosted by an episode on 'gifting' and 'regifting' in the TV series 'Seinfeld'.[38]

New words are invented daily, especially in talking of technical matters: 'twitterverse', 'unfriend', 'selfie', 'e-mail', 'iPhone', and the verbs 'to text' and 'to sext'. A new word enters the language roughly every 100 minutes, and other words disappear unnoticed, though one never knows whether a word that one has not heard for years will be heard tomorrow.[39]

Phrases also drift: 'Anniversary' used to mean a date exactly one year after an event, or more years later when accompanied by a modifier such as 'tenth anniversary'. 'Anniversary' now means a period such as a week, a month or a year after the event, needing modifiers such as 'one month anniversary' in contrast to 'one-year anniversary', while 'second anniversary' has changed to 'two-year anniversary'. "So I said…" began to change to "Then I'm like…" some thirty or so years ago, at least in my part of the world. Is this change 'forever'? Perhaps, but it is more likely that some other phrase will soon usurp its place. In the Lord's Prayer that I learned as a child, there is a line "Deliver us from evil." Now "deliver us" would be more likely to imply putting us in a box and taking us as a package to some defined location.

---

38   See https://www.merriam-webster.com/grammar/yes-impact-is-a-verb; https://www.merriam-webster.com/grammar/gift-as-a-verb

39   See https://languagemonitor.com/

The introduction of new words for new concepts and the use of slang for in-group identification, for example by teenagers and some criminal organisations, differs substantially from the drifts that change one language into another over centuries and millennia, though they may contribute to it. We now consider some less obvious but more important long-term drifts and failures to drift in language.

Over a very long time span, some aspects of language change very little. Proto-Indo-European (PIE), the language from which English, Russian, Latin, Greek, Sanskrit, Swedish, French, Romanian, Gaelic, and many others derive, was spoken maybe 5000 years ago, and yet many of its features can still be heard in its distant descendants. Here are a few samples of things that have changed and things that have not, over that long time span (based on Watkins 2000).

The PIE root terk^w- meant 'to twist'. Some English words descended from it, possibly by way of 'twerk', are queer, thwart, torch, torment, torque, torsion, tort, torture, truss, contort, distort, extort, nasturtium, retort, torticollis. But the English word 'twist' does not descend from terk^w-. 'Twist' comes from dwo-, along with two, twelve, twilight, twill, twine, twice, between, twin, twig, diploma, duet, dyad, double, duplex, doubt, dubious, redoubtable, and many more.

The words that have drifted from terk^w- form distinct families. Most of them have a common sound shift in the vowel from 'er' to 'or', but some do not. Most have kept the initial t, but in many the k^w– is softened to a 't' or 'sh'. In their meanings, some (torque, torsion, torticollis) retain the sense of a mechanical twist, but for most the 'twist' is mental or metaphoric. In some cases the twist has become an obliqueness, or a sense of crossing, as with 'truss', and 'thwart' (of a boat), 'athwart' meaning a barrier across some passage, or 'thwart' (blocking someone's ability to do something). Twisting parts of the body can be painful, a sense that is retained in 'tort' (a legal hurt), 'torment', 'torture', and 'extort'.

Despite the wide range of meanings, for example from twilight to twig to duet to duplicity in the descendants of dwo-, the sound pattern associated with the meaning 'two-ness' has shifted very little over the thousands of years since the early Indo-Europeans roamed the shores of the Black Sea. Watkins (2000) even is able to present a table of consistent sound drifts of 35 phonemes between PIE and twelve ancient Indo-European languages spread over a region from India to northwest China to Ireland, and from old Germanic to five more recent Germanic languages such as Old English. Similarly, Hogben (1964) provides a table illustrating the patterned sound shifts within and between Indo-European language families, using four Germanic, four Romance, and four Gaelic languages as examples.

As an example, Watkins says: "...Proto-Indo-European initial **p** remains **p** in Latin, but it is lost entirely in Old Irish and becomes **f** in Germanic and consequently in Old English; this Indo-European ***pəter-**, meaning 'father', becomes Latin pater, Old Irish *athir*, and Common Germanic ***fadar**, old English *fæder*." And **dyeu-pəter** (God the Father) drifted into classic Greek *Zeus* (omitting the 'father' aspect of the god, despite his sexual reputation) and Latin

Jupiter (which has a sound very like its ancestral form). And English prayer may refer to "Our Father, which art in Heaven", omitting the **dyeu** (French 'Dieu', God) part of the Proto-Indo-European progenitor.

The drifts mentioned above have happened over a time span of perhaps five thousand years, and yet in many cases the modern forms in the various languages have an acoustic resemblance to the ancient forms. You can hear the family resemblance, though it is doubtful that you would immediately have recognised them as 'the same' word. The same is true of the different meanings of the descendants. In many cases but not in all, you can see the two-ness of the descendants of dwo- or the twisting effects in the descendants of 'turkey'. The k of kaito is softened into the h of 'heath', but one can see this same softening even now in the comparison of 'ski' (pronounced 'she' in some English dialects) and the Scandinavian 'skip' for English 'ship', or in the the English words 'skirt' and 'shirt', derived by different routes from PIE (s)ker-. (These two words are interesting in that an intermediate form of 'shirt' apparently was 'skirt', and vice-versa.)

Such drifts and non-drifts do not occur as a result of decisions by some authority. They happen in the course of using language in ways that may subtly differ from one person to another. Ohala (1992) provides evidence from many languages in different language families that the source of the drifts is to be found in perceptual confusion, not in production ease.

What happened to the PIE **p**? When it becomes **f**, pretty well all that changes perceptually is the duration of the aspiration, the puff of air that accompanies initial p in many English words (though not in some other Indo-European languages), although the easy ways to produce the two sounds differ in where the lower lip closes off the air flow — against the upper lip or the upper front teeth. Changing that mode of production can cause subtle changes in the acoustic representation of neighbouring vowels, they may influence the perception of other consonants, and so forth. The drifts are all interconnected.

By all these criteria, 'language' or 'a language' is hard to define. Defining it is even harder when one considers the non-verbal ways people communicate with facial gestures, 'body language', and the like, all of which can, in appropriate circumstances, be used almost interchangeably with words. While some gestures seem to be shared with other animals, as Darwin (1899) noted, others differ substantially across cultures, and are as arbitrary as the sounds we associate with the concepts we communicate using words.

Language in words or gestures is changeable, but does it exist outside the actual things that are written or spoken? Is there now an Assyrian language? or Latin? Was there ever? Does the Minoan Cretan language exist that was written in Linear A and that nobody can now understand, and that nobody has spoken or written in over 3000 years? How about a computer programming language such as FORTRAN or C++? We do not claim to provide a definitive answer to such questions; rather, we would argue that there can be no definitive answer without presupposing the answer in some arbitrary definition.

I.3.3 Culture

What do we mean by 'culture'? In this work we mean the whole network of behaviours and tangible artefacts that distinguish one group of people from another, whether the group be a family, all those who profess a particular religion, those who support a particular sports team, or in any other way participate in a common set of protocols and rituals for how they interact with one another. The use of a particular language, dialect, and even accent can be an aspect of a culture, as can whether one keeps both knife and fork in hand when dining or cuts the food before transferring the fork to the other hand for actual eating. As with language, many, if not all, of the elements that distinguish one culture from another are, on the face of it, completely arbitrary.

The OED sense of 'Culture' as raising and training is clearly involved in the development of 'culture' in the sense of behaving appropriately to group membership, since a baby born into one group will be acculturated differently from a baby born or even adopted into a different group. This may be obvious, but how and to what extent it happens is not always obvious.

'Culture' need not require 'language', but to be viable a language needs a culture. We can talk of a general North Atlantic culture, because the peoples of the regions near the North Atlantic from Finland to North America have a lot in common about the ways they deal with each other, even though they speak very different languages. To say that there is a 'North Atlantic culture' is to use the word 'culture' very broadly; at the other extreme, we can say that just about every family within the North Atlantic region has its own specific culture, which a house guest must learn in order to live congenially with the family. Between these extremes of group size we can talk about a 'teen culture', a 'biker culture', or other patterns of interaction within the group and with people outside the group.

Across the spread of possible definitions of cultures, we see that people in the 'same culture' can use different languages, and people that use 'the same language', such as English or Finnish, may belong to different cultures. One person may belong to several different cultures, just as a person can speak several different languages. However, if we talk only about the most highly restricted versions of 'culture' and 'language', then there is a tighter link between a culture and the way language is used in that culture. Using the jargon of a technical or professional culture defines a person as at least being able to participate in some interactions within that culture, and participating in such interactions requires a person to be able to use at least some of the technical jargon.

The word 'jargon' is used deliberately here, as signifying language unintelligible to most people who do not belong to that culture. It is a pejorative word used by those who do not understand the sublanguage for a restricted domain, as distinguished from those who do. The users of a jargon may be perceived as controlling a perception of themselves as belonging to an exclusive and somehow superior cult or club. Perceiving someone to be controlling some perception is the core concept of what we will describe as a 'protocol', and the concept of a protocol developed in Volume II is at the heart of our later analyses of language and of culture.

Part 2: Simple Perceptual Control

In the next three chapters, we look a little more deeply into the single control loop, before going into wider-ranging issues such as the control hierarchy that is the central feature of Powers's HPCT. Extensions of the simple control loop, the hierarchy and the central reason for it, are described in Chapter I.5, while some practical issues that may arise in everyday control are mentioned in Chapter I.6.

Chapter I.4. Basic Aspects of Control

As we will do several more times in this book, we again consider the relationship between what we perceive and the world around us that contains the things we think we perceive. This problem becomes central when we start dealing with cultural and political issues that often hinge on differences between what various people and groups of people believe to be true of the world. On this occasion, we do not start with PCT but with a roughly contemporaneous theory that was developed in a book published a decade before Powers's *Behavior: The Control of Perception* (1973). The earlier book (1962) was called *The Behavioral Basis of Perception*, by J. G. Taylor.[40] The main body of this book is based primarily on Powers, but we start here with a few words on J. G. Taylor.

Powers and J. G. Taylor operated in ignorance of each other's work, but their ideas mesh well. Taylor argued that what we perceive depends on our need to behave in relation to it, in a feedback process that can alter the relation between our senses and our perceptions. Powers argued the other side of the same feedback loop and used the concept of control, which Taylor did not. For Powers, the critical point was the ancient understanding that what we do has the objective of bringing our perception of the world nearer to the way we would like the world to be. Hence our actions change what we perceive, and if we are to act effectively, what we perceive must have a close relationship to what is in the world.

I.4.1 Perception, Control, and Reality

It has long been a philosophical puzzle that although all we can ever know is obtained through our senses, nevertheless because our perceptions are for the most part sufficiently coherent as to persuade us that there is a 'real world out there' in which we can act with reasonably consistent results, we believe that our perceptions show us the real world more or less accurately, and that it continues to exist when we look away. J. G. Taylor (1962) argued and showed experimentally (e.g. J. G. Taylor, 1966) that what we actually perceive by way of our senses depends largely if not entirely on feedback from behaviour to perception. His

---

40   No relation to the present author. However, I was asked to review *The Behavioral Basis of Perception* on its first publication, and ten years later I contributed to a Festschrift on the occasion of the tenth anniversary of its publication, in the *South African Journal of Psychology*.

theoretical feedback loops were not control loops in the PCT model, being based in Hullian reinforcement theory, but his results fit well with PCT.

Experimentally, Taylor showed that if the relation between the outer world and the senses was changed, such as by wearing prism spectacles, the perception of the world was corrected only in those aspects which influence and are influenced by behaviour. This could lead to the subject (often Taylor himself) experiencing some weirdly non-logical perceptions. For example, if someone acts purposefully while wearing inverting spectacles, they soon learn to perceive the world as being right way up in some aspects related to active movement, but the smoke from a cigarette might be seen as 'rising downwards' to a ceiling perceived as above the smoker's head. When Taylor, who used a cane when walking, wore prism spectacles, initially the floor in front of him appeared to be sloped left-right, but after some walking, his perception changed so that a narrow flat path appeared ahead of him on a floor that otherwise remained sloping (J.G. Taylor, 1962b).

Many blind people 'see' the world by echolocation, an ability often thought to represent some kind of compensation for blindness. Taylor (1966) trained both blind and blindfolded sighted subjects to detect the locations and material of vertical rectangular panels of plywood or tinplate (12' high by 6', 4', 2' or 1' wide) set at arm's length on a table in the middle of a large room, by speaking at them and reaching to touch them. Without the reaching to touch, the subjects appeared not to learn, but when they were asked to touch the object, many of the subjects learned to perceive its location quickly and easily by sound alone, and could often identify its size and material. Taylor (personal communication, 1966) asserted that he could find no difference in this ability between his trained sighted subjects and blind people who used echolocation in everyday life.

When locating the targets, Taylor's sighted subjects reported widely different subjective experiences, which Taylor labelled 'visual', 'cutaneous', and 'somatic'. Many of them reported no subjective perception at all other than a clear knowledge that they would touch the target when they reached for it. It seems as though the subjects were in the process of creating a new sensory modality. If so, the echolocating ability of blind subjects should perhaps not be called 'blindsight' even though it performs much the same function as does vision for sighted people. It is a different and sometimes not subjectively conscious perceptual type, as dependent on the feedback from sensory data that accompanies action as is any other kind of perception. When the same technique is used by bats in avoiding obstacles and chasing prey in the dark, we call it 'sonar', which suggests the action of sound ranging rather than the perceptual world of the bat.

Powers looked at the same issue from the opposite side. All we can know is contained in our perceptions, and we control those perceptions by seeming to act on the supposed real world, our Perceptual Reality (PR). If our actions consistently enable us to control a perception, then it is reasonable to treat the perceived world 'out there' as being real, at least in respect of whatever corresponds to the controlled perception and the aspects of the world used in controlling it.

'Reasonable' in the foregoing does not mean the result of a logical reasoning process; it means that if the world is real, then our perceptions must be based in large part on how that real world affects our senses (or how in the past it has affected our senses or our ancestors' senses). Starting in Chapter I.11, we will discuss 'reorganisation', the process by which this match between perception and reality is achieved. We may not be able to say what is 'really out there', but we can say that the assumption that it corresponds somewhat to our perceptions has allowed us to control perceptions well enough to survive over both evolutionary time and the lifetimes of individuals.

In most of this book, we ignore the philosophical question, and simply assume that there is a real world 'out there', though at the same time we assume that not every perception accurately corresponds to reality. Some may be illusory, as is a mirage that is later perceived to be a lake, or a specially constructed 'Ames Room' (https://en.wikipedia.org/wiki/Ames_room) that looks rectilinear but is not, and in which a ball seems to roll uphill and people seem to change sizes when they walk.

Some, such as the monsters perceived by some children to be under the bed or the horrors we hear about in the daily news. may have little or no relation to direct sensory experience. They might be real, but we have not had direct sensory experience of them. In all these cases, we assume that there is a reality, even if that reality may not correspond to our perception of it. In Volumes III and IV of this book, we will consider issues of truth and trust in what other people tell us, leading to some social implications of the uncertainties inherent in the possibility of dissociation between perception and reality.

In order to understand PCT, we must distinguish between the conscious perceptions that form the colourful world of which we are aware, and the 'perceptions' that are the meat and potatoes of PCT. These perceptions or perceptual signals exist in the brain. They consist of firings of neurons or the combined effects of the firings of bundles of neurons. Presumably the conscious experiences we call perceptions are generated by such firings, but the interest of the theory is in the mechanism, not in the experience itself.

In Chapter I.12, we will address the relation between mechanism and conscious experience (including conscious thinking), after we have examined the important principles at work and their implications in a variety of areas of individual control, and before we launch into the implications of perceptual control in the functioning of small and large groups of people and other organisms.

Each perceptual signal that we control by acting on the environment is created in the brain by some operations on incoming sensory data (possibly together with imagined data, as we will discuss in various places). Although we call the net effect of these operations a 'perceptual function' that generates the perceptual signal in a control loop, most such perceptual functions change over time as systems adapt to changing or unusual conditions and as we learn new things (e.g., J. G. Taylor's experiments mentioned above, or the ordinary visual system adaptation to changing light levels). In most of what follows, however, unless otherwise indicated, we assume that the perceptual function of a control loop is stable enough that we can ignore the inevitable slow changes.

A perceptual function in the PCT hierarchy does one thing. The processing that executes the function defines a perceptual category and produces a single scalar value, the degree to which its input matches that perceptual category. When we control a perception, we are controlling a scalar value of an instance of the category.

The mechanism of the processor is irrelevant. It could be a structure of gears and levers, of interacting fluid flows, of changing chemistries, of neural interactions, among other possibilities. But since we are most interested in living things, humans in particular, we limit ourselves to biochemical and neural possibilities, with our initial emphasis strongly on the neural. In these early chapters, we largely or entirely ignore non-neural possibilities, even though many forms of life have no neurons. Their perceptual control depends on other mechanisms, such as biochemical and hydrostatic systems. Volume II begins with suggestions about how neural processes could have evolved from purely chemical ones, but non-neural processes are largely ignored in Volume I.

Powers (2005, pp. 23-24) treated the firings of individual nerve fibres as being contributors to a construct he called a 'neural current':

> *The level of detail one accepts as basic must be consistent with the level of detail in the phenomena to be described in these basic terms…. No one neural impulse has any discernible relationship to observations (objective or subjective) of behavior. Even if we knew where all neural impulses were at any given instant, the listing of their locations would convey only meaningless detail, like a halftone photograph viewed under a microscope. If we want understanding of relationships, we must keep the level of detail consistent and comprehensible, inside and outside the organism…. As the basic measure of nervous system activity, therefore, I choose to use neural current, defined as the number of impulses passing through a cross section of all parallel redundant fibers in a given bundle per unit time.*

As Powers was well aware, this definition of neural current is loose in several ways. Both 'per unit time' and 'parallel redundant fibres in a given bundle' are constructs that sound sharply delimited, but they are not. It is unlikely that any two neurons, with their hundreds or thousands of input and output connections, respond identically to identical inputs, because their biochemical environments will be different. It is even less likely that any two have the same set of input or output connections.

The meaning of 'redundant' therefore is no more than 'correlated' in an information theoretic sense; if several neurons of the bundle increase their firing rates, so will many others. The 'bundle' therefore has a core constituency of neurons whose firing rates are often similar, and peripheral members which often fire rapidly when the core members do, but not always. Different 'bundles' will have overlapping memberships, and the memberships will drift over time as the trillions of synapses among the hundred billion neurons strengthen and weaken. Such drifts may be at the heart of the reorganisation process that we discuss later. The fuzzy input patterns that result in changing 'neural currents' determine the category for which a function, such as a perceptual function, produces a meaningfully increased neural current as its output.

For now, and for most of this work, we can ignore such looseness of definition and agree with Powers that the construct of 'neural current' is a useful one, in the same way that the concept of 'a brick' is more useful when building a wall than would be a description of all the inhomogeneities of the clay of which the brick is formed. The 'value' of a signal is taken simply to be the strength of its 'neural current', which we take to be a continuously variable number rather than a discrete integer. All the function inputs and outputs we discuss are neural currents. When we refer to a 'perception', for example, it means the neural current from one specific 'bundle' (Figure I.4.1a)

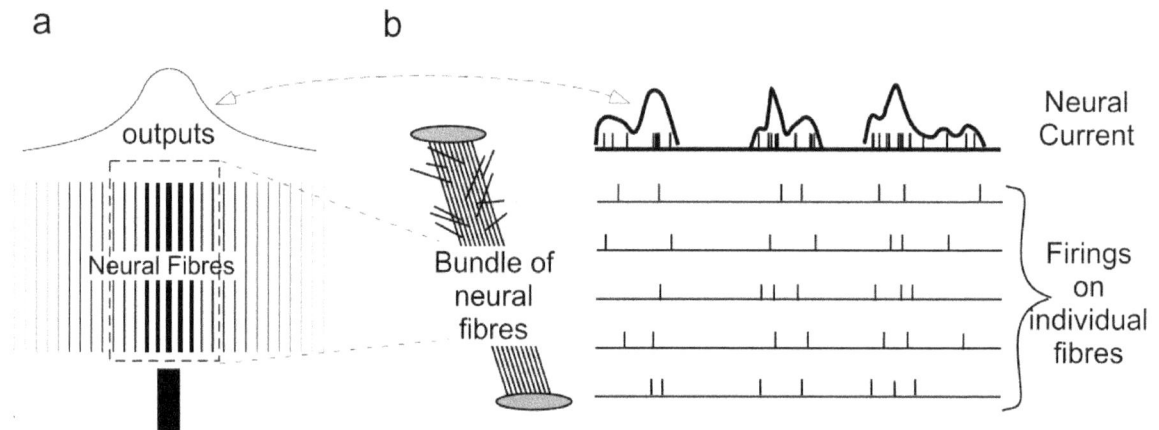

Figure I.4.1a Schematic to suggest the nature of Powers's "neural current" concept. A bundle of nerve fibres connects more or less the same source to more or less the same target (while of course branching out in myriads of other ways). The firings on the bundle are summed and smoothed to create the "neural current". (a, left) a "space" view in which the "outputs" graph represents the average firing rate per fibre when a specific input is present. (b, right) a "time" view of how firings of fibres in the bundle add to form a time-varying "neural current" in the bundle. The dashed outline in (a) delimits the fibres taken to belong to the specific bundle. More correctly, the summations would be weighted according to the sensitivity of each fibre to that input pattern.

The perceptual input function (PIF) of an Elementary Control Unit (ECU) of Figure I.1.2b takes as input a variety of values of properties, themselves the result of perceptions of categories of the real world created by lower-level PIFs.[41] If components of the real world have such properties, the perceptual function defines a specific complex of them, which is the 'Corresponding Environmental Variable' (CEV) for that perceptual function. As a concrete example, suppose we have already created perceptual functions for such CEVs as 'seats' or 'legs' (of a chair), and so forth. When a seat, four legs, and some other properties are appropriately presented to the relevant perceptual function, we perceive 'a chair', and we say that the corresponding configuration of aspects of the real world — the CEV — is a chair, a real, solid, chair on which one can sit. But it could be an illusion, which we would not know without trying to sit on the chair that we perceive.

All of these perceptual functions intrinsically define categories rather than instances of environmental variables. The variable output by a perceptual function may be the degree to which a pattern of sensory values matches the pattern to which the perceptual function is tuned, or it might be the amount of that pattern in the sensory data. In Chapter I.9 we will introduce 'lateral inhibition', but here we should note that one effect of lateral inhibition is to tone down general increases and decreases in value, such as happens to the brightness of everything in a room when a light is turned on at night. An increase in the output of a category recogniser will signal its quantitative value more strongly if related categories do not show increases, because they might otherwise inhibit the output of the one in question. Although we will discuss lateral inhibition and its effects in more detail in Chapter I.9, I introduce it here only to pre-empt possible objections to the treatment of perceptual variables in control loops later in this chapter and the next two chapters.

The attribution of existence to things we perceive to be 'out there' extends to things about which we change our mind, and later perceive not to be there, like the desert lake that turns out to be a mirage when we approach it to get some water. If we sit on what we perceive to be a 'chair' and instead of then resting in a comfortable position we fall directly to the ground, the failure of perceptual control suggests that what we saw as a chair was not a chair in Real Reality, but an illusion of some kind that formed part of our Perceptual Reality. Such illusory objects usually lose their perceived reality when perceptions of them fail to be controlled as they 'should be', or when they are not useful as atenfels in controlling other perceptions in the way that they 'should be'.[42]

Things exist for us while we perceive them, and as long as they serve as atenfels or CEVs for effective perceptual control, they continue to exist for us. This is true at all levels of perception, as suggested in Figure I.4.1b for an imaginary case in which someone is controlling his perceived financial state by making money by means of writing reports for someone else who is out of the picture. It

---

41 Although in humans and some other species, as we shall see later, some may include components derived from imagination, conscious analysis, or memory, especially at higher levels of the hierarchy.

42 We introduced the word and concept 'atenfel' and related concepts in Section I.2.4. For our purposes here, an adequate paraphrase might be "means of controlling some perception."

is important to remember that the Mirror World is only what we perceive to be 'out there', not necessarily what is 'out there' in Real Reality. It is a constructed Reality created by our perceptual input functions from data those functions have received from Real Reality or from imagination.[43] In later volumes, we discuss many cases in which our perception of a state of the world depends on what someone else tells us, rather than on our direct sensory experience of the state.

The means of control (atenfels) for the controller of financial state in Figure I.4.1b is to write a report that includes pictures and text. To do so, the person must control perceptions of the picture elements and words which have their own CEVs, using a computer to select and place them with a mouse and keyboard, again controlling perceptions that define their own CEVs in the world where anyone else could see them if they wanted to look. So long as the control actions do serve to control the perceptions, the perceived objects exist, and the lower-level perceptual control processes (or the perceptions that they control) can serve as atenfels for the ones at higher levels. The reason financial state is itself controlled is that it supports some higher-level perceptual control, such as the controller's perception of self-worth, which may increase with the amount of available money.

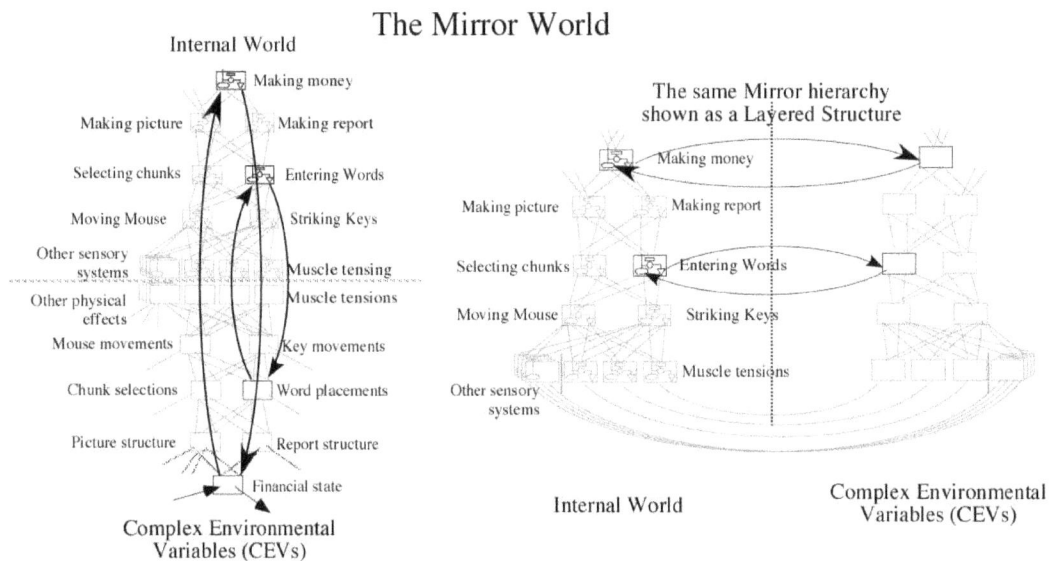

Figure I.4.1b A suggestion of how the perceptual functions at different levels create the equivalent complex environmental variables, and how the perceptions are controlled through the control of intermediate level perceptions. The example is of someone earning money by writing a report. The labels on the upper part of the left diagram represent the controlling behaviours, whereas the labels in the lower part represent perceptual entities that are mirrored into the environment. The two diagrams are the same, the right-hand diagram emphasizing the one-to-one relationship between the hierarchy of control and the equivalent hierarchy of CEVs in the outer world that are defined by the controlled perceptions. (Figure 1. from Taylor, 1993b)

---

43   In Section I.7 and in Volume II we will consider theories of how imagination and control interact.

Fig I.4.1b graphic detail enlarged for clarity

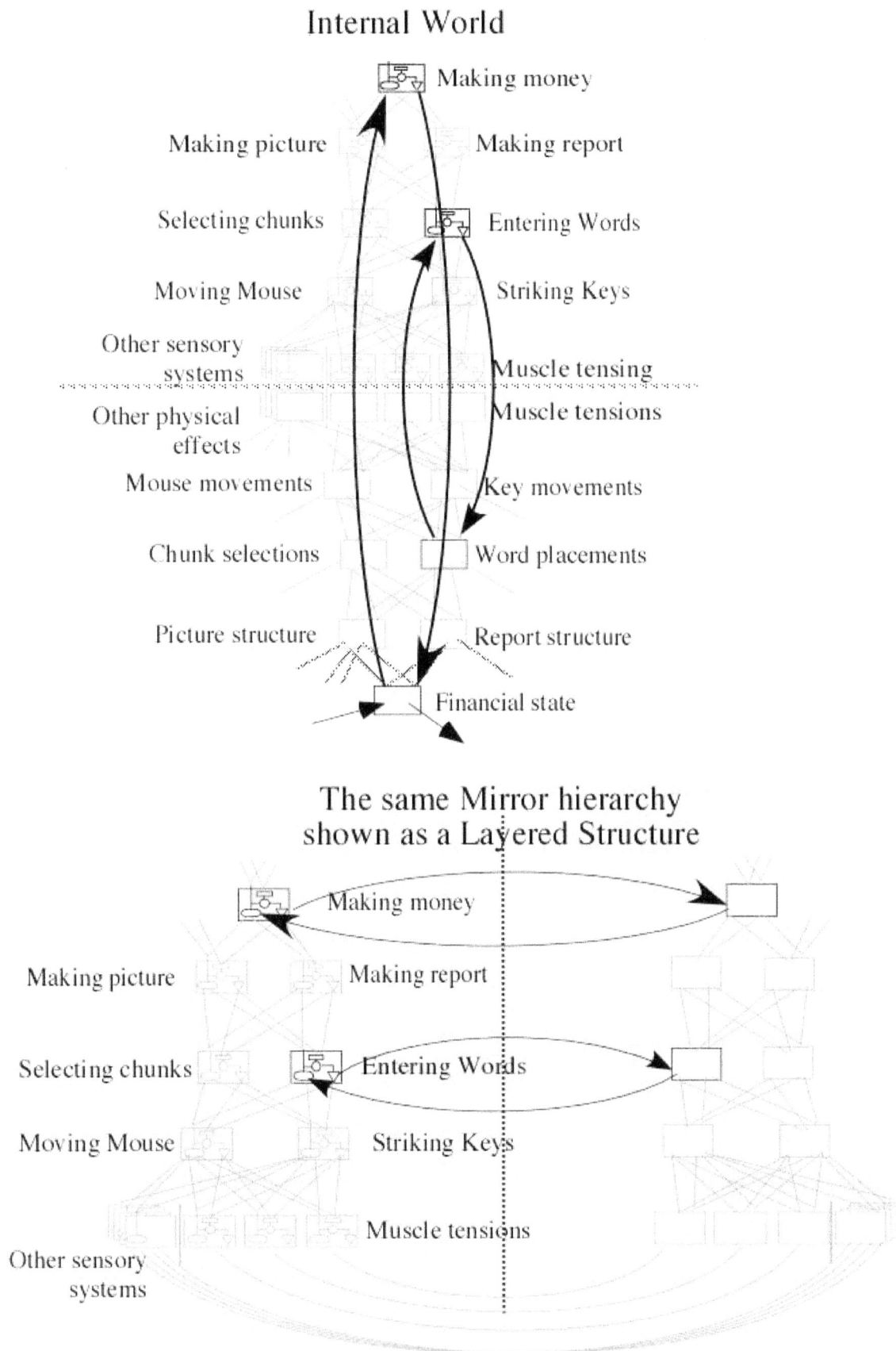

Internal World

Making money

Making picture   Making report

Selecting chunks   Entering Words

Moving Mouse   Striking Keys

Other sensory
systems   Muscle tensing

Other physical
effects   Muscle tensions

Mouse movements   Key movements

Chunk selections   Word placements

Picture structure   Report structure

Financial state

The same Mirror hierarchy
shown as a Layered Structure

Making money

Making picture   Making report

Selecting chunks   Entering Words

Moving Mouse   Striking Keys

Muscle tensions

Other sensory
systems

Because CEVs are created in the world by perceptual functions working with attributes provided to the senses by whatever is now in the real world (or once was and is now recalled or imagined), the CEVs have a hierarchic structure that exactly mirrors the structure of the perceptual hierarchy, because the CEVs are constructed from the perceptions produced by the hierarchy. Every perceptual function in the hierarchy defines one CEV in perceived reality, and approximately in reality as well (after effective reorganisation). CEVs are artefacts. CEVs that are perceived as concrete are easily observed by other people; abstract CEVs exist as artefacts to be influenced only in the environment of the Controller, but artefacts in the perceived environment nevertheless, even if the perceptual values are supplied partly or completely from the perceiver's imagination, as they are in dreams or in the hallucinations of schizophrenia.

Figure I.4.1b shows the relationship between the Perceptual Input Functions (PIFs) of a hierarchy of control systems and the corresponding CEVs. It represents the Analyst's viewpoint. The controller sees only the things that are depicted as being in the 'Mirror World', and only the controller sees that specific Perceptual Reality. Outside observers might well see something different, because the controller has access only to the 'Internal World' half of the diagrams, while an external observer may, at best, have access only to their own perceptions of the environmental properties that the controller is using to form those perceptions. Different people in a common environment are liable to construct different Mirror Worlds in their respective Perceptual Realities, and these differences can lead to conflicts from family spats to widespread wars (Volume IV).

If the observer does have perceptual functions similar to those of the controller, then the observer might be able to perceive a structure similar to that of the Mirror World of the CEVs corresponding to the controller's controlled perceptions. Skill is important; it might take a trained Certified Public Accountant (CPA) to perceive the subject's financial state when presented with all the paperwork! But no external observer could perceive accurately a Mirror World that is constructed from CEVs that include components from the observer's imagination or memory. The writer of a document such as this book cannot know what additional perceptions a reader will bring to the reading of it.

I.4.2 The Basic Control Loop

Figure I.4.2 sketches again the components of a simple control loop (Figure I.1.2a repeated, except that the effect of the output on the CEV is now shown as a unitary path in which all the different influences are combined, in the same way as neural impulses are combined into a single 'neural current'). The control loop consists of an Elementary Control Unit (above the horizontal line in the figure) plus an environmental feedback path (through the grey area of the figure).

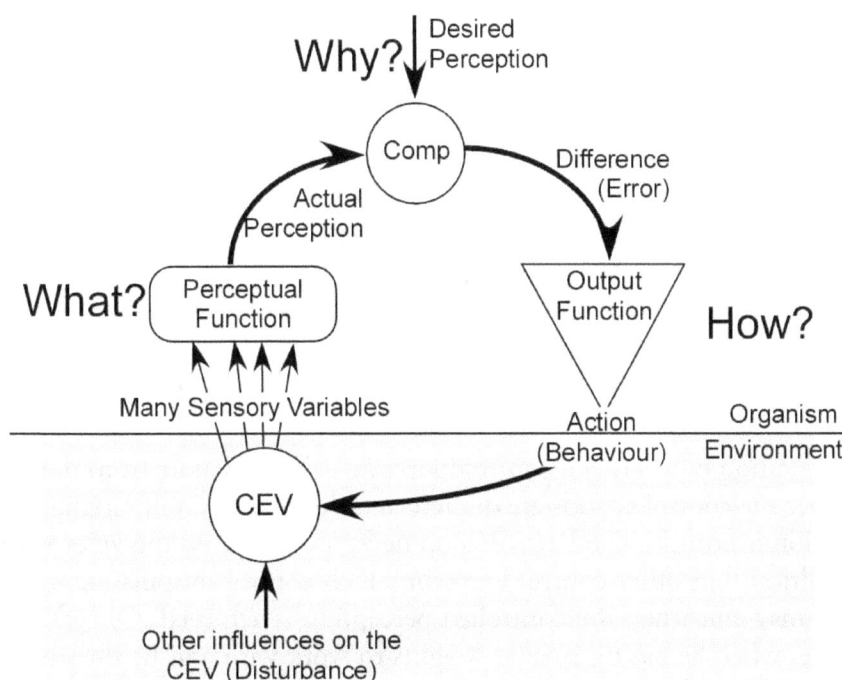

Figure I.4.2. A basic control loop showing the major constituent elements. The "Elementary Control Unit" (ECU) is in the white upper portion of the diagram. The "External Environment" is external to the ECU but much of it is likely to be inside the organism.

Powers's hierarchic structure consists of an indefinite number of such elementary control units (ECUs), each consisting of a Perceptual Input Function (PIF), a Reference Input Function (RIF), a comparator, and an Output function. As in Figure I.1.1 or Figure I.4.1b they are arranged in several levels, so that the outputs of ECUs at one level contribute to the reference values at the next lower level, and perceptual signals at one level contribute to the perceptual input functions at the next higher level. We consider the hierarchy in much more detail later. For now we are interested in the basic control loop itself

The most important thing about a control loop is so obvious that it is easily overlooked — that it should be seen as a continuous loop in space **and time**. At every instant the actual perception has some value, the difference has some value, the influence on the CEV has some value, and the signals between the CEV and the Perceptual Function have some values. All these values may be changing all the time. This is fundamentally different from the TOTE (Test, Operate, Test, Exit) loop of Miller, Galanter, and Pribram (1960). In that scheme, one event goes cycling around the loop, now as the perception, then as an error, then causing output which influences the CEV, before showing up once again as the perception.

In the generic control loop, everything is happening concurrently all the time all around the loop. Nevertheless, there are circumstances in which TOTE may appear to be a special case of Perceptual Control. When Oliver is using his scales to measure the weight on the 'rock' pan, before he makes his next addition or subtraction of the weight in the scale pan he has to wait after he adds a scale weight until the scale pointer stabilises on one side or the other of vertical. The democratic election cycle is an example at a quite different perceptual level. Each potential elector 'Tests' the performance of the existing government. On Election Day the elector 'Operates' by casting a vote. Over the next few years the elector 'Tests' the government, then 'Operates' again, and so the loop continues. Rather than calling these examples special cases in which TOTE is the correct model, we should recognise that for various reasons, such as availability of atenfels, control outputs can be 'episodic' or 'sporadic' even as the variables in the loop have continuously *variable* values. Neither Oliver nor the voter stops controlling the outcome during the waiting period.

The question of whether a control loop can 'exit' is separate from the question of whether the control actions are discrete and dependent on observing the effect of an action over time. The election loop never 'exits' (unless the voter withdraws from politics), but other control loops in which control outputs are episodic or sporadic may end when the controlled perception is achieved. Oliver's weighing terminates when he has no smaller weights to work with and he does not expect the 'rock-pan' weight to change thereafter, or when he uses the rock he weighed for some other purpose. Using the TOTE language, Oliver's perceptual control of the rock weight 'exits'. Shaving a piece of wood that must tightly fit into a shaped hole is another example. If too much is shaved off, the whole process must start again with a new piece of wood, so after each 'shave-operation' the joiner Tests whether the piece will fit and is sufficiently tight. When it is, the loop 'exits'.

The second thing to be aware of about the loop is that although all the variables may be changing continuously and simultaneously, yet it takes a finite time for the effect of an event at one place to begin to influence the values of the variables elsewhere around the loop. The time it takes for an event at, say, the CEV to begin to influence the effect of the action on the CEV is called the 'loop delay' or 'loop transport lag', among other names. Furthermore, the delay between a sharp event and the return of its influence around the loop may be affected by different properties of the components of the loop. For example, if, as is common in simulations, the output function includes an integrator, the influence of any event may last forever, but if that integrator is leaky and the components of the loop are 'linear', the influence will decay exponentially over time.[44]

---

44   'Linear' is a mathematical term which implies properties such as $c(Y+Z) = cY + cZ$.

Between clearly continuous control and purely TOTE-like episodic control lies a continuum of possibility, in which, for example, continuous control brings the perception near its reference value or the momentum of the environmental variable is perceived as bringing it nearer its reference value, at which point the controller ceases acting on the changing environmental variable, though it may continue to observe it. Controlling the landing point of a rock rolling down a slope might be an example. All of these, including TOTE, are examples of perceptual control.

On the same continuum, beyond TOTE, lies what we might call 'fire-and-forget' which is not perceptual control. A cannoneer may be able to watch the cannonball as it flies to its target, but cannot influence it. The ball falls where the laws of physics demand. The cannoneer can do no better than to adjust the cannon for the next shot in the hope that the next ball will fall nearer its target. Firing the first ball is 'command' rather than 'control'. The ball , as opposed to a laser-guided missile, is 'commanded' to land at a particular place, but is not controlled to do so. The cannon's aiming direction is controlled, but that control is not control of the flight of the ball.

The control loop works by acting on the environment as the loop's perceptual function sees it, changing the environment so that the perceptual value approaches and stays near the reference value for the loop. Any change in the environment is likely to affect some other property of the environment. Moving a rock from there to here may leave a trail or expose something the rock was hiding, and any such change may affect what is perceived by a perceptual function in another control loop. This effect of controlling a perception is a side-effect. Side effects will become important in many places in later chapters. Here, it is just something the reader should keep in mind when considering anything about control that involves more than one control loop.

I.4.3 The Behavioural Illusion and Model Fitting

We have mentioned some things that are true of the entire loop structure but not of its parts. We will soon consider the loop's components, namely a Perceptual Function, an Output Function, a Reference Input Function, and a Comparator Function. Before we do that, however, we will consider an important illusion that underlies one significant difference between PCT and a class of theories of behaviour that consider actions ('responses') to be determined or instigated by environmental events ('stimuli').

From the viewpoint of the Observer, and especially from the viewpoint of the Experimenter, there is a very seductive illusion that we call the Behavioural Illusion. The Observer sees that the controller detects some change in the environment (a 'stimulus') and then acts ('responds') in some more or less predictable way. The Experimenter introduces some such change in the environment and then measures some aspect of the 'subject's' action. The illusion is that the relation between the 'stimulus' and the 'response' depends only on the internal structure of the subject.

The 'Behavioural Illusion' (BI) is easily described, but less easily analysed. It is the illusion that the form of the 'response' that follows a 'stimulus' is determined by the processing that occurs inside the organism. Of course, it is true that internal processing produces the output, and even the form of the output, but the environment determines what that form must be, because in order for the subject actually to control a perception of the aspect of the environment which was disturbed by the 'stimulus', the internal processing must previously be accommodated to the environment.[45] This illusion is seductive because obviously if the organism was not there, the response would not occur, so it appears that something about the internal structure of the organism, something that makes it different from a chair or a rock, must be shaping the 'response' to the 'stimulus'.

This is the underlying thought behind the predominant paradigm for psychology in the first half of the 20th century, 'Behaviourism', which is based on the idea that if you could specify exactly all the sensory inputs and could measure correctly all the behavioural outputs, you could determine everything that goes on inside an organism. Since PCT suggests that this simply is not true, and is the basic 'Behavioural Illusion', it behooves us to examine that illusion and explain why it is an illusion.

But, you may say, does not a spring produce a 'response' of lengthening when you pull on it? As Shakespear's Shylock in *The Merchant of Venice* says: "If you prick us, do we not bleed?" Yes these are true, and experiences like that amplify the strength of the illusion each time the same kind of thing seems to happen when a person is disturbed by a 'stimulus'. When the front door bell rings, do I not get up and go to see who is there? No, that's not the same thing at all! The words are similar, but that's as far as it goes. I may go to the door, and usually do, but on this occasion maybe I believe I know who is there, and do not want them to know that I am at home. There's a difference between me and a spring or a rock. I control, springs and rocks don't.

The main feature of the Behavioural Illusion is that it occurs only when some perception is well controlled. When physical inanimate objects change as a consequence of applied influences, their changes are in principle completely determined by their material and structural properties, and can be calculated in advance. But a person who is presented with a 'stimulus' may 'respond' in different ways, depending on what they want to do (what perceptions they are controlling with what reference values), how they perceive the situation, and what intervenes between their muscular output and the thing in the environment which they might wish to influence.

Everyday observation suggests we act differently depending on the values of perceptions which we do not control. If we want to go out, we may take an umbrella if we perceive the sky to be dark grey and we do not want to get wet,

---

45   'Reorganisation' is the PCT concept accounting for the accommodation of the internal processing to the properties of the external environment. We examine the process of reorganisation in several parts of these volumes, beginning in Chapter I.11.

but not if we see the sky a clear blue or if we do want to get wet. We cannot control the colour of the sky, but it influences what we do. Going out is an action in the control of a perception of our location. Taking an umbrella is one way of going out, not taking one is another. The difference allows a person to control for an imagined perception of future wetness, using an uncontrolled (and uncontrollable) present perception of an aspect of the current environment.

Taking or not taking an umbrella when we control for our location to be somewhere that requires going outside depends both on our reference value for our perception of wetness and on the uncontrolled perception of the sky. It looks like a simple stimulus-response: *see grey and rain* (stimulus) ➜ *take umbrella* (response), but doing so is part of a more complex control loop. After all, on another day the person may want to get wet, and would go out without the umbrella.

If the subject of an experiment has no way to influence the perception that is disturbed by the 'stimulus', as is often the case in psychophysical tests of detection or discrimination, why would she ever act in a way that depends on the stimulus? The guiding principle of PCT is that all intentional behaviour is the control of perception, and if pressing the appropriate button in an experiment isn't intentional behaviour, what is?

Why press a particular button when doing so will affect nothing about what is presented? Pressing the button doesn't influence what was presented, but it might influence some other perception the subject controls, such as the experimenter's goodwill toward the subject. The means of controlling that other perception is to press the proper button. It is not enough to press any old button. It must be the right button, insofar as the subject is able. The 'right' button is the one that is related to the stimulus in some way known to the subject. So the button that gets pressed is related to the presented stimulus even though the stimulus is not influenced by pressing the button or anything else that the subject does.

In such cases, the Experimenter has no guarantee that the relation between stimulus and response indicates anything about the subject's internal organisation. If, however, the assumptions about what perception the subject controls are correct, the commands to 'respond' act as atenfels for them; the behavioural illusion then ceases to be an illusion and the relation between 'stimulus' and 'response' does tell the experimenter about something internal to the subject's brain and body.

One of my colleagues told me a story long ago about participating in an experiment when he was an undergraduate. He had been asked to keep a stylus as long as possible on a sensitive area on a rotating disk, so, perceiving (incorrectly) that it was an intelligence test, he dismounted the disk and laid the stylus on the sensitive spot. The experimenter, on his part having the 'stimulus' of an impossibly perfect result in his test of mechanical skill, produced the 'response' of expressing anger at my 'stupid' friend for not producing the correct 'response' to the 'stimulus' of the rotating disk. Looking through the lens of Perceptual Control Theory, we can see that each of them was actually controlling perceptions that were not what the other perceived them to be controlling.

In many experiments, however, the subject actually controls a perception directly disturbed by the 'stimulus'. The subject might be asked by a doctor in an annual checkup to look at the doctor's fingertip while the doctor moves his finger up, down, and sideways. In such cases the 'response' turns out to be related to the stimulus mainly by the characteristics of the environment, with little contribution from the details of the subject's internal processing, except insofar as the subject controls inaccurately. The subject still controls for the experimenter to be pleased, or for maintenance of a 'competent' self-image, but the interpretation of the relation between 'stimulus' and 'response' is different. Some information can still be gleaned about the internal processing of the subject, but only to the degree that control is imperfect.

Before analysing the complex interactions between the experimenter and the subject (the underlying principles of which are examined in Chapter II.14 when we deal with 'protocols'), we follow Powers (1978) and consider the Behavioural Illusion for a single control loop.

Figure I.4.3 shows a simple control loop, but complicates it by introducing an unspecified function labelled 'f..' into each of the connecting 'wires'. In the environmental feedback path they represent the combined effect of atenfels in the path segment. Of these, f_{CP} represents ways in which the ability to perceive the CEV may be influenced, such as by telescope or microphone, whereas f_{OC} represents ways in which the CEV can be influenced, such as by using a lever to move a rock or by voting to change the policy of the government.

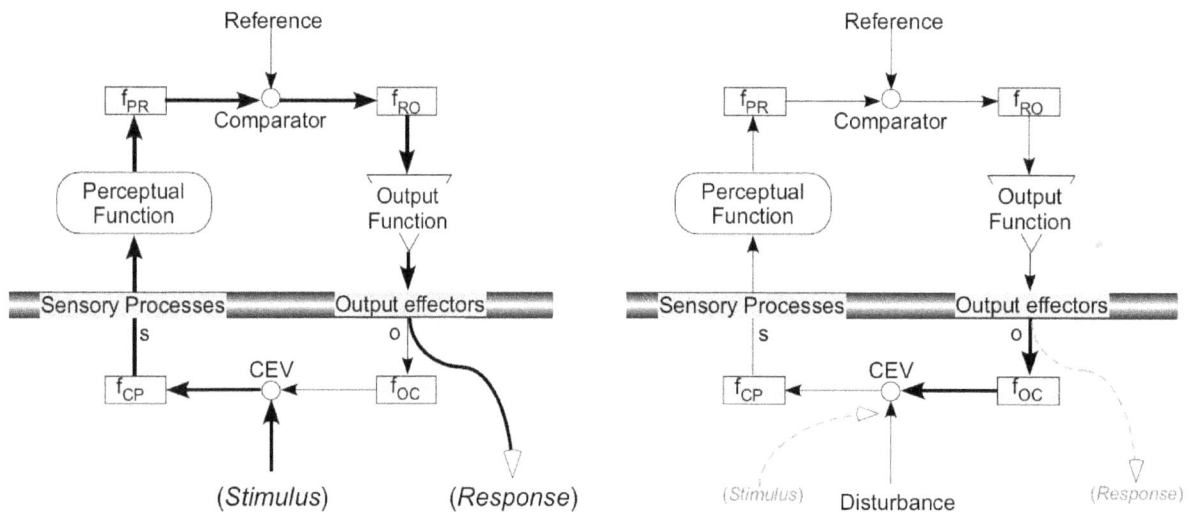

Figure I.4.3 A simple control loop, showing possible functions that might occur at different places around the loop. (left) The interpretation of the loop as a "Stimulus-Response" process. (right) According to PCT, the "Stimulus-Response" interpretation simply inverts the effect of the environmental feedback path f_{oc}. In a typical experiment, the "stimulus" is what a PCT experimenter would call a "disturbance", while the "response" is what the PCT observer would call "output to the environment" (not the output of the control unit's output function).

Along with the Perceptual Function and the Output Function, f_{PR} and f_{RO} are the internal processes that the 'Stimulus-Response' analyst wants to describe. For the purposes of the present discussion, however, we will suppose only that they are necessarily monotonic, meaning that for every possible input value there is only one possible output value, and if the input value increases by a small amount, so does the output value. As may become evident, most of these functions have very little effect on what can be seen by an outside observer or experimenter when the perception is well controlled.

Following Powers (1978), conceptually but not in detail, we observe that if control were perfect (a physically unrealizable condition), the output of the function f_{OC} consequent on the disturbance would be exactly opposite to the stimulus (the change in the disturbance). The value of the CEV would remain perfectly constant, no matter what the other functions and internal processes of the controller. Every relationship between the 'stimulus' (disturbance) and the 'response' (input to f_{OC}) the experimenter would observe would be a property of f_{OC} and nothing else. The Behavioural Illusion is that what the experimenter perceives to be a property of the internal organisation of the subject is actually a property of the environment.

No physically realisable controller can control exactly perfectly, but it takes very subtle measurements to distinguish very good control from the ideal of perfect control. Simulations done to evaluate parameters of a hypothesised control model usually do so in situations in which the properties of the atenfels are very simple, contributing minimal lag and influencing the CEV in a linear way. A rare exception is Powers's 'Circle-Square' demonstration in LCS III (Powers 2008), which is a dramatic demonstration of the fact that we control input rather than output.

At the other extreme, if f_{OC} were a switch that had been turned off so that there was no feedback connection, no influence of output upon the perceived CEV, then the response would be a combined function of f_{CP} and the internal processes of the subject organism. The experimenter would be fully justified in claiming that the relationship between stimulus and response provided evidence about what processes occurred within the organism.

For many experimental cases, we need no such complexity. The simple control loop of Figure I.4.2 is nearly enough. The 'subject' (who we can call Sean) controls some perception (the output of f_{OC} in the figure) to some value. The experimenter (who we call Ethel) does something that affects the CEV (without necessarily being precise about exactly what Sean will perceive), and observes something about his action (without necessarily being precise about observing all and only those action components that affect the CEV). Some part of what Ethel does will influence the CEV, some part of what she observes will be Sean's actions to influence the CEV as part of the control loop, and

Sean will be only partially successful in controlling the perception.[46] All of these imprecisions affect the experimenter's observations.

Ignoring these imprecisions, although they are almost certain to occur, let us assume that Ethel is absolutely accurate both in influencing Sean's CEV and in observing what he does in controlling his perception of it. In any physically possible case, a subject cannot control perfectly. This means that the 'response' input to f_{OC} does not produce exactly the opposite of the 'stimulus' (change in the disturbance). It is not quite the 'right' response, but it is the response produced by the internal processes that include all the other f_{XY} processes that may or may not exist, as well as the Perceptual Function, the Output Function, and the comparator process. Only this failure of perfect control allows any experimenter to infer anything at all about what happens inside the organism in an experiment. The success of control enforces the Behavioural Illusion, but control imperfection allows at least some access to the processes in the rest of the loop.

A PCT researcher would propose a model of these processes, generate a software simulation, run the software model, and compare it with what the subject actually did. In many tracking simulations, the simulation model treats these internal processes (except for the Output Function) as simple pass-through operations, even though it is obvious that a Perceptual Function is likely to be extremely complicated, and nowhere in the loop in a real live organism are any of the f_{XY} processes likely to be trivial.

The reason the simple model works is that when control is good, the Behavioural Illusion (BI) leaves little scope for distinguishing among different complexities and non-linearities that may actually be there. Many different organisations that implement control would fit just about as well, so the Occam's Razor approach is to accept the simplest. Because of the BI, the zero'th approximation simple model works so well that it is very difficult to discriminate more precisely the processes that may be operating in the path from CEV to Output.

The 'pretty good success' of such simple models is a testament to the power of the Behavioural Illusion when subjects control well. The models cannot have enough resolution to distinguish among different functional forms such as non-linearities. For example, most researchers in perception think that perceptual magnitudes tend to bear something like logarithmic or power-law relationships to the corresponding physical magnitudes. (We will say more about this below.) The simple models can't distinguish such varieties in the internal processing because the fact of control simply inverts the effects of any nonlinearity. Experiment 3 in Powers (1978) demonstrates this.

---

46   Remember that in Figure I.1.2a the CEV is shown as being influenced by many separate paths and as influencing several different perceptions. In trying to influence the CEV, the Experimenter is likely to manipulate some or all of these paths, but unlikely to do so in exactly the way the Subject combines them to form the CEV. Similarly with the experimenter's view of what the Subject senses.

More information may be extracted from a tracking experiment. This is best done under conditions where the human controls relatively poorly, or in which feedback of some component of a controlled perception is impossible, analogous to the case of controlling for not getting wet by taking or not taking an umbrella depending on whether the uncontrolled perception of sky colour is grey or blue. In this case, by observing the correlation of 'umbrella-ness' with sky colour, the experimenter could determine that the person was able to perceive the sky colour, or something closely associated with it.

An interesting example of the Behavioural Illusion was described by Marken (2014:133-142) following earlier discussions with Powers. A famous psychophysical 'law' is Stevens Power Law (e.g. Stevens 1957), which says that the perceived magnitude of a 'stimulus' such as the intensity of a light or the loudness of a sound is a power function of the physical magnitude of the stimulus. Stevens (1966) found also that when another sensory dimension rather than number was used as a match, the result was again a power law with the exponent that would be predicted from the exponents of the individual dimensions when each was compared to number. Marken notes that the subject is likely to be controlling for the perceived magnitude of the reported number to be equal to the perceived magnitude of the 'stimulus'. The same presumably would be true when a sensory magnitude is compared to another sensory magnitude. Indeed, when you look through PCT glasses, it is hard to see any other way the subject could do the task.

The issue, then, is how the magnitude of a number or another sensory dimension is perceived. If both the number and the 'stimulus' are perceived as logarithmically related to their physical magnitude (as Fechner's Law would suggest), then Marken shows that the output value (number or physical magnitude) will necessarily be a power function of the 'stimulus' physical magnitude. In this case, as with Sean's umbrella, the Behavioural Illusion allows the experimenter to probe the perception of the uncontrolled variable (the 'stimulus') because a higher-level variable that incorporates it (the difference between the number and the 'stimulus' magnitudes) is being controlled. The same experiment could be done by specifying the number and asking the subject to adjust the stimulus.[47]

---

47 One should be aware, however, that Garner (1958) long ago demonstrated the unreliability of experimental methods used to generate the power law, at least in the case of perceived loudness. His criticisms do not affect perceived equality judgments, although Garner had previously shown that when people were asked to judge whether sounds were louder or softer than a standard sound, their 'equality to half' judgement was very close to the midpoint of the loudness of the set of sounds offered for comparison (Garner, 1954). Garner's assessment in 1958 included: "...*it is clear that we are on very dangerous ground in assuming that the loudness scale proposed by Stevens has any real meaning in the experience of normal observers.*"

All actual control exists somewhere between the extremes of unachievable perfect control and no control. Control can be very poor, as in the case of an average voting-age citizen trying to control the perceived policies of the government, or very good, as in a tracking study with a slowly varying disturbance. The better the control, the stronger the Behavioural Illusion and the less possible it is to use the relationship between the control actions and the disturbance to say anything about the interior processes of the controller. But on the other hand, the better the control of a higher-level perception that incorporates an uncontrolled perception, the more one can discover about the uncontrolled perception because of the Behavioural Illusion at the higher level.

Next we turn to the individual components of the ECU. We will begin with the Output Function, and ignore the Perceptual Function until the end of this chapter. The Perceptual Function simply provides a variable that the rest of the loop serves to control. How it does control may be very complicated, as is illustrated by the amount of research dedicated to finding out how, for instance, we perceive someone's face. Complex as it may be, only the resulting variable, called in PCT 'a perceptual signal', is of interest.

I.4.4 The Output Function

The job of the output function is to accept as input the *perceptual error* and produce output which (through the loop) brings the perceptual error near zero and keeps it there.[48] The Analyst sees that the output acts through the environment to influence the CEV which is defined by the perceptual function, but the ECU knows nothing of this. It simply produces more or less output depending on the state and history of the error.

In most simulations of human control in the context of PCT, the output function does not follow the immediate error value, but integrates it using a leaky integrator. If the disturbance remains constant, the error will reduce asymptotically toward some value dependent on the ratio of the integrator's gain rate and leak rate. In most of these simulations, the environmental feedback path is treated as a simple connection that does not contribute to the loop dynamics.

In these simulations, not only is the feedback path often treated as a simple connection, but so are the other paths in the loop. The Perceptual Input Function, despite its complicated character in practice, is treated as a simple unity multiplier that converts a measured value of the CEV to an equal perceptual value, and the comparator simply inverts the perceptual value and adds the result to the reference value. In these simulations, then, the loop gain around the loop is represented only in the output function.

---

48   The 'perceptual error' is often simply called 'the error'. It is the output of the Comparator Function, which is typically considered to produce the simple difference between the reference and perceptual values, but which we will suggest may take other forms.

The concept of 'loop gain' is important but is easily misunderstood. In a stable loop of any kind that has variable values around its different parts, the apparent gain has to be exactly 1.0, because if you evaluate a variable at one point and then go around the loop, on average that variable will have exactly the same value. So 'loop gain' means something else. It means the gain that would be observed if you simply multiplied the gains of the different individual components considered in isolation all around the loop. For example, if in a control loop such as that of Figure I.4.2, the perceptual function had a gain of 0.5 (the variability at its output had half the amplitude of the variability at its input), the output function had a gain of 10.0, and the path through the environment and sensors back to the perceptual function had a gain of 3.0, the loop gain would be 15.

In a control loop, the loop gain must be negative, so these example numbers would not produce a control loop, or even a stable loop. But there's a more serious issue with them. Suppose the perception is of the movement of a dot in a screen, as it has been in many studies based on PCT. The movement is measured in centimetres, while the perceptual value is measured in neural firings per second. It makes no sense to say that if the movement is 1 cm. and the firing rate changes by 20 impulses per second, the perceptual function gain multiplier is 20. It is not. It is 20 impulses per second per cm. Since the impulse count is a pure number, this value is succinctly written as 20 sec-1-cm-1.

A similar issue arises at the output, where we ask the question how far would the dot move if the output firing rate changed by one impulse per second and the person could not see the movement. That value might be, say, 100 cm., in which case the output gain would be 100 cm per impulse per second, or 100 cm-sec. If you multiply these two values, and if there were no other multipliers around the loop, the loop gain would be 2000, the impulse rate and the movement distance cancelling out. Such a loop gain is unrealistic for human control, but is easily exceeded in mechanical systems, even human guided ones. Simple naked-eye tracking experiments usually result in an apparent loop gain in the neighbourhood of -10 or -20.

This description assumes that the loop gain is a simple constant, but that is not usually true. To get good fits to the actual traces made when a human controls in an experiment, the output function is most often best if it is a leaky integrator, which has no simple multiplier gain (though it does have an asymptotic gain, as we discuss elsewhere). If the simulated output function is a leaky integrator, the loop gain is a function of frequency that depends on the relation between the integrator gain rate and the leak rate. At zero frequency (the effect an infinite time after a step change) the gain is the ratio of these two gain rates.[49]

---

49   However, when the results of such simulations are published, the gain rate is often called simply 'Gain', while the leak is called a 'slowing factor' for historical reasons. The confusion between Gain and gain rate can sometimes cause difficulties in interpreting the simulation results because the time-base of the rate measurement is left implicit, sometimes as 'gain per sample', sometimes as as 'gain per second', either of which is reported as 'Gain', because that was the number that was plugged into a formula.

Output functions need not be leaky integrators, though a leaky integrator does have certain statistical advantages when the environmental context is unknown and variable. In the real world the feedback path is not a simple connection, because the muscular output to the external environment is a force, objects have mass, and friction has its effects. Force applied to an object with mass produces acceleration, not location. Acceleration integrated is velocity, and velocity integrated is distance. Forces applied to springs produce counteracting forces, and masses on springs bounce. In an everyday environmental feedback path, several such dynamical effects may be in play, influencing the timings and manner in which the output affects the controlled perception.

Powers (1994) described an adaptive output function that would compensate for the effects of such dynamical issues in the environmental feedback path. He called his device an 'Artificial Cerebellum' (AC), and demonstrated it in controlling a disturbed spring-loaded mass, which dynamically is the same as a swinging pendulum. If appropriate for a particular control loop, the AC would adapt to perform a leaky integration, but it would also adapt to compensate for a wide range of dynamical issues in the loop structure.

Though more complex in its implementation, the AC was in principle an adaptive Finite Impulse Response (FIR) filter in the form of a shift-register multiplier. A shift register holds in a succession of registers the value of its input at successive sample moments, and the multiplier component adds a different multiple of the value at each register to form the final output. If the multiplier values are all the same and the shift register has infinitely many registers, the system is a perfect integrator. If the multiplier values decline exponentially as the register values age, it is a leaky integrator. If older values are subtracted from more recent ones, it is a differentiator. But the multiplier values can be anything at all, and can compensate for natural 'ringing' in the loop. Seen as a FIR filter, the spectral response of the AC reduces the peaks and raises the valleys of the loop response spectrum.[50]

It seems not improbable that naturally evolved control systems might incorporate something functionally or even possibly structurally similar to Powers's Artificial Cerebellum, because they have been exposed for billions of years to the same problem Powers addressed. However, whether this kind of module exists in the biological cerebellum is quite another question, a question not addressed here. It is, however, addressed by a review article (Popa, Hewitt, and Ebner, 2014), the abstract of which could serve equally well as an abstract for Powers's description of his AC, except that Powers uses neural currents rather than spikes from individual cells. Here is the Popa et al. abstract:

---

50   This process is called 'whitening' or 'prewhitening', because noise with a flat spectrum is known as 'white noise'. However, because transport lag may cause positive feedback at higher frequencies (as happens when a loudspeaker feeds back into the speaker's microphone), the AC will usually adapt to produce a 'pink' rather than a 'white' spectrum.

Historically the cerebellum has been implicated in the control of movement. However, the cerebellum's role in non-motor functions, including cognitive and emotional processes, has also received increasing attention. Starting from the premise that the uniform architecture of the cerebellum underlies a common mode of information processing, this review examines recent electrophysiological findings on the motor signals encoded in the cerebellar cortex and then relates these signals to observations in the non-motor domain. Simple spike firing of individual Purkinje cells encodes performance errors, both predicting upcoming errors as well as providing feedback about those errors. Further, this dual temporal encoding of prediction and feedback involves a change in the sign of the simple spike modulation. Therefore, Purkinje cell simple spike firing both predicts and responds to feedback about a specific parameter, consistent with computing sensory prediction errors in which the predictions about the consequences of a motor command are compared with the feedback resulting from the motor command execution. These new findings are in contrast with the historical view that complex spikes encode errors. Evaluation of the kinematic coding in the simple spike discharge shows the same dual temporal encoding, suggesting this is a common mode of signal processing in the cerebellar cortex. Decoding analyses show the considerable accuracy of the predictions provided by Purkinje cells across a range of times. Further, individual Purkinje cells encode linearly and independently a multitude of signals, both kinematic and performance errors. Therefore, the cerebellar cortex's capacity to make associations across different sensory, motor and non-motor signals is large. The results from studying how Purkinje cells encode movement signals suggest that the cerebellar cortex circuitry can support associative learning, sequencing, working memory, and forward internal models in non-motor domains.

It is also suggestive that the mammalian cerebellum contains between 2 and 5 times more neurons than the cortex (Herculano-Houzel, 2010), and that Powers's AC is more complex than other parts of the control hierarchy.

In the rest of this book, we simply assume that the output functions of all control units are adaptively tuned, either on an evolutionary timescale or in the course of learning to function in a particular environment.

I.4.5 The Reference Input Function

Variation of functional type across levels could be accomplished if the Reference Input Functions (RIFs) at different levels had as different characteristics as do the PIFs at the different levels. Although Powers (2008) demonstrated that in a uniform environment a multilevel control structure with simple weighted-sum between-level reference connections can learn to control a complex structure such as an arm with 14 individually variable joint possibilities, that demonstration was performed in a uniform environment that provided no opportunity for different kinds of perception or of action other than in the arm itself. Overall, it seems more probable that differences in effect would be accomplished at different levels by different kinds of RIF. At lower levels the RIFs might well be nonlinear weighted summations. At higher levels, they might be context-addressable associative memories of perceptual values, or even complete logical programs.

In his book *Behavior: The Control of Perception* (B:CP), Powers came to the conclusion that every RIF would be an associative memory, writing: "We will assume from now on that *all reference signals are retrieved recordings of past perceptual signals*. This requires giving the outputs from higher-order systems the function of address signals..." (Powers 1973:217; 2005:219). We need not go so far, and indeed, in his many simulations of multi-level control systems in the 40 years following the original publication of B:CP, Powers apparently never used associative addressing. Even in demonstrations such as the *Little Man* (Section II.12.1, in Volume II) he used a linear weighted summation of the higher level outputs as the RIF for every ECU.

With Powers, though, we assume that at least some Reference Input Functions are associative memories. Lateral connections within levels of the hierarchy can, among other effects, construct associative memories. Lateral connections are an explicit extension of and deviation from the Powers hierarchical structure. In Chapter I.9, we demonstrate their functioning for associative memory and provide other arguments for introducing them.

In B:CP, Figure 15.2 shows a reference input function as an associative memory that takes input from only one higher-level control unit's output. However, since in the Powers hierarchy every ECU is likely to take its reference value from a combination of higher-level outputs, the single connection shown in Powers's figure must be seen as representing a vector of outputs from the higher level systems which contribute to the reference value of the one depicted.[51]

When several ECUs at the next higher level of the HPCT structure output a particular set of values to a content-addressable RIF, that vector of outputs acts as an address into a memory, which produces a particular reference value for the perception controlled by this ECU. Powers suggests that at some prior time, a

---

51   A vector is an ordered set of values of any length, as in {1, 2, 3, ...} or {..., x_{n-1}, x_n, x_{n+1},...}. A vector of length 1 (a unitary value) is a 'scalar' value, but can be used in computations in the same way as any other vector.

condition not represented in his Figure 15.2 had signalled that the current value of the controlled perceptions should be stored at that address. In effect, the RIF says: "When in this situation before (as perceived at higher levels), things worked when we managed to produce this perception." Setting the reference to that value requests the ECU to provide the same perception again. Of course, this does not imply that the same actions will be used to produce that perception, or even that the same yet-lower-level perceptions will be used.

Those same output values from the higher level which as a set were associated with successful control in the past may be distributed to possibly many ECUs at the next lower level, but resulting in a different reference value for each of them. The effect is to produce a vector of lower-level perceptual signals with values that resulted in low errors in the higher level ECUs in the earlier context. It is worth emphasising that word 'context', because if many higher-level outputs influence the reference value for an ECU, the perceptual value to be obtained depends entirely on the higher-level context. *Perceptual control is contextual.* In other words, the set of ECUs at one level reproduces a coordinated pattern of reference values at the lower level, one of four *profiles* of control.

Figure I.4.5 illustrates three of the four profiles for a trivial structure of three high-level control units providing reference values for three lower-level units in which the RIFs are simple summations. The error profile (the vector of error values) is not shown, but is analogous to the other profiles.

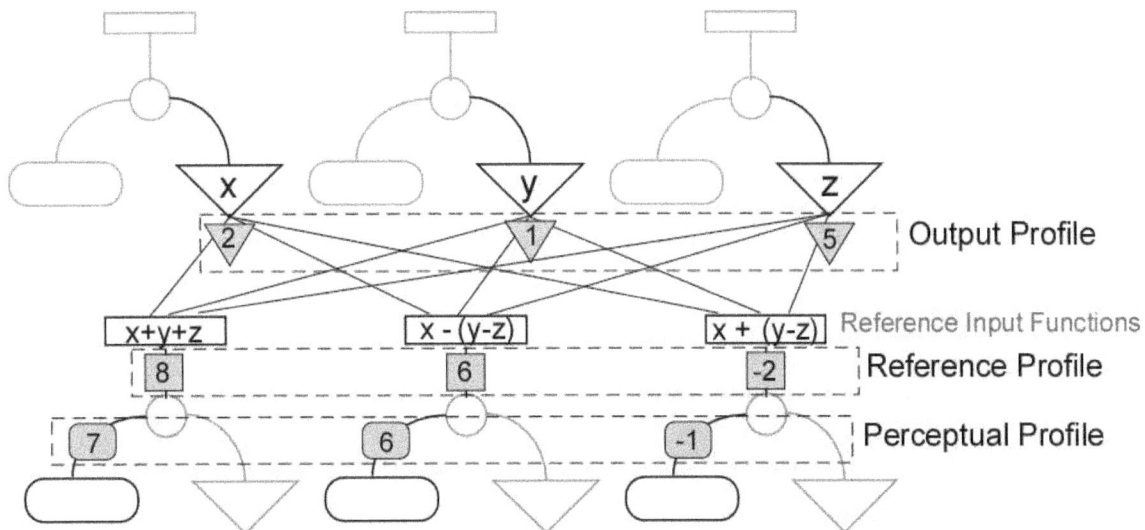

Figure I.4.5 Profiles of control: Three lower-level Elementary Control Units (ECUs) have different reference input functions, and three higher-level ECUs provide input to them. An illustrative set of values are shown, for three profiles (or vectors) based on these RIFs. The output profile at the upper level is {2, 1, 5}, which produces a reference profile of {8, 6, -2} at the lower level, for which the current perceptual profile is {7, 6, -1}. These values lead to an error profile (not illustrated) of {1, 0 ,-1}. For reasons of clarity, the error profile is not shown, but it is analogous to the others.

I.4.6 The Comparator Function

The final key component of an ECU is the Comparator function. Every ECU that has a potentially variable reference value must have a comparator.[52] In most PCT simulations the comparator of every ECU is taken to be a simple subtractor, subtracting the perceptual value from the reference value provided by the Reference Input Function. The comparator could, however, be more than that, and the ways it might differ from a simple subtractor are the reason the comparator is given its own section here instead of being taken for granted.

One of the many aspects of the conventionally accepted and simulated version of PCT that are unsupported by direct evidence is assumption that the value of the error signal has a linear relationship to the difference between the perception and reference values. This relationship is sometimes called the 'error function' for an ECU. Figure I.4.6a shows a few possibilities for the error function, all of which have been mentioned in informal discussions of, or in writings on, PCT. The last possibility, (d), is 'tolerance', which we consider at length below.

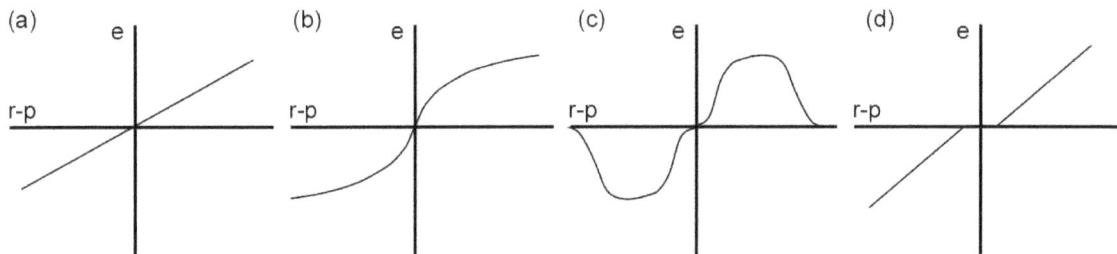

Figure I.4.6a A few plausible error functions. In each panel, "e" is the value of the error signal, "r-p" the difference between reference and perceptual values. (a) Simple linear, which is almost always used in simulations; (b) power-law or quasi-logarithmic; (c) "give up" if error is too large; (d) tolerance zone: the error is zero if the perception is "close enough" to the reference.

Figure I.4.6a panel (a) shows an ordinary subtractor comparator function. A function like that of panel (b) is appropriate if the precision of small errors matters relatively more than that of large ones. Panel (c) represents a case in which the controller stops controlling if the error is too large, while panel (d) represents a controller that is linear like a subtractor only outside a tolerance zone. Regarding (d) and tolerance, hardware control systems are designed with a tolerance zone to reduce or eliminate high-frequency jitter when the error is

---

52   In the Powers hierarchy that has no loop-backs from a higher level to a lower, the top level has no variable reference value and therefore the top-level comparator function is trivial. It is as though the ECU were supplied with a fixed reference value, conveniently assumed to be zero because any bias can be subsumed in the Perceptual Input Function (PIF) of the ECU. The comparator effect is simply a sign reversal for the perception, which also can be done by the PIF.

near zero, and it is easy to imagine that physiological control systems might solve the same problem the same way. Also, the four possibilities can be combined in different ways to produce other possible error functions, such as 'give up with tolerance', or 'low-error precision with high error give-up'.

Experimentally, it is very difficult to distinguish different curvatures of the error function for any particular control task, for two reasons. Firstly, if control is good, the error seldom deviates much from zero, and the different possible curvatures can all be approximated by a straight line over very small spans. Even 'tolerance' can be soft, with a curve that moves smoothly from a zero-error region around zero r-p into a locally linear function, rather than the sharp break between zero and linearly rising error shown in Figure I.4.6 panel (d). Secondly, if control is not good, the traces are relatively noisy and many models that differ only in degree of curvature fit the human tracks equally well.

A second aspect of the error function that is hard to test independently is the slope of the function. Suppose the function is linear, as in Figure I.4.6 panel (a). In analyses and simulations, the slope is usually taken to be 1.0, the value of r-p being sent 'raw' to the output function, which contains a gain factor. If the slope of the error function were to be multiplied by a factor K, the result would be the same as if the output function's gain factor were to be divided by the same factor. This point may seem trivial, but it has consequences if the form of the error function is not linear, as we will see shortly.

Next, we suggest a possible way in which a comparator might work with negative as well as positive values of r-p, a problem that must be resolved in a neural system in which the 'neural current' can never be negative.

If the comparator subtracts the perceptual value from the reference value, it will produce negative error values whenever the perceptual value exceeds the reference value. This would pose no problem for a designer of a hardware control system, but we are dealing with a system in which the values are assumed to represent neural firing rates, and no neuron can fire a negative number of impulses per second. Accordingly, we must assume that Nature has some trick up her sleeve.

One possibility is that a value of 'zero' is represented by some resting firing rate, a rate that is depressed by inhibition to represent negative values and enhanced by added excitation to represent positive values. Since the subtraction is presumably done by comparing inhibition from the perceptual signal to excitation from the reference signal, the suggestion that 'zero' is represented by a resting firing rate is not unreasonable. But continual firing is energetically wasteful, and one of the problems of a brain is to get rid of excess heat, so if this solution is ever used, it cannot be the main way the brain deals with negative values.

Figure I.4.6b is a schematic of how a comparator function that provides positive and negative error values might possibly be implemented with neural currents that can never be negative. (Arrowheads indicate excitation of their target, whereas circles indicate inhibition).

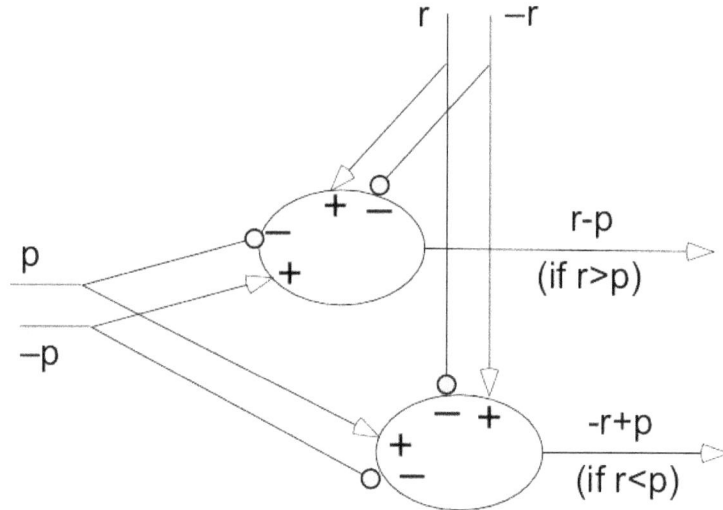

Figure I.4.6b Possible implementation of a comparator. All signal values are positive. "-p" and "-r" are signal paths for the absolute values of perceptual and reference values that would be negative, such as "above" when the perceived position (or reference) relationship is "below".

When r and p have the same sign, sometimes one is greater, sometimes the other. Two different outputs are required for the two cases. When r > p, the error is positive and serves to excite the output function of the control unit, but when r < p, the error is negative, which we assume inhibits the output function. When r > 0 and p < 0 or the reverse, only one of the output possibilities exists, as illustrated for a numerical example in Figure I.4.6c.

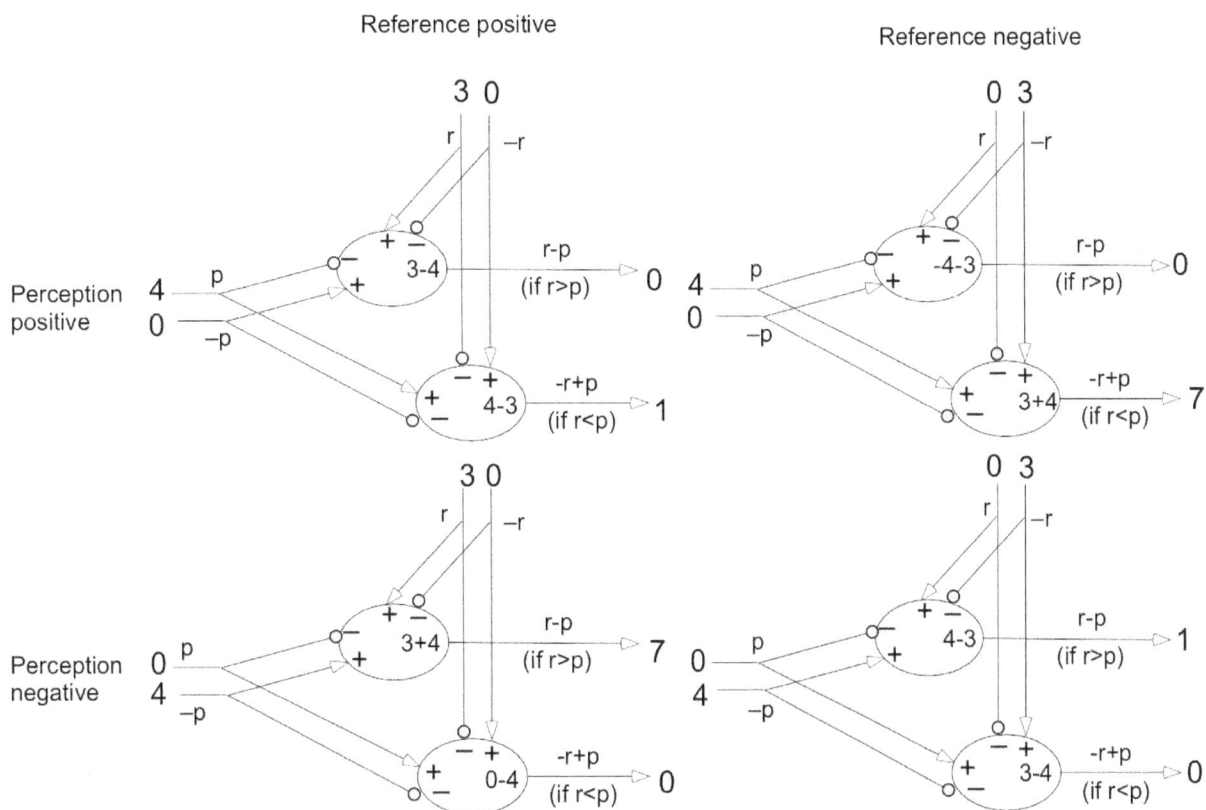

Figure I.4.6c The comparator with numerical examples for all four possible cases. Assume that a positive perceptual or reference value represents "X above Y", a negative value "X below Y". Signal values that are neural firing rates can never be negative, so negative perceptual values are reported as their absolute value on a different signal path. The Figure I.[Martin???] shows reference values of plus or minus 3, paired with actual perceptual values of plus or minus 4.

The two outputs, one for r-p < 0, the other for r-p > 0, separately produce one-sided error values that when supplied to the output function act as though they form a single two-sided function as in Figure I.4.6d. The p-r pathway would present an inhibitory error value to the output function when r < p, while the r-p pathway would provide an excitatory input when r > p.

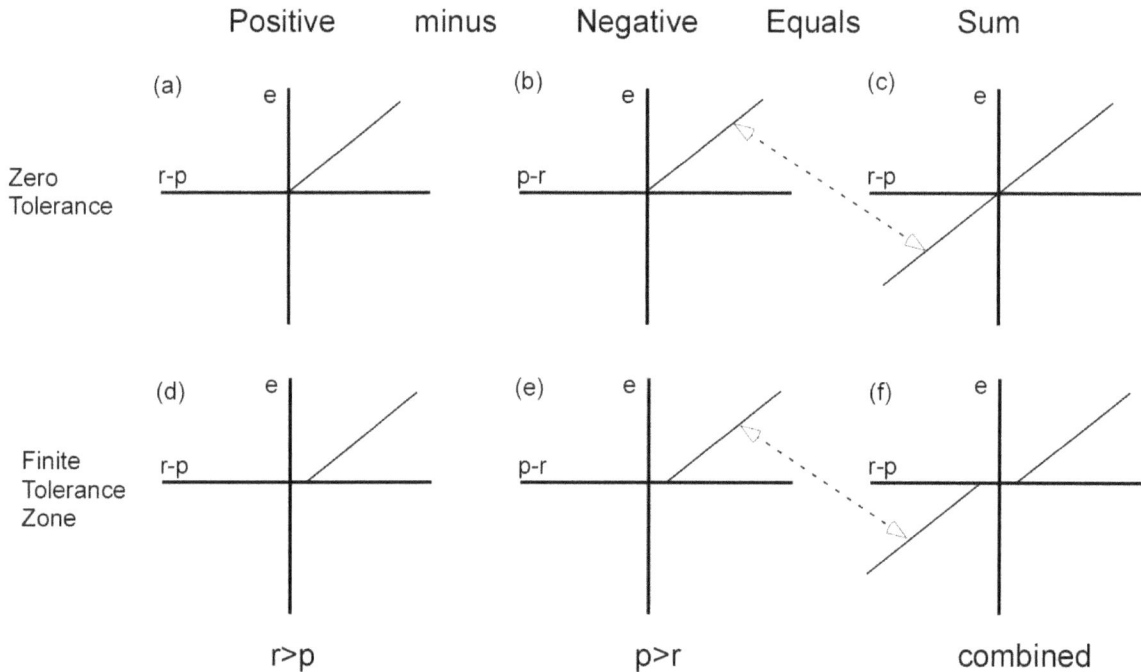

Figure I.4.6d Formation of a two-sided error function from two one-sided functions that use only physiologically feasible firing rates greater than zero. The two-sided function is purely conceptual, and cannot exist as a function with just one output value that is a neural firing rate.

It is also possible for the positive and negative outputs of the comparator to be fed to independent output functions, analogous to flexor and tensor muscles that cannot push, but must always pull against each other until they come to a pair of tensions that result in a particular joint angle. We will consider this possibility further when we deal with 'stiffness' and the control analogy to physical tensegrity structures. In most of the rest of the book, we will ignore the neurological implications of positive and negative values, and simply assert that we can perceive and control the difference between 'above' and 'below', 'left' and 'right', 'more' and 'less', 'before' and 'after', and the like.

What we will not ignore is that between these opposites, there is often a middle ground, suggested by Figure I.4.6d panels (d) to (f). 'The same height' is between 'above' and 'below', but 'the same height' means not that the heights are the same within picometres, but that they are close enough for the purpose for which the comparison is made. In the context of perceptual control, 'Close enough' is a rough definition of 'tolerance', or rather, of 'within tolerance bounds'.

I.4.7 Tolerance

In PCT, tolerance means that although a controlled perception may differ somewhat from its reference value, nevertheless the error value that enters the output function is zero, as though the perception matched its reference. No change in action output is needed to improve the match between perception and reference values. If two things are 'the same height', the plank laid between them will serve its purpose at some higher perceptual control level. Different purposes demand different levels of tolerance.

Perhaps a plank laid between two supports of 'the same height' is level enough for you to sit on when the heights differ by a couple of inches, perhaps it is level enough that a ball laid on it does not roll to one end, perhaps it is level enough to allow you to walk from one end to the other without slipping because of a perilous slope from one end to the other. We are talking about 'tolerance', and for different purposes these 'same heights' have different ranges of tolerance. Any time you say to yourself (or to others) something like 'That's close enough' or 'I can live with that', you are demonstrating tolerance.

'Tolerance' has a meaning in engineering. A part-length may be specified as, say, 5 cm, but a difference of 1 mm either way may have no effect on the usefulness of the part. Suppose I want something to prop open a window. A 1 cm stick doesn't leave enough gap, and a 1 m stick won't fit, but anything between 30 cm and 50 cm will do. My tolerance zone is roughly 20 cm wide. In the early days of electronics, a component such as a resistor could be specified as having 1%, 5%, or 10% tolerance, which was shown by the gold, silver, or black colour of a band on the resistor body. Suppose I want to create a standard clock to test some prediction of relativity theory. A clock that loses one second per millennium is too inaccurate. It must be much more precise than that. My tolerance zone allows the clock to gain or lose perhaps one second per billion years.

In everyday speech, the word 'tolerance' often implies an ability to accept that someone else may do something you perceive to be wrong or misguided, without trying to correct them. You may think a religion other than your own is wrong-headed, but you 'tolerate' people who profess it — or you don't, and may try to convert them, kill them, or keep them out of your neighbourhood. We will turn our attention to this everyday meaning later. In all these cases, social, mechanical, electrical, or perceptual control, 'tolerance' implies a difference that does not matter so long as the difference is small enough that the value in question remains within a 'tolerance zone'. For now we restrict ourselves to a discussion of tolerance as a technical term that applies to analogue control loops.

We now begin to show that, paradoxically, the existence of a finite tolerance zone in perceptual control does matter even for a single ECU. It can improve the ability of a control loop to react rapidly but stably to a sharp change in reference or disturbance value. In common language, it helps an organism to 'turn on a dime' when circumstances change abruptly.

The presence or absence of a tolerance zone can be tested, at least in tracking studies. In an unpublished aspect of a tracking study conducted as part of a study on sleep deprivation, I compared the fit of models of several different control structures to over 1,300 human data tracks for two different kinds of perceptual control (Taylor 1995) and found that including a small tolerance zone always improved the fit. Although the fits with the linear error function of Figure I.4.6a panel (a) were good, those with the function of Figure I.4.6a panel (d) were better, sometimes appreciably. This result should not be taken as definitive, because adding a parameter to a fitting process is always likely to improve the fit. Nevertheless, everyday experience suggests that we encounter a wide range of situations in which we see that something is not exactly as we would wish it, but do not act to bring our perception of it closer to its reference value. And in socio-political discourse as well as pop psychology, it is often said that a little tolerance is a good thing — in other words, "Don't sweat the small stuff."

One reason a little tolerance is a good thing is that it can allow for conflict-free control of more independent perceptions than would be possible in a strict 'zero-tolerance' regime. If control of one perception slightly disturbs another controlled perception in a 'zero-tolerance' regime, the disturbed ECU will act to correct the error, thereby probably further disturbing the original and leading to an escalating conflict. If each has a small tolerance zone, such conflict over trivial error values can be avoided. The effect is, for well controlled variables, as though the available degrees of freedom had been greatly increased. In an ecology of many interrelated perceptual control systems, the ecology with zero tolerance is rigid like a crystal, and can be altered only with some intermediate disruption, whereas a system with tolerance allows for a certain degree of smooth modular reorganisation.

A tolerance zone falls out naturally from the no negative neural firing rate representation of the comparator (Figure I.4.6b), as it depends only on the balance between excitation and inhibition in the individual physiological components. The tolerance zone could be implemented by adding a signal that connects to an inhibitory (subtractive) input to the two half-comparator units.[53]

---

53   Compare Bill Powers (920722.0800) in the CSGnet archive. — Ed.

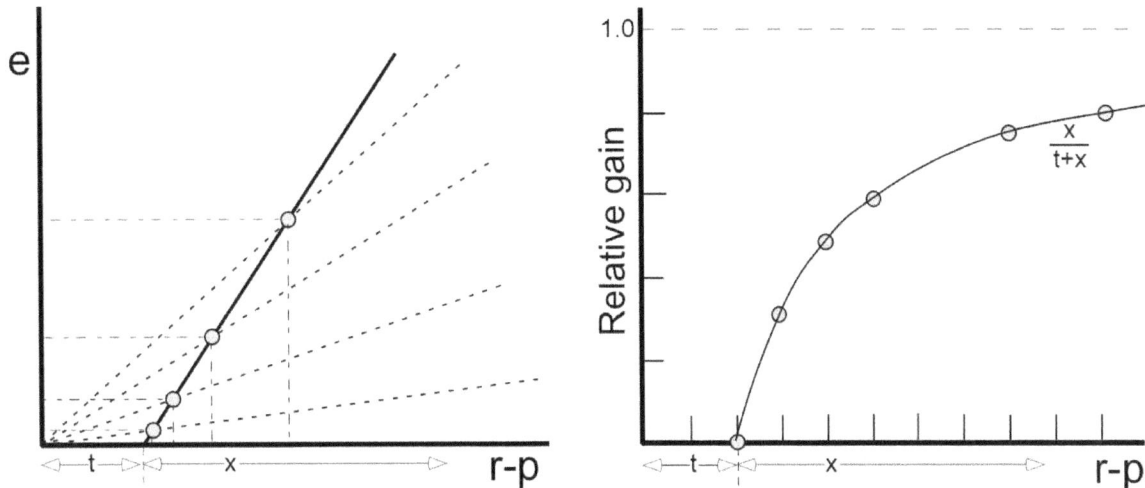

Figure I.4.7a (a, left) The error value "e" as a function of the difference between reference and perception if there is a tolerance zone of half-width "t". The dashed lines show "intolerant" error functions that would provide the same value of the error for a given value of r-p when r>p; (b, right) The apparent increase in instantaneous loop gain for different magnitudes of error if there is tolerance. "x" is the excess of |r-p| beyond the tolerance zone.

The left-side graph of Figure I.4.7a shows one side of a linear error function in a system with a tolerance zone of half-width t. When the error is greater than t, the instantaneous gain is equivalent to that of a zero-tolerance system with a gain represented by the slope of a dashed line in the figure; the larger the error, the higher the instantaneous gain. The right-side graph of Figure I.4.7a shows the result. When error is large, the effective loop gain is large, as it would have been if the error function was a linear function with a high slope, but when the error is small, the loop gain is near zero. The relative loop gain is given by the function x/(x+t) where t is the half-width of the tolerance zone and x is (|r-p|)-t.

Tolerance has an unexpected effect on the speed and stability of control in situations such as tracking a moving target that sometimes shifts abruptly. All physically realisable loops have some transport lag, and when there is transport lag, a sufficiently high gain will send the loop into oscillation, and lower but still high gains will cause it to oscillate around its asymptotic value after a step change in the reference or disturbance value. A change in the slope of the error function has the same effect in the control loop as a change in the gain multiplier of the output function, so what this analysis suggests is that control loops with a tolerance zone should tend to correct large errors faster than a linear model that is a good average fit, but correct small errors more slowly or not at all. The effect of this shift is a reduction in overshoot after a step change in a disturbance, as compared to the linear model, thus allowing an increase in the gain of the output function without loss of stability, as suggested in Figure I.4.7b for an idealised situation.

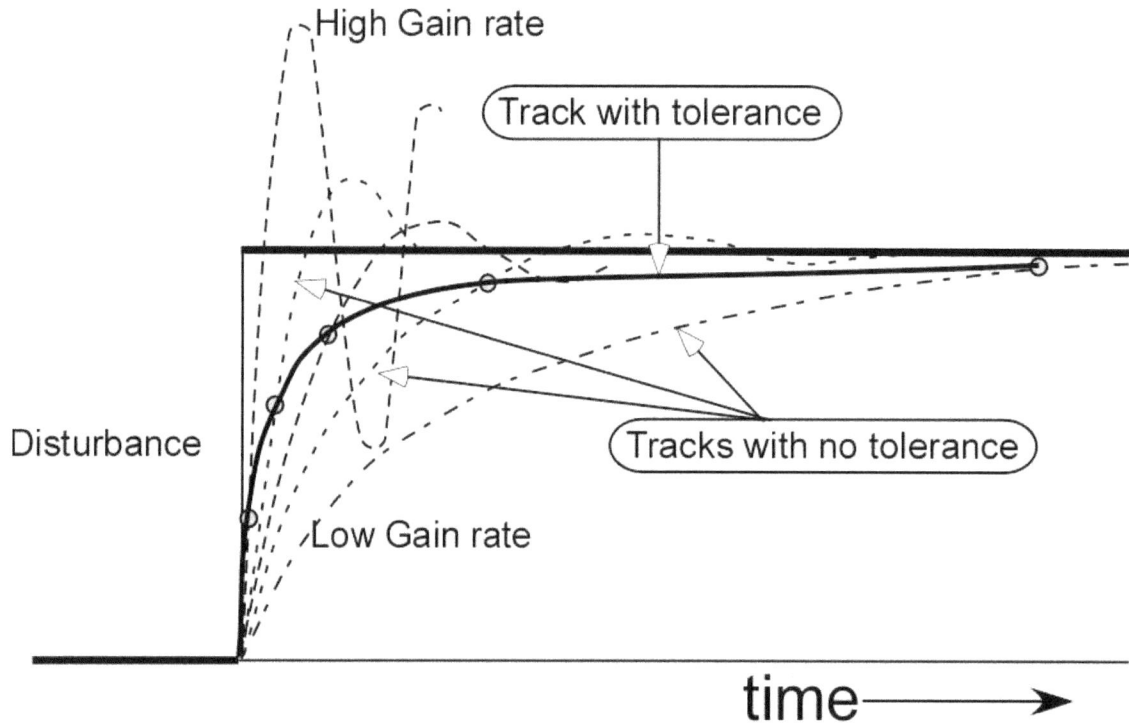

*Figure I.4.7b The effect of including a small tolerance zone on the speed
and stability of control in the presence of loop transport lag. Dashed curves
suggest the response of loops with no tolerance and various gains if the output
function is a leaky integrator with the indicated gain rate. The heavy curve
suggests the response if the loop has a high gain rate and a tolerance zone in the
error function. So long as the error is large, the effective gain is high and the
track nearly follows the "High Gain Rate" function. As the error continues to
decrease, the effective gain rate also decreases.*

Figure I.4.7c shows the effect of a tolerance zone for an illustrative example
condition simulated in a Microsoft Excel program provided by McClelland.
Two controllers are compared, one with zero tolerance, and one with a tolerance
zone. The left panel of Figure I.4.6c shows the speed-up effect illustrated
schematically in Figure I.4.7b, whereas the right panel shows the 'High Gain
Rate' overshoot that occurs without, but not with, tolerance when the gain rates
of the two controllers are set identically.

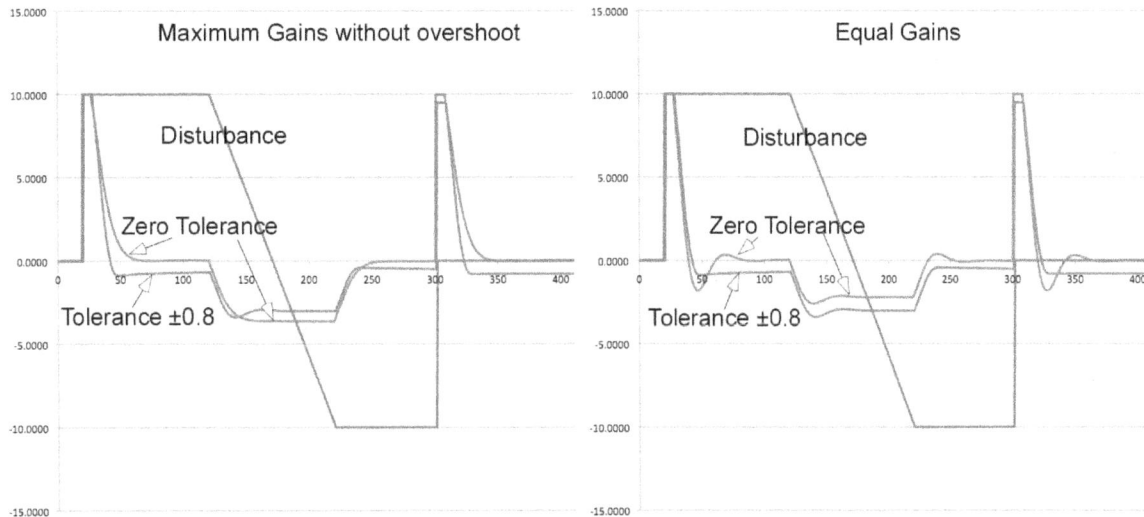

Figure I.4.7c Examples of the effect of the tolerance zone on overshoot and on speed of approximate correction of rapid disturbance changes in a simple simulation (Excel program kindly supplied by McClelland). Graphs show the effects of changing disturbance values on the CEV of the controller with and without tolerance. The controller is as in Figure I.1.2a, in which all functions except the output function are simple pass-through operations in which the output is the same as the input. The output function is a leaky integrator and the comparator provides an error signal that is the amount by which the difference between reference and perception exceeds (falls below) the upper (lower) tolerance bound. The only difference between the other properties of the two controllers is that in the left panel the Gain rate of the zero-tolerance controller has been reduced to the highest value consistent with avoiding overshoot.

A controller with a fixed tolerance zone will never bring the error to zero, which might seem to be a reason to discount it as a normal component of a control loop. On the other hand, the ability of the controller with tolerance to reach a close approximation of the reference value relatively quickly may often have more survival value than the ability eventually to reach a closer approximation to a reference value that might well have changed before that close approximation could be reached.

Just as with the leak, the size of the tolerance zone presumably has some optimum value in any specific circumstance. Both are non-zero in fits of simulation models to human performance under tested conditions, though the sensitivity of the measurement is seldom good enough to track any changes in their value as the testing parameters change. All we can say from current experiments is that if the output function of a control loop tracking an analogue variable has the form of a leaky integrator, then the leak is almost certainly greater than zero, and so is the tolerance zone.

When we deal with alerting in Section I.7.5 and more when we deal with social conflict, tolerance will be seen to be much more significant than it is for a single isolated control loop. We will deal with social, political, and religious implications of 'tolerance' on the interactions of many people well beyond individual control, as well as implications for some mental health issues internal to the individual.

Chapter I.5. Further Aspects of Perceptual Control

I.5.1 Perceiving Magnitude and Perceiving Place

Much of the explicitly PCT-based experimentation and simulation uses tracking of place, in the form of making a cursor follow a target or making a target stay at a chosen location regardless of a disturbance that would move it if the controller stopped acting. Some, however, do control magnitude, a rather different proposition. Why is controlling magnitude different from controlling place? Because magnitude could be the measure of firing rate in one neuron (or a bundle of them), whereas changing place is necessarily a change of which neurons are most sensitive to the moving entity. It is a question of controlling Mackay's logon values (e.g Mackay 1950, 1953) as contrasted with controlling his metron identities (Figure I.5.1).

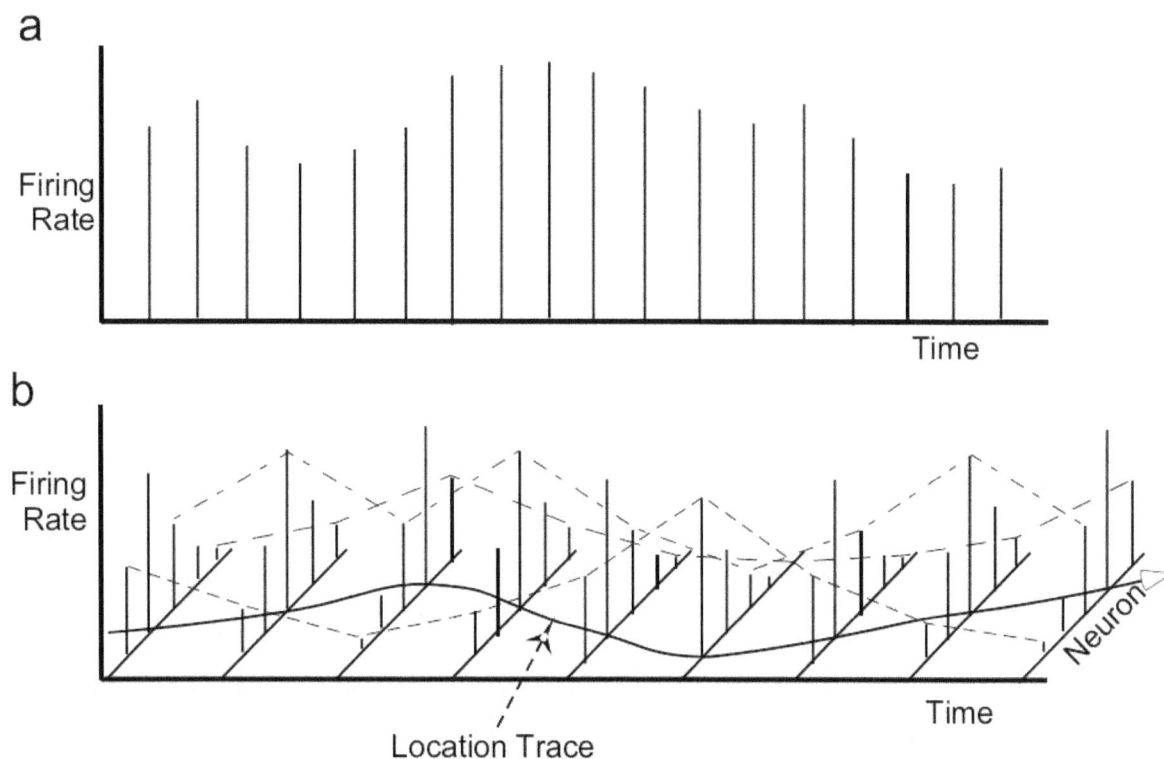

Figure I.5.1 Changing magnitudes versus changing place. (a) Changing magnitude can, in principle, be tracked through the firing rate of a single neuron. (b) Changing place cannot, because it involves the change of relative firing rates across several neurons. The solid curve represents the movement of the "target" across the set of neurons, while the three light dashed curves represent the changing firing rates of the first, third, and fifth depicted neurons as the target moves.

When tracking a magnitude, the firing rate of a reference neuron or 'neural bundle' could, in principle, be compared with the firing rate of a corresponding perceptual neuron or bundle, but this is clearly not possible when tracking a place. In the first case the error value is simply the difference between the two firing rates, but what is the error value in the case of place tracking? Conceptually, it is the distance between the neuron that has the maximum firing rate in the reference profile and the one that has the maximum firing rate in the perceptual profile. Could this difference be converted into a single neural firing rate, thus converting control of place into control of the magnitude of a difference between target and perception, the usual job of a comparator?

This turns out to be a wrongly posed question. At the interface to the environment there are not myriad independently moving deployers of force. The movements of individual arms, fingers, legs, and so on each create effects in the environment, some of which change the values of perceptions produced by perceptual functions. Although these perceptions are treated as unitary, they are the results of myriad neurons that form what Powers called a 'neural bundle', and the results are a 'neural current' in a particular bundle. To treat the neural firings individually, as the wrongly posed question does, is to ask at a level of fine detail a question that can only be answered in the aggregate.

The answer parallels in the aggregate the answer that is used in detail when discussing the comparator. The place on the sensory surface (e.g. the retina or the skin) is reported by the various connections made by the neurons that are differentially excited by the event, the place being the location of the maximum effect. The place in the environment uses the internally localised place plus similar aggregates that report posture (e.g. head and eye pointing direction) to form some of the inputs to a perceptual function that reports location in space.

We need not go into the details of how place is computed in the brain. All that we need to know is that we perceive where things are relative to one another, without perceiving any absolute location for any one item. The relevant perceptions are distances, not locations. Distances are magnitudes. By reducing the problem of perceiving place to one of perceiving magnitudes, the apparent problem of accounting for success in the usual 'cursor-tracking' task disappears.

I.5.2 Energy and Entropy

This section is likely to seem like very technical physics, but if you can get even the gist of its theme, it will help you to gain a better understanding of control. If not, you should be able to understand the rest of the book without reading this section, but at a shallower level, yet still sufficient to justify the intrusion of PCT ideas into the socially oriented research domains indicated in the subtitle of the book: *An Inquiry into Language, Culture, Power, and Politics.*

Even the smallest of organisms that we generally accept as living, such as bacteria, need food if they are to stay alive. Why should this be? Because the organism needs to expend some energy in order to act in any way. According to basic physics, any realisable process requires energy to be expended, and that includes any action performed by any living thing. The expended energy can never be completely recovered. The energy used to stretch a rubber band seems to be recovered when the band relaxes to its unstretched state, but the energy is not all recovered during the cycle of stretching and relaxing, as you can tell if you feel the band after doing it a few times in a row. It gets hotter. That is a sign of increased entropy in the band, entropy that is released into the air as the band cools down.

Entropy is often confusing because it sometimes seems to be the opposite of what you at first think it should be, especially in the dynamic situations (non-equilibrium conditions) that are necessarily one aspect of control. So let's think about that for a moment. One point about entropy is that it is additive. If you add some entropy from somewhere, it doesn't change how much was originally there, it just adds its contribution as an arithmetic sum. This is true in the static or quasi-static conditions which engineers of steam pumps and railway locomotives considered and which were considered later by the developers of the theory of heat engines, and it is true in the dynamic 'non-equilibrium' conditions which are of interest in control systems.

Entropy calculation is a mathematical operation on a set of real-valued variables. If they are variables in a control loop, entropy can be calculated only from the external analyst's viewpoint. From that viewpoint, a disturbance to an environmental variable that corresponds to a perception adds entropy to the affected control loop and to the body that contains the loop. The external omniscient analyst can see all the variables concerned, but the organism cannot.

One way of looking at the entropy problem is that it is the survivalist job of control to stop entropy introduced by external disturbances from continually increasing within the body, and to get rid of it to the environment somehow. Most mammals get rid of it at least in part by heating up water and dissipating the heat into the environment by evaporating the water either as sweat like humans, or by expelling moisture-laden breath that dissipates into the cooler air. They also take in well-structured (low-entropy) food and excrete less well structured (higher entropy) material as urine and faeces, these being another way of exporting entropy to the environment in the process of extracting from food some of the energy used in creating it, whether that might be the solar energy taken in by plants or the energy used by animals in building their bodies from the structured material of plants or other animals.

There's an old student mantram that says "Heat can't pass from a colder to a hotter." This may be true of cold and hot bodies in contact. Two otherwise similar bodies, one hot and the other cold, will tend to equalise their entropies per unit quantity (volume or mass) if they come into contact, usually but not always by passing heat (entropy) from the hotter to the colder. However, the mere existence of a refrigerator gives the lie to the idea that one cannot transfer entropy (heat) from the cold interior of the fridge to the warmer environment. A fridge does this as its main job. But it needs an energy supply if it is to work. A fridge whose workings are neither plugged into an energy supply nor provided with a local energy source equalises the temperature of its contents with that of its environment. In such cases, the environment leaks entropy through whatever insulation the fridge may have into whatever it contains, until the specific entropy per unit of the contents matches that of the environment — and the contents which were being kept cold start to increase their internal entropy, in the form of decay rather than heat.

Entropy is not a property just of a single variable over time, or of a collection of variables at a single moment. It also depends on how the variables relate to each other. If two variables always covary, their combination is of very low entropy, perhaps zero, but as a pair their variation contributes directly to the entropy of the collection. More interesting, however, is what we might call 'structure'. Later in this chapter (Section I.5.5) we use as an example of two-level control a wooden chair that has four legs, a seat, and a back. When they are just parts not yet assembled, each can be moved separately in three dimensions, but when the chair is complete, they all move together. It is the chair that can be moved in three dimensions, not the no-longer independent parts. The chair is (or has) a 'structure'.

The relatively high entropy of the initial arrangements of the six parts becomes very low or zero as an assembled structure. Looking at a chair as a question of "Where did the entropy go?", the answer must be that the energy used in assembling the chair is dissipated eventually as heat to the local environment. In practice, this includes the work of some person or machine which picks up the pieces and puts them together. Energy and entropy are closely linked and are often discussed together, but they are not the same concept. We will see this in many places through the book, such as when we consider the use of money for maintenance of some physical structures such as roads, bridges, and buildings. But it also applies to the maintenance of less concrete items such as subscriptions to periodicals or club memberships.

Throughout the book we will encounter situations that could usefully be viewed by considering entropy relationships and changes, but in most such cases we will not. All the same, it might help if one sometimes imagines a control system as a tool for getting rid of the entropy which the continual actions of the environment on any lower-entropy structure adds and which would otherwise tend to destroy the structure. In this view, the essence of an organism's survival, and in one way the essence of perceptual control, is export of entropy by means of a through flow of energy.

I.5.3 Quality of Control and Introduction to Rattling

When one looks at a control loop from the Analyst's viewpoint (Section I.2.5), one can imagine the different signal values (the loop variables) at any moment as time passes. The two inputs to the control loop, the disturbance variable and the reference variable, vary with no relationship to the other parameters of the loop. As each varies, so do the other variables of the loop — the perception, error, and output variables, and what we call the Complex Environmental Variable (CEV), that which is jointly affected by the external disturbance variable and the opposing 'output variable' or Action.

The Analyst can determine the variance of all these variables over time. The better the control, the lower the magnitude of the error, given a particular variance of the disturbance. (Henceforth, I will omit the word 'variable' in this context.) When we compare the disturbance variance and the error variance,[54] their ratio is a measure of how good the control is. In much PCT discussion, this ratio, 'disturbance variance divided by error variance', is a measure of the Quality of Control, or QoC. We shall use the QoC notation frequently throughout this book.

In Chapter I.10 we will talk a lot about measures of uncertainty and information. 'Uncertainty' is a non-parametric measure that can be numerically related to variance by a fixed multiplicative constant if the distribution of values is the common Normal (or Gaussian) bell shaped curve. Unlike the variance, the uncertainty of a distribution can be precisely and usefully calculated no matter how different it is from the standard bell-shape. Uncertainties and variances both can be simply added if you want to compute the uncertainty or variance of a variable that is the sum of independent variables 'a, b, c, …'. The opposite of uncertainty is 'information'. Numerically, 'information' is the change of uncertainty, positive or negative, following some event or observation. Uncertainty as a measure should not be confused with personal uncertainty.[55]

At this point, we have all we need in order to discuss uncertainty-based Quality of Control measures, especially one new measure, 'Rattling', that we will use increasingly as we move to discussions of multiple interacting control loops, occasionally in Volume II and more often in Volume III. The 'rattling' measure applied to entire organisation was, so far as I am aware, first described by Chvykov et al. (2021). Here, we will apply it to the control loop as a structural organisation.

---

54 One cannot compare the immediate magnitudes of the disturbance and the error directly. Their variances are measured over a stretch of time called a 'window', which must cover a time period appreciably longer than the loop transport delay. This is because of the loop transport lag that delays the effect on the error of the change in output opposing the changed disturbance.

55 Uncertainty is a calculated measure conversely related to probability: higher probability is lower uncertainty. Probability is calculated as the ratio of favourable outcomes to the set of all possible outcomes. Personal uncertainty is in practice seldom derived from calculating probabilities, not only because these sets are seldom fully identified or even considered but also because of the relation of personal uncertainty to conflict. Such confusions are the source of the many metaphorical abuses of the technical notion of information by people who should know better. — Ed.

Before we introduce rattling as a concept and measure, we need to look at a couple of other concepts, 'amplitude modulation' and the relationship between an amplitude-modulated waveform defined as x(t) and its derivative dx(t)/dt. A waveform is uncorrelated with its derivative, but if you magnify the waveform by some magnifier M, creating a waveform Mx/t, the derivative also is magnified by M, making Mdx(t)/dt. If M varies over time, both the original waveform and its derivative are amplitude modulated by the waveform of M(t).

AM radio is an example of amplitude modulation, in which the signal being transmitted is the modulation waveform M(t), and the base waveform, the 'carrier', is the signal frequency to which the radio dial is tuned. The waveform and its derivative are uncorrelated, but the modulation of the waveform is precisely correlated with the modulation of its derivative.

The modulation M(t) is a waveform that has some variance or uncertainty, namely, that of the program to which you are listening.

Let us imagine that the signal in question is the disturbance to a control loop rather than the program heard on an AM radio. The disturbance value has a variance or uncertainty from moment to moment, but its amplitude cannot be measured until at least a full cycle of its lowest frequency has affected the controlled environmental variable (CEV). To follow its modulation waveform, M(t) requires several cycles of the lowest frequency in the disturbance, and to determine the modulation variance requires following the modulation for several times that duration.[56]

What matters here is change in the uncertainty of the modulation of the disturbance to a control loop, and the effect such change has on the Quality of Control (QoC) of the loop. As an example of such a change, imagine a radio announcer delivering an ordinary message, and at some point his voice is overridden by a loud bang and the sound of alarm bells. The uncertainty of the modulation would be greatly changed for any measurement period that included the bang, and a control loop controlling to distinguish the announcer's voice would have an instant rise in its error variable.

If we now consider the uncertainty of the error variable, it rises relative to its normal value faster than the disturbance does, because the disturbance is largely opposed, but the bang is not. The control loop is limited by its transport lag in how fast it can control against rapid changes in the signal it is controlling.

Another viewpoint is always useful when you want to understand something. The 'rattling' measure introduced by Chvykov et al. (2021) provides another viewpoint on stability of control and on reorganisation processes that restore stability. Rattling is a measure over an organisation of interacting entities. The central point of the paper by Chvykov et al. was that organisations trend (but are

---

56   Given enough time, a receiver could in principle produce a signal that represented the changing variance of the modulation signal, and this 'modulation of modulation' process could be carried on over an indefinite number of levels, but this is not the point of the present discussion.

not driven) to structures which are calmer (have lower total rattling) more often and more strongly than toward structures that are more rattled. They investigated properties of groups of entities which are not perceptual controllers, so they do not use these words, but in terms of the discussion above we can see rattling as a measure of the uncertainty of the derivative of the modulation of a waveform. It applies not to individuals but only to the structure of the entire set of entities and how much effect they have on one another. Like thermodynamics it is a general property which helps to delimit the functional scope within which control systems can operate, in this case especially their capacity for reorganisation.

We will start to pursue the social implications of this finding by Chvykov et al. in Chapter II.5, and then as we progress through the rest of the book we will more deeply develop wider-ranging consequences of the tendency of organisations of individuals to reorganise to calmer, less rattled structures.

I.5.4 Control Stability

The effects of time are often ignored in discussions of Perceptual Control, in favour of the discovery of equilibrium states of variables in the loop, but time-effects are important. It takes time for neural impulses to travel along the axons, and it takes time for perceptions to form, but these millisecond-level times pale when compared with the time it may take for the influence of the output signal to appear at the CEV and return to the perceptual signal. If you are in a restaurant, for example, and after perusing the menu you control for perceiving (want to see) a plate of fish in front of you, you actually see that there is no plate of fish currently on your table, and your control action is to ask a waiter for a plate of fish. Ten or twenty minutes later, you may perceive your plate of fish in front of you, but it may have taken only ten or twenty seconds between choosing your meal from the menu and acting to correct the perceptual error of the empty place setting by telling the waiter what you wanted.

The time it takes for the effect of a change somewhere in the loop to return around the loop to the same point in the loop is called the 'loop transport lag' or 'loop delay'. In the restaurant example, the loop delay is ten or twenty minutes, even though the action delay might have been only ten or twenty seconds. As is often the case, the rest of the delay is in the environmental feedback path. We will examine the implications of path delay much more closely in Volume IV, Appendix 9, on 'relativistic networks'.

Suppose that during the long delay before the waiter returned with the food you had decided that you had been forgotten or the waiter had had an emergency, and you had summoned another waiter to order again, because your perception of the state of your place setting was still different from its reference value. A few minutes later, your original waiter emerges with your original order and, before you finished, the second waiter appears with your second order. Now your place setting is in error again, differing from its reference value by one extra plate of food. Had the loop transport delay been shorter, this would not have happened.

Consider another example, one that most people have experienced: the howl, scream, squeal, name it what you will, of a public address microphone sensitive to the sound from a nearby loudspeaker broadcasting its user's voice. The loudspeaker sound returns to the microphone after a delay determined by the distance between them. Whatever waveform was induced by the speaker's voice initially returns after this delay and is added to the current voice input. Then that addition comes back again, and again, and again, always after the same delay. If some initial input, no matter how small, is added again and again to itself and the re-amplified return exceeds the original input for some frequency whose period is an exact multiple of the delay, the result will be an exponentially increasing output at that frequency — the squeal.

The same is true for any feedback loop. Loop transport delay always exists, because even at the speed of light, it takes time for events at one location to have any effect at another location. If at some frequency there is positive gain greater than unity, the loop will be unstable. In the case of the waiter it looks different because the events are discrete. Let's imagine that you ate none of the meal brought by the first waiter before the second waiter arrived with his contribution. Now you want a plate to be taken away, but the same delays happen, and both plates get taken away. Because you again have no food, you again place an order. This kind of oscillation does not escalate, but it could continue until something about the control loop changes.

In his 1979 tutorial articles in *Ised* magazine, Powers demonstrated that by adding a prediction component to the perception of the current state, this kind of loop instability could be reduced. Adding some multiple of the derivative of the perception is equivalent to asserting that the perception will be the current perception a certain time in the future if it continues to change at the same rate. Almost always, its rate of change will change over time, just as does the actual value of the perception, but on average, the effect of using the predicted perception at the time when the output will have its effect on the CEV, rather than using the current perception, offers an improvement in control stability and accuracy.

We return to the relationship of delay to instability in the more complex case of conversational interaction, where the problem is to avoid instability in the loop between the conversational partners while maintaining accuracy in their communications.[57] Accuracy demands long delays, stability demands short loop delays. We will discuss how these conflicting requirements are addressed by the structures of language, and how they relate to language drift between cultural groups and over time.

---

57   R. Kennaway's Annex in Powers (2008) provides a careful discussion of stability in single control loops with delay.

I.5.5 Perceptual Complexes

For simplicity, we have so far been concerned mainly with the operation of a single control loop, but in most of this book we will be dealing with structures of several, sometimes very many, control loops which sometimes work together, sometimes interfere with each other's operation, and sometimes operate entirely independently of each other.

The basic structure with which we start is the perceptual control hierarchy as described by Powers in his many writings. We do not develop it further until Chapter I.8, but here we offer an answer to the question of why perceptual control in an organism should be organised hierarchically, as Powers argued it must be and as experimental demonstrations suggest it is (see for example Marken 1986; Marken and Powers 1989a; Marken, Mansell and Khatib 2013). We approach the question differently from the way it is usually approached, because we start by assuming that the 'real world' in the environment has certain coherences that disallow some patterns of perceptual control while supporting others. We will go further into environmental structure when we introduce uncertainty analysis in Chapter I.10.

First, we consider a row of undifferentiated control systems, each of which independently controls a perception of the orientation of a different part of the environment. Then we consider the consequences if these independent control units happen to be looking at different parts of one rigid object. One of the consequences is the probable development of a second level of control.

A control loop controls a perception that has a single scalar value, and all the variables everywhere around the loop are scalars.[58] However, control loops in real organisms do not exist in isolation. Control loops interact, and they do so in many ways. One of the ways is Powers's hierarchy of control. Another is through the side-effects of their actions on each other. A third is resource-limitation conflict, and yet another is in a mutually useful arrangement we call a 'protocol loop', which is discussed mainly in Volume II and in my Chapter in *LCS IV* (2020). A final possibility is lateral inhibition between control loops which control somewhat similar perceptions. We discuss some implications of this last way in Chapter I.9, and implications of all of them in scattered places throughout the book starting in Chapter I.11.

At this point, however, we will take an Analyst's view of a set of control loops that act simultaneously in a common environment, but which are not otherwise interconnected (Figure I.5.5a). They interact only through the side effects which the controlling by each has upon the ability of others to control effectively.

---

58   A scalar value is one that can be described by a single number, as opposed to a vector value represented by an ordered set of numbers, or a matrix value which is described by a rectangular array of numbers.

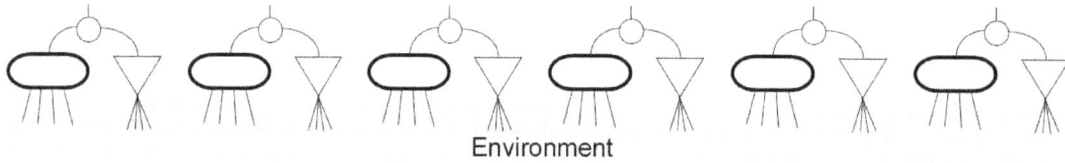

Figure I.5.5a A set of six elementary control units independently controlling their perceptions in a common environment. The Perceptual Input Functions (PIFs) are emphasised.

Each of these control units knows nothing but the scalar value of its perception. Each perceptual value is independent of all the others, but the Analyst can see all of them at the same time. To the Analyst, the set of control units controls a vector-valued perception. From time to time we will use that way of looking a complex control structure, notably when we talk about control 'profiles' such as are shown in Figure I.4.5, but for now we can take the vector description or leave it, because we will be dealing with consistencies in how these independent control systems see the environment.

Although the values of these perceptions are independent of each other, each is determined by the part of the environment being sensed, and the different parts of the environment being sensed may perhaps not change independently of each other, as suggested in Figure I.5.5b. In Figure I.5.5b, every perceptual function gets its input data from a different part of the environment, but the Analyst sees that they are all looking at different parts of a chair. Let's see how the orientation -perceivers change their perceptions when the Analyst turns experimenter and moves the chair slightly in different ways. A caution: it may be difficult for you to set aside the privileged Analyst view which the image of the chair encourages you to take, and to consider only the limited input of each controller, the orientation of one part in X-Y-Z coordinates.

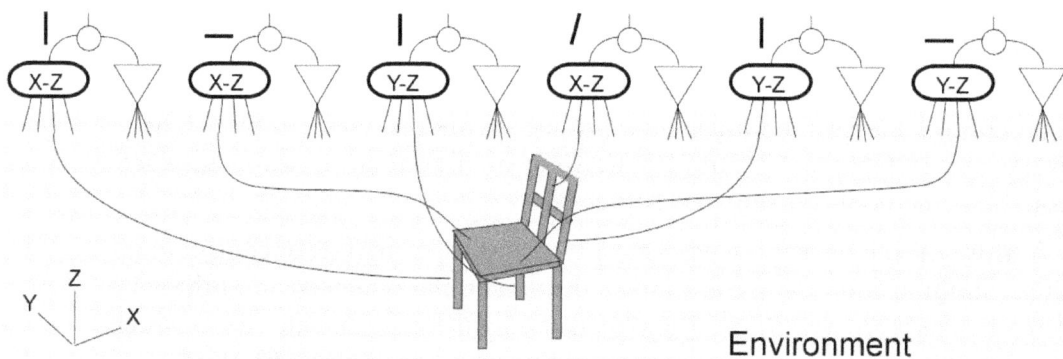

Figure I.5.5b Six independent perceptual functions "looking at" six different parts of the environment may not have completely independent values of their perceptions, if the parts of the environment that they depend on are in some way coherent. The notation "X-Y" means that the perception is the orientation in the X-Y plane, or rotation around the Z axis (no X-Y units are actually shown).

First, suppose the Experimenter turns the chair, rotating it in the X-Y plane so that if someone were sitting squarely on it they would face a bit more out of the picture. If we label the control units 1 through 6, reading left-to-right in the figure, the perceptual values of units 4 and 5 that perceive the orientation of the back slats change, but the others do not because their very limited inputs from other parts of the chair are unaffected. Of course, any units that could perceive rotation about the Z axis would change their perceptual values, but there are none in the example set.

After putting the chair back in its original position, the Analyst-Experimenter next tips the chair backward. Now the perceptual values of units 1, 2, and 4 all change together, but the others do not. (For example, tipping the chair along the X axis does not change the orientation of the back leg from the point of view of controller 3 'looking' along the X axis.) If instead, the Analyst-Experimenter tips the chair to its right side, units 1 and 2 do not change, but all the others do. And so it goes. However the chair is moved, always some group of the perceptual functions change together, and sometimes all of them will change. The six perceptual values are constructed to be independent by virtue of perceiving different aspects of different parts of the environment, but their data in practice are not independent. The Analyst can see why they are not independent; it is because the different ECUs happen to perceive regions of the environment that the Analyst knows to be physically linked, even though none of the ECUs is individually affected by their non-independence.

Even if we define the chair as simply as in the cartoon form of Figure I.5.5b, there are far more than six apparently independent orientations of its different parts. Each 'leg' has three possible rotations, as does the seat and each element of the back. Even if we say the cartoon 'back' is a single slab, and allow only one perceiver of each of the three rotations for each part, the chair parts have at least 18 different 'independent' orientations that might be perceived. All of these orientation perceptions change in consistently coordinated patterns as the Analyst-Experimenter rotates the chair this way and that.

If we now consider the perceptual values of the original six ECUs in the example set, there may be special ways the chair can be rotated so that only one of the perceptual values is changed, but in general, this is not possible. The Analyst, who sees the vector of perceptual orientations together with the chair as a coherent object, can see that there are three independent axes for rotating the chair, which means that at most three of our chosen six orientation perceptions can be independently and freely changed. The other three, whichever they might be, must change in ways determined by the values of the independent three. The same is true if we include all 18 of the orientations of the chair parts. Only three can be independently changed, and the rest must follow in a coherent pattern.

Coherence of orientation perceptions does not define a 'chair'. It is a fact about any rigid object. Another fact about any rigid object is that no matter how many different parts of it are perceived to be in particular places by independent location-perceivers, only three of the location perceptions can be independently changed without also changing at least some of the orientation perceptual values. If you move the chair centre-of-gravity up-down, left-right, and forward-backward, you can independently move, say, the left front leg (a little) up-down, but only by tilting the chair around a diagonal axis. Together, although there may be 36 controllers of perceptions of location and orientations, their common linkages to the chair allow only six independent patterns of location and orientation of the parts of the chair, three of orientation and three of location.

Now imagine for a moment that each of these 36 little controllers of orientation and location had its own muscles to influence the location or orientation of whatever bit of the chair it was looking at, and acted to control its perception to its own local reference value. What would happen when the Analyst-Experimenter introduced disturbances like rotating the chair or moving it to a new position?

Many or all of these 36 controllers would find that their perceptions no longer matched their reference values, and would act to reduce the error. Behind the scenes all 36 controllers would push and pull at their bit of the chair, but they might get nowhere, because they would be in conflict with other controllers pushing and pulling at other parts of the chair. None of the individual controllers could 'see' this conflict, but to control their perceptions they would need to push and pull ever harder on what would seem to be a remarkably heavy object stuck in trembling jelly. This is not very effective control, at least for the individual controllers.59

Despite the conflicts among the 36 controllers (and if their escalating conflicts did not break the chair apart), they might tend to move it back toward its original position, since the 36 different controlled perceptions were all affected by a coherent set of six disturbances. How could these little controllers come to work together to move the chair instead of possibly breaking it?

So far, the system has been described as consisting of 36 individuals that interact only through the object. Nothing has been said about the source of their reference values, which have been assumed to be independently determined. These 36 values cannot be all satisfied at the same time if they vary independently. The only way that they could be all satisfied is if they are somehow made to be not independent. Some common source or sources must supply no more than six independent reference values, from which all 36 reference values can be constructed.

These six reference values would have to come from the outputs of six controllers, each controlling a perception that was some function of all 36 individual perceptions. In Power's terms, we are talking about a second perceptual level above our original little controllers. We are starting to build a hierarchy of control levels (Figure I.5.5c).

---

59   As we will see later (mainly in Volume II), the set of controllers is exercising 'collect-ive control', which, to an external observer, looks as though a single controller that we call a 'Giant Virtual Controller' (GVC) is operating with a reference value (observ-able in the environment) which may differ from the reference values with which the six independent units is controlling.

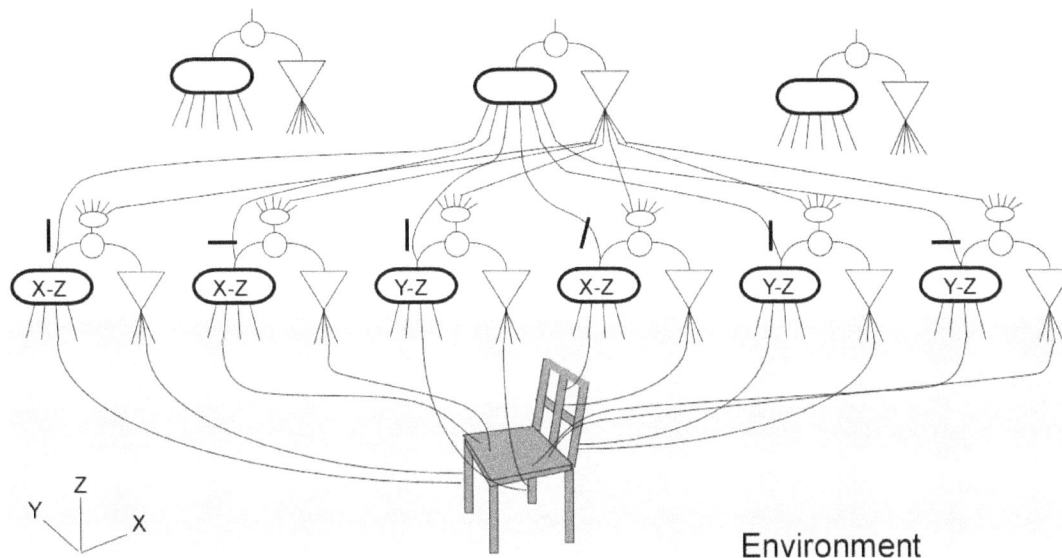

Figure I.5.5c A smaller number of control units each of which controls a perception built from the perceptions of all the lower level units and influences the reference values of all the originally independent units could, in principle, avoid conflict among the lower-level units.

These higher-level units would not miraculously come into being fully formed, but they might start as more or less random multi-way interconnections that become self-organised by virtue of the consistencies induced by the fixed configuration of the chair's parts. Generalising from this example, we can see that environmental consistencies of any kind can result in the development of a corresponding set of perceptual functions, and that consistencies among these consistencies might easily lead to a hierarchic set of levels of perceptual control.

'Reorganisation' is a process by which these consistencies may build on one another, and by which they may change when the environment changes. Reorganisation is the primary form of learning in the Powers hierarchy, corresponding to 'Procedural Memory' on the output side, and 'Semantic Memory' on the perceptual side of the hierarchy. Reorganisation as a concept is the topic of a later section. We will discuss a plausible mechanism at the level of synapses when we talk about Hebbian and anti-Hebbian learning, and will investigate reorganisation more closely when we consider the development of mutually useful control structures which we call 'protocols' in the 'Story of Rob and Len'.

It is also possible for the development of particular types of detectors at different levels of the control hierarchy to be genetically programmed. If so, the organisation would exist for the same reasons, but its form would have been found by natural selection over evolutionary time rather than by reorganisation during the life of an individual.

For now, it suffices to point out that if six higher level control units develop for 'chair-object' control, and are eventually connected appropriately to the perceptions and the reference input functions of the lower 'chair-part' level units, each higher-level unit connected to all of the lower-level ones, then the six higher level units could independently control perceptions of three rotations and three locations of the entire chair object. The lower-level controls would be operating in a coordinated way, rather than being in perpetual conflict. After the Analyst-Experimenter moves it, the chair would be returned to its reference position smoothly and without internal conflict among the 36 little control units. We may not yet have a 'chair' perceiver as such, but we do have a 'solid object' perceiver that is able to control perceptions of the object's position in space.

The development of levels of control could also go in the other direction, starting with control of 'chair' perceptions, the chair being initially perceived as a unitary object, the legs, seat, and back elements only later being perceived as potentially separable, with locations that could be individually controllable.

Taylor and Taylor (1983) proposed that exactly this does happen along with bottom-up combinations of parts when people learn to read. A string of letters is a visual pattern, which can be learned as a word. 'Whole word learning' advocates propose that children should start learning to read at this level, because the 'words' have meaning, and the strings of learned words make sense to the reader. On the other hand, advocates of phonetic learning believe children should learn to read by sounding out sequences of letters and discovering words in the sound patterns.

Taylor and Taylor suggested that the typical untutored way of learning might be an amalgam that they called 'three-phase learning'. In the initial phase, the child might learn the visual forms of some words as whole entities. When they knew enough simple words, phonetic similarities might be pointed out, such as that the 'c' in 'cat' sounded like the 'c' in 'cow'. With the phonetic patterns of the visual symbols established, the child might be able to perceive patterns and sound out unfamiliar words, a task impossible to a child who learned only whole words. In the third phase, large (perhaps syllable-level) sequences of phonemes would be perceived as units that recur in words of similar meanings, allowing the child not only to sound out new words, but also to perceive their likely meanings while perceiving them optionally at any of the three levels that might be labelled letter, syllable, word or phoneme, morpheme, word.

I.5.6 The Control Hierarchy

Although the core of Perceptual Control Theory is the observable fact of control, implemented by the single Elementary Control Unit, the essence of the Powers version is the control *hierarchy*. The hierarchy has been described in many publications by Powers (Powers 1973/2005, a.k.a. B:CP, is usually cited most prominently) and others, including chapters in LCS IV (Mansell et al. 2020). Here I review the most critical properties of the Powers hierarchical model, HPCT, and indicate how the structure on which the present chapter is based deviates from the strict hierarchy envisioned by Powers.

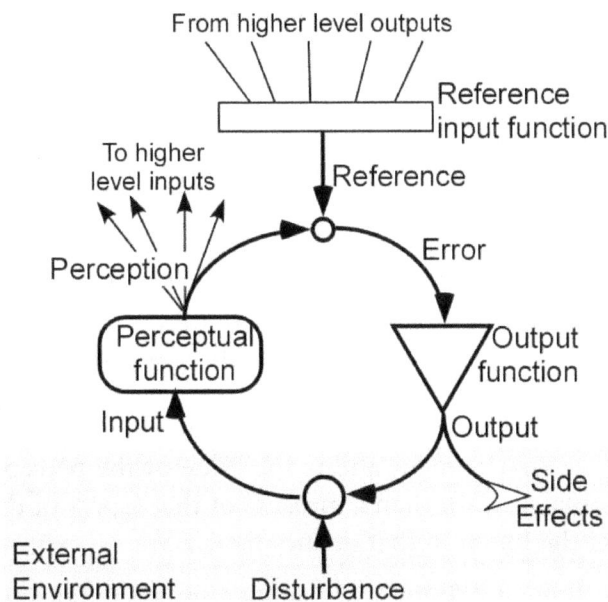

Figure I.5.6a. A basic control loop showing the major constituent elements. The "Elementary Control Unit" (ECU) is in the white upper portion of the diagram. The "External Environment" is external to the ECU but much of it is likely to be inside the organism. In the external environment, the CEV (Corresponding Environmental Variable) is the small circle at the head of the arrow marked "Disturbance".

The HPCT structure incorporates an indefinitely large number of ECUs, connected in a series of 'levels', distinguished by the type of perceptual function that produces the perceptual signal. Powers proposed the actual nature of each successive level based on his own introspection, and regarded them as provisional.

The key, however, is that only at the lowest levels do ECUs contact the environment outside the organism, through sensors such as retinal rods and cones, auditory hair cells, taste and smell receptors, touch receptors and so forth, and through effectors such as muscles, chemical emitters and material waste. All other ECUs interact only with ECUs at the neighbouring levels above and below ('below' meaning toward that interface with the environment). Most perceptual functions receive inputs that are the perceptual signals of ECUs at the level below and send their perceptual signals to perceptual functions at the level above. The outputs are distributed to the Reference Input Functions (RIFs) of units at the level below and each ECU's RIF receives values from units at the level above.

Figure I.5.6a shows an arrow labelled 'Side Effects'. These are effects on aspects of the environment other than the CEV. These side effects may influence other variables for which the corresponding perceptions are being controlled by the same person or another, but they have no effect on the controlled perception of the CEV. The side effects are unimportant to the performance of the given controller but are important for interactions treated in later discussions, and in Section I.5.5, above, we saw examples of side effect interactions among the ECUs that controlled perceptions of the parts of the chair, and argued that these side effects were likely to result in construction of more complex ECUs that would form a higher level in a growing hierarchy.

The environment through which the feedback path of a control unit passes between its output and the input to its perceptual function can be very complex, but much of the complexity can often be conceptually compressed into a single CEV (Corresponding Environmental Variable) that is defined by the perceptual input function. In Figure I.5.6a, the CEV is represented by a small circle at the head of the arrow where the Disturbance enters. The rest of the environmental complexity is in the atenfels of Figure I.2.4 by which the output influences the CEV and those by which the CEV influences the perceptual signal. If some change in the environment affects the value of the perceptual signal (the 'Controlled Variable' or CV), then a measure of that aspect of the environment is an argument of the function that defines the CEV.

Figure I.5.6b shows a hypothetical example of hierarchic control. As in the example at the beginning of Chapter I.1, a person is ringing a doorbell. What perceptions does she control? There's a reason she wants to hear the doorbell ringing, a perception, perhaps, of seeing the door opening. She had several available actions to create this perception, ranging from knocking on the door to having a strong friend bring a battering ram. But she chose to ring the bell, which implies that she has a reference to hear the bell ringing. So long as she can hear it ring, she does not need to change whatever actions she is performing, but if she does not, she must act differently to bring about the 'ringing' perception or otherwise see the door open.

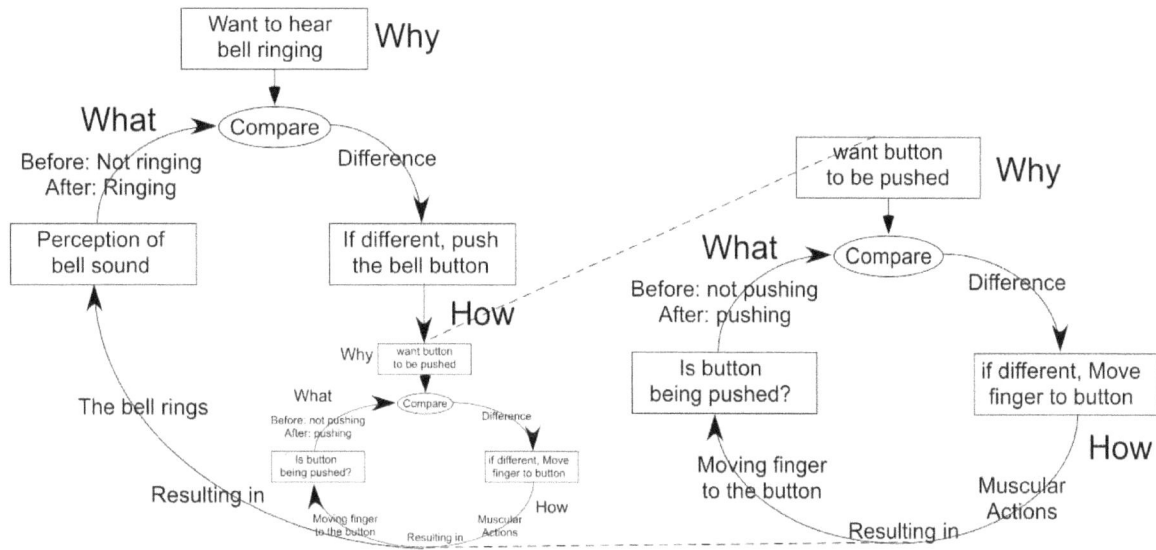

Figure I.5.6b A two-loop control structure. A person wants to perceive the sound of a doorbell ringing, and to bring about this perceptual value, wants to push the appropriate button, so acts to move a finger to push the button, which causes the doorbell to ring.

She knows only one way to generate the perception of hearing the doorbell ring, which is to find a button that looks like other buttons which she has learned will ring bells when pushed. Such buttons often provide atenfels for control of perceptions of doorbells ringing. Having found one, she has an action available, which is to perceive her finger to be pushing the button. She moves her finger to the button, pushes it (a sequence not shown in the figure) and hears the bell ringing. The figure does not show the higher-level control loop whose perception matches its reference value when the door opens, nor a control loop above that, when her perception of being welcomed into the house matches its reference value, nor ... we could continue with the small child's infinite recursion of 'why', but we refrain. You probably get the picture.

Figure I.5.6b shows one perception being controlled by sending a reference value to one other perception, but things are seldom so simple. In general, the output of the higher ECU is sent to the reference input functions of many ECUs at the level below, and many ECUs at the level below contribute their perceptual values to the inputs of any one ECU at the level above (shown by Figure I.5.6c and Figure I.5.6d). Various uncontrolled perceptual values also enter PIFs, such as the present time of day and possibly values obtained from memory and imagination. The connections between levels are many-to-many, but as with the doorbell example, there is often a dominant one-to-one connection supporting some specific controlled perception.

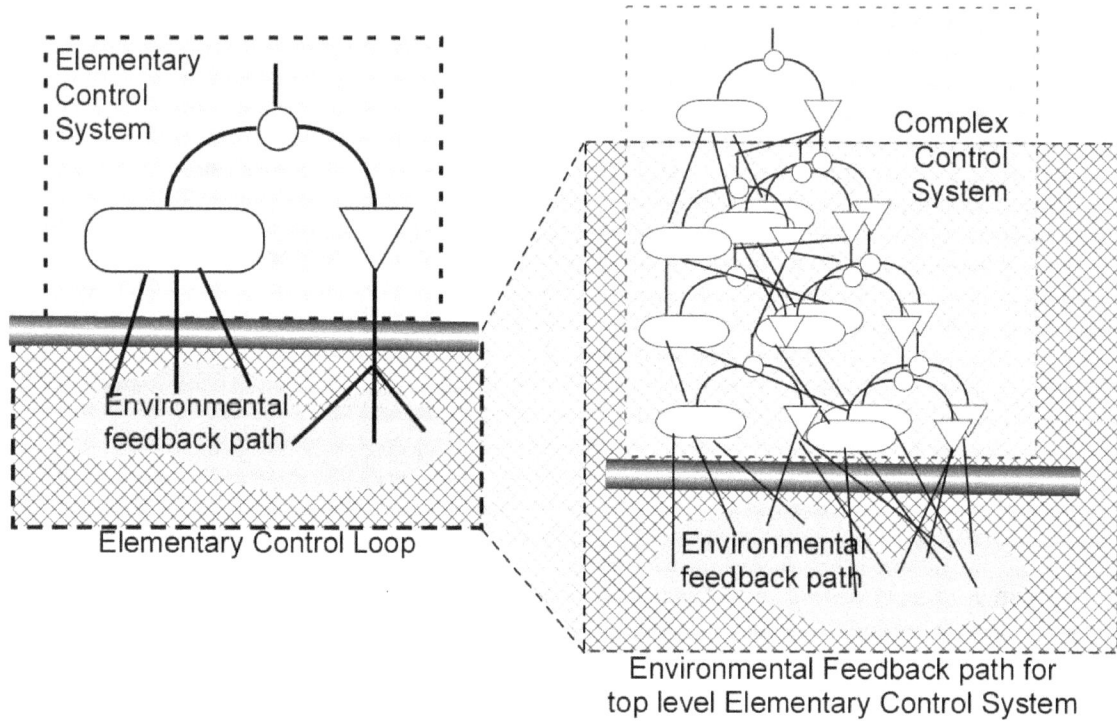

Figure I.5.6c The Environmental Feedback Path of a control system passes through the levels of control below it before the effects reach the environment outside the organism.

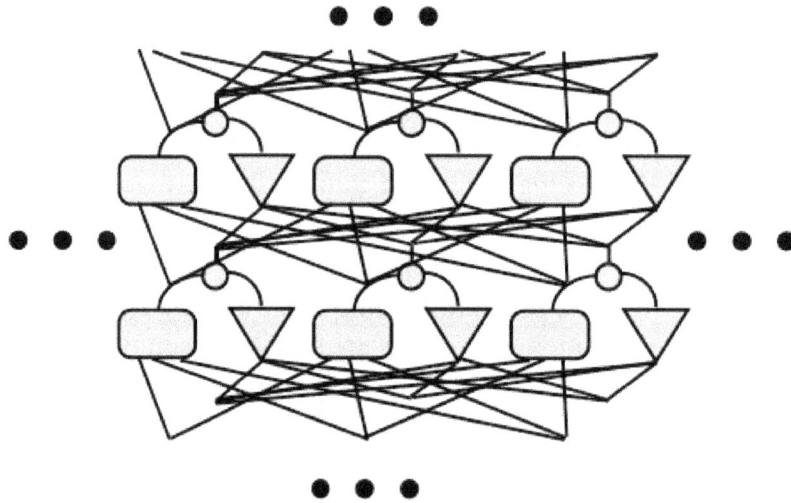

Figure I.5.6d A fully connected segment of two layers of a control hierarchy that might extend in all directions. The hierarchy is often shown as fully connected, with all lower-level loops connected back to the upper level loops that contribute to their reference signals, but reciprocal connection is not necessary, so long as an effective feedback loop exists through the environment.

The Powers HPCT model does not require such complete connections between neighbouring levels. As we shall see, modular sets of connections with some overlaps among modules are more likely in practice. Furthermore, in the Powers HPCT model, there are no lateral connections from the output of one ECU at a level to the reference input of another at the same level, and no internal feedback loops that such lateral connections would permit. In Chapter I.9, I will argue that such connections and feedback loops are likely to exist, and that their presence accounts in a straightforward way for several phenomena that are otherwise less easy to explain. For the present, however, we explore the strict HPCT model, ignoring the possibility of within-level lateral connections and feedback loops.

Just as the Perceptual Input Functions (PIFs) of the ECUs must almost always be more than simple weighted additions, so also should we expect the reference values at the different levels of the hierarchy to have different functional relationships with the patterns of higher-level outputs that influence them. An ECU must have a Reference Input Function (RIF), just as it has a Perceptual Input Function. The RIF, however, is different from the PIF, in that it is not part of the ECU's control loop through the environment. Instead, an RIF at level N is in the environmental feedback path of each level N+1 controller whose output contributes to the input of the level N's RIF.

I.5.7 Atenfels and Molenfels

In Section I.2.4, an 'atenfel' was described as a link in the environmental feedback path of a control loop. An atenfel is not simply an observable property of an environmental entity, it is a skill used with an environmental property in control of a specific perception. The skill to draw a line is part of an atenfel of which another part is the drawing instrument used to create the drawing. One cannot properly talk about something as an atenfel without specifying the perception controlled in the loop, at least implicitly.

An atenfel should not be confused with an 'affordance' (Gibson 1966). An affordance is a view by an outside Analyst, who imagines what something such as a pencil might be used for by the right kind of user, whereas an atenfel includes an affordance like that when it is incorporated in the control of some perception such as to perceive a shaped mark on a piece of paper. The atenfel includes the user's skill and the nature of the perception to be controlled. For example, a steep snow-filled gully might seem to be an affordance for a skier to get down a mountain, but it would be incorporated in an atenfel only for an expert skier who was controlling a perception of being lower on the mountainside, perhaps as well as a perception of self with a reference of being able to overcome a challenge.

Being able to ride a bicycle is not much use if you do not have a bicycle, as was demonstrated powerfully in the 1948 Vittorio de Sica movie *Ladri di biciclette* (*Bicycle Thieves*). Nor is having a bicycle much use if you do not know how to ride it. The atenfel includes both the skill and the environmental requirements for exercising the skill.

As shown by Figure I.5.6c, the environmental feedback path of a control loop consists of everything, every function or entity between the output function of an Elementary Control Unit (ECU) and its Perceptual Input Function. Some of that is internal to the skin of the individual, some of it outside, but if something is to be an atenfel for control of some perception, the individual must have the ability to use it, which means to manipulate it in some way that appropriately affects the CEV corresponding to the controlled perception. An atenfel therefore consists of an entire lower-level control loop, not just a property of an object that an outside observer might be able to perceive. The building of atenfels and molenfels is a major part of reorganisation of the perceptual control hierarchy.

Some atenfels may be changed by the side effects of the actions used in control. A cook may control for the taste of what is cooking by adding a little salt. At that point, the cook is not controlling for the quantity of salt available, but the act of adding salt to the food changes how much salt will be available for cooking the next meal. If there is no salt left, the 'adding salt' atenfel is of no use for control of any perception for which it might otherwise have been available. The salt is a resource that is affected by side-effects of the cook controlling a perception of taste, and remains as a part of an atenfel only as long as some salt is left available for use.

On the other hand, the cook uses a spoon as an atenfel when controlling for the taste of what is being cooked. The use of the spoon for that purpose does not change the spoon. This is analogous to a catalyst for a chemical reaction, which eases the reaction without being itself changed by the reaction. We can call atenfels that are essentially unchanged by being used 'catalytic atenfels', as opposed to 'resource atenfels' which are changed or depleted by use. The loss of a few molecules off the surface of a spoon does not change its value as a cooking atenfel, at least not during the lifetimes of several generations of cooks.

A single atom or molecule of catalyst in chemistry can be used in only one reaction at a time, so the amount by which the bulk reaction is speeded by the catalyst depends on how much catalyst there is. Likewise, if the cook is using the spoon to taste whether his food has enough salt, another cook cannot use it at the same time to taste whether his dessert has enough sugar. How many perceptions can be controlled at once depends on how many spoons are available as catalytic atenfels. In that sense the bulk atenfel 'spoons' is a 'renewable resource', like physical wind power, of which only a limited amount can be supplied at any one time, but for which the usable future amount available is not affected by how much is used now. We call a catalytic atenfel that can support a restricted number of independent feedback paths at a time 'limited'.

Here is a small taxonomy of atenfel types:

1. The CEV itself.

2. Path atenfels from output to CEV or from CEV to
   the Perceptual Input Function.

   a. Resource: Is changed or depleted by use, possibly useable only once.

      i. renewable: the resource supply is regenerated or resupplied over time.

      ii. non-renewable: the supply of the resource is permanently depleted
          by being used.

   b. Catalytic: Remains unchanged by being used, and can be reused
                   indefinitely.

      i. Limited: Only a restricted number of simultaneous uses, perhaps
                     only one.

      ii. Unlimited: Can be used simultaneously in the control of any
                      number of perceptions.

These different types of atenfel have different consequences in social interactions, as we shall see when we discuss psychological and social power. The renewable Path Resource type, for instance, is the type that features in The Tragedy of the Commons (Hardin 1968, discussed in Chapter III.8). The different kinds can be combined in what are called 'atenexes' with little if any restriction. For example, a hammer could be a Path-Catalytic-Limited atenfel for hammering nails, a Path-Resource atenfel for perceiving oneself to be warmed by burning it in a fire, and a CEV when someone is choosing the best tool for a job. Not all those atenfels could be used simultaneously, but until the hammer is used as a resource, all of them remain available for use. The hammer is an atenex, a provider of a variety of potential atenfels. Almost all objects are atenexes, even if they were designed to be tools for one specific purpose.

McClelland in LCS IV (McClelland 2020) lists a different set of properties of objects or artefacts that provide potential atenfels, coming at it from quite a different angle. His list is

- Durability: How long the atenfel may endure.

- Portability: Whether the object can easily be moved, with its potential atenfels.

- Accessibility: How many different control loops can the atenfels of an object serve simultaneously.

- Versatility: How many different kinds of atenfel the object is designed to provide.

- Malleability: How the object may be reshaped to provide different kinds of atenfels by design.

Most of McClelland's types refer most obviously to atenfels provided by concrete objects and to the objects themselves, but they can also refer to the more abstract artefacts we call language and culture and to any other stable structures created by interacting control systems. Chapter II.3 contains an extended quote from McClelland, repeated in the Introduction to Volume III, which shows how many different kinds of CEVs, both abstract and concrete (sometimes literally the building material, concrete), provide atenfels for control of perceptions in people unknown to those whose work creates and maintains those stabilities.

Whereas the proposed taxonomy above defines categories of atenfel, McClelland's list refers to a measurable property of any member of a category. When such a measure is useful we can use it in place of the 'Limited-Unlimited' categorical distinction. For a Path atenfel of the 'Limited Catalytic' category, for example, his Accessibility measure indicates how limited it is. His Portability, Versatility, and Malleability measures, however, apply to atenexes rather than to a specific atenfel.

We might add to McClelland's list a measure such as 'Design': the degree to which an object has been purposefully selected or created to provide an atenfel for control of a particular type of function. A sharp blade is a 'Design Atenfel' for cutting, and is less likely to be used for pounding a nail or for reflecting one's face while shaving. The blade is an atenex and could be used for those purposes, but the Design measure is the relative likelihood of it being used for a purpose other than that for which it was intended by its selector or creator.

In complex situations, where verbal descriptions may become hard to follow, a formal notation may be used (described in Volume IV: Appendix 3). Tom{P[Bridge]} indicates that Tom uses a bridge that provides an atenfel for controlling the perception P. If P is a perception of his location with a reference value of perceiving himself to be on the other side of the river, a more complete notation is Tom{Location@other side[Bridge]}, which should be read as "Tom controls a perception of his Location with a reference value of 'other side' using a Bridge."

Most objects in the world provide potential atenfels for controlling many different kinds of perception. Sometimes the control is possible only by using more than one object in the environmental feedback path. Blades that can be mounted into a razor are not usable for shaving until they are in a razor. Conversely, the razor cannot be used for shaving unless a blade is mounted in it. A book can provide a firm backing for writing a note with a pen (Nevin 2020, in LCS IV). The notation for this structure would be A{P[Pen:Book]} 'Person A controls perception P using a pen and a book together'. When a complex of atenfels works together to provide a possibility which no individual atenfel can offer by itself, we give the complex the name 'molenfel', as in Figure 1.4.1b. The name 'molenfel', for a complex of atenfels that provides a feedback path that none of its atenfels permit singly, comes from 'MOLecular ENvironmental FEedback Link' (Volume IV: Appendix 2).

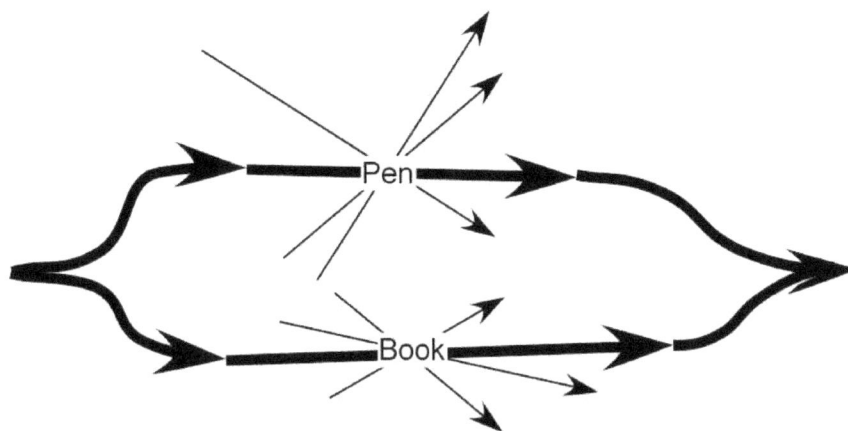

Figure I.5.7. A simple molenfel. The pen and the book each can serve in the control of many different kinds of perception, but by using the book as a backing for the paper the writer is able to use the pen to control a perception of writing a note on a sheet of paper while resting in a comfortable chair. Without the book, the pen would be useless for writing, and without the pen the book would not enable writing.

As the divergent arrows in Figure I.5.7 suggest, both the pen and the book could provide atenfels for other controlled perceptions. If that fact is relevant in any particular situation, the possibility of conflict arises, in that it is quite likely that if an item is being used as an atenfel for one perception it will not be available to serve the control of a quite different one. For example: "Could I borrow your pen to write a note, please?" "No, sorry, I'm using it to try to stab this spider that keeps running across the table". As briefly indicated before, when an object is being actually or potentially used in service of more than one perception, we give the name of 'atenex' or 'potential atenex', a term derived from 'ATomic Environmental NEXus'. Almost any object, perhaps every object, is an atenex for some perception someone is capable of wanting to control.

Not all atenexes involve conflict. Many cars can use a stretch of road at the same time. However, there is usually a limit on how many cars can use that bit of road without causing a traffic jam in which nobody uses it very satisfactorily. Many of the cases in which an artefact can serve in controlling several different perceptions are of this type. It can simultaneously serve more than one perceptual controller, but there is a limit to how many can be satisfactorily served at the same time. That limit might be hard — for example, traffic controls might prevent more than n cars from entering a highway stretch — or soft, as is normally the case with traffic, when at some point the density results in traffic slowing and more density brings it to an almost complete stop.

Sometimes, what we need in order to control a perception is the participation of another person, who provides the required atenfel. Money is often useful in getting another person to help us control a perception, so money is an abstract atenex with atenfels that appear in many different molenfels. Very seldom, if ever, does the physical molecular form of money provide a useful atenfel — in contrast to the mental representations of money in the form of coins or pieces of paper, which can provide atenfels unrelated to their representation of value for trade. Later, when we discuss the catalytic effect of the invention of money, we will suggest that money has much the same role in the development of culture as carbon does in the development of biological structures.

I.5.8 "You Can't Tell What Someone Is Doing ...."

A catch phrase of Perceptual Control Theory is "You can't tell what someone is doing by watching what they are doing." What does this apparently self-contradictory statement mean? Consider the doorbell ringing example of Figure I.5.6b. You can watch a person standing outside a house, pushing a small knob beside the front door. What is she doing?

She is certainly pushing a small knob. You may know from past experience that such small knobs beside external house doors usually cause a bell to ring inside the house, so you may guess that another thing she is doing is ringing the bell. But what else is she doing? Is she 'casing the joint' to see if anyone is home before burglarising the house? Is she canvassing for votes for an upcoming election? Is she visiting the residents for a cozy chat? Is she testing the doorbell circuitry to see if a failure has been fixed? Or what? Without other information, you have no idea.

If it matters to you, you might ask her, but if she was casing the joint, might she not answer that she was just visiting, but nobody seems to be home? How would you know? We will explore such situations in Volume II when we discuss the General Protocol Grammar, which implements a kind of Test for the Controlled Variable (Section I.2.5) at many levels in a dialogue.

A rather dramatic demonstration of that catch phrase, devised by Powers, is the 'rubber band' task.[60] Two rubber bands are knotted together and an experimenter E puts a finger in one loop and asks a subject S to put a finger through the other. E has put a mark on a table, and without letting the audience know, asks S to keep the knot over the mark as closely as possible. When E pulls on one loop, S must pull equally on the other to keep the knot over the mark, so if E traces a particular pattern over the table, S's finger must trace the mirror image. If S's trace is visible to the audience as it happens, it seems that S is trying to trace a particular shape. Very few naïve audience members ever suggest that S is trying to keep the knot in a fixed place.

How might a third person, an observer 'O', determine that the subject had been asked to keep the knot stationary over the mark, rather than being asked to draw a particular pattern? One way is to look for stillness where it should not be expected. If E pulls on the rubber band, or relaxes the tension, ordinarily the knot would move, but it doesn't, because the subject produces an equal and opposite pull.

---

60 See https://www.youtube.com/watch?v=zgXqsP0uEbY for a video demonstration of this by Warren Mansell.

If O notices that the knot tends to move first in the direction in which E moves but always returns to the region of the mark, E clearly is not trying to keep the knot stationary. Why then does it always stay close to the mark? It must be because S wants it there. But S appears to be trying to mirror what E is doing. Could that not be what S is doing? If S is good at mirroring, the knot would stay over the mark as a side-effect. How could O tell whether S was controlling to move the opposite way from E or to keep the knot stationary? In each case, the other is a side-effect.

O might attach another elastic band at the knot, making a three-leafed clover pattern, and pull on it. O's pull would move the knot away from the mark if S is not controlling for the knot to be over it, but would have no effect on S's ability to control the mirroring relation between the two patterns.

O might obscure S's view of E's finger movements. That would make it impossible for S to mirror E's pattern, but if S is controlling proximity of the knot to the mark, obscuring E's finger movements would have no effect.

When O has, to O's satisfaction, determined what S is controlling, O perceives at one level the failure of the corresponding environmental variable to change, but at another level the actions S uses to counter O's disturbance. These two levels will become important when we discuss dialogue in Volume II, but, for now, let's look at them a bit more closely. Suppose O is not really interested in just what perception S is controlling, but rather is interested in some effect, possibly a side-effect, of S's actions in performing the control. O may be controlling for S to perform some specific action. Indeed, in the basic form of the elastic band demonstration, E could easily control for S to draw a circle, a duck, or any connected shape at all, because E expects S to be controlling for the knot to stay over the mark. Imagine an entirely different scenario, in which Ingrid wants Charles to produce some food (*action*), so Ingrid tries to get Charles to perceive that she is hungry (*controlled perception*), on the assumption that Charles controls for Ingrid not to be hungry.

The so-called 'coin game' is another frequently used PCT demonstration. E gives S five coins and S lays them out in a pattern that conforms to a description S keeps private. E's task is to discover S's private description by moving one coin at a time and observing S's resulting action in moving a coin or not. S guarantees to E that the pattern after S's move conforms to S's hidden description. Figure I.5.8 illustrates one possible sequence of moves. The reference pattern for S in this example is not a specific arrangement of coins, but any arrangement that could be described as "Four coins forming a rough square, with the fifth outside the square."

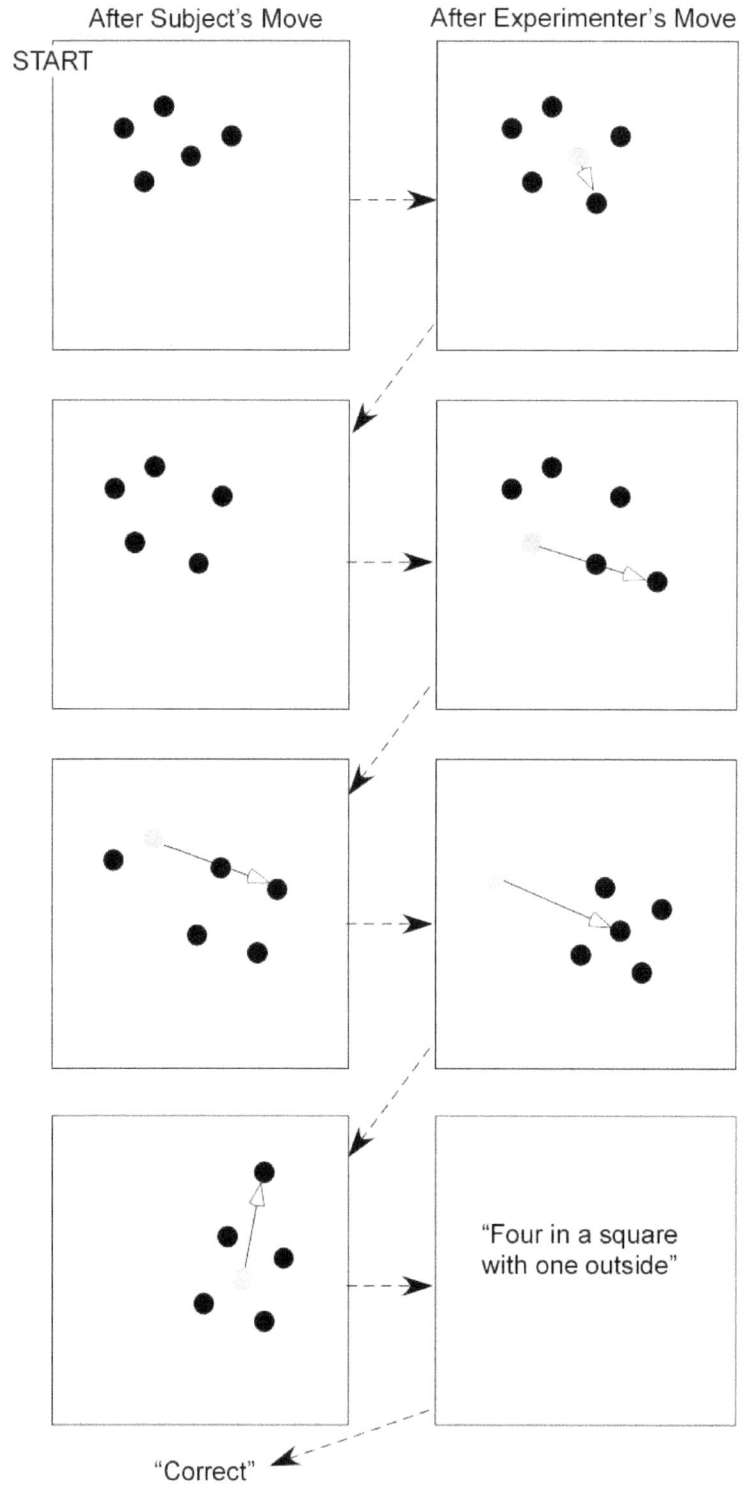

Figure I.5.8 A possible play of the coin game. In the Subject's second turn, the coin moved by the Experimenter creates a pattern that satisfies the Subject's reference value for it, so the Subject does nothing. Notice the variety of shapes in the left column, all of which satisfy the Subject's reference.

The coin game looks very much like a turn-taking dialogue without words, apart from the final guess and confirmation. But E is actually performing the Test for the Controlled Variable (TCV) by disturbing in various ways a perception S is controlling. In this TCV, E believes that S is controlling a perception of a pattern, because they have agreed that this will be so. The controlled perception is not in question, but its reference value is, and that is what this particular TCV is trying to find out.

In the elastic band demo, if E asked S to choose some perception to control by moving the finger that is stretching the band, that exercise could also be an analogy of a dialogue, but not a turn taking dialog.

Compare this coin game process with the larger 'game' of teaching — say, teaching a complex technique to an apprentice. The master has in mind a reference for the technique, and the apprentice wants to learn it. In trying to perform it, the apprentice tries things that do and things that do not conform to the master's reference perception for the pattern that is the technique. Just as in the example coin game where more than one placement of the coins satisfies S's reference for the pattern, so the successful apprentice may not perform the technique exactly as does the master, but the results satisfy the master's reference for what he would like to perceive.

As we will discuss in Volume II of this book, when we go more deeply into the interactions of two communicating partners, much of the work of communication is the discovery by either partner of the other's intentions, the controlled perceptions and their reference values. Language makes the task easier if the communicators are cooperative. For example, the Experimenter's job in the coin game would be much easier if the Subject were allowed to tell the Experimenter that the pattern was "Four coins in a square, with the fifth coin outside the square". But even with language, and especially if the communication is deceitful or non-cooperative, the coin game serves as a reasonable analogy to some of the testing that interactors do as part of the process of communicating.

I.5.9 Avoiding, and the Perception of 'Not'

We have been dealing with situations in which the act of control moves the CEV and thus the perception toward a single reference value or a specific reference category. Only at the reference value of the perception (within tolerance limits in the case of a quantitative perception) does the error value go to zero. When we want to not perceive some particular value of the perception, however, the situation is different. The CEV can be almost anywhere so long as it does *not* create the undesired perceptual value — the 'anti-reference' value, to coin a word.

Avoidance represents a situation in which the perceiver controls so that the 'anti-reference' value of a perception will not arise during some future time period. Imagine the following situations, all of which might plausibly complete a sentence that starts with "I want to avoid …":

- bumping into anyone in a crowd.
- falling into the old mineshaft in the field.
- falling over the balustrade on the seventh floor balcony.
- seeing the wine glass too near the edge of the table.
- hearing foreigners talking their disgusting language in the bus.
- offending that person with whose policies I disagree.
- having that wall red when we redecorate.
- being in the same room with Jack.
- having Rachel see me with Dora.
- making a foot-fault when I serve in tennis.
- having Rachel be within talking distance of Dora.
- making a burning smell when I cook.
- being served a food to which I am allergic
- seeing the present government re-elected.
- being near someone smoking.

All of these have one thing in common, that there is no specific preferred alternative to the environmental condition that is to be avoided. For example, your avoiding being near someone smoking is not the same as your being far from someone smoking, because that presupposes that you expect to perceive someone smoking, whereas not perceiving anyone smoking might be even better. Not falling into a mineshaft is not the same as falling into something else, or as doing something else with a mineshaft you perceive to exist. Not offending that person does not imply offending someone else, or ingratiating oneself with that person. And so on. For none of these is there an obvious reference value toward which one's actions might influence the corresponding perception, although the tennis example might be an exception, if the only alternative to a foot-fault is a fair serve.

These environmental conditions can be restated as perceptions to be avoided, by starting each with "I want to perceive myself not ..." and then using the same completions listed before. In the examples, with this modification, the perception currently is already at its reference value. Although you may at this moment be in a crowd, you are not in the process of bumping into someone. You may be in a seventh floor apartment, but you are not at this moment falling off the balcony. You are not currently playing tennis, so you are not making a foot fault. In all the cases, what your control is 'not doing' is bringing the controlled perception toward its reference. However, if you are in the situation to be avoided, it may already be too late; yet that is not always so — you may be able to escape.

Some of the examples, those for which the perceptions have a quantitative value, also could be completed "I want to perceive myself far from...." Categorical perceptions may not seem right when used to complete this last kind of sentence, though there are exceptions.

For at least some of these examples, to 'avoid' can be considered the same as to control the perception in imagination with a reference value of zero.[61] What, then, is the difference between "I want to avoid..." and "I want to perceive myself not..."? 'To avoid' seems to imply present action so that the unwanted perception will not occur, whereas 'to perceive myself not' seems to imply that if the perception were to occur, I would act to change it, to escape the situation. 'Avoid' implies an action now to control an imagined future perception; or, to say the same thing in another way, 'avoid' is an action output for control of a higher-level perception that includes the imagined unwanted perception as one of its components.

I can avoid falling into the old mineshaft that I know of by not walking in that field. If, however, I walk in the field not knowing the mineshaft is there, I cannot avoid the mineshaft, but I can control for not falling into it when I see it ahead of me. In the 'Rachel–Dora' examples, the reader can easily imagine a scenario into which one might control for those perceptions not to happen — a perceptual profile with a corresponding reference profile. If the 'Jack' example is added into the same profile, the reader's imagined situation probably becomes a little more precise, as Jack might be imagined to be Dora's 'Significant Other'. But if the 'Jack' example is combined instead with the smoking example or the 'offending' example, the reader is likely to imagine an entirely different complex of perceptions — a different scenario.

'It is not A' means something quite different from 'It is B', although 'It is B' may well imply 'It is not A'. 'Not A' could mean any other letter, or no letter at all. The English language use of 'not' is quite vague. "John did not give a book to Jane" could mean that John had no interaction with Jane, that he gave a book to someone else, that he gave something other than a book to Jane, or that he sold a book to Jane, among other possibilities. The one thing that is certain is

---

61   Control in imagination is taken up in Section I.7 and in Volume II.

that the triple relationship specified did not happen. No alternative event is specified, as is also the case for avoidance, but "John avoided giving the book to Jane" means something rather different from "John did not give the book to Jane". The former seems to suggest that at some point someone, even perhaps John, had been controlling for perceiving John to be giving the book to Jane, and John knew it. The latter has no such connotation.

To 'avoid', in addition to implying control of an imagined future state, also implies that the avoided state would have been more likely to have occurred if the avoidance action had not been taken.

Whereas a perception of 'X' defines a small region of the space of possible perceptions of the environment, a perception of 'not-X' includes the whole of that space except for the specified small region. But do we ever actually perceive 'not-X' alone, out of context? Probably not. Rather, we perceive an absence of some X that we might well have perceived in that context. Looking at a table set for a formal dinner, someone brought up as a hunter-gatherer would probably not perceive 'the wine glasses are missing', but a properly trained butler would immediately perceive 'not wine-glass', the absence of the expected wine glasses.

Earlier, I noted that avoidance perception can sometimes be equivalent to controlling a perception with a reference value of zero.[62] This is true if the perception is of a quantity of something that you do not wish to have, but it is not true of, say, avoiding having two quantities being equal, since the perception of the difference between the two things could satisfactorily be anything except zero. Control of avoidance requires some more general mechanism.

In his 'Crowd' demonstrations, Powers (2008, Chapter 10) finessed how to control for perceiving 'not close', by defining as a perceptual function a 'proximity detector' with two essential properties: firstly it could never go negative, and secondly it was monotonically related, inversely, to the distance, so that any positive value could be used as a reference. If the reference value for proximity was zero, the controller tried to get as far away as possible from the avoided location. This worked very well for that situation, avoiding bumping into anyone in a crowd. But is it reasonable to suggest that two different functions, one for approach and one for avoidance, are required for each perception that might sometimes be wanted and at other times avoided, such as, say a dip in the ocean, which may be wanted in summer but not in winter?

Furthermore, any functional inverse would be likely to fail in most situations of the types listed. What would be, for example, the inverse of falling into the old mineshaft or of being served the peanuts to which one is allergic? The location to be avoided is the whole empty surface area of the mineshaft, but anywhere outside a few metres from its edge is fine for a casual walk, even with one's dog. In another context, one might want to be right at the edge so as to see whether there is a ladder down the shaft, but definitely not over the edge so that one would fall in. In the case of avoiding peanuts, taking any other food or

---

62   Thanks to a personal communication from Warren Mansell 2015/07/15.

no food at all leaves the error value at zero, so what then counts as 'proximity'? A different kind of nut, perhaps, might be conceptually close, but if one is not allergic to it, the other kind of nut might just as well be an onion or a steak, both of which leave the 'not peanut' controller with zero error.

It seems more reasonable to think that the perception to be avoided is exactly the same as the perception that in another context would be approached. In Powers's Crowd, it would be conceptually quite reasonable to expect any one 'person' to try to avoid all the other generic 'persons', while wanting to be as close to a particular 'person' as possible. The person and distance perceptions seem to be the same whether the person is to be approached or avoided. One tries to stay close to a friend in a milling crowd, while bumping into as few others as one can manage. What seems to be required is a mechanism for reversing the sense of the error function, so that the comparator could produce either of the two functions shown in Figure I.5.9 depending on whether the reference value is to be approached or avoided.

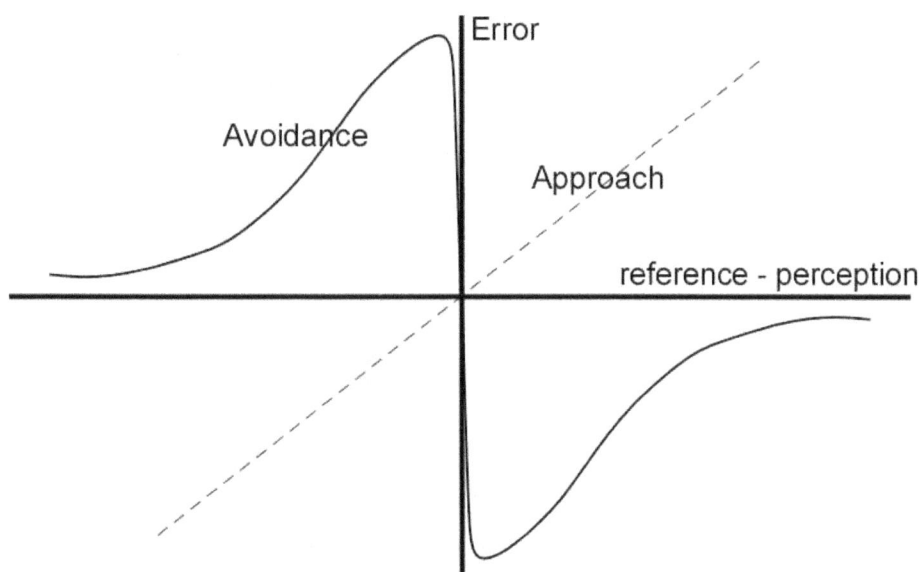

Figure I.5.9 Possible relationships between the reference-perception difference and the error value sent to the output function. The dashed line represents the "classical" linear comparator function for normal "Approach" control that brings the perception close to its reference value, while the solid curve represents a function that would take the perception further from the reference if the rest of the control loop were unchanged.

When we avoid perceiving the CV at the to-be-avoided reference value, the push-force away from the avoided point decreases the further from that point is the current perceptual value, in the same way that in the usual approach control system the force toward the reference value decreases the closer the perception is to the reference — allowing for any integrators in the output function or the environmental feedback path in both cases.

The avoidance curve produces positive, not negative, feedback, but if the difference between reference and perception is large enough, the loop gain eventually falls below unity, ending the 'explosive' escape of the perception from the to-be-avoided reference value. Since this control is happening in imagination when one is avoiding a possible future perceptual value, the positive feedback does not influence the current environment. Only when the results of the planning process are executed does this influence the current environment.

A general principle of Perceptual Control Theory, whether Powers's Hierarchic version or any other type, is that the 'Why' of control is imposed from outside the control loop itself. Figure I.1.2a shows the 'Why' as the reference input to the control loop, and this is the normal case. But now we have to set not only the value of the reference 'Why' but also whether that value is to be approached or avoided — two independent input values for the loop, one of which is a binary switch that changes the operation of the loop.

A mechanism to do this is a topic for a later place. For now, we must continue to recognise that people can and do sometimes control to approach and sometimes control to avoid a given perception with a given reference (or 'anti-reference') value. The importance of this will be elaborated when we introduce 'tensegrity structures of control', which may be important in stabilising the structures of complex systems within and among individuals.

Chapter I.6. Practical Control Issues

Perceptual Control Theory is not just an abstract theory, interesting because of its simplicity and beauty. It is a useful theory, with wide-ranging application to practical issues, as the subtitle of this book indicates.

I.6.1 Opportunity and Attention

When you are walking alongside a brook, thinking that it might be nicer to walk on the other side, you are controlling a perception of your potential walking path, but are doing so in imagination only. You are not actively inventing a bridge and getting materials to build it. That would be too much trouble, and you might not even have the skills to design the bridge or the strength to move and tools to work the materials.

Many different atenfels are required for control of many perceptions in the course of designing, building, and then using anything, let alone a load-carrying bridge over water. But if you see a suitable-looking plank, place it across the brook, and cross over, you are now controlling the same perception in reality.

What changed between your walking along the bank, wanting to be on the other side, and actually arranging the environment so that you could safely achieve that purpose without getting wet? Nothing much. Perhaps you had even imagined finding a plank that could serve as an atenfel for controlling your perception of your walking path. A plank wouldn't be too much trouble to use, if you did find one and it wasn't too heavy for you to move. And then you found such a plank, and having imagined using such a means of crossing over, you could implement in the real environment what you had imagined doing.

Another example. You are out for a walk on a hot day. You see an ice-cream truck and say to yourself "That might be nice", then go and buy a cone to enjoy while you walk. You hadn't been thinking of ice-cream and were not controlling any perceptions that included ice-cream. But an atenfel appeared in your real-world perceptual field that you could imagine enabling you to control for enjoying an ice-cream cone, and suddenly there you were, controlling a perception that moments ago you were not controlling even in imagination.

What happened? One doesn't start controlling a perception out of the blue, at least not in any of the versions of PCT we are exploring. A reference value has to be supplied, and in the HPCT hierarchy it can only be supplied from higher levels. If the day had been cold, would you be as likely to start controlling for enjoying an ice-cream immediately after you saw the truck? Probably not. You would not be actively controlling for tasting an ice-cream whether or not you saw the truck. Only if it was hot would you start actively controlling for it when you saw the opportunity.

What makes the difference? Your reference values from higher levels, of which there are many, only a few of which are used at any moment in active control. When the day is cold, you do not actify a reference value that would result in acting on the opportunity to have an ice-cream, so when you see the truck, you do not take advantage of it.

If you were wearing a business suit and walking with a potential client, would you start controlling for enjoying an ice-cream as soon as you saw the truck? Perhaps, but what would decide the question for you? Surely it would depend on some perception you might be controlling with respect to the client's perception of you. We will talk about controlling perceptions of other people's perceptions when we come to discuss 'protocols' in Volume II. The point for now is that 'ice-cream availability' does not make you control for perceiving the sensation of ice-cream in your mouth. Rather, it provides a possible atenfel for controlling some other perception that is perhaps less easily controlled by other existing means.

As noted above, the ice-cream truck is not itself an atenfel. Acquiring ice-cream is only one of the many things that can be done by means of that same truck, so it is an atenex. In your view, however, the view of the ice-cream truck has previously allowed you or someone you have observed to acquire ice-cream. We will talk about association in Section I.9.6, but for now let us note only that an object can be associated with an atenfel that it can provide. Any atenfel that has been perceived while it is being used in control of some other perception is likely to become associated with that perception, which implies that the ice-cream truck is likely to be associated with a perception of ice-cream in one's hand.

The perception 'ice-cream-in-hand' was not being controlled before the ice-cream truck was spotted, not even with a reference value of zero (which would mean controlling for not having an ice-cream). However, memory associates other perceptions in context of an ice-cream truck. One context was previously associated with perception of having an ice-cream and another with not having an ice-cream. Current perceptions of the environment are associated with one context more than with the other, including a set of higher-level perceptions which provide outputs to the reference input of the 'ice-cream-in-hand' perception, so that it is controlled with a high or low reference value depending on the context. The choice is likely to have commanded conscious attention, which we consider later.

One question I have sometimes used in conversation is "In what coin do we pay attention?" It is not a frivolous question, though I often use it as though it were. 'Attention' does not have a place in HPCT, though Powers and others have made a variety of suggestions about it. Most of these suggestions hinge on changing something about the control process.

For example, we may pay attention to some perception we are trying but failing to control well, or to a difficult decision as to which path to take in some situation. We may also pay attention to a perception we were not trying to control but which might indicate something we perhaps should start to control, such as a flicker of motion seen out of the corner of the eye. We might want to control some perception because of that flicker, perhaps ducking to avoid a flying rock, something we were not controlling for earlier. In addition, the counselling technique called 'Method of Levels' is partly based on the idea that where we pay attention is where reorganisation may occur in the control hierarchy.

The theme that all these ideas have in common is that control at that moment is not as good as it should be, whether as a transient issue or on a longer time scale. But who or what perceives, and moreover controls, the quality of control?

To say 'attention' is simply to give a name to an ill-understood phenomenon. Powers gave the name 'Dormitive Principle' to this kind of naming, illustrating it with a hypothetical assertion that we go to sleep because of a buildup of a 'dormitive principle' over time.[63] One does not avoid a problem by naming it. If there is a 'dormitive principle', the act of naming it demands a search to find out whether it exists. Likewise, to say that 'attention' accounts for the various occasions that result in our being conscious of some control problem demands a search for a mechanism, a statement of what coin we use to 'pay attention', supposing that we pay it only when control is not good using the means of controlling our various perceptions that we are actively using at the given moment.

"Less easily controlled by existing methods" means something in everyday language, but what does it mean in PCT, where we have only elements that change the relation of perceptions to their reference values? The word 'relation' suggests control of a relationship perception. We perceive A, separately we perceive B, and we perceive the difference between their current values. That difference perception is a perception of a relationship between their values.

Do we perceive how 'easily controlled' some perception might be? Consciously, we can often perceive in imagination the state in which we would like something to be — our reference value for it. Consciously, we often also perceive that something is not as it should be and that our actions are not making it much better. However, HPCT as described by Powers (Figure I.4.2) does not allow for direct perception of the value of the error output from a comparator.

Powers intuited that any conscious perception must correspond to some perceptual signal in the hierarchy, or perhaps a combination of signals which is not (yet) represented by a perceptual function that produces a perceptual signal. If his intuition was correct, then in some way error values as well as reference values must be available to perception. In the next chapter we shall see how a nearly equivalent alternative form of the hierarchy does permit this, as illustrated in Figure I.7.3b.

Although it does not necessarily follow directly, a suggestion consistent with the foregoing is that 'attention' signals the provision of temporary links that generate reference values for existing control units, to allow control of conscious perceptions. Conscious perceptions then would be the ones for which 'attention' provides this service when the currently active part of the hierarchy does not produce good control.

---

63   Powers (931003.0030 MDT) introduced the term on CSGnet, citing Bateson. The ultimate source is Molière's play "The Imaginary Invalid," in which a group of physicians explain that the virtus dormitiva of opium induces sleep. Bateson (1976:xx) translated virtus as 'principle' and popularised the term as a critique of specious claims in science.

Since according to Powers's HPCT proposal the rate of reorganisation varies inversely with the quality of control, this suggestion is consistent with the Method of Levels (e.g. Carey, 2006), a psychotherapeutic procedure that works by directing the patient's attention to a region of the hierarchy that includes conflicted control systems (which by the nature of conflict do not control well). Clinical experience suggests that such attention results in reorganisation forming and consolidating new linkages.[64]

If 'attention' is connected with possible change of the roster of perceptions being actively controlled,[65] the 'ice-cream truck' opportunity then makes sense. If one might start controlling an 'ice-cream taste' perception, one would need to have paid attention to the ice-cream truck, whether or not one actually started to control having an ice-cream in hand and on tongue. 'Opportunity' then can be translated into the perception of a potential atenfel for control of a perception that either is not being controlled or is being controlled less well than it would be by using that new atenfel.

I.6.2 Cost and the Perception of 'Worth'

Opportunity requires attention to be paid, so now we must consider the currency in which attention must be paid. That currency is not money; indeed we shall argue later that money is just one possible way of accounting for cost in suitable situations, and that as a measure of 'cost' money is no more fundamental than time or inconvenience.

In discussions of mental and especially physical function, cost is often accounted in terms of energy — physical energy that can be measured in ergs. We have thus far not mentioned physical energy much, if at all, in our treatment of perceptual control, but we will do so extensively later. Energy is intimately connected with the way we will address 'cost' and 'worth'. We start with systems in which the CEV is the location of a mass that can be moved by a force applied either directly or to an atenfel provided by an atenex such as a computer mouse.

Disturbances introduce energy into a negative feedback loop. The disturbance moves the CEV or would do so in the absence of the countervailing force applied by feedback. In an equilibrium system such as a spring, a ball-in-a-bowl, or a pendulum, the energy supplied by the disturbance is the only energy source involved in the feedback loop (Volume IV, Appendix 1). That energy is returned to the environment, possibly in the form of frictional heat, when the disturbing force is removed.

A control system, on the other hand, is supplied with energy by an independent external source, such as its food, an electricity supply, or something else that it

---

64   When we deal with the 'tensegrity' properties of control hierarchies, we will see that conflict, by itself, is not the issue. The issue is with conflicts that do not allow for the kind of cooperative control that is the fundamental basis for the existence of the hierarchy itself.

65   Or of the means of controlling them, which is the same thing in the PCT hierarchy.

can take advantage of. The external source enables the control output to apply a force that opposes the force of the disturbance. If control were perfect, the mass would not move (unless the reference value for its location changed) and with no movement there would be no loss of energy to the environment due to friction. The energy that is input is used to continually extract entropy from the CEV, at least in those degrees of freedom that are controlled (as we discuss in Volume II, Chapter II.1), and is exported to other parts of the environment.

Control is never perfect, however, which means that in the process of acting on the CEV, some of the energy (from the disturbance as well as from the power supply for the control unit) is dissipated into the environment. It is hard to say how much will be dissipated, because that depends on many factors, including the quality of control. Side-effects of control also dissipate energy, and that energy is lost to the environment without any direct benefit to the controller. We will treat side-effects at length later in this four-volume book. For now, we treat them simply as wasters of energy that coincide with control. Although side-effects do not have any influence on the quality of control, they do have consequences for the thermodynamic efficiency of control.

Why do I concentrate on energy? Surely there isn't much force involved in turning a switch or talking on the phone, is there? And isn't the main energy we use provided by power plants that supply motors with electricity and gasoline? All of that is true, and yet one of the big problems evolution has had to solve is how to dissipate all the heat produced by our big brains. Energy conversion from one form to another is always accompanied by the generation of heat, and the firing of a neuron is no exception. The more spikes there are, the hotter the brain would get if the excess heat were not carried away.

The maintenance of internal body temperature is very important for any mammal, and even for a reptile which controls it by moving between shade and sun. Every neural impulse generates some heat — not very much, to be sure — but we have billions of neurons and trillions of synapses. Every little bit of energy saving helps the heat problem, so we should not be surprised if evolution has resulted in ways to conserve energy. One of these ways is the improvement of control to reduce the number of neuron firings required for effective control and to reduce unwanted side-effects. For a given method of control using a constant set of atenfels, the tighter the control, the greater the energy cost. To control better without increasing the energy cost, something has to change, whether it be using different atenfels, increasing skills so that the actions increase their fit to the perception being controlled, changing the suite of perceptions being controlled in support of the one in question, or something else. In a word, '*reorganisation*'.

Earlier we argued that in reorganisation to improve the stability of intrinsic variables, the quality of control should be treated as though it were an intrinsic variable. The reason was that the stability of low-level control is needed in order for actions to produce consistent effects upon higher-level perceptual variables. This was something of an ad-hoc argument, but now we have a reason to say not that good control should be treated as though it were an intrinsic variable, but rather that it *is* an intrinsic variable for the purpose of directing reorganisation.

These last two paragraphs can be re-interpreted to say that any reduction in the efficiency of control incurs a cost, an energy cost rather than a monetary one, but a very real cost nevertheless. I will argue that more familiar 'costs' can be attributed at root to the energy cost of control.

For example, if you have a lot of money, then buying a fancy watch little changes your ability to control other perceptions using money as an atenfel, but if you have very little money, to buy that watch might make control of many other perceptions more difficult, thereby incurring the energy cost we have been speaking of. To a poor person, buying the watch costs a lot in everyday language and in the language of control efficiency, whereas to a rich person it does not. Even though they both may spend the same number of dollars on the watch, their costs are very different.

Here we return full circle to 'opportunity' and 'opportunity cost'. To perceive an atenfel that would improve control of something which is currently not well controlled is to perceive a way to reduce the energy cost of control. If the atenfel would assist control of a perception which is not currently being controlled, and which is unlikely to be controlled even in imagination, it is not a real opportunity. In economics, 'opportunity cost' refers to the benefits of choice B which are lost when choice A is selected. We translate that as the energy cost incurred by using a means of control which is less effective or efficient than would be possible using the best available atenfels.

Now we can answer the question posed earlier: "In what coin do we pay attention?" The coin is energy usage in the brain. One pays attention to perceptions that might need to be controlled but are poorly controlled or are not controlled. Some non-PCT psychologists may call 'attention' an executive function, and even in PCT that might not be an unreasonable term to apply. But it is a function, and as such it implies increasing numbers of neural spikes — neural spikes, for example, which are used in different kinds of brain scans to study what parts of the brain are involved in different functions.

Taylor, Lindsay and Forbes (1967) found that attending to two or four simultaneous short visual or auditory events, compared to attending to just one, reduced the conscious perceptual total capability in informational terms by about 20%, regardless of which or how many perceptions needed attention in order to perform the task. We might expect attention more generally to have some such level of cost. It can be a high price to pay, but effective reorganisation can reduce the cost greatly by building an effective hierarchy of perceptual control subsystems to which no attention need be paid, controlling perceptions that are not normally consciously perceived.

Tolerance can also reduce this cost. If every perception that departed even marginally from its reference value required correction, only a few perceptions could be controlled at any one moment and all would require a full repertoire of neural firings. By allowing a tolerance zone around most, and perhaps all, controlled perceptions, it is much less likely that at any one moment many of them would be outside their tolerance zone. The 20% added cost of attention required for switching among perceptions to be controlled might be avoided and far offset by the reduced cost of controlling each of them.

We now see that 'value' is a personal perception. The value to you of a 'thingamajig' you do not now have is the quantitative improvement of control (the 'worth') you might enjoy if you had it. If, like an illness, the thingamajig reduces your ability to control well, then its value to you is negative. An omniscient analyst might be able to put absolute numbers on this 'value' to you, but your perception of it is more likely to be a perception of the relative value of a thingamajig as opposed to something else, a mumblybob.

Is that object more valuable to you than $20? To have the object might ease control of some perception (increase your worth), whereas to lose $20 reduces your worth (your overall ability to control other things that you might be interested in at some future time). Do you want to swap that mantel ornament for a picture your friend owns? The same applies. Which one has more value for you, in terms of changing how much control of perceptions relating to, say, artistic appreciation? We will return to the question of 'value' as differential quality of control several times in this work, notably when we talk of trade and barter, and of an economy that includes money. In Chapter I.8 we deal with 'motifs' of control. Trade is one such motif (Chapter III.9), built on the idea of comparative value we have just described.

The essential point in this section, however, is that 'value', as a differential of 'worth' has a close connection with 'reorganisation', and that this suggests that Powers was correct in suggesting that reorganisation proceeds more quickly when (and where in the hierarchy) control is poor, than it does when and where control is good. Increasing worth is equivalent to improving control and reducing rates of reorganisation. Decreasing worth has the opposite effect. Hence, an effective trade — one that increases the total worth of both parties — decreases the rate of reorganisation in both. Emotionally, the qualitative result is often a feeling of satisfaction.

On the other hand, a feeling of dissatisfaction accompanies a trade that is not effective, in which one party perceives a reduction in total worth, having been cheated or coerced into making a trade that now seems to have been a bad one. Deception, too, we will discuss further toward the end of Volume II, and in many places when we deal with power and politics in Volume IV.

It is probably a reasonable first approximation to suggest that most people control for ever-increasing worth, which entails an increasing ability to control. A child's maturation increases both the number and complexity of the perceptions it can control, both of which enhance the total 'worth' of the growing child. For any one or group of those perceptions that it can control, if the child practices controlling them, its skill increases, augmenting its worth in that dimension as well. Worth, then, is an increasing function of the number of different perceptions that can be controlled and the speed and precision with which they are controlled, together with the greatest disturbance that can be fully countered. Conceptually, though not numerically, we could write 'worth' as

$$worth = \sum\nolimits_{perceptions} (speed \times precision \times range)............(6.1)$$

This is effectively the same equation as is implied by the mathematization of Ockham's razor (Appendix 10 in Volume IV) to measure what Einstein (quoted in the Preface) called the 'impressiveness' of a theory.

I.6.3 'Worth' and the Perception of Self

Do we perceive our own worth? Do others? We seem to and so do others, though probably with little accuracy. We recognise as special those people who do a lot of things fast and well, as well as those who are clumsy and ill-coordinated or who think slowly, and those 'idiots savants' who do one thing extremely well but many other things rather less well than average.

How these people perceive their own worth is another matter, but one thing we can assume for most people — setting aside those who take vows of poverty, perhaps — is that their reference value for their perception of their own worth is higher than their actual perception value for it. In plain language, most people want to better themselves in one way or another.

We are talking here about the difference between control of 'self-self-image', the perception of one's own self, and control of 'other-self-image', one's perception of how others perceive one. These do not necessarily have similar actual or reference values. For example, if Quentin is afraid of spiders, that fear is part of his self-self-image, but if Agnes shrieks "Quentin, there's a big spider. Please get rid of it." and Quentin is controlling for others not to perceive that he fears spiders, he has a conflict between showing his fear to Agnes and his control for being far from any spider. The error in his Agnes-related other-self-image perception will be increased if he shies away from the spider, but the error in his control for avoiding spiders is increased if he acts to get rid of it for Agnes.

If Quentin does remove the spider for Agnes, he is likely to perceive himself as having more 'worth' than before, because he now can control various spider perceptions that he perceived himself as unable to do before the event. If he does not get rid of the spider, he will probably not change the value of his self-self-image perception, but, presuming he controls for Agnes to perceive him as worthy, that perception will be more in error. Similar contrasts can apply in many situations which, like the spider removal, can be traced to a conflict between self-self-image and other-self-image.

Robert Burns said in his "To a Louse": "*O wad some Pow'r the giftie gie us/ To see oursels as ithers see us.*" He points out that we cannot directly perceive how others see us, so our other-self-image is in no way guaranteed to reflect the environmental reality of how others actually see us. But then neither is any perception guaranteed to correspond to reality, as visual illusions, mirages, hallucinations, dreams, and the like vividly attest. For other perceptions, the fit to real reality is probably improved by experience based on the success of control — reorganisation of the perceptual functions included — and so it must be with other-self-image perception.

The measure of worth in equation 6.1 is applicable to any living thing, not just to humans. But is it complete? The simple number of perceptions controllable is not enough, even when weighted by the 'speed-precision-range' triad. To control some perceptions may be more useful than to control others. There is little point in carefully controlling, say, the exact position of a place-mat on a dining table while not controlling to relieve a perception of hunger. You would die of starvation.

What matters is whether the perceptions controlled affect the intrinsic variables important to the well-being of the organism and, in evolutionary terms, the maintenance of that well being for long enough to allow the organism to produce descendants. We could call the influence of varying the perceptual value (and hence the corresponding environmental value, the CEV) on the intrinsic variables the 'import' of the perception, and augment our conceptual equation 6.1 as follows.

$$worth = \sum\nolimits_{perceptions} (speed \times precision \times range \times import) \dots\dots(6.2)$$

In later volumes of this book, we will see that the import variable depends to a large extent on social factors, but for now we can ignore that complication. When we consider equation 6.2 rather than 6.1, we see that the ability to control the position of a place-mat with speed and accuracy contributes less to one's worth than, say, the ability to ride a bicycle fast and precisely, or the ability to acquire money that allows one to buy machinery and hire people that help control of a wide variety of perceptions that one cannot control with one's own senses and muscles unaided. In other words, one's 'worth' has three possible values as seen from three different viewpoints.

I.6.4 Frustration and Reorganisation

We come to frustration. Considered as a term in physics, 'frustration' loosely implies the inability to achieve some optimum over a set of variables. For example, if X, Y, and Z are arranged in a triangle and there is a rule that neighbours must be of opposite sign, any two of X, Y, and Z could obey the rule if the other were removed. In colloquial terms, 'frustration' loosely implies the inability to do what one wants to do even when it seems that to do it should be possible. Here is the opening of the Wikipedia page on 'frustration':

> *In psychology, frustration is a common emotional response to opposition, related to anger, annoyance and disappointment. Frustration arises from the perceived resistance to the fulfillment of an individual's will or goal and is likely to increase when a will or goal is denied or blocked. There are two types of frustration: internal and external. Internal frustration may arise from challenges in fulfilling personal goals, desires, instinctual drives and needs, or dealing with perceived deficiencies, such as a lack of confidence or fear of social situations. [...] External causes of frustration involve conditions outside an individual's control, such as a physical roadblock, a difficult task, or the perception of wasting time.[66]*

Within the PCT framework, "outside an individual's control" applies equally to internal and external frustration, and we need not distinguish them. They both wind up with an inability to control a perception that 'ought' to be controllable, creating a perception of one's worth as being lower than it might be. The decline in worth happens because 'ought' implies either that the perception had been controllable, at least to some extent, or that in imagination an available means of control has been found to be effective.

We can suggest this in a diagram, Figure I.6.4.

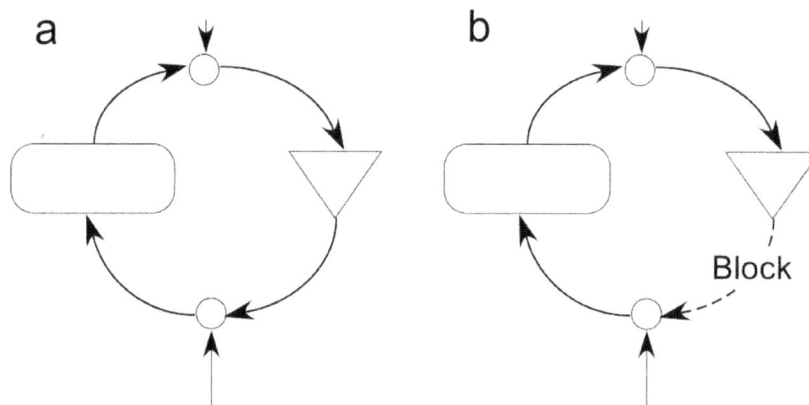

Figure I.6.4 The canonical condition for frustration. In (a) a perception was controllable, at least to some extent and possibly only in imagination, but in (b) the previously available environmental feedback path is blocked, leaving no alternate means of controlling the perception.

When control by a previously available means is blocked — frustrated — the error in the perception that should be controlled can become indefinitely large. In this situation, the rate of reorganisation increases, and while no effective means of control is found, the e-coli method of reorganisation (trial-and-error) would keep changing parameters in the hierarchy erratically. The part of the control hierarchy to which this control unit sends reference values would start behaving wildly, as many people do, especially immature or autistic ones, in a fit of anger sometimes called a temper tantrum.

---

66   Retrieved 2019/01/23.

What is the *e-coli* method of reorganisation? It is a type of hill-climbing optimization process, based on a study of how the bacterium *e-coli* navigates in a fluid medium, which optimises some criterion value by changing the values that describe the parameters of the control loops and of the connections among them within some part of the perceptual control hierarchy.[67] The parameters are represented by individual dimensions — axes — of a space of description, and the values of the parameters by locations on these axes. Taken together, these values locate the current structure as a point in the 'parameter space'. The criterion value, often taken as overall Quality of Control (QoC) of this part of the hierarchy in the current external environment, is a property that can be evaluated for any set of parameter values, represented by any point in the parameter space.

The *e-coli* method consists of arbitrarily choosing a direction within the parameter space, and moving the structure to a new point some distance along the chosen direction. If the criterion result is better than before, the structure is moved further in the same direction. This continues until a point is found to be worse than its predecessor, when the simulated e-coli (the structure) 'tumbles' and choses another arbitrary direction.

Always if a checked structure description point has a worse criterion value than its predecessor, the direction tumbles to a new arbitrary direction. Otherwise it continues in the currently selected direction. There are various implementations, and lots of details differentiate one particular e-coli procedure from another, but those details should not be allowed to obscure the basic process.

At some point, while frustration persists and no way of controlling the variable is found, most frustrated people are likely to control for other socially related perceptions that would conflict with the ones 'de-controlled'. The anger, if that emotion persists — and it may well persist — no longer results in the externally visible tantrum. Furthermore, as more complex perceptions and their means of control are built into the control hierarchy, among them may well be included mechanisms for finding ways to bring frustrated perceptions back under control.

I.6.5 The 'Bomb in the Hierarchy'

The '*Bomb in the Hierarchy*' is a phenomenon related to frustration, but with a different and possibly more dangerous source and consequent effect on the hierarchy itself. The Bomb is an important aspect of any self-organised complex control hierarchy, which we now discuss.

Control hierarchies are usually discussed as though the sign of each link were adjusted so as to ensure that the feedback from output through the world to perception was always negative. This criterion can be met in a fully designed system working through a predictable world, but not in a system that develops through its varied interactions with a complex and changing world.

---

67   Marken & Powers (1989b).

Like frustration, the Bomb begins with control that was or should have been possible becoming impossible, but unlike frustration, the feedback loop is not broken. Instead the negative feedback loop is turned into a positive feedback loop by some environmental event. The effect is not a random variation in the perceptual error, but a directed and possibly explosively exponential increase in the error, which is more damaging, hence the term 'Bomb'.)

In PCT diagrams, signal paths are usually shown as simple arcs, but in practice many of them consist of multiple paths with different dynamic characteristics. Inside the brain, the 'neural current' represented in the diagrams as a single line is a simplifying concept. A better representation might be a braided wire representing the combination of myriad firings on different nerve fibres. We will look at the effects of this simplification many times through this book.

In the external environment, there may also be a variety of direct and indirect paths through which an action influences the controlled perception. Although reorganisation has ensured that under normal circumstances these different paths combine to create negative feedback, reorganisation cannot ensure that all of the sub-paths leading through the environment from the output function individually influence the controlled perception in the direction that would oppose a change introduced by a disturbing influence. It is quite possible for some of these actions, taken individually, to have undesirable positive feedback effects on the error. But any such positive feedback sub-loops are ordinarily overwhelmed by the negative feedback sub-loops in any Elementary Control System (ECS) that maintains good control. All the positive sub-loops do is to reduce the Quality of Control for the loop. The only thing that can be assured is that for a control loop functioning well under normal conditions, the total loop gain has come to be negative.

Conditions in the world may change, blocking the effect of some of the desirable negative sub-loops, as in the 'frustration' situation. The previously hidden positive feedback sub-loop then are unmasked, as in the top row of Figure I.6.5a, and the overall feedback gain may then become positive. The loop begins to produce actions that increase, rather than decrease its error. It 'loses its temper' due to the non-correctable failure to control a perception that had been perfectly well controlled.[68] The path may be blocked because something fails that normally works, or because another independent control system is acting on an atenfel normally part of a negative feedback sub-loop, or for any of a number of reasons.

We here consider the case in which there is at least one sub-path which by itself would create a positive feedback loop. Normally the influence of this positive feedback path is hidden, because the other negative feedback paths together overwhelm it. If something in the environment blocks the action of a negative feedback sub-path, the previously hidden positive feedback sub-path may dominate, turning the overall loop into a positive feedback state, and destroying control, as suggested in Figure I.6.5a.

---

68   In such a case, the e-coli method of reorganisation creates many tumbles, each unsuccessfully trying to reduce the error in this perception and failing while having real effects in the environment.

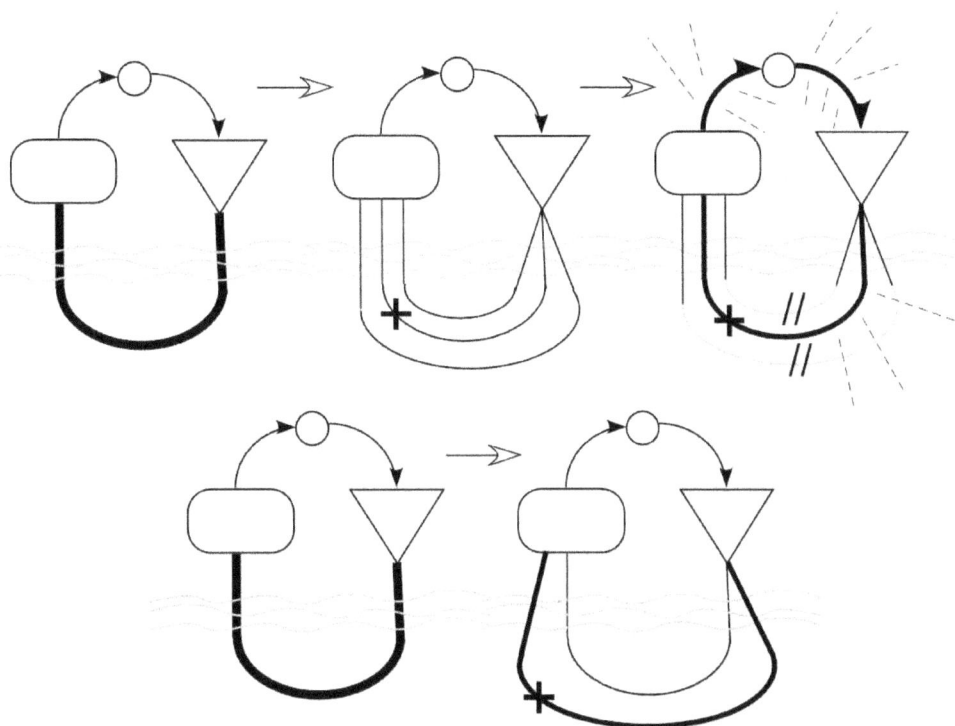

Figure I.6.5a Positive feedback in an environmental feedback path with several channels. (Top) Many environmental feedback paths consist of several parallel influences on the controlled perception. Not all of these necessarily have the same direction of influence on the perception, and if those that contribute to negative feedback are blocked more than are those that provide positive feedback (signified by a "+" sign on the path arc), the loop may "explode" exponentially. (Bottom) A new positive feedback path might be added through the environment. This could happen, for example, if another control system tried to control "the same" perception to a different reference level. This is a "classic" conflict situation, the result of which is a positive feedback loop that passes through both conflicted control units, often resulting in increasing output from both until some limit is reached.

Control works on the principle of "If at first you don't succeed, try harder." At its core, that is what is done by the integrating component usually considered to be part of the output function in lower-level control systems. At higher levels, it is a principle taught to children (in Scotland, using the parable of Robert the Bruce and the spider whose web he repeatedly broke). And it is a principle used almost exclusively by politicians whose ideologically driven policy failures lead them to continue doing the same thing, only more so.

Reorganisation, on the other hand, works on two other principles: "If it ain't broke, don't fix it", and "If at last you don't succeed, try something else." "At last" here often means "after trying harder with the same actions." Those same actions will have produced powerful but ineffective output that necessarily has enhanced any side-effects that would disturb other controlled perceptions.

To "try something else" means to vary the action consequent on the error signal, and to do that means changing the interconnections that convert output values into reference values for lower level controlled perceptions. If these changes are random, it is highly probable that they will interfere with the control of other perceptions, leading to 'Try something else' for those perceptions as well. Propagation of 'trying something else' might develop into an explosive chain reaction of disrupted perceptual control, depending on how well those other control loops have earlier been reorganised to control against severe disturbances in varying contexts.

Conditions in the world may change, blocking the effect of some desirable negative sub-loops, as occurs in the 'frustration' situation. The overall feedback gain may then become positive, the previously hidden positive feedback sub-loop having been unmasked, as in the top row of Figure I.6.5a. The loop produces actions that increase, rather than decrease its error. It 'loses its temper' due to the blockage of a normally available path to its goal. The path may be blocked because something fails that normally works, or because another independent control system is acting on an atenfel normally part of a negative feedback sub-loop, or for any of a number of reasons.

Positive feedback in one control loop could conceivably propagate up to higher-level ones that it supports, creating an avalanche of error in the hierarchy. In such a case, some event in the world causes a hidden positive feedback sub-loop of some control loop to become manifest, and as a result the overall feedback gain of some higher-level loop that incorporates the failing loop also becomes positive. If the other paths that serve it are not strongly enough negative, any of these higher ECSs may succumb, and go into a 'bombed' positive feedback state. The Bomb can in this way propagate upward through the hierarchy like an inverted avalanche, causing maladaptive behaviour at any level of abstraction.

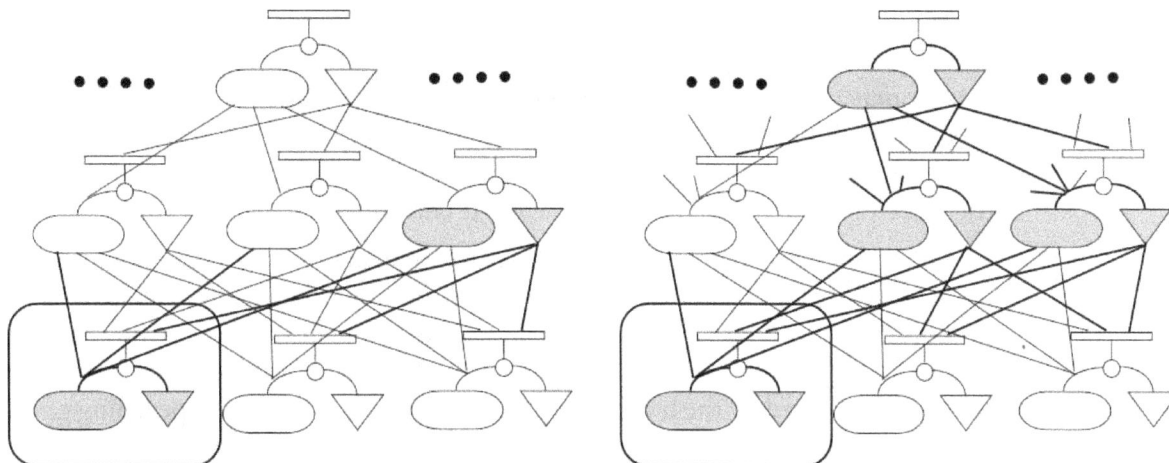

Figure I.6.5b A Bomb avalanche can be small or large. In both panels, the control system at the lower left is experiencing positive feedback. In the left panel, this control system destabilises one control system at the level above but the problem goes no further, whereas in the right panel, two control systems at the level above are sent into positive feedback by the positive feedback of the first, and they in turn destabilise at least one at a yet higher level, perhaps together with units above and beside those shown in the diagram. Such avalanches will induce strong reorganization, and the structure will subsequently be less susceptible to an avalanche propagating through the same control units.

A milder form of the Bomb can exist, in which the damaged sub-loop path does not contribute significantly to the controlled perceptual signal. The output of the loop, which overall still moves the percept closer to its reference value, causes through this sub-loop additional irrelevant side-effect actions — wasted effort or superstitious behaviour. These actions will be eliminated only if the wasted resources affect the ability of the hierarchy to control other perceptions, and may be retained for the life of the organism.

When a control unit or a part of the control hierarchy does not control well or at all, reorganisation will happen more quickly than it otherwise might. If the organism has not died or become severely damaged in the tantrum, reorganisation will tend to eliminate or at least hide the positive feedback loops that actively led to the loss of control. Subsequently, that particular exploded bomb no longer endangers the structure, which has become more resilient and controls well under a wider range of circumstances than before. Other unexploded bombs may lurk in the hierarchy-environment interaction, but as a whole, the control structure has become more mature and less prone to erupt in a temper tantrum.

The propagation of the explosion through the hierarchy has many of the characteristics of a sandpile avalanche. In the simplest version of a sandpile avalanche, sand grains fall one by one from a stationary aperture onto a flat table. As a sand grain falls, one of two things may happen. Either it stays where it falls, on top of the pile, or it bounces off and lands somewhere else. If it stays on top of the pile, the pile gets taller and the likelihood that the next grain will stay on top is reduced. If the new grain bounces off, it lands somewhere else where it may stop, it may continue with another bounce down the slope, or it may dislodge a precariously lodged grain and the two of them may continue downslope. At the end of each bounce, the grain may stick, continue with another bounce down the slope or dislodge further grains while it continues.

The steeper the slope, the more likely it is that the impact at the end of each bounce will dislodge another grain and that the downslope flow will grow. The result is that the sandpile will experience a series of smaller and larger flows called avalanches which maintain its average slope as the pile gets larger and larger. The slope is determined by the balance between the energy of a downslope falling grain and the energy required to dislodge a previously placed grain that it may hit.

One way to reduce the distribution of avalanche sizes in the physical sandpile is to keep shaking the table, adding energy to every sand grain and thereby reducing the slope of the sand pile. Small avalanches become more likely, but large ones become rarer.

The equivalent of the energy required to displace a sand grain involved is the strength of the individual positive or negative feedback path, and the energy of the bouncing grain corresponds to the side-effect feedback loops between the new control system (or old ones affected by an environmental change) and existing control loops. Such side-effect loops, in which each unit's actions influence the other's perceptual value, are much more likely to be positive than negative feedback loops.

Strong Bombs probably cannot last unexploded very long in a hierarchy that is exposed to a moderately disturbed world, just as a large avalanche is unlikely in a sandpile subjected to continual shaking, but they can persist in a 'coddled' hierarchy, one that is seldom stressed by exposure to unfamiliar or difficult circumstances, like a child who is overly protected from experiencing difficulties and dangers to be overcome.

In such a 'coddled' hierarchy, a Bomb is likely to be particularly dangerous and to cause a large avalanche when it explodes. This is the situation faced by a teenager brought up in a very ordered and especially in a very pampered environment, and then exposed to the wider world in which his wishes are no longer catered to and where the rigid mechanisms of his youth no longer function. People 'out there' do things differently, and often that difference leads to conflict (a situation of positive feedback through the actions of another control system). The introduction of positive feedback into a functioning control loop is precisely the situation described above as being likely to trigger The Bomb.[69]

---

69   For more detail on avalanches like those of the Bomb, and in general on the development of self-organised critical structures — which the control hierarchy is likely to become — see the Wikipedia entry "Self-organised criticality" and the links and references therein, many of which refer to neural avalanches.

A hierarchy reorganised in a too stable environment is always vulnerable to 'The Bomb' if accustomed negative feedback paths are blocked or absent due to a change in environmental conditions, revealing a positive feedback path which had been hidden. The result might be as mild as a slight reduction in the quality of control at some low level, or it could be a catastrophic and possibly fatal failure of control up through many levels. Control systems which are 'idealistic', in the sense of having reorganised in a complaisant environment to have only one rule of behaviour for the control of their perceptions, will be most vulnerable to 'The Bomb', since the increase in error caused by the revealed positive feedback path could not be reduced by 'going the other route'. If you know the byways, a roadblock on the highway may slow you, but will not keep you from your destination. If you know only the highways, you might be stuck for hours.

As we noted above, the outward manifestation of the Bomb, at least in idealistic children whose control systems have not been sufficiently reorganised to control well against a variety of disturbances, is sometimes a temper tantrum. In idealistic older people, the lower level structures are likely to have been reorganised so that 'Bomb Avalanches' seldom occur. In people in their teens and twenties, the Bomb phenomenon may be initiated at a moderately high level in the hierarchy, as evidenced by destructive purposes, apparently aimed at 'the system' rather than at parents, being employed in a more coordinated way to control higher-level perceptions. Extreme versions of such behaviour are often called 'rebellion' or 'terrorism' by those whose stability is affected by them. All these cases may be a consequence of inability to control important moderately high-level perceptions, either from lack of atenfels (means and skills for an unpredictable world) or because of conflict at supporting levels.

This shift of target between childish tantrums and mature rebellion corresponds to a change in the part of the person's environment that had provided most of the means by which the person had previously been able to control perceptions — the constellation of available atenfels in family interactions. Since earlier reorganisation had resulted in the effective use of atenfels in the family part of the environment, reorganisation in the newly accessible area has a greater range of options for random rearrangement of the means of control than were available previously. We shall return to this after we have developed the concept of 'rattling' in Chapter II.5, and even more in Volume IV.

I.6.6 Resource Limitation Conflict

So far, we have talked mainly about the operations of a single control loop. However, control loops never act in isolation, and most of this book is concerned with things that happen when two or more control loops act in ways that affect one another's operation. Part 3 deals with interacting control loops. In preparation for that, we look now at what can happen when two or more control systems act on or through the same part of the external environment: conflict.

The most obvious kind of conflict occurs when two control systems control perceptions of the 'same' aspect of the environment but have different reference values for their perceptions. John wants to paint the room blue, while Jane wants it to be red. Toby wants the plush teddy bear and so does Alexandra. These are 'classic' PCT conflicts, and the form of this conflict is our first 'motif'. A 'Motif' is a recurrent structure of control loops and forces; that is, it recurs much more often than it would if control loops impinged on each other at random. A control loop itself is a motif when considered as a non-random arrangement of its components.

Consider a somewhat less obvious kind of conflict. Ken and Joy want to live together, but Joy wants to live in the country while Ken wants to live in the city. Each seems to be controlling a different perceptual variable, and relinquishing control of either would permit control of the other. Is the conflict between the two of them, or is it inside each of them? One can read it either way or both ways, but if one considers the perceptions as a higher-order 'Us living together in the place I want', they really are seen to be controlling 'the same' environmental variable, the higher-level perception of a triple relation among the partners and the location in which they live.

The classic conflict is a special case of a more general class of conflict in which a number n of ECUs control their perceptions but their loops pass through some part of their environment with fewer than n degrees of freedom. A frequently encountered constraint might be that there are not enough effector channels to do the job — "I haven't got three hands, Joey." For example, suppose a circular disc on a screen moves randomly and changes the lightness of its neutral grey colour under some exterior influences. Using only the x-y movement of an ordinary computer mouse, one can control either its colour match to the ring or its location centred in the ring, but not both simultaneously (Figure I.6.6a).

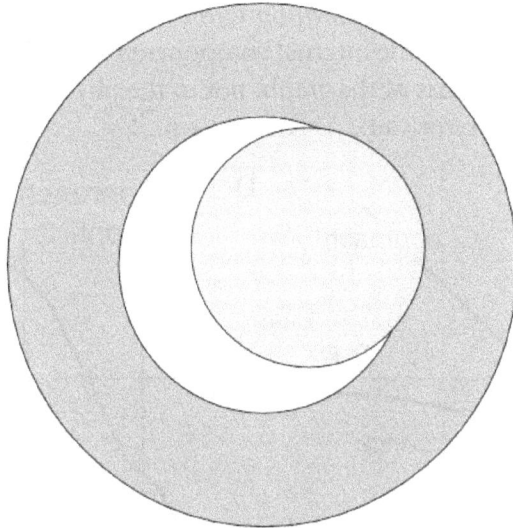

Figure I.6.6a The central disc moves and changes its lightness under external influences. The tracker's task is to keep it centred in the ring and matching the ring in lightness by moving an ordinary mouse. There are three degrees of freedom to be controlled, but the mouse can move only in two dimensions. The task is impossible without ceasing to control one of the three for a while before returning to it.

It doesn't matter where in the control loop the resource limitation exists. If three different controllers control different functions of these variables and there is no combination of variable values that satisfies all three reference values at the same time, a conflict situation exists. Since all the controllers are trying to 'pull' the CEVs toward their reference values, we can call this situation an 'approach' conflict.

The opposite of the 'approach' conflict is the 'avoidance' conflict. From the viewpoint of the observer looking at the influences on the CEV, there is no difference between two forces pulling in opposite directions ('approach conflict') and two forces pushing in opposite directions ('avoidance conflict'). In one dimension, the avoidance conflict looks very like the approach conflict, except that the conflicted controllers are trying to 'push' the CEV away from their 'to be avoided' reference value. Of course, if the conflicted controllers can themselves move away from the CEV, there is no problem. The conflict occurs when they cannot.

Powers implemented 'avoidance' by having the perceptual function return some inverse function of distance, so that large distances corresponded to small perceptual values, and a reference value of zero for that perception implied that the CEV was far from the 'to be avoided' value. Different reference values would correspond to particular preferred distances from the fixed point.

The same kind of result could be achieved with an inverted error function, such that small values of reference-minus-perception produced a large error value to be fed to the output function of the controller. There is a difference, however, in how the strength of the force acting on the CEV changes as the deviation between the CEV values R and P (the environmental values that correspond respectively to the reference and perception values r and p) increases (Figure I.6.6b panel a). Figure

I.6.6b is shown from the Analyst's viewpoint, in that the values and forces are those associated with the CEV, not the internal components of the ECU. 'Leftward' and 'Rightward' refer to the x axis of the graph, not to the physical environment of the CEV, and the 'apparent error value' is R-P, not r-p.[70]

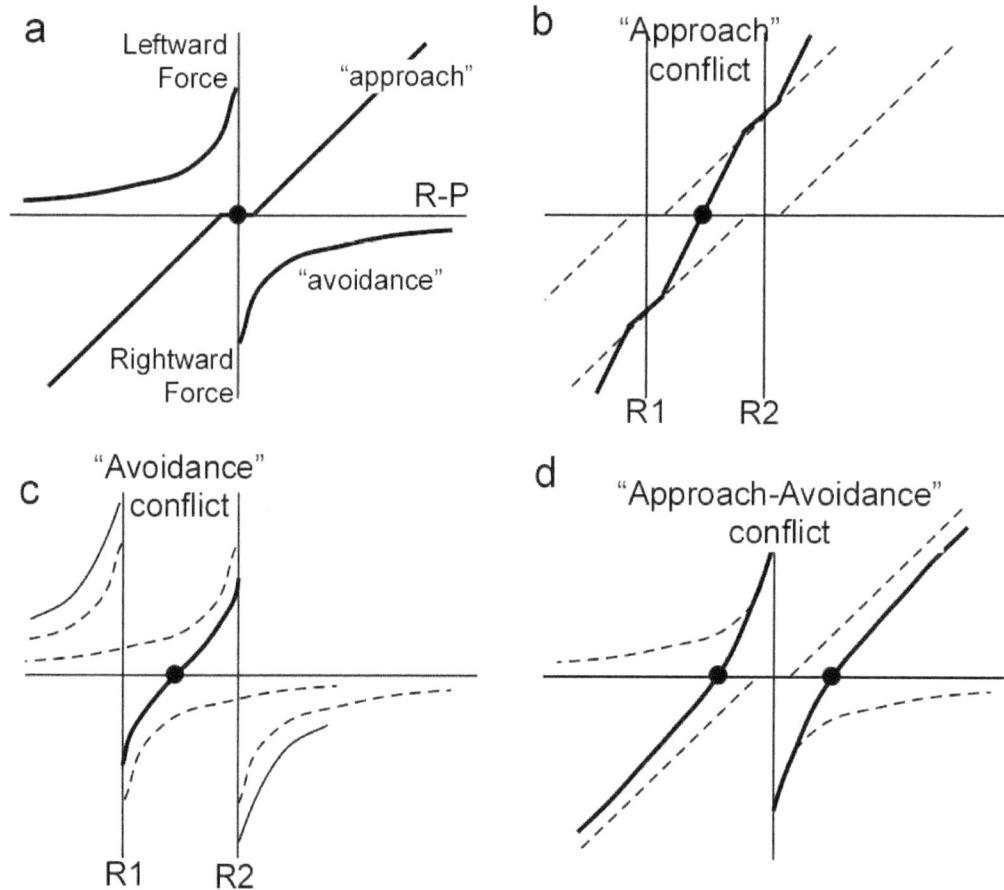

Figure I.6.6b Analyst's views of how the apparent error (R-P) influences the value of the CEV (P) that corresponds to the current perceptual value (p) in Approach control and Avoidance control. R is the value the CEV would have if the perceptual value equalled the reference value. In each diagram the solid circle indicates a stable position of the CEV, and in b, c, and d, the slope of the heavy curve shows how the apparent error signal R-P relates to the directed action output of the collective controller created by the two competing control units. (a) Individual approach (P moves toward R) and avoidance (P driven away from R); (b) Resource limitation ("Approach") conflict in which two control systems both try to draw P toward their different reference values R1 and R2; (c) Mutual repulsion ("Avoidance") conflict in which two control units each try to move P as far from their reference values R1 and R2 as possible; (d) Approach-Avoidance conflict in which two control systems have the same apparent reference value R, but one tries to bring P toward it while the other tries to Move P away from R.

---

70   This is necessarily also the case when an experimenter performs a 'Test for the Controlled Variable'.

Panel (a) of Figure I.6.6b shows both approach control and avoidance control. Approach control, which moves the CEV toward an apparent reference value (P is influenced to approach R), is shown with a tolerance zone; if there were an equivalent tolerance zone for avoidance, the zero-error region would include everywhere 'far-enough' away from the avoided value, not just a small region around the apparent reference value.

The other panels of Figure I.6.6b show what happens when two control systems independently influence the same CEV, creating a conflict. A conflict occurs for all values of the CEV for which the dashed curves lie on opposite sides of the X axis, so that the two control units work against one another. For the stable points shown by the black discs in panels (b), (c), and (d), the conflict will probably escalate with the controllers continually increasing their outputs, if the output function integrates the error, as is the case in most PCT simulations.

What also happens, as McClelland (1993) showed, is that the CEV moves exactly as it would if it were being influenced by a single virtual controller that had a loop gain the sum of the gains of the individual controllers and a reference value proportionately between the individual ones in proportion to their relative gains. This is a minimal instance of a concept we will use at length and in varying degrees of complexity in Volume III of this book — 'Collective Control'. We would say that the CEV influenced by the two conflicted controllers is the 'CCEV' (Collective Corresponding Environmental Variable) of a 'Giant Virtual Controller' (GVC) that has only two members. Some GVCs have many millions of members, others have only a few, but never less than two.

The approach-avoidance conflict of panel d often occurs within a single individual, when, say, stealing a cookie is a way to bring a perception of taste in the mouth toward its reference value, but at the same time is a way to bring a perception (imagined) of punishment nearer to a value that is to be avoided. Stealing a cookie is both to be approached and to be avoided, a situation that can lead to a physical approach to the cookie jar that stops a certain distance away from it.

Conflicts are not always resolved by the opponents increasing their output until one of them reaches a limit, allowing the other to control their perception. Often a more effective approach is to alter the opponent's ability to control, either by making it harder for the opponent to perceive the CEV corresponding to the controlled perception, or by making it more difficult for the opponent to apply the output to the CEV. In a military conflict, destroying the enemy's transportation infrastructure reduces the force that the enemy could apply to attack you or to defend against your attack. In an example that we will see in Section I.7.4 John tries and fails to break a branch off a tree, and Jill helpfully provides some tools to cut it. If Jill did not want the branch removed, she might hide all the cutting tools.

Chapter I.7. Consciousness and Imagination

This section reframes a number of concepts in terms of perceptual control, to lay the groundwork for later developments. The first group of concepts involves our imagination and planning, and the second group deals with what we might call 'difficulties', things that either are to be avoided or that go wrong. Although you might have assumed at first, as most readers do, that we control perceptions of which we are conscious, the previous chapters should have made it clear that in the main the perceptual control hierarchy operates quickly and non-consciously, and that the perceptions which we are conscious of controlling are those which we have not yet fully learned how to control.

In Volume II and beyond, we shall be looking into the relationships between conscious thought and non-conscious control. We do not yet maintain this careful separation in this chapter, because the underlying question of 'control' is the same in both worlds. The distinction becomes important later when we discuss details such as the speed of a control loop in the (fast) reorganised non-conscious perceptual control hierarchy as opposed to the (slow) thought-out control done consciously.

I.7.1 The Conscious Perceptual World

In your currently perceived world, there is much more than what your senses tell you at this instant. Much of what your senses tell you is unconscious, but even the unconscious perceptions are available for use in control, and some of them are currently being controlled. Think about the tensions in the muscles that hold you up if you are standing or walking. Were you conscious of them? Probably not, but if they were not being controlled — sensed, compared with reference values, and adjusted to compensate for problems that might lead you to fall down — would you not be flat on the floor? Think about the sounds around you at this moment. Are there any sounds of which you were not conscious? If one of them happened to have been your baby crying or someone calling your name, would you not already have been conscious of it?

Think about the building you are in at this moment. Maybe it is your house. What can you see without moving your eyes or turning your head? Not much, I think. But in your conscious mind can you not see a lot more if you want to? Do you not perceive the bookshelf or the wall that is behind your head? If it is your house, do you not see rooms, doorways, furniture, all of which are hidden from view until you move? And if you have moved your head, have the objects you saw at first now vanished from your world? They have not. They remain available for control and many are probably being used in perceptual controls of which you are as unaware as you were of their nature before they were called to your mind by the query you just read. Your house continues to be a part of your perceived environment, whole and entire. The room you were in does not vanish into a mysterious void when you move to another, though it may vanish from your conscious awareness.

If your house has two stories, can you see how to get from one story to the other? In my house I see in my imagination a stairway. If I imagine myself going upstairs in my house 0when I am down, I can see in my mind the pictures on the wall as I pass them, and the view into a couple of rooms as I near the top of the stairway. The world available to my conscious perception contains those stairs and those rooms, whether I am in my house or not. They are part of my 'Perceptual Reality' (PR) or (in conscious perception) my 'World Model'. This is not the 'Real Reality' (RR) in which we all must live, but the two worlds, PR and RR, must be related in some way. In Chapter I.11 we will illustrate how it is possible that we could find to the smallest detail what functions RR can perform, without ever getting the slightest clue about the mechanism by which it is able to do it.

The 'World Model' concept is developed in Section I.7.7. In AI, a 'model' is a program which detects specific patterns, using a collection of data sets. In PCT discussions, a 'model' is a generative artefact, a functioning system which produces effects like those which would be produced by the thing modelled, under the same boundary conditions and given the same values of variables. We extend the term here to perceptions which are imagined on the basis of their reference values. Imagination emulates perceptions that would be produced if those reference values were to be used in generating action on the external world. This is the basis for planning, in which we may imagine diverse possible disturbances and vagaries of the environmental feedback — as exemplified in the common expression 'Plan B'.

My Perceptual Reality contains the things that are now and have recently been in my field of view, as well as those of which I am or have recently been conscious. But it contains much, much more. It contains things and people I may not have seen for a long time, and events too. It contains people I have never met and who may not ever have existed, such as Presidents and Prime Ministers about whom I have been told. It contains events that might have happened long before I was born, such as the Diamond Jubilee of Queen Victoria, and the Battle of Actium in which (so I am told) Octavian defeated Antony and Cleopatra. It contains abstractions such as the policies of political parties and the beliefs of various religions. Like any other perceptions, whether or not they correspond closely to the present or historical outer world, they are my perceptions, and when I act, some of them might contribute to the perceptions I control by acting. What is available to my perception of '*the way the world is*' is large, much, much larger than the tiny part of it that meets my senses right now.

My Perceptual World is not limited to static objects such as stairways, doors, and furniture, abstractions such as political policies, or remembered and imagined entire events. It includes the dynamic consequences of controlling perceptions in that world of objects, abstractions, and historical events (We treat this kind of thing as 'Narrative' in Chapter II.10). I can imagine walking up my stairs and arriving at the top, successfully sitting on a chair, putting a pot of water on the stove and watching it boil. I can imagine throwing a stone into a pond and watching the expanding rings of ripples, or slipping on an icy sidewalk and falling. I can imagine putting some long thin object across a little brook and stepping on it. If I have imagined something as thin as a drinking straw, I may perceive it bending and breaking, perhaps under its own weight but certainly under mine, but I might also imagine miraculously walking across it to the other side, like a tightrope walker.

I can imagine myself floating through space, unaffected by the 'breathing vacuum' until I land on the Moon, and then watching a lunar eclipse from the other side. I can imagine '*the way the world works*', which, like '*the way the world is*' may not correspond to reality, though in several places in this book we examine why my imagined '*way the world works*' is likely to trend toward reality on timescales both of evolution and personal learning.

Often I can imagine that if I did various things in the world I perceive to be 'real', they both would and would not work the way I imagine them. This is a different level of imagination, the first being of the world as I believe it to be, in which the long thin straw would certainly break under my weight, and the second being of a world in which I can shape the way that it works. By imagining in this second world in which I am able to do things that are not possible in the 'real' world, I may also imagine creating entities that might provide atenfels to allow me to control those perceptions in reality. If '*Necessity is the Mother of Invention*', invention is a baby gestated during control in imagination.

In my imagination, I can compare what would happen if I walked, or tried to walk, across the long drinking straw with what would happen if I placed a sturdy plank across the stream instead of the weak straw. I can imagine that the plank might wobble, but not break, whereas the straw would break, but not wobble. I can imagine that I could manipulate a thin plank to lie across the stream, but a thick plank might be too heavy for me to move. Would a plank light enough for me to handle be liable to break if I used it? My experience with the strengths of different planks will usually allow me to answer such questions correctly if my imagination is tested by trying to do what I imagined doing.

I call this combination of perceptions — my present, remembered, and imagined perceptions of '*the way the world is*' — my Perceptual World, which together with my imagination of '*the way the world works*' constitutes my 'World Model'. My 'Perceptual World' and my conscious 'World Model' overlap, but they are not the same.

And I can imagine '*the way the world does **not** work*'. When I imagine myself floating to the moon while remaining completely aware of my surroundings, I imagine at the same time that this could never happen in the world in which I perceive myself to exist. It is not part of my World Model but is in yet another perceptual space that I might call a 'Fantasy World Model', as separate from my 'World Model', as is the more mundane imagination of tightrope-walking over the drinking straw placed across the brook.

Perceptual Control Theory seldom deals explicitly with consciousness, but we will, especially in the series of Chapters II.7 to II.10. In those chapters we will try to relate conscious, rational thought to the non-conscious, rapid parallel processing of the perceptual control hierarchy. Here we will talk about the everyday experience of conscious perceptual control. This is what people actually experience. Non-conscious perceptual control is more fundamental and more extensive, but usually ignored. Even Powers did not explicitly distinguish conscious rational thought from non-consciously produced intuition and action, despite the considerable difference in speed between serial rational thought and the highly parallel processing of the reorganised perceptual control hierarchy.

Consider the example of placing a plank across a brook in order to cross dryshod. When you see the plank and go to pick it up, you can consciously imagine its weight and probable strength, but would you take the time to do so if you had done this many times before using the same plank? When I try the thought experiment on myself, I find that the result is that I would perhaps be conscious of where I was attempting to place the ends of the plank, but I would just be using its properties in controlling the perceptions at various levels that are involved in getting across the brook. Over enough similar occasions using consciousness when you were unfamiliar with that plank, you may have reorganised its potential as a 'bridge' atenfel into your normal control hierarchy. We discuss how this may come about in Chapter II.8.

Perceptions in my Fantasy World seem in one way to be as real as the perceptions in my World Model, but in a different way they do not, because Fantasy World perceptions always are conscious and include a component that marks them as belonging to '*the way the world does **not** work*' (unless I am asleep). In the same way, a perception (memory) of having dined at the Ritz includes a component that marks it as different from a memory of having dined at Aunt Martha's — assuming I have done both (which I have not). The Ritz and Aunt Martha's are different worlds, and so are the World Model and the Fantasy World.

All of the above is a report of subjective phenomenology, without yet making any suggestion of how these phenomena might occur. They are effects to be explained by some mechanism, even if they are entirely subjective and unique to me. I simply assume that other people have similar experiences, having no evidence for or against that assumption other than what other people say and write. We can crudely call them all 'imagination' in contrast to direct perception, though different kinds of imagination may distinguish the Fantasy World from the World Model. Powers offered suggestions as to one possible way imagination might be used in perceptual control. We examine these next.

I.7.2 When Am I? Past, Present, and Future

When is 'now'? How long is 'now'? How does the perception of 'being now' relate to the perceptual control hierarchy, if it does at all? 'Nowness' isn't a perception we can control, after all, but we do perceive some things as being in the past, others as perhaps being in our future, such as a graduation ceremony in which years ago we did participate or in which we might participate next year. Somewhere between past and future is 'now'. But when is it? That is a question philosophers have long asked, without agreeing on a stable answer that was, is, and will be satisfying and unchanging. Let's see if the PCT hierarchy of perceptual control can help untangle the mess.

A PCT way of looking at a distinction between what was, what is, and what may come to pass, depends on what perception we are controlling and on the time scale of the control loop involved. For example, imagine that on a dark night in the First World War (1914-18) you are in the middle of watching a morse code message being sent by flashing a light across a body of water. Perhaps you are on a boat that is a member of a group planning to land on an enemy-controlled shore. When is 'Now', what is in the past, and what is still to come? Let us consider a very short message (SOS), which could be translated as 'Send help quickly'.

The entire 'SOS' Morse code 'emergency' message (Figure I.7.2) is clearly in the 'Now', as is each dot, each dash, and each space between them — and the waiting time until help arrives, if it ever does. But so also is the landing operation on which you are engaged. After you have executed whatever the message told you to do, understanding the message was clearly 'Then', in the past, but the instructions it conveyed are still being acted upon, in the 'Now'. These several 'Nows' are simultaneously experienced, but take very different times to expire. While you were in the middle of seeing the message being transmitted, some individual flashes were clearly in the past, and some were yet to come, as you presume because so far the message makes no sense.

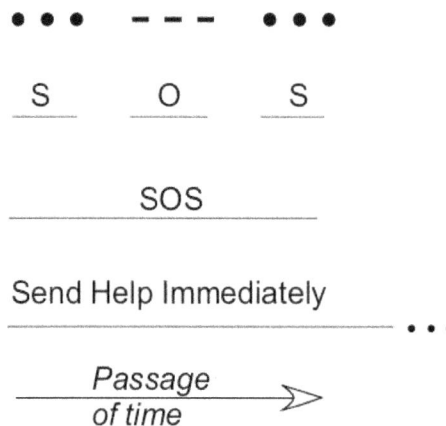

Figure I.7.2 A hierarchy of "Now", from the duration of a Morse "dot" or "dash" to the waiting for help in an emergency. The length of an underlying line indicates a possible duration of "Now" for the unit in question, whether a Morse dot or a lengthy message for which "Now" may extend until the message has been answered.

Are any flashes individually 'Now'? Probably so long as you are still determining whether a particular flash is a dot or a dash, it is in the 'Now', and so is the alphabet letter composed of a sequence of from one to five dots and dashes. As parts of the letter, so long as a pause between flashes has not exceeded some threshold duration, all the contributory flashes are 'Now', but as flashes, some are past and some may be yet to come. The same relationship exists between the letters and words, between the words and the message, and between the message and the arrival of help, all are Nows, nested one within another like a set of Matryoshka dolls — or lower-level perceptual input categories within higher-level ones.

Returning to the longer message being flashed from shore to a hopeful landing craft, at the operations level, everything affecting control of its success, including all of that Morse message, is 'Now' from the start of actions until the success or failure of the operation can be perceived. What tactical or strategic reasons might have led to the operation are reference values at yet higher levels of perceptual control, changing more slowly, and therefore being perceived by their controllers as 'Now' for even longer periods of time, as are the political and social changes that might exist within a 'Now' that lasts for centuries. We are Now in the 'Western' Industrial Era, for example, and we are 'Now' in a time of rapid technological change, just as much as I am in the short 'Now' of typing this sentence or an even shorter 'Now' of typing a word that forms part of the message.

I think the point has been made that 'Now' is a perception, a conscious perception, but not a controllable perception, and not something that is true or false of the environment. How long may be the Present, before which was the Past and only after the conclusion of which is the Future, depends on the perceptual hierarchy level of what you (and possibly nobody else) are consciously thinking. Your 'Now' actions affect the Future, the Past affects your Present perceptions, and the length of 'Now' is the effective duration of the Loop transport delay for control of any perception — 'effective' because the output function and the environmental feedback path may (actually 'must') smooth out the effects of sudden changes in the disturbance, so the actual Loop transport delay is ill-defined and is sometimes taken to be how long it takes for a change at one place in the control loop to undergo half the stabilising effect of feedback control. 'Now', even within a single control loop, is not precisely bounded, even though it is always nested within some longer 'Now', presumably to the duration of the controller's life.

I.7.3 The Imagination Loop

How did Powers treat imagination? As a scientist, he always required that a mechanism be available for any concept that might be developed into a PCT construct. Imagination was no exception. Accordingly, Powers treated imagination one perception at a time, not as a whole world or set of possible worlds. Just as complex molecules are built from simple atoms, and complex control hierarchies are built from simple Elementary Control Units (ECUs), so also may complex fantasies or realistic plans be built from simple unitary imagined perceptions.

In B:CP, (Fig. 15.3 in the 2005 edition) Powers illustrated what he called the 'imagination mode' connection that completed an 'imagination loop'. In the imagination loop, the reference value for an ECU is short-circuited to its perceptual output, as though the lower-level ECU had produced actions that brought the perception to its reference value. The Powers' imagination short-cut circuit has no relation to consciousness, any more than do any of the perceptions controlled in the hierarchy. Figure I.7.3a (left panel) shows the essential components of this form of the 'imagination loop'.

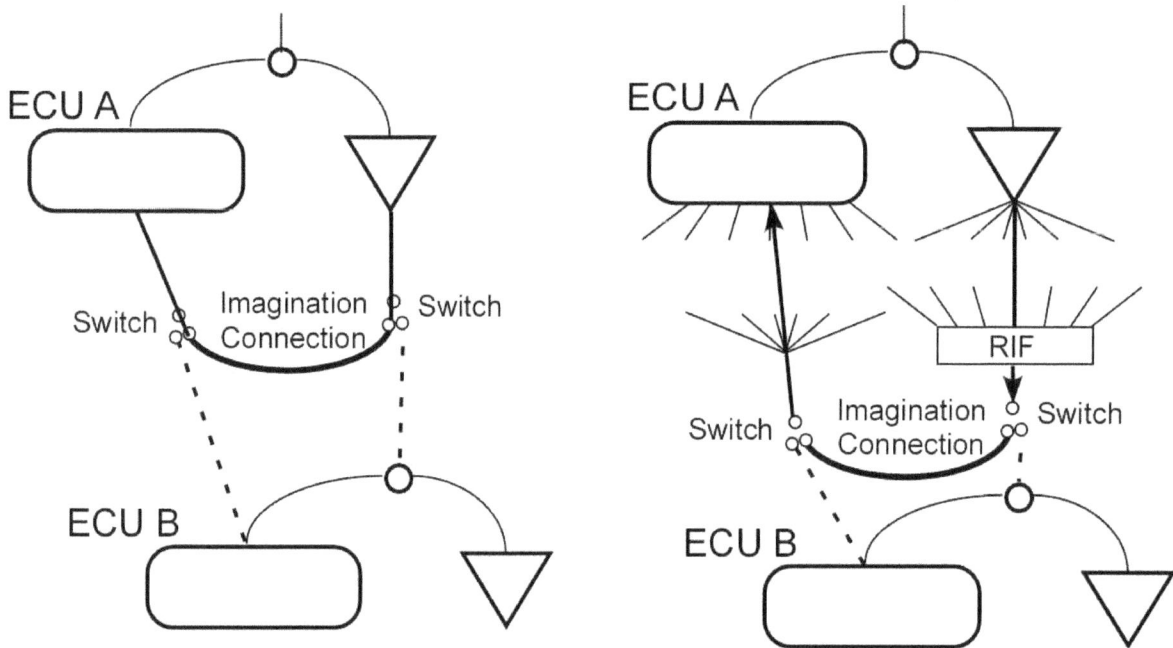

Figure I.7.3a The basic imagination connection for ECU B. Instead of actually producing an output and controlling its perception, ECU B supplies as its perceptual value its reference value, as though it had controlled perfectly. (Left) Connection as in B:CP Fig. 15.3. (Right) The same, allowing for the fact that a reference value may depend on more than one higher-level output.

The effect of the 'imagination loop' was to restrict imagined perceptions to ones built from perceptual types that had been developed by reorganisation. At the lowest imagined level, imagined perceptions would be replays of experienced perceptions. At higher levels, imagined perceptions would be built from such lower-level replays, but not necessarily in any previously encountered constellation, since the imagined lower-level perceptions would be inputs to the higher-level perceptual functions that also had other inputs.

When one is imagining, one does not want the output that creates the imagined perception to cause actual behaviour in the external environment. That is the reason for the switch at the right side of each panel of Figure I.7.3a. While we dream, our bodies do not usually perform the actions we dream that we are performing, even if those actions are physically possible. The contribution to lower level reference inputs must be blocked before they reach the musculature, but that block should not also block ongoing control of perceptions from the senses. We should be able to walk on the sidewalk while imagining that we are flying like Superman or swimming across a lake. How can the necessary separation between imagined and on-line control of perceptions that would use similar musculature be achieved? How can we walk across a street while imagining we are hiking on a mountain trail using the same walking muscles?

In a human brain there are billions of neurons that have trillions of interconnections called synapses. At our periphery we have about 600 muscles that can influence the outer world, and a few million individual sensors such as retinal rods and cones, auditory hair cells, and so forth. The numerical discrepancy between what is inside and what has access to the outside is enormous. At any one instant, we could not possibly be controlling more independent perceptions than we have muscles, and since many sets of muscles work in coordination, the actual number of perceptions independently controllable at any one moment is even smaller.

To control the many perceptions we do over time, we must be able to switch control quite fluently from one perception to another. We don't lose control over the others, because we can switch back to controlling them actively if they stray outside their tolerance zones, but neither do we control them as tightly as we would if we kept controlling them actively all the time with no tolerance. The consequence of this is that at any one moment we probably have far more uncontrolled perceptions than controlled ones, but many of the uncontrolled ones could quickly become controlled and the controlled ones uncontrolled, so that we can keep most of them tolerably close to the way we want them most of the time. If a pencil starts rolling off the desk I quickly restore it without losing control of the sentence which I resume typing.

To switch control from one perception to another, either the physical connections must be remade so that a few ECUs are transferred to perform new duties, or the ECUs remain associated with their individual Perceptual Input Functions and Output Functions but are switched in and out of connection to the musculature. The same issue must apply to sensory input, since the number of things we can perceive through our senses over time as we move around the world vastly exceeds the number we can directly perceive at any one moment. Because of the thousands or millions-to-one ratio between the internal and external functional possibilities, the likelihood is strong that any one ECU remains associated with its own perceptual function and is switched in and out of operation, rather than being transferred around control of an ever-changing repertoire of possible perceptions like the central processing unit of an ordinary computer being time-shared among several processes.

This argument suggests that any ECU may be disconnected from the peripheral musculature while still controlling its particular perception, as in the left panel of Figure I.7.3a. It further suggests that when the perception created by the PIF of an ECU is currently uncontrolled, the output of that ECU is likely to be disconnected from lower-level reference inputs, at least when those lower-level ECUs are actively controlling higher-level perceptions. It follows that, for most if not all perceptions that can be controlled through actions on the environment, the living brain is likely to have a functional equivalent to the right-hand switch in either panel of Figure I.7.3a. Such a switch could be implemented by reducing the gain of the output function to zero.

The question, then, is whether a neural connection such as the 'imagination-mode' connection back to the perceptual input shown in each panel of the figure is likely to exist or to be common if it does exist. What we can note is that the imagination connection constitutes a minimal environmental feedback path for control of ECU A's perception. If the connection exists and the loop gain is negative, the perception will be kept near its reference value. In the next section, however, we will complicate the imagination connection by re-introducing the World Model or the Fantasy World.

In the right panel of Figure I.7.3a, the other inputs to the Reference Input Function (RIF) of ECU B could require the output of ECU A to keep changing to compensate. They can be interpreted as disturbances to this feedback loop. The 'disturbance' created by the other inputs to the ECU B RIF in the right panel of Figure I.7.3a is not, however, as arbitrary as disturbances from external sources. The changing input to the RIF reflects the changing action context of other ECUs controlling perceptions at the same level as that of ECU A. In other words, the changing output of ECU A that controls its perception 'in imagination' could be interpreted as 'maintaining situation awareness' in a changing outer world, being ready to contribute its output to the ongoing action when required, avoiding the need to 'bring it up to speed'.

When one is not actively controlling some perception but has 'situation awareness' of it in its context, one's perception of the whole situation has the same sense of reality as do the parts of the room that we cannot at the moment see with our eyes. The same cannot be said of imagining one is walking upstairs while one is actually sitting in an easy chair. Imagining walking upstairs has no feeling of reality, however vivid the imagining and despite its feeling of realism. We perceive ourselves as 'really' sitting in the chair, and perceive ourselves to be imagining walking up the stairs. Can the imagination connection of either panel of Figure I.7.3a accommodate this kind of imagination as well as awareness of the currently unsensed part of the world? Yes, it can. The left-hand switch in either panel of the figure accommodates this requirement, isolating the imagined perception from other inputs.

The PIF of ECU A does have other inputs that might come from ongoing sensory data (or from an imagined world, as we shall see presently). but the imagined perception that concerns us now is the perception which would be derived from environmental inputs by ECU B, were it not short-circuited by the imagination connection. What ECU A does with the imagined perception is another question entirely.

There is an alternate hierarchic organisation of controlled perceptions functionally equivalent to that proposed by Powers, but with some conceptual advantages. The idea for the alternative circuit was inspired by the following passage from Seth and Friston (2016), taking 'prediction' to be equivalent to 'reference value':

> *... descending predictions are compared with lower-level representations to form a prediction error (usually associated with the activity of superficial pyramidal cells). This mismatch or difference signal is passed back up the hierarchy, to update higher representations (usually associated with the activity of deep pyramidal cells).* [71]

In the Powers hierarchy, the perceptual signal value from one level is sent as an input to the perceptual functions of control units at the next level above. But since the error value is the reference value minus the perceptual value, the same result could be achieved by sending to the next higher level not the perceptual value but the reference value and the error value (Figure I.7.3b).

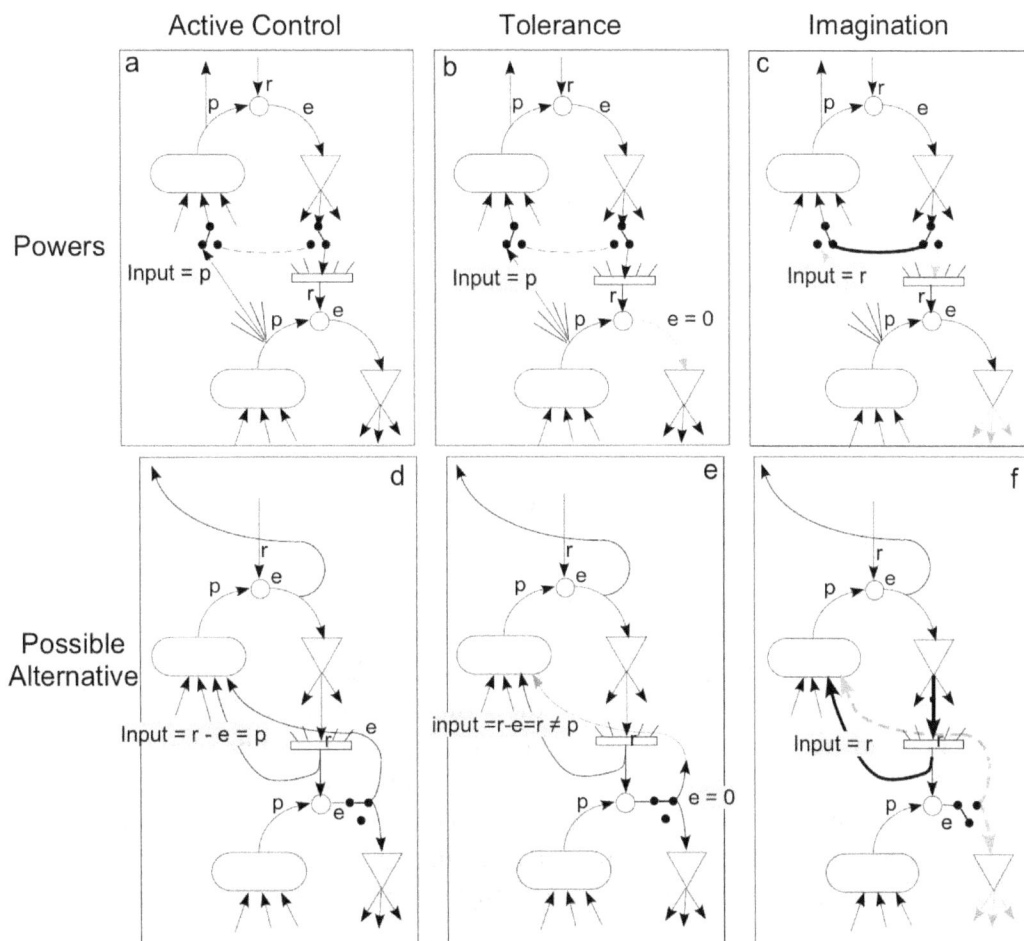

Figure I.7.3b The subtly different effects of the Powers connection between the perceptual functions of two neighbouring levels and a possible alternative connection inspired by Seth and Friston (2016). (Upper row) Powers, (Lower row) alternative connection. (Left column) When controlling with zero lower-level tolerance, (Middle) When the lower level perception is within its tolerance zone so that the error value is zero. (Right) The imagination connection; the error connection to the next level is switched together with the reference connection from above so that the lower unit produces no output.

---

71   Pyramidal neurons map many dendritic inputs to a single axon in specialized ways. — Ed.

In the alternative circuit, the Powers 'Imagination connection' is never switched out, but is always active. The error signal is ordinarily taken to be Reference minus Perception (e = r - p). This equation is equivalent to p = r - e. Consequently, the higher level function could treat the combination of the two separate signals, error and reference, the same as the perceptual signal of the Powers hierarchy.

In Figure I.7.3b, the simpler form of the Powers connection (Figure I.7.3b panel a) is shown for visual simplicity, but the following arguments apply equally if the more complex version is used (Figure I.7.3b panel b). If the tolerance range of the lower control unit is zero, the actual perception is available to the upper perceptual input function with both connections.

With the alternative form of inter-level connection, other forms of perceptual function would be possible, including ones that ignore the reference value and work only with the error value, and others that do the converse, using only the reference value. In the Powers sense, this last kind would be always imagining that it had produced the desired perceptual value from the lower level, regardless of what actually occurred as a consequence of issuing that reference value to the lower level. It would be a 'Command-only' unit.

Circuits that ignored the reference value but that used the error value would be good candidates as components of a reorganising system, or as control units whose outputs influenced parameters such as the gains of lower-level units rather than their reference values. They might also serve as alerting units that monitored when an uncontrolled lower-level unit's perception moved out of its tolerance zone, so that the unit might be brought into control once more, in a short-term form of reorganisation. We will not pursue these possibilities here but should keep them in mind as possibilities.

If the tolerance zone is finite, and the perception is within it, the upper-level perceptual input function treats the lower-level perceptual value as if it were actually the reference value, because the error output is zero. Anyone who has, for example, played the piano as an amateur will know the phenomenon of hearing what you intended to play rather than what you did play when playing alone, but being painfully aware of your errors when playing for an interested listener such as your teacher. In the latter situation, your tolerance zone for error may be drastically reduced as compared to when you are just playing for fun. In the Powers connection, this does not happen, as the actual perceptual value is reported to the higher level.

When one is controlling in imagination, the Powers circuit disconnects the lower-level system entirely from the upper one, providing the upper with the reference value that would have been sent to the lower system. Of course, the more complex form would provide the reference value that in the specific context would have been provided to the lower level. The alternative connection always provides the lower reference value to the upper perceptual input. What changes in the three conditions is the role of the error value, which is r-p if the tolerance zone is zero, zero if the perceptual signal is within the tolerance zone, and disconnected if the upper unit is operating in imagination mode.

This alternative connection, inspired by Seth and Friston (2016), resolves a few possible issues with the Powers circuitry. Firstly, ability to perceive the reference value means we are consciously able to perceive what we want to achieve, and independently we are able to perceive the margin by which the current state differs from it, the error value. Powers assumed that any perception of which we can become conscious must exist somewhere in the control hierarchy of perceptions, but in the standard HPCT hierarchy reference signals and error signal are not perceptions. With the alternative connection, higher levels can perceive both. Furthermore, with this interconnection circuit the higher-level perception can take more nuanced values than simply a choice between the actual sensory perception and the reference value; instead, it can attenuate or augment the perceived error signal from the lower level, so that the higher level perceives the lower level to have exaggerated error (as when one plays the piano for one's teacher) or diminished error (as when playing the same piece for fun).

So far, we have been ignoring the fact that 'switching' is an all-or none process, and that the requirement for a 'switch' depends on the signal path being connected or disconnected. In a designed electrical circuit, switches are often replaced by attenuators such as variable resistors which can vary the degree of connection continuously between perfect (1.0) and disconnected (0.0). Could the nervous system do something similar without destroying the pretty picture we have been describing? Indeed it can, and not only is it possible, it is likely.

The reason is that we have been treating signal values as 'neural currents' analogous to the electrical current on a wire. The neural current was defined as an abstraction based on the activity of many independent neurons that happen usually to respond similarly to the same inputs. 'Similarly', however, does not in practice mean 'identically'. Though the neurons that contribute to a 'bundle' that carries a neural current may have a lot of overlap in their input connections, they also have their differences, so that as an input pattern changes slowly, first one and then another of the fibres in a bundle becomes the one that responds most strongly, and at the same time other fibres respond less and less, while yet others are recruited into the bundle.

If the pattern to which they are responding is noisy — consciously we might say the pattern was 'faded', 'washed out', or 'uncertain' — then this shifting of responses among the fibres of the bundle would also be noisy. Averaged over time, this would not affect the value of the neural current carried by the bundle, but looked at fibre-by-fibre, the time average across the bundle would be flatter than it would be with the same pattern seen more clearly (Figure I.7.3c). Fewer of the fibres would have a persistently strong response, and more would have a generally weak response. The bundle would seem wider but flatter, if measured by commonality of responses. The effect on downstream neural currents would be similar to that of an attenuator in an electric circuit.

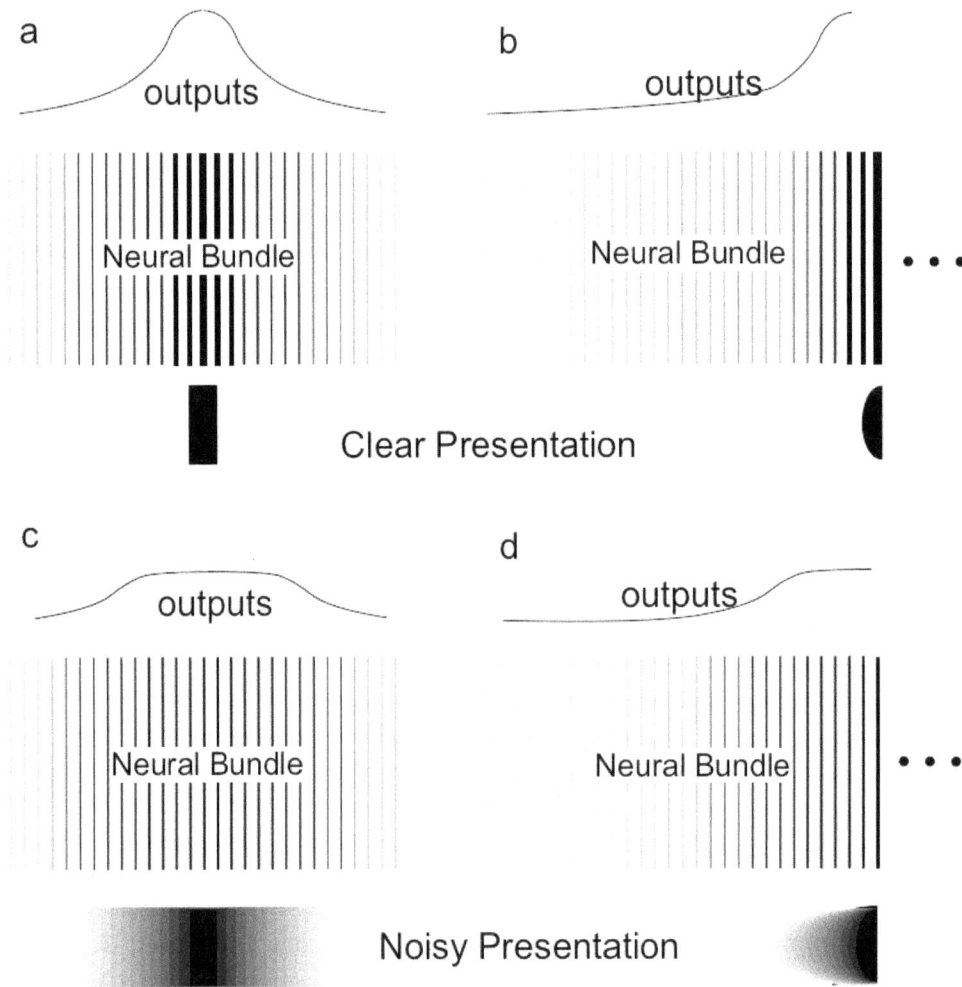

Figure I.7.3c Outputs from individual fibres of a "neural bundle" that sum to produce a "neural current". The left-right dimension represents changes in any property, not necessarily related to location in space. When the input is clear, the more central fibres in the bundle produce a concentrated output, but when the input is noisy, the same total "neural current" is produced by a less concentrated set of central fibres (top and bottom panels). The right hand pair of panels suggest that individual fibres may belong to more than one "bundle" and can respond to somewhat changed inputs as well as to the original.

In the Powers 'Imagination Loop' the switches would be replaced by reciprocal attenuators, one side having a generally stronger connection as the other side gets weaker. By deconstructing the fibre bundle and the 'neural current', it is possible to see how imagination and sensory perception might both always contribute to the higher-level perceptual input function, with weighting corresponding to their relative clarity. The same applies to the alternative connection, which instead of switching from one mode to another would transition continuously from, say, being fully based on sensory evidence to being fully based in imagination, with all mixtures of sensory and imaginative evidence being possible.

We will return to this point several times in various sections of the book, particularly in Chapter II.10, but for the most part until then we will ignore the complexities we have just been discussing. The structure as described by Powers is adequate for most purposes. We will explicitly note departures from it.

I.7.4 Teaching, Imagination, Learning and Invention

Reorganization is a kind of learning. Whether random or e-coli, reorganisation as described before alters only the internal structural relations and functions of the control hierarchy. It takes advantage of the environment as it exists, in order to improve control of perceptions that are to be controlled and to adjust the control actions so that their side-effects do not disturb intrinsic variables beyond ranges compatible with survival and propagation of the genes. Reorganization is an important kind of learning, but it does not take advantage of learning by being told how to do something — learning by being taught — nor by inventing a new way to control some perception.

To move an overlarge rock, you would not imagine using a lever unless you had in your conscious World Model the properties of levers and of the material requirement for something to be a lever — length and bending strength, for example. Could your World Model include a lever-atenfel if you had never used a lever? Perhaps it would, if someone had explained levers to you in the past. That someone would be a 'teacher'. If you needed no such atenfel when you were given the explanation, however, why would you consciously remember it later when you could use it? This is the conundrum of all schooling. The student is being taught about atenfels that are not currently useful, and for the most part would not have been useful in the student's past, but that the teacher imagines will be useful in the student's future, a future not imagined by the student.

So why should the student want to learn anything that would be useful only in some unknown future? What perceptions would the student be controlling by storing the lesson in her World Model? What imagined perceptual control other than passing a future exam could be improved by remembering that Confitia is the political capital of Haveristan, though the commercial capital is Fincifitia, which was overrun by the Mongols in early 1342 while Confitia held out for another month? At the time the 14-year-old is taught these amazing facts, she might well imagine that the only perception for which learning them could be an atenfel would be marks on a future exam. When the student, ten or twenty years later, has secured a lucrative contract in Haveristan, that part of her World Model might be critical because of the pride the Confitians still hold despite the more recent rise of Fincifitia in regional banking. But would her World Model still contain those perceptions of the way the world is?

Learning about levers has more obvious uses for the student to imagine. Those facts can be associated with perceptions which are already in the student's

World Model, and if parts of the lesson also involve the student using levers to control perceptions, those aspects of the way the world works are likely to both be incorporated in the World Model in imagination and be reorganised into the hierarchy in a way that, for example, walking in low gravity can only be if the student visits a place like the dwarf planet Ceres.

Learning is different from teaching. As frequent variations and controversies in the philosophy and practice of education attest, the student does not necessarily learn what the teacher wants to teach. The teacher controls her perceptions by teaching, whereas the student controls her perceptions by learning. The perceptual controls by teacher and student may work together, or they may conflict, and either way the experience can be productive and influence reorganisation in both parties, but reorganisation is seldom amenable to direction, and the student might well reorganise in ways that disturb perceptions the teacher controls, and may control perceptions which are not part of the course material. In the later Parts of this work, we consider how the control of different perceptions by different people may work together, both in dyadic interactions and in creating social structures and movements. For now, however, we return to examining control by individuals in a passive environment.

No single controller (ECU) can perceive anything other than the signal its Perceptual Input Function produces, which is the perceived state of its CEV. An ECU cannot perceive anything of its atenfels, the means by which it senses and acts upon its CEV. Other perceiving systems can, however, perceive 'relationships' (one of the levels of the Powers hierarchy). At the end of Section I.7.7 is a quotation on the World Model in which Powers points this out. In particular, some perceptual system X could, in principle, perceive a relationship between an atenfel of controller Y and the state of Y's CEV or any other signal in Y's control loop. Controller X could then control that relationship perception, perhaps by altering the atenfel so as to assist or inhibit the ability of controller Y to control its perception. In the notation of Appendix 3 (in Volume IV), X{Y{.@.[.]}@Y-success[.]}, read out as "*X controls for Y to be successful in controlling something by some means, and X uses some unspecified means to make this happen*".

If that sounds a little too abstract, consider this scenario. John, who is up a ladder in a tree, tries unsuccessfully to break off a small branch. With no word or sign from John, Jill hands him some cutters. John uses them to sever the branch. What perception has Jill controlled? One cannot tell for sure, but a possibility is that she controlled for seeing John achieve what he seems to be trying to do (perceive the branch as separate from the tree) by providing an atenfel which eases his control of that perception. In the above formula, Jill is taking the part of 'Controller X' to John's 'Controller Y'.

In our notation, *Jill{John{branch-tree relation@branch severed[.]}@John perceives branch to be severed[Give John cutter]}*. The dot in the first square brackets indicates that Jill perceives John's atenfel to be unspecified (she perceives John not to know how or have the means to achieve his objective). The brackets '{ }' indicate that Jill is controlling something (as she perceives John also to be doing), which can only be John's atenfel relation, the only unspecified item within the brackets that define her control loop. She controls her perception of John's ability to control by giving John the cutter, providing him with the missing atenfel. This kind of control of atenfels of one control loop by another requires no innovation to Powers's control hierarchy, though it is seldom addressed explicitly in the PCT literature.

Controllers X and Y need not be in separate bodies. Another controller within John could just as easily have perceived that the cutters would provide the atenfel that would allow him to perceive the branch to be severed, and he could have gone to get them himself (controlling a perception of the location of the cutters to a reference value of 'in hand'). To perceive the branch severed, John might control a sequence perception consisting in part of 'perceive cutters in hand' 'perceive branch between cutter blades' 'perceive cutter blades closed'.

John would not, however, control such a sequence perception if he had not previously reorganised to use the cutters to cut things. Had he never used cutters to sever a branch or seen it done, he might not have been able to imagine that scissoring the cutters was the action that would result in the branch being severed. If we now present a similar situation as a puzzle to be solved (John wants to perceive the branch as separate from the tree, and twisting it by hand has failed to control this perception), John must imagine a sequence that produces a useful atenfel. He has a reference for perceiving 'branch severed', but must produce in another control unit a reference relationship between 'cutter in hand' and 'branch severed', and then control the corresponding perceptions that generate the 'cutter in hand' atenfel.

This simple example illustrates a theme that pervades the study of conscious control — finding the means to influence a perception. Earlier we treated reorganisation by approximation (e-coli) and by random 'try something else' approaches. Here we have another approach, putting together in imagination an atenfel and a so far uncontrollable perception. Indeed, this is the primary function of a formal 'teacher', who often provides 'students' with the means to control perceptions they have not yet had to try to control, as well as assisting them to control perceptions with which they have difficulty. The result is a new component of the reorganised hierarchy, in which provision of that atenfel becomes a perception controlled below the now-controllable one for which this means of control was invented.

Looking from outside at a child trying and failing to control some perception, an adult may sometimes be tempted to say "If you do it this way, it will be much easier", or "Suppose you put that in there, then it would work." This is the same control of perceived relationships between atenfels and assumed perceptions being controlled that we discussed with Jill's provision of cutters for John to use. The adult imagines what perception the child is trying to control and what actions might improve that control, a perception the adult controls perhaps at a level of the hierarchy which Powers called the Program level.

The different perception ascribed above to the adult could equally well be in the child. That perception might include the actions used by the problem control system as well as a larger view of the environment than is available to the first control unit. Just as the adult might imagine how the child might control, this high-level system within the child might in imagination perceive that 'If I put that in there, then it will work', and then act so as to put 'that' in 'there', providing a new atenfel for the problematic perception.

Returning to the 'branch-severing' perception, it may be that what John learns is the 'cutting' atenfel rather than the 'cutters' tool. The next time John wants to sever a branch too big for his hands, he may use a different tool that includes a 'cutting' atenfel among its properties, such as a knife, a saw, or pruning shears.

Considering the problem at yet another level, John may reorganise so that 'searching for a new atenfel' becomes part of his hierarchy, to be used when he is unable to control some perception. In this way, imagination using the World Model may be a direct contributor to reorganisation, and (from the viewpoint of an external Observer) to invention.

An internal Observer may have a more privileged position than an outside Observer, in two respects. Firstly, the internal Observer may have direct access to the perceptual function or the comparator output of the problematic system, in which case it would not need to imagine what perception was being controlled poorly. Secondly, the internal Observer control system might, as its action output, induce localised reorganisation in the output connections of the problematic system so that the newly created atenfel would be permanently available as part of its environmental feedback path.

The adult version of the active internal Observer is the Inventor-Constructor, which in imagination controls an imagined perception together with an environment changed in imagination along the lines of 'If that were across there, then I could control this perception', or in less abstract terms 'If I can find a long enough and strong enough plank, I could put it across the brook and control for perceiving myself to be on the other side.' If there is no plank at hand, this Inventor-Constructor might imagine 'If I put together this and that, it would act just like a plank' and so forth.

Resource limitations that exist in the actual environment do not exist in the imagined environment unless the imaginer imagines them to exist. When the imaginer does imagine a limitation that prevents control of some imagined perception, it might lead to a feeling of impotence, or to further imagining of how to create an atenfel that would remove the limitation. We will explore these and related issues in various places throughout this book.

I.7.5 Alerting

Section I.6.6: 'Resource Limitation Conflict' highlights a situation that can occur when one tries to control more degrees of freedom than are available at the narrowest part of the parallel control loops involved in the conflict. The popular aphorism "Too many cooks spoil the broth" colourfully describes the problem. Since we may want to control many more things at once than we are able — we could be controlling for being seated at the piano, driving to a friend's house, playing golf, polishing the cutlery, and myriad other things, of which we can do only one at any moment — how do we avoid being in perpetual conflict?

The answer to this question comes in two forms. One is that we don't actually avoid the perpetual conflict entirely. In the next chapter (Section I.8.10), we will introduce the concept of 'tensegrity', and argue that mild, non-escalating conflict helps in stabilising our control hierarchy. (This will be further developed in Section II.1.7). Here, we take a different tack and base the discussion on the concept of tolerance (Section I.4.7) in a simple control loop. Tolerance is a variable parameter of the loop, not a property of the loop in the same way as its loop transport lag. The concept is closely related to our everyday concept of tolerance, as 'allowing' other people to do things we don't much like, but it is not the same.

Consider Figure I.4.7a, which shows the relative loop gain of a control loop as a function of how far outside its tolerance bound is the current perception. In the absence of a tolerance bound the loop gain is constant whatever the perceptual value, but when there is tolerance, the absolute magnitude of the negative loop gain rises from zero to its maximum value.

For some part of this rise, the absolute loop gain is less than unity, and the loop does not actually control. It only cushions and slows the effects of changes in the disturbance value. Only when the perception has gone far enough outside the tolerance zone to increase the absolute negative loop gain appreciably above unity will the loop's output function begin to apply much force in a conflict with other loops that simultaneously try to use the same limited degrees of freedom.

Before any of the various perceptions have gone very much outside their tolerance zones and while their loop gains are all low, the low-intensity conflict induces the helpful kind of tension that maintains the tensegrity structure (Chapter I.8) of our control hierarchy. If, however, one or more of the controlled perceptions in the conflict goes far enough out of its tolerance zone to bring its absolute loop gain above unity, that loop will begin to dominate, effectively eliminating the beneficial tensegrity-assisting tension.

Here is where a hypothetical alerting system comes in. It may be, but need not be, conscious. There is no effective limit on the number of perceptions one may monitor if one is not controlling them. Perceptions involved in the tensions of minor conflicts because they are just outside their tolerance zones are not being controlled. In principle, they could be monitored simultaneously with the myriad other potentially controllable perceptions that exist in the hierarchy. In practice there is likely to be a limit, at least if consciousness of the perceptions involved is required. Later, we suggest that the 'Magic Number Seven' (Section II.8.8; Miller1956) may be relevant.

There is, however, a criterion that does not require monitoring perceptions involved in conflict, however mild that conflict might be. Observation of the error trend may be sufficient. The alerting system need only monitor the trend of the reference minus perception value as compared to the tolerance bound value. The comparator necessarily computes this difference in order to determine whether to report a zero or non-zero error value. If (r-p) is trending downward, any conflicts will be reduced in strength, so no action is necessary, but if (r-p) is both trending upward and the error value passed to the output function is non-zero, then that perception is a candidate for active control.

With such an alerting process, active control may be switched off or remain unchanged for most perceptions, but those with upward-trending error (which could lead to serious problems) might be reported for further evaluation, perhaps consciously, to determine whether to discontinue controlling some other perception in favour of the one that triggered the alert, or to ignore it for a while, leaving the ones currently being controlled to continue being controlled. In everyday life, you may consciously experience this phenomenon when a movement sensed in the corner of the eye has led you to flick your eye momentarily in that direction and back again when the rapid non-conscious system has very quickly determined that the movement did not signify a change that might need a change as to which perceptions you were currently controlling.

I.7.6 Exploring and Searching

Figure I.7.6 suggests how a 'World Model' might be a part of the imagination loop. (We consider a 'Fantasy World' in the next section.) In this proposal, the World Model acts as a simulator of what might actually happen to the perceptual signal of ECU B if B's output were to be distributed to its various lower-level RIF connections. In this way, ECU A would be able to control its imagined perception — trying out a variety of possible plans — much more quickly than would be possible in the real world, and without the drawback that the actions involved in a bad plan can irrevocably change the world in which a good plan might otherwise have been effective.

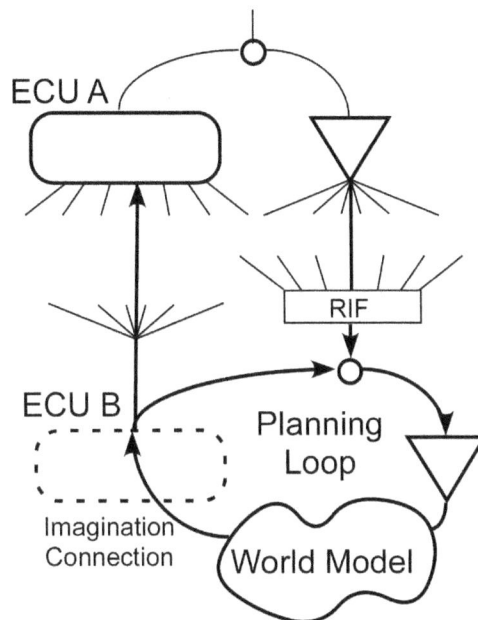

Figure I.7.6. A 'Planning Loop' in imagination that uses a World Model to imagine the potential effect of different values of the output of ECU B.

We can divide the World Model's perceptual part into two distinct components, one based on current sensory input, the other based on memories of past sensory inputs. If the latter were recently sensed, they were as likely to correspond to reality as were any perceptions, but as time passes the environment changes, and memories no longer correspond to what would be perceived if the opportunity arose to re-view that part of the environment. In this sense, the World Model decays over time, or rather its unrefreshed associations with direct sensing of the external environment do.

We are all explorers. We learn things about the way the world works and about the way the world is when we control any perception. If control is not effective, we learn by reorganising or by changing our action to control that perception. If it is effective, we learn that our reorganised system is still compatible with the world. The world has not changed so much that what used to work now fails. And that is the point of exploration. Some things in the world change quickly, and some stay the same for long periods, even lifetimes. Useful exploration finds the latter, which we can include in higher-level perceptions which we want to control without having to go and use our senses to perceive their current state. If the value is unlikely to have changed much since we last found it in the real world, then to bring the perceptual value from imagination is much quicker, if less reliable, than to seek it out again.

Actions that long ago would have had one effect on a controlled perception might now have a different effect. If you want to control a perception of yourself writing a note on a scrap of paper, you may remember that you saw a pencil in your desk drawer yesterday, and opening the drawer, you get the pencil and start writing. But if it was last year that you saw the pencil there, it may very well not be there when you want to use it.

If you had recently observed the pencil while controlling something else entirely, and now you needed the pencil as an atenfel, the fact that you had seen it greatly speeded up your control of the writing perception, as compared to searching your house for a pencil. In general, the more potential atenfels we can perceive in our World Model, the quicker we can control perceptions that we were not controlling when we observed the potential atenfel — provided the environment still contains that atenfel and still enables that means of controlling the perception.

In the example of the pencil, you did not explore for the pencil when you saw it yesterday. Indeed to say you 'explored for' something is a little strange. You explore to see what is there that might be useful for future perceptual control. If you have a current perception you are failing to control because you lack an atenfel in the environment though you have reorganised to use such an atenfel, you don't 'explore for' it, you 'search for' it. You want a pencil in order to write, but you have no pencil in a specific location in your World Model, so you search for a pencil. You want to read but you put your glasses down somewhere and forgot where you put them, so you search for them. In the process, you may see other potential atenfels, such as the pencil. They are irrelevant to your current control of the perception of having your glasses in front of your eyes, but they may become part of your World Model and be used tomorrow.

The externally observable actions involved in searching may be the same as are involved in exploring. The actions are controlling perceptions, in the one case the perception of the atenfel now in position to be used by the corresponding control unit, in the other case the perception of updating the contents of the World Model; in the one case "Ah, there's the pencil," in the other "I know where the pencil is if I need it in the near future." How near is the 'near future'? Close enough that by the time the pencil is needed it has a reasonable probability of still being in the drawer. How close is 'close enough' is determined by the information rate of disturbances that would influence the perception if it were currently being controlled.

Exploration to find *how the world is*, so that it can be used in controlling future perceptions, is no use if the wanted atenfel has changed before it is wanted. Half a millennium ago, Magellan found a passage to the Pacific Ocean without rounding Cape Horn. It is still there, though it may not be there a few million years from now. I found that today there are tulips in the garden, but next week they may have finished. There is no value in storing 'tulips in the garden' as a World Model perception useful in controlling a perception of table decorations that may be wanted next month. On the other hand, if you are searching for a means of controlling for having a table decoration now, 'tulips in the garden' might be a suitable atenfel, a perceived fact quickly forgotten if not changed by your actions in controlling the 'table decoration' perception.

Exploration serves planning; searching does not, except as a side-effect of observing unsought things during the search.

I.7.7 Planning and World Models

We make the imagined situation a bit more of a fantasy and imagine being an astronaut walking on the surface of a dwarf planet such as Ceres, which has very low gravity compared to either the Earth or the Moon, on both of which people have actually walked. If the person doing the imagining knows nothing of low-gravity celestial bodies, the imagined world may not include this change in the physical constraints on walking. But if the imaginer does include the effects of low gravity in the imagined world, it will work differently from the real world, since in the everyday world a moderate jump does not result in flying off the earth into space at more than escape velocity. The perceptions of leg muscle tensions when walking on Ceres must differ substantially from those involved in everyday walking, a difficult thing to imagine correctly, never having experienced altered gravity and seldom consciously perceiving one's leg muscle tensions when walking on this Earth.

What does it mean 'to imagine correctly'? Surely we can imagine anything we want, can we not? The answer depends on whether we want the imagination to perform as the real world would if we were placed in the imagined situation. Our imagined walk on Ceres is an obvious fantasy. Is it imagined in our World Model, where consequences of action match the way we have reorganised our control systems, or is it imagined in a Fantasy World in which we can have any consequences we want? If it is imagined in the World Model of the World in which we perceive ourselves to live, our imagination is called 'planning', even if we never expect to be placed in the situation for which we plan.

Planning is control in imagination at Powers's Program level of perception, the perception of 'if-then' selection of the reference values to be provided to lower-level ECUs. In Chapter II.10 we look at the consequences of the fact that such conditionals of which we are aware are conscious. Those that are non-conscious are built into the hierarchy by fragmenting the relevant category so that the conditional is built into different sub-categories. For example, "If *the bridge is broken* then *we will take the detour*" incorporates two different categories, "unbroken bridge" and "broken bridge", though the actual categories might not involve "broken" as a reason for the bridge not being usable, such as being blocked for maintenance operations, which might have the same effect on 'our' choice of route. If the bridge is unusable often enough when we want to use it, the sub-categories might have been reorganised into the hierarchy.

If *Daphne doesn't want to come,*

    then *maybe Rosalind will be able to join us*

    but *if Rosalind does not come,*

        then *the three of us will go as we are.*

This conditional describes a set of possibilities that are unlikely to have occurred so often as to have been the cause of fragmenting any existing categories (perceptual input functions). In each case, the clause following the 'if' describes a consciously imagined perception, while the clause following the 'then' describes a reference value for a course of action in control of a perceptual instance. A possible failure of control is also described: the action to ask Rosalind is to control for Rosalind joining the group, but if Rosalind does not want to join the group — a perception — then reference values will be supplied for a different set of controlled perceptions.

When one plans for future control in the 'real' world, a useful imagined world should behave as the real world would. This does not mean that one's World Model contains an explicit replica of the real world, but it does mean that the control systems built by reorganisation should function as they would do in active control, even if the reference values with which they are supplied are derived from control of imagined perceptions at some high level. If one imagines moving a large rock to a new place, one might imagine picking it up, but might also imagine that one does not have the strength to do so; one might then imagine using a lever, which in one's imagination might leave unwanted gouges in the ground; one might then imagine hiring some equipment, and so forth. If one is planning rather than imagining an idealised world, one has to imagine the effects on the world realistically. One cannot correctly imagine disturbances that *will* occur when one executes the plan, but one could imagine a variety of *possible* disturbances and imagine the Program so that the disturbances foreseen as being possible are included among the if-then choices.

Plans are Program-level perceptions controlled in imagination. Perceptions controlled in imagination are easily brought to their reference values, but if those plans are executed in the real world, with feedback coming not through the imagination loop but through the external environment, there is no guarantee that they will work as imagined. As Rabbie Burns put it, "*The best laid plans of mice and men gang aft agley*", and in words of General Eisenhower[72] (when he was President of the United States), "In preparing for battle I have always found that plans are useless, but planning is indispensable." There is no guarantee that 'the way the world works' has been correctly imagined, let alone the inevitable disturbances that will be likely to influence the various perceptions the plan uses as lower-level stabilised values. Planning to move the large rock on Ceres is very different from planning to move it on Earth, because although the inertial mass of the rock is the same in both places, the weight on Ceres is very much less. The rock would be imagined as easy to lift, but difficult to move accurately to the desired place. But the plan, a controlled perception itself, would never be tested against reality and never be controlled using environmental feedback.

The imagined walk on Ceres contains no perceptions from the immediately present external world, but the world in which it occurs is no less complex than

---

72 Remarks at the National Defense Executive Reserve Conference, November 14, 1957" t.ly/LFSlY.

the world supplying the perceptions currently controlled by muscular action. The imaginer probably has never even experienced the perceptions that are involved with moving arms and legs while wearing a spacesuit. However, the perceptions controlled in imagination are almost always, if not always, of the same kind as are the perceptions controlled in 'real life'. A colour-blind person probably does not dream in colour, and a person with normal senses does not easily perceive radio waves in imagination; if they do, the radio waves are likely to be imaginatively perceived as though they were visible light. Walking on Ceres requires control of the same perceptions as does walking on Earth, though the magnitudes of the reference values sent to the lower level systems are likely to be quite different.

When we imagine walking on Ceres, we are imagining a situation far removed from anything we have encountered in 'real life'. While we are imagining this fantastical walk, we might in 'real life' be walking the dog in the park. Imagining a fantastical walk does not block control of the perceptions used for actual walking. It does, however, require separating the perceptions resulting from actual walking so that they do not influence the perceptions of imagined walking, and vice-versa. The imagined perception of black, starred Space contains no components of the perception of the park and the dog, and the perception of the park, the dog, and the various perceptions controlled while walking on this Earth contain nothing of the very low gravity of Ceres.

We again have to assume either that planning uses only some fibres from the bundles that are shown as single lines in PCT diagrams, or that a separate world model exists outside the active hierarchy. Later, starting seriously in Chapter II.10, we will be careful to distinguish this separate conscious experience from the non-conscious perceptions represented in those PCT diagrams. At that time we will suggest why concentrating on planning is likely to interfere with the precision or speed of active control, and why active controlling should be expected to interfere with planning. I know of no experimental evidence either way, but subjectively I personally experience interference. When I am trying to solve a problem, I have difficulty acting in any complex manner with speed and accuracy, and vice-versa. Subjective impressions are not science, but in this instance they have to suffice.

As everyday experience tells at least some of us, we may daydream of an imaginary world to such an extent that we fail to control important real-world perceptions. Conversely, we may be so involved with something we are doing in the real world that we have no room for imagining. The implication is that normal perceptions occur by multiple parallel paths, and that some of these paths may be devoted to an imaginary world while others are used in active control. So it is reasonable to suppose that individual perceptual fibres that collectively form a perceptual path might be used at some times for active control and at others for control in imagination. The Powers switch of B:CP Figure 15.3 (2005, p. 223) or Figure I.7.3a above is likely to be applied not to the entire perception, but to individual contributory threads.

Suppose now that the imagined world is not completely dissociated by switches from the real world. Suppose real world perceptual values contribute to an imagined world and output values derived from the imagined world contribute to reference values in the real world. This is not situation awareness. Indeed, it is almost the converse, actions in the real world being determined in part by the comparisons between imagined perceptions and their reference values.

In the absence of a pathological condition such as schizophrenia, if the perception derived from current sensory input is clear, fibres carrying its signal will dominate those carrying a signal derived from imagination, but if the perception is absent or unclear, the impulses from the imagination connection may substitute or support the absent or unclear current sensory input, providing an appropriate perceptual value for the next-level perceptual input functions.

In the extreme, we might be talking about schizophrenia, but the same can happen in more normal conditions. Sometimes it causes surprise, when the real and imagined perceptions of what 'should be' the same thing are appreciably different. One imagines that the door was locked when one turned the key, but is surprised when a push on the door opens it after the handle is turned. Section III.7.4, entitled "The Tail of the Invisible Rabbit", discusses the implications of cultural myths, imagined facts about the real world which are believed by many people to be as true as that the sun rises in the East in the morning. An imagined perception can have the same effects in real-world perceptual control as do any perceptions derived from the senses.

A strong caveat must be mentioned in respect to much of the foregoing. All the perceptions assumed to be divisible between real and imaginary control are conscious. However, we do not ordinarily assume that perceptions controlled in the Powers hierarchy are conscious. It may be that the division mentioned above has nothing to do with dividing the strands involved in active control, but is instead a division between perceptions that are conscious and perceptions that are not. Whatever the resolution of that question turns out to be, the imagined perceptions are controlled in an imagined world, which might have physical and social laws quite different from the ones with which we live.

Imagination is not all about what we might be able to do, but if plans are to be effective when deployed in actual practice, they should also consider what we should not do or would not be able to do based on the contingencies of the environment. The converse of an atenfel is a block. A block either prevents the use of an atenfel or prevents the development of one that might have been constructed in the absence of the block. In a CSGnet message commenting on a draft of McClelland (1994) Powers talked about 'contingencies'.[73]

---

73   In the CSGnet archive, Bill Powers (931210.1145 MST).

A contingency is a cause-effect relationship imposed by the environment. If you drive your car into a tree, the car will be damaged. That is, the condition of your car is contingent on where you drive it. Likewise, if you want to drive from Durango to Denver, you will not arrive at Denver unless you drive on the roads. So achieving the goal of driving to Denver is contingent on driving your car where the car is capable of going. And again, if you want to drive from First Avenue to 30th Avenue along Main Street in Durango, your success is contingent on driving at considerably less than 50 miles per hour; if you drive too fast, you will be arrested.

The first of these contingencies is in the class of natural law: nobody can drive a car (at speed) into a tree (of large size) without damaging the car. There's nothing personal in it; that's just how the world works. The second contingency is man-made, because the roads were built by human agents. They were built along certain routes and not others; they provide access by car to certain places and not to others. By building roads where they are, the builder in effect said, "Here are the ways a person can go by car." Driving to a certain place is contingent on staying on the roads that already exist. Nobody can just build a road to go to any arbitrary place, so the choice of places to drive to is limited, as is the route. There was nothing personal in the choice of where to build the roads; that is, the builders were not thinking of the convenience of any particular person (normally).

The third contingency is also man-made, but it is not a physical thing: it is a social rule. It says that anyone who drives in that place above a certain speed will be arrested. Still nothing personal: the rule isn't aimed at you or President Clinton, but at anyone who exceeds the cutoff speed. Driving from A to B successfully is contingent upon following this rule. In this case, the contingency is not implemented by a physical arrangement of the environment, but by the actions of a person.

The special property of a contingency in relationship to behavior is that it does nothing but create links between actions and consequences. It does not say whether a person should seek or avoid those consequences, or that the person must do or not do the act that leads to them. It just says that if the act is performed, the consequence will follow. A Skinner box is set up so that for every n presses of a lever, a piece of food will fall into a dish. This box in no way says that anything or anyone has to press the lever, or that the appearance of the food in the dish is of consequence to anything or anyone. It just says, "If you do this, that will happen."

In this sense a contingency is an enabling factor for either an atenfel or a block. If you want 'that' to happen, you could do 'this' if you know how. In that case, you would be controlling a perception with a reference value of 'that', and to do so you could use 'this' as an atenfel. But at the same time, you can imagine that a 'bad thing' might happen. It would be a 'bad thing' because it would increase the error in some perception you planned to control, or which perhaps you already control. In Chapter II.5, we will begin to discuss 'rattling', a measure of velocity of change in the values of variables, which tends to be reduced in organisations of active entities. Including one of Powers's contingencies in the message would probably increase rattling and would be unlikely to be included in a final plan, unless it reduced rattling elsewhere in the organisation (of the individual body or the social group that might be affected by the event).

Powers's example of building a road illustrates both atenfel and block. The road builders are expected to control for linking two towns by building a road that runs between them, but not by building a road in some random place. Powers's third contingency, the speeding ticket, is a block that prevents one from controlling for arriving too quickly at a target location. In interpersonal interactions, with which most of this book is concerned, the creation of atenfels and blocks for other people is a major distinction between cooperation and conflict. Everything imagined in Powers's message was part of his personal World Model.

In 2010, Powers had come to think of the 'World Model' as follows:[74]

> *We can sense output force because the tendons have sensors that report how hard the muscles are pulling, and we have pressure sensors all over that detect how hard a hand or foot is pressing against something else. We have sensors to tell us if a joint angle is changing as a result of the force, and of course we have vision to give us a different spatial view of the result. So by experimenting with output forces, we can build up a set of control systems for controlling the immediate consequences of applying forces. We can get to know how much consequence a given amount of force produces. Years later we will learn that the ratio of force to consequence is called 'mass'. But if we integrate the force to produce a velocity, we can discover empirically what the value of this ratio is for different objects, without calling it anything.*
>
> *That's all we need to do to build up a model of the external world. It's not even that; it's just a model of the world. The idea that there's also an external world that we don't experience takes a while to develop. At first, it's just the only world there is....*
>
> *When we examine that external [world][75] in order to model it, we are already looking at the brain's model. It lacks detail, but as we probe and push and peer and twiddle and otherwise act on these rudimentary perceptions, new perceptions form that begin to add features and properties — like mass — to the model. ... Why we have to act one way instead of another to get a particular effect is unknown, but we learn the rules. When we don't get the effect we want, we alter what we are doing until we do get it.*

74   CSGnet archive (Bill Powers 2010.12.23.2300 MDT), "Insight into PCT Models".

75   Here, instead of 'world' Powers wrote 'plant', the conventional term in engineering control theory. — MMT

Powers is not talking about a model of *The way this world is*, he is talking about a model of *The way this world works*. *The way this world is* consists of the set of all the current values of perceptual variables, but that World is the World of Perceptual Reality (PR), not necessarily that of Real Reality (RR, Section I.2.1).

The World Model concept will be used in what follows, without further enquiry into its mechanism. It is just an aspect of conscious control, not of the reorganised hierarchy of non-conscious perceptual control. Later we will enquire a little further into social influences on World Models, with many discussions about aspects of conscious control. If we are planning possible action sequences in the real world, we just assume that imagination can use the lower-level structure that has reorganised (see next section) to work for control through the exterior environment to plan how to alter perceptions in different ways which might be useful in future. When we imagine walking on Ceres, we use imagination of the way the world might work if some Natural Laws happened to be a little different.

Part 3: Interacting Control Loops

Thus far in Parts 1 and 2 of this Volume we have been considering individual control loops, each of which controls a single-valued variable called a perception. The perception is the output of a perceptual function that has inputs both from the outer world through the senses and from the imagination or memory of the individual. It communicates with other control loops, but in a limited way, with loops sending their outputs downward to contribute to the reference values of one or more lower-level loops and receiving its own reference value from one or more higher-level loops, the whole structure resembling a braided stream rather than a hierarchy. Nevertheless, the word 'hierarchy' has become embedded into the vocabulary of Perceptual Control as proposed by Powers, and we shall continue to use it.

In Chapter I.8, the first chapter of Part 3, we discuss some simple motifs, on which more complex ones may be built. A 'motif of control' refers to a pattern or interconnection structure that is frequently seen, and has a particular function involving an emergent property of the motif. For example, what we may call a minimal hierarchy is a control motif. The control hierarchy is constructed from multiple minimal hierarchies. Each minimal hierarchy consists of only two levels of control in which the perceptual function of a controller has as inputs the perceptual outputs of at least two of the subordinate controllers and contributes its output to the reference input of at least one subordinate controller.

In Chapter I.9, we depart from the austere Powers hierarchy in which lateral interconnections among control units are disallowed, and allow them without otherwise altering the hierarchy as so far described. In particular, we propose that lateral inhibitory interconnections may be developed by reorganisation, and that these serve to enhance the discrimination between perceptions of similar aspects of the environment, even in some cases to the extent of producing perceptions of categories.

Category perceptual control is one of the levels of the Powers hierarchy. We, however, recognise that every perceptual function necessarily describes a category, and the perception that is controlled at any moment is a property of an instance of a category. What Powers describes as a Category is a consciously observed type with clear boundaries between related categories (Section I.9.7). Lateral interconnection allows category perception to be displaced so that it is not a level on its own, but is a possible way of perceiving at any level, whether it be of a colour that is the category 'red', and not 'purple', or of a political party that is 'socialist' and not 'communist'.

Precisely defined categories are conscious perceptions in contrast to the categorising done by perceptual input functions in the non-conscious perceptual control hierarchy. It is tempting to assign category perceptual control preferentially to (simplistically speaking) the thoughtful operations of the left half of the brain and analogue perceptual control preferentially to the pattern-preferring right half, but this book explores the functional implications of perceptual control, and does not go into the neurophysiological correlates of control. We leave that to others who are much better informed on the matter than this author.

Chapter I.8. Motifs of Control, Stiffness, and Tensegrity

When we have multiple controllers that interact, as they do in a hierarchy, there is an opportunity for new 'Motifs' to show up. What do I mean by a 'Motif'? The Oxford English Dictionary (OED) definition includes "*In painting, sculpture, architecture, decoration, etc.: A constituent feature of a composition; an object or group of objects forming a distinct element of a design;...*". In this book, I mean something very similar, but I include something extra, which I think is implicit in the motifs mentioned in the OED. Each Motif recurs relatively frequently compared with the myriad other possible structural patterns, and is likely to use the emergent property associated with it for its particular function within its context. That function is the same each time the structure recurs. Many Motifs in this book, especially the more complex ones in later volumes, have very precise functions. I give some of these Motifs labels such as 'Protocol' or 'Trade'.

The function of a Motif is that it has an emergent property that is used elsewhere in a way that is stable throughout reorganisation, a topic we discuss beginning in Chapter I.11. I do not mean that a Motif is immune to being reorganised out of existence, because that can happen no matter how useful the structure is. I mean that when components are organised into a Motif, the probability that they will be discarded or modified during reorganisation is low compared to when similar components are differently organised. A Motif has an 'emergent property' (an 'emergent') that is unique to that structure or more complex ones which are elaborations of it or which use it in its entirety.

The Motifs that will be important later in this book are all built from one basic Motif — the control loop itself (the only Motif not constructed from control loops). 'Control' is an emergent of a particular structural relationship among component parts that have specific functions. Nothing interesting is created by simply plugging randomly together a comparator, a perceptual function, a time-binding function that uses an external energy source to amplify its input, and an agent that acts on the external environment, unless that random plugging of bits and pieces together by chance happens to create a very specific circuit, the control loop. 'Control' is an emergent property of the *structure*, not of the components of the structure.

The structure is the motif, and one motif can be incorporated into another with an unrelated emergent function or property. We will see this often, as the motifs with which we are concerned become ever more complex in the later volumes where some motifs of control extend across individuals. But the simple motifs with which we start here occasionally have surprising emergent properties.

Structures can have both static and dynamic emergents. 'Stiffness' is a static emergent property of a motif consisting of opposing forces. Such forces may exist within a control loop because of the fact that neurons cannot fire at negative rates, or it may be a property of interacting independent control loops. For 'stiffness' to emerge requires tension between two opposing forces. In control, stiffness requires persistent error in two controllers acting on the same environmental variable but with different reference values for their perceptual values. This may also produce 'conflict', another dynamic emergent of the same structure.

'Conflict' can also arise when two controllers influence the same environmental variable to the same reference value, but perceive it differently so that their error values are opposed and they act on it in opposite directions. The structure and its static and dynamic emergents are the same, but the mechanism is different.

A small to moderate tension or persistent error is necessary if a structure is to be able to act rapidly to counter extrinsic disturbances or changes in the reference values in question. In more homely terms, you act more quickly and accurately if you are wide awake than if you were just roused from a relaxed slumber. The small tension matters.[76]

The third emergent property we discuss quite often in this book is 'tensegrity', which depends on some of its parts resisting bending. Tensegrity is a property that can be realised by a hierarchic control structure of at least two levels, in which the 'stiffness' motif occurs at least three times if the control hierarchy acts in a 3-D environment, which I treat as minimal. As we discuss toward the end of this chapter, persistent error becomes important simply to maintain the viability of the two-level hierarchy of a minimal 3-D control tensegrity structure. Since a 'dimension' of control represents the possible outputs of an arbitrary perceptual function at any level of the Powers hierarchy, there is no inherent limit to the number of dimensions involved in the tensegrity structure of a control hierarchy.

This will all be explained as we go along, but the key point is that if a control hierarchy is a structure of tensegrity modules using tensegrity modules as building blocks, it will be resistant to many kinds of insults because the stresses on a disturbed control loop will be distributed throughout the structure rather than all the corrective action being the responsibility of a single output function. The rest of the structure increases the ability of the disturbed system to control and the power available for it to do so. We will call the tensegrity of a control hierarchy 'control-tensegrity' to distinguish it from other domains in which the elastic resilient strength property of tensegrity is found. We mention five of these domains later in this chapter, including 'psycho-tensegrity', a conscious manifestation built on 'control-tensegrity'.

I.8.1 Control 'Stiffness'

'Stiffness' is usually thought of as a resistance to bending or buckling. In control, however, what is there to be bent? A control loop can be thought of as 'pulling' an environmental variable towards a point at which the corresponding perceptual value equals its reference value, as though the perception were being pulled at the end of a wire toward its reference value.

---

76   See Powers' model of muscle tone, Bill Powers (920722.0800) in the CSGnet archive.

Something can be metaphorically bent in a 'classic conflict' (McClelland 1993), where two control loops pull the environmental variable in opposite directions. Imagine a tug of war. Two teams pull on a rope in opposite directions, let's say Eastward and Westward, each trying to move a marker on the middle of the rope to their side of a neutral zone. They are not allowed to move sideways, North-South orthogonally to the rope, so if some force such as a wind would tend to move them sideways, they resist it. The same is true of a North or South force applied at the middle of the rope. Such a force would move them in its direction if they did not resist it (Figure I.8.1a).

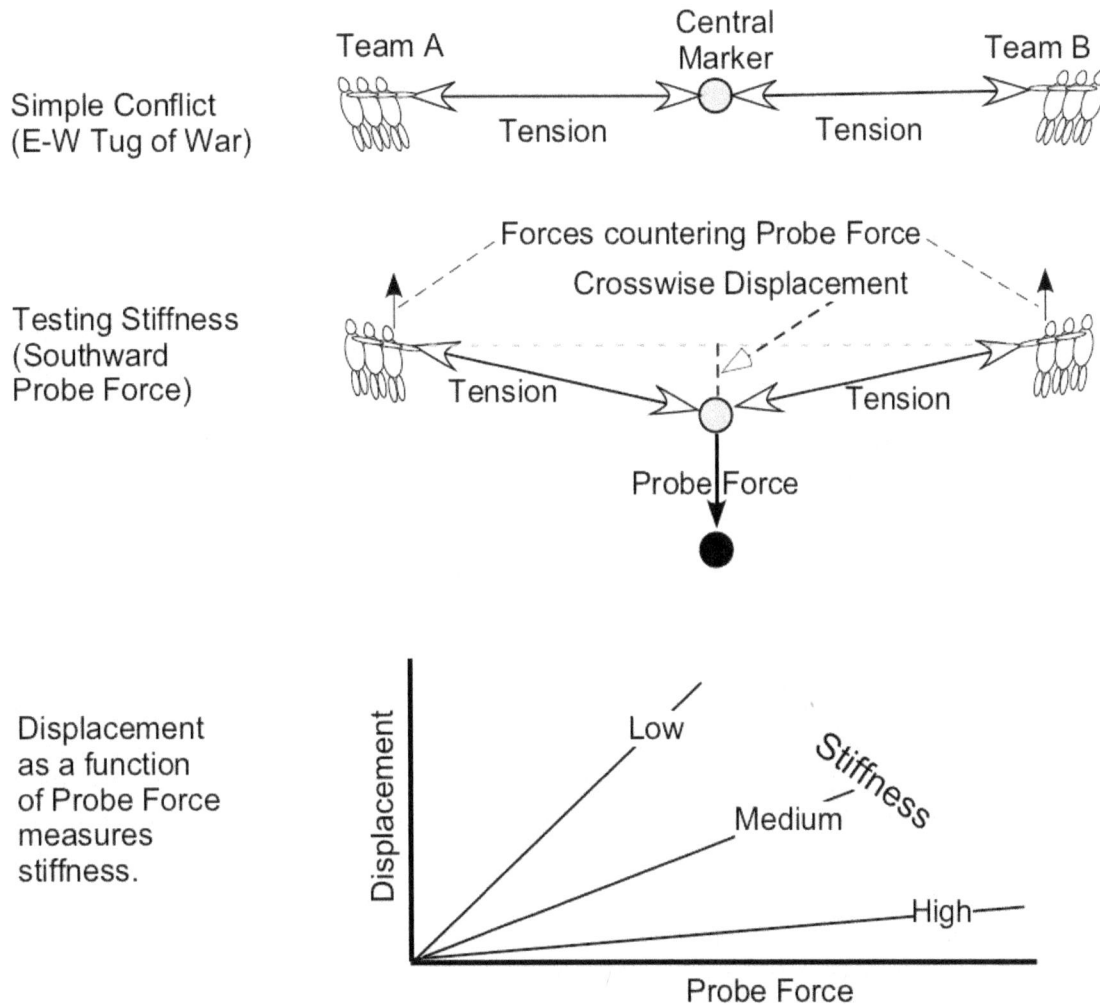

Figure I.8.1a Testing stiffness. A tug of war stiffens the rope so that it resists bending at the point where a lateral probe force is applied, just as a stiff rod would. The lateral force is balanced by the force each team uses to prevent it from being displaced laterally. The opposition to the probe force is distributed between the two teams. The more strongly the teams pull on the rope, the less will the lateral force displace the point where it is applied.

When a connection, such as a rope or wire, is pulled from both ends, it will get a little longer. It stores some energy. In basic mechanics, force times the distance over which the force moves whatever it does move is a measure of the energy used, which in this case is energy stored in the rope. How much longer does the tug-of-war rope get when the two teams pull with how much force? That is a measure of the capacity of the rope to store energy per unit of force applied (called its Elastic Modulus).

Now a side force is applied while the competitors are pulling against each other. If the tug of war teams maintain the force they are applying to the rope, and do not move when the side-force is applied, the side force will add length to the rope, increasing its length and its energy storage. How much length is added? The crosswise displacement C, half the rope length R, and half the distance D between the teams together form a right-triangle, so $R^2 = C^2 + D^2$, and R/D is the proportionate increase in the length of the rope. The added energy to be stored in the rope is the crosswise force times the distance C over which it is applied in order to displace the rope centre as far as it moves (assuming the opponents do not move).

In the discussion of the comparator (Section I.4.6, Figure I.4.6b, and Figure I.4.6c), we noted that many comparators have to deal with both positive and negative values of r-p (reference minus perception), and possibly of r and p separately. Values of variables in a control loop are carried by the firing rates of neurons in a neural bundle, and firing rates cannot be negative. To handle the negative values, we postulated that inhibition might be treated as having the effect of subtracting from excitatory inputs when the firing rates are conceptually combined into the 'neural currents' often used in diagrams and analyses of biological control loops. We now extend those thoughts further around the control loop, and examine what might happen if the two separate error outputs of Figure I.4.6c fed distinct output functions whose outputs had opposing influences on the CEV in the way that the extensor and flexor muscles have opposing influences on the angle of their joint. Figure I.8.1b sketches such a loop structure.

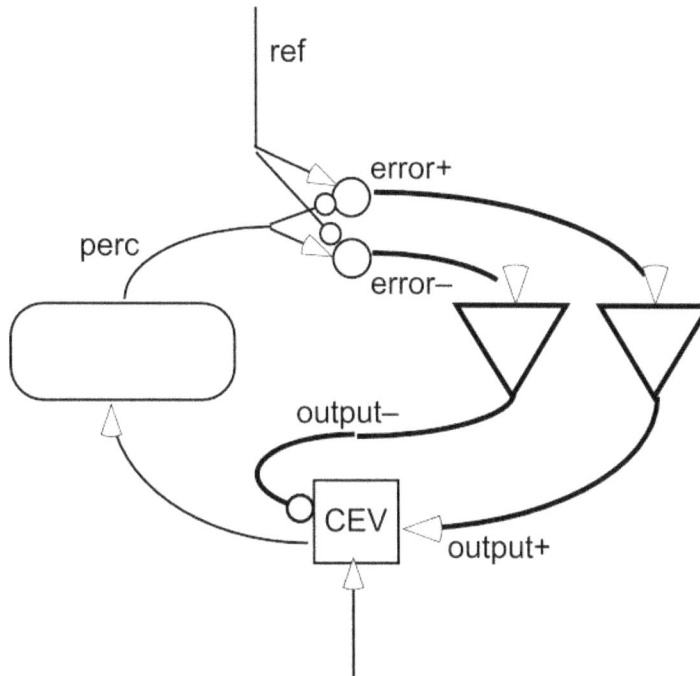

Figure I.8.1b The two error outputs serve separate output functions, one for influencing the CEV one way, the other for influencing the CEV the other way. The two outputs serve opposing muscles, such as the elbow tensor and flexor. For simplicity, the comparator is shown as it would be for perceptions and reference always positive, but the same two outputs would exist for a comparator with the full complexity of Figure I.4.6c.

The two outputs might drive tensor and flexor muscles, but how they are actually used depends entirely on the nature of the CEV and the atenfels of the environmental feedback path. All that matters for what follows is that there are two output functions rather than one, and that they have opposed influences on the CEV. One question we will consider is whether such an arrangement is likely to be the normal case for a control loop, and the implications if it is. For example, it does seem that most of the skeletal muscles in the body, if not all, act in opposed pairs.

First, we consider the degrees of freedom around the loop over time. At any one moment, all the signals such as the perception, the reference, and the disturbance are single-valued variables. The error seems to have two values, one that is greater than zero if the error is positive, the other that is greater than zero if the error is negative. So there is really only one value, which passes smoothly between positive and negative. This single value, however, cannot be expressed as a rate of nerve firing that passes smoothly from positive to negative. It must be expressed as a difference between two positive rates of nerve firings.

What about the output? Are its two values independent? To answer this question, we must consider 'degrees of freedom' more closely, though in a simplified manner. The error has only one degree of freedom shared between its two signal paths, because only one at a time can have a non-zero value. Over time, the values vary up and down, but those changes are smooth over a long enough time-scale to allow the impulses to be averaged into a rate and not infinitely fast.

If one observes a value at time t_0, that value represents one degree of freedom. The signal value very shortly after (whether shortly means nanoseconds, minutes, or years) will be almost the same. After a slightly longer period, the value will have changed unpredictably, but will still be somewhere near the value at t_0. If you wait long enough, though, the value at t_1 will have no relation to the value at t_0, other than that they both belong to the same overall probability distribution of values. When this is the case, t_1 represents a new degree of freedom. Waiting long enough again results in a new third value that represents a third degree of freedom, and so on and so on.

In a time interval of duration T, N of these independent measures can be packed together, but no more. The value of N depends on many variables, some of which we will encounter later, when we discuss uncertainty and information. If you try to add another, its value could be computed from the N tightly packed original values. All the values at any moment in that interval can be exactly calculated using just the values of these individual degrees of freedom.

The packing might vary, but however they are packed, there are always the same maximum number N of independent values that can be used to describe the waveform over that time interval. We say that the independent sample rate is N/T samples per second, and that there are N/T degrees of freedom (df) per second. Every signal waveform can be characterised in this way. So the error function has, say, N_e df/sec, the perception has N_p df/sec, and so forth. These values may seem to differ because the error degrees of freedom are split between the two signal paths. The total remains the same, however, because when one of them is varying, the other is a steady zero, which has zero df/sec.

The output from the two Output Functions (OFs) is different. Let us suppose that each branch has its own leaky integrator. If the error has ever changed sign, each OF will have received some positive input some time in the past, though it may not be doing so at any particular moment. Indeed, only one can be receiving non-zero input at that moment. So we know that one output function integrator is currently only leaking, while the other is also integrating incoming positive values. (As noted, all incoming values are positive; both have a non-zero positive value at all times.)

Can we predict the value of one OF output from the other? No we can not, because both of them depend on the entire history of when and by how much the error values were positive on their respective inputs. Those outputs look as though they are independent, though we know that they cannot have more degrees of freedom over time than the total of their inputs, which is N_e df/sec. If we accept that the two outputs at any moment are truly independent, where did the extra apparent degrees of freedom come from?

The answer is that they were stolen from history. By simultaneously specifying the two outputs at a given moment, one degree of freedom less is available for independent specification of their values over deep history. Knowing the history of the error values, one could compute the values of both OF outputs, so knowing the current value of one together with enough history, the value of the other could be computed. Without the history, the two outputs are effectively independent, and sometimes, including in the following discussion, I will treat them as though they were truly independent.

The two outputs might be considered as the outputs of two separate controllers in conflict. The CEV moves according to the difference in their momentary influences. The greater the integrated value of their historical inputs, the greater the output values, since they will not have been able to leak to as low values as they would if the disturbances historically had been easier to oppose and the perception had been kept closer to its reference value. The implication is that strong opposing forces on the CEV from the two outputs is a measure of control difficulty due to rapid variation in the disturbance. The combined strength of the two outputs is an indicator of recent control difficulty, whereas their difference is a measure of the current influence on the CEV. Algebraically, $OF_1 - OF_2$ is the effective output in the corresponding 'canonical' control loop of Figure I.1.2b.

So what is $OF_1 + OF_2$? It is a measure, which we might call instantaneous 'stiffness', which does not exist in the canonical loop. It is a new emergent property that requires a structure that includes opposing forces before it emerges.

The following examples may clarify the concept of 'stiffness' in this context of muscular opposition. In Figure I.8.1c panel d, the usual approach-avoidance conflict is represented in a single dimension, a single scalar environmental variable being influenced in value by the conflicting controllers. However, for a situation in which the controlled perception is always the distance of an object from some reference location, both object and reference location lying on a plane surface, the object whose property is being controlled as the CEV may allow changes in two or more dimensions that interact, as suggested in Figure I.8.1c panels b, c, e, and f. In the top half of the Figure, the situation is the control conflict with the reference locations marked by dots; the bottom half shows an analogous mechanical configuration.

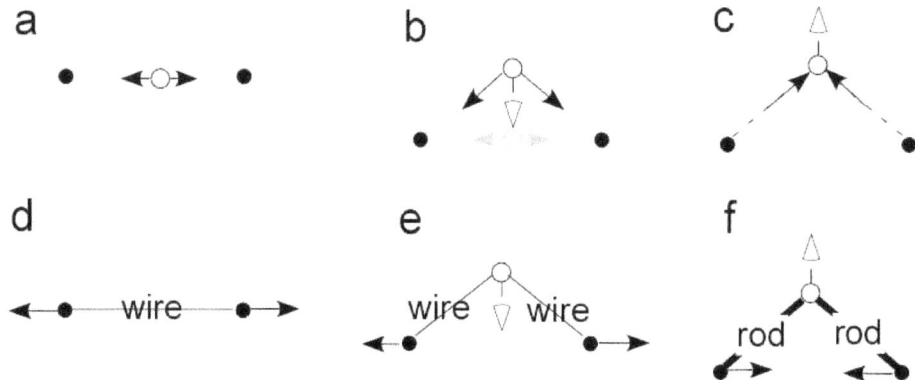

Figure I.8.1c Mechanical analogy to approach and avoidance conflicts on a plane (a, b, c, conflicts, d, e, f, mechanical). In the "conflict" panels, the solid black dots represent reference values (locations in some abstract space) toward which controllers try to move the perceived value of the object property represented by the open circle. (a) Approach conflict when the object is located directly between the controllers; (b) Approach conflict when the object is off the line between the controllers, but the control action pulls the object into line; (c) avoidance conflict when the object is off the line. In this case the control actions cease to be in conflict because both influences increase the distance of the object from the controller; (d) two engines pulling on a wire in tension; (e) two engines pulling on a slack wire bring it into tension; (f) two engines pushing together, with a rod that had been holding them apart now bent at the centre and no longer preventing their motions.

Panels a-b and d-e of this figure show control and mechanical stiffness, respectively. The greater the gains (panel a) or the stronger the pull (panel d), the stiffer the CEV is against orthogonally oriented disturbances. Panels b and e also show the stiffness manifest as a tendency to align the CEV with the reference values in the control case or the fixed points from which the force is applied in the mechanical case. Although the forces (or the control actions) are only in the X direction, if the points from which the force is applied are fixed in the Y direction, the pair also diminishes the ability of a disturbance to move the CEV in Y. In all these cases, the existence of conflictive collective control in the X dimension induces a stable equilibrium state in the orthogonal direction (or directions in a higher-dimensional situation).

The interesting panels of Figure I.8.1c, however, are the right-hand pair, c and f. In c and f the stiffness is *negative*, and the induced equilibrium is unstable. Any disturbing force orthogonal to the direction of the opposing forces will result in collapse of the structure, with the CEV being pushed ever further from the line between the points from which the push is directed. If the object is exactly aligned between the two influences or forces, the situation is not going to change except for the escalation of output due to any integration of error in the control units, but if, as shown in c and f, the object is at all misaligned, the avoidance output forces in c or the compressive force in f will push the object even further out of line, thus eliminating the conflict or the impediment. The collinear arrangement is in an unstable equilibrium, like a pencil balanced on its point.

None of the panels shows control in two dimensions, but they do show how control in one dimension can affect the nature of the equilibrium in another. The strength of that equilibrium — the force required to move the CEV a given distance from the stable point — depends on the total of the forces applied in the opposed directions. That equilibrium strength is the stiffness of the complex.

To make the 'stiffness' property less abstract, let's consider a couple of examples. First, think of the opposed muscles that influence the joint angle of Figure I.8.1d panel c, and imagine a varying force on the 'hand' at the end of the 'arm'. As the disturbing force varies direction, first one and then the other muscle is needed to oppose it and keep the joint angle constant. But when the disturbance switches direction, the muscle that was opposing it does not relax instantly. It responds not to the joint angle or the disturbing force, but to the output from the Output Function to which it is connected. That output is from a leaky integrator that reduces its output slowly, so the muscle retains some tension. The same is true for both muscles, so the more often and the more wildly the disturbance changes direction, the poorer the control, the greater the error in each direction, and the more tension remains in both opposed muscles at all times. The joint 'stiffens'.

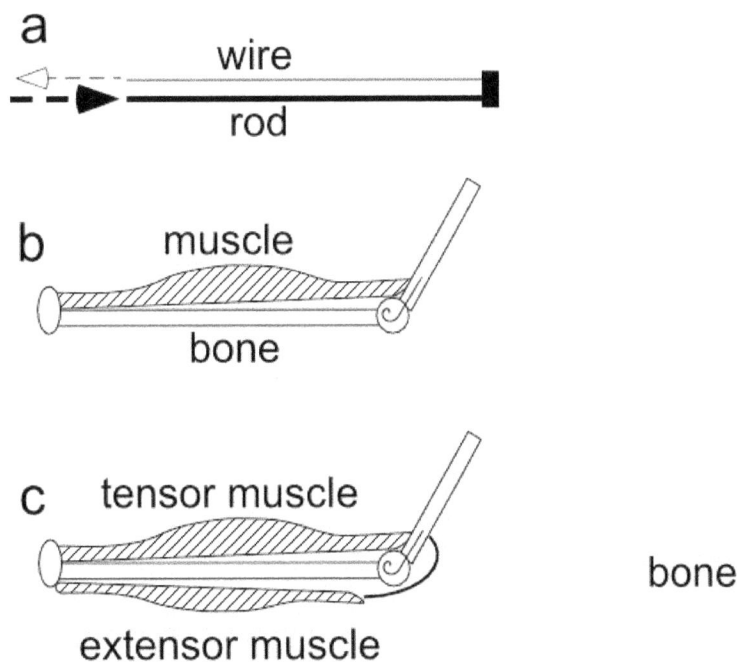

Figure I.8.1d A mechanical push-pull structure used extensively in organisms with skeletons. This is the simplest possible one-dimensional tensegrity structure. In (b) a spring opposes the force of the muscle by increasing amounts as the joint is flexed, whereas in (c), a second extensor muscle (wire) is used to pull against the tensor muscle. The latter is more efficient, since with the spring, the muscle always has to be tensed to keep the joint at a particular angle, whereas with the two-muscle system, both muscles can be relaxed at any joint angle.

This artificially simplified elbow joint may still be a little abstract, and it has only one angle through which it can move. For a more real-world example, imagine trying to carry a flag at the head of a parade on a gusty day. You are likely to hold the flagstaff tightly with both hands, as the gusts blow this way and that. The wind does not simply blow from the East and then from the West. It may be predominantly westerly, for which you can easily compensate by applying a steady easterly force with little tension, but along with this predominant direction are variations in all directions, and against those, there is little you can do in a relaxed manner. Your tight hold makes it harder for a northerly gust to tear the flag from your grasp, even though you are primarily controlling against a westerly disturbance. Contrast this with how you would hold the flagstaff on a calm, sunny day. Probably you would apply only enough force to hold the flag against gravity with one hand, allowing you to wave to the crowd with the other if you wanted.

Stiffness is an emergent property of a motif of control that requires a minimum of two forces in opposition. This arrangement of controllers is also a minimal system that illustrates conflict between the controllers over the variable in the environment that corresponds to the perceptions being controlled by the conflicted parties. The stiffness property, on the other hand, is relevant to a quite different variable, whose variation is uncorrelated with that of the conflicted variable. The two independent variables are properties of what is perceived as a unitary object, such as the flagstaff held by a parade leader or the ribbon tied in the middle of a tug-of-war rope.

We move on to a more complicated motif, a control hierarchy with a new emergent property, tensegrity.

I.8.2 Five Domains of Tensegrity

The word 'tensegrity' was introduced by Buckminster Fuller to describe physical structures whose structural integrity was enabled by the tensions in wires that connected the ends of otherwise free-floating compression elements, which I will call 'rods', because in many physical tensegrity structures, that is what they are. These compression elements must be stiff, so that they do not buckle under compression. Figure I.8.2 shows a physical tensegrity structure, placed as a piece of public art in Buffalo, NY, USA.[77]

---

77   More photos in Snelson and Heartney (n.d.), p. 57, which provides a wealth of illustration and informed discussion of tensegrity structures. See also the Wikipedia article "Tensegrity". — Ed.

Figure I.8.2 A tensegrity structure on a plaza in Buffalo, NY, USA.
(Credit: Snelson & Heartney, n.d.)

Tensegrity is ordinarily considered to be a mechanical property of a certain kind of physical structure built from compression elements I call 'rods' and tension elements I call 'wires'. A tensegrity structure has several properties that are worth noting up-front. A physical tensegrity structure is very light for its strength, compared to bulk material. It is able to withstand forces from outside by distributing the imposed energy throughout its structure. It can store energy and can return it to its environment slowly or explosively, depending on the structure. It is resilient, bending rather than breaking under stresses, like a tree that can withstand a hurricane.[78] If the tensions on its wires can be changed individually, a tensegrity structure can move in ways a rigid structure cannot, and it also can be made more or less rigid in whole or in part by coordinated changes of its wire tensions.[79]

---

78   The tree actually is a biotensegrity structure (Swanson 2013, Scarr 2014, Levin 2015).

79   NASA's 'Super Ball Bot' (https://www.nasa.gov/image-article/super-ball-bot/) is designed to crash-land on a planetary surface and move about by changing the lengths of tension units ('wires') in its structure. — Ed.

These are a lot of desirable properties, and we will be claiming throughout this book that they are properties to be expected of hierarchic control structures as well as of social structures such as culture and language. The implication will be that these structures would not be expected to be more than approximately amenable to a single formal description such as the 'Grammar' that children may be taught in school.

Rods must be stiff to avoid buckling under compression, while wires are stiff by virtue of the opposition of 'pulling' forces, as described in the last section. In a classical tensegrity structure, rods do not contact other rods, their ends only being connected by wires. Each rod end has at least as many wires attached to it as there are dimensions in the space in which the structure lives (typically three in a physical structure), so that the rod end can be moved in any spatial direction by changes in the tensions in the attached wires.[80]

Tensegrity structures not resting on some surface will hold together under external forces such as gravity only if they are properly constructed (by design or by evolution) and if there are adequate tensions in the wires and balancing compressive forces in the rods. If your legs are totally relaxed, you cannot stand. In this and the rest of this chapter, we will, however, be increasingly concentrated on intangible tensegrity structures built of control loops. These are implicit in the hierarchic perceptual control structure proposed by Powers, and also, as we shall see in Chapter I.10, in an engineering reorganisation process suggested by Wiener in 1963 for discovering the internal processes of an impenetrable 'Black Box'. For these, 'gravity' does not apply, though as we shall see in Volumes II through IV, similar 'globally attractive' analogues to forces can occur within community structures.

There is no reason why the rods and wires of a physical tensegrity structure must be solid metal or other material. Each rod and each wire could itself be a long linear tensegrity structure, thin for the 'wire' elements, thicker for the 'rods' to keep them from buckling. Such elements would be lighter and less brittle than the corresponding solid elements. Furthermore, these component tensegrity wires and rods could themselves be built from microscopic tensegrity components, and so forth down to to scale of individual molecules such as proteins, which could be shaped to have internal attractive and repulsive forces arranged to create tensegrity effects that may have influences on the loops formed by interactions involving hormones and enzymes.

Scarr (2014) argued that our physical bodies are built just this way, down to and including the structure of the molecules used as the material of flesh and bone. He entitled his book Biotensegrity: The Structural Basis of Life. Furthermore, as Scarr pointed out and as we described in the last section,

---

80 You may notice that the three rods that form an inverted pyramid at the base of the structure in Figure I.8.2 have only two wires connected at the upper end. There is, however, a virtual wire in the form of gravity, which pulls the upper rod end downward just as would a wire in a structure floating in space.

'stiffness' is a property not only of solid rods, but also of pairs of wires pulling in opposing directions. We use this property when we discuss a minimal 3-D control tensegrity structure below.

Biotensegrity differs from other examples of tensegrity only in that the rods and wires are parts of living systems. This leads one to enquire what really matters about a tensegrity structure. The material clearly does not matter, so long as at any particular scale the components can come in two linear forms — a form that holds its ends apart against applied forces that would push them together (rods) and a form that tends to pull its ends together if they are pushed or pulled apart (wires). The opposition of these two forces enables the construction of stable tensegrity structures.

We have mentioned two kinds of tensegrity that differ only in the materials of which they are made and the context in which they are deployed: physical tensegrity or simply 'tensegrity' and 'biotensegrity'. We will add three more kinds of tensegrity, which I will call 'control-tensegrity', 'psycho-tensegrity', and 'socio-tensegrity'.

'Control-tensegrity' refers to structures created by control loops that have been incorporated into the Powers hierarchy by reorganisation. It is unlikely to occur in conscious perceptual control, because (as is argued in Section II.8.8) even a minimal tensegrity structure would contain too many components to be accommodated in a single thought. Control-tensegrity deals with tensegrity effects that can be attributed to control of perceptions in the reorganised non-conscious perceptual control hierarchy. 'Psycho-tensegrity' applies in the domain of consciousness, where words such as 'thinking', 'imagining', and 'planning' are typically used. These are two aspects of the mental life and behaviour of an individual. 'Socio-tensegrity' applies in the domain of interpersonal interaction. It is a catch-all phrase because this domain has been very little studied within PCT, and further study may well suggest that it should be subdivided into different domains, either in parallel, depending directly on psycho-tensegrity or less complex levels of tensegrity, or forming new levels, one depending on its predecessor as happens in the other four domains of tensegrity.

Just as bio-tensegrity can be seen as an instance of tensegrity in general, so each of the others is closely related to its antecedents, though not always being an example of it. The important thing to note, however, is that while control-tensegrity loops through the internal or external environment that contributes to the perceptions controlled by its component control loops, neither psycho-tensegrity nor socio-tensegrity need directly involve the material world at all. Psycho-tensegrity structures may be built entirely in a fantasy world in which the fantasist uses entirely different Laws of Nature, whereas for example socio-tensegrity structures might use antagonisms between hostile groups as 'rods', and affiliation to political parties as 'wires'. We will leave the effects of socio-tensegrity and much of psycho-tensegrity for later volumes.

I.8.3 Basic Tensegrity

How can a control hierarchy be resilient — strong yet flexible — so that it can cope with an ever-changing natural and social environment? One answer we shall explore in some depth is that it has the same characteristics as those of a mechanical tensegrity structure. So let us examine what that means, simply at first, and then in more detail.

Having loosely described tensegrity structures in five different domains which are distinguished from each other by the constitution of the rods and wires, we must return to the question of what the word 'tensegrity' might mean that is common to all these domains. Clearly, it cannot be material, so it must be functional, in the same way that one does not ask "Two of what" when told "Two plus two equals four." The function 'plus' by design ignores what, if anything, is being counted, in the same way and for the same reason that Claude Shannon designed his measure of communication channel information rate capacity to be independent of what kind of meaning the message might have.

In tensegrity, the function of a rod or of a wire is to resist an influence that might change the 'distance' between entities at its two ends. The rod acts against forces that might bring its ends nearer to each other, while the wire acts against forces that might separate its ends further. This can make sense only if there is some concept of 'space' within which the concept of 'distance' becomes meaningful. In the Powers hierarchy, it could perhaps be the magnitude of a relationship perception, a controllable quantity. Among a group of people, 'distance' might represent dissimilarity between two people in their opinion on some subject, or the dissimilarity of the official position of two political parties on some item of policy. A 'wire' in this last example might be some pressure that tends to bring their official positions closer, whereas a 'rod' would be some tendency to increase their difference. Socially, efforts by one party to prove to the public that their position is 'better' than the other are likely to create such a 'rod' between the parties.

It does not matter what the domain is or what the influence is that tends to increase or decrease the 'distance', but for the idea of 'distance' to make sense the two ends of a rod or wire must represent perceptions of the same kind. 'Distance' might be frequency of texting between two people, perceived difference between two perfumes, political balance among several parties, average positivity of reviews of two movies, or anything else that might, in principle, be measured quantitatively. Again, it does not matter whether the distance measure represents a perception that is controlled by an individual or collectively, influenced by side-effects, or otherwise altered when some other perceptions are controlled. Whether the situation is structured as a tensegrity structure must be determined for each occasion. Functionally, just as in the world of material rods and wires, tensegrity potentially stabilises the interactions among control systems within people or among groups of people. So let us look more closely at tensegrity in the abstract, without committing it to any particular environment.

The structure in Figure I.8.2 is static. It is strong in that if it were struck or pulled from any direction, it might bend slightly, but so long as its 'feet' stayed firmly anchored, to break it would require forces substantially stronger than would break any single rod or wire. It could hold a lot of weight if a heavy object were placed on it in some kind of basket or sling.

Despite its strength, the structure is very light. If the rods happened to be thin-walled aluminium tubes, one person could probably lift it. (They are in fact stainless steel.) Figure I.8.3a illustrates in a 2D diagram how the redistribution of tensions among several wires can balance a laterally applied force on a tower structure. Weight-bearing structures can be built with non-touching straight rods, as attested by the twelve-metre tower shown in Figure I.8.3b.

Figure I.8.3a illustrates another feature of tensegrity structures.[81] They can be reconfigured from within, by changing the tensions on the wires. Suppose that the structure were not subject to an externally applied lateral force, but instead the wires up the right side were tensioned a little each, while the wires up the left side were slightly slackened. The tower would bend to the right, exactly as in the illustration.

Scarr (2014) noted that living bodies do this all the time, changing the tensions on muscle fibres to make the parts of their bodies move. In PCT, these tensions are the environmental variables of loops that control their perceptions, as Powers noted early in B:CP (Powers 1973/2005). Their control is the way we move, no matter how ballerina-graceful or clodhopper-clumsy we might be.

---

81   The 2-D diagram of Figure I.8.3a is not of a conventional tensegrity structure, since the 'rods' are bent, and therefore sustain forces at the bend that are not present in a standard tensegrity structure. It nevertheless illustrates the point at hand.

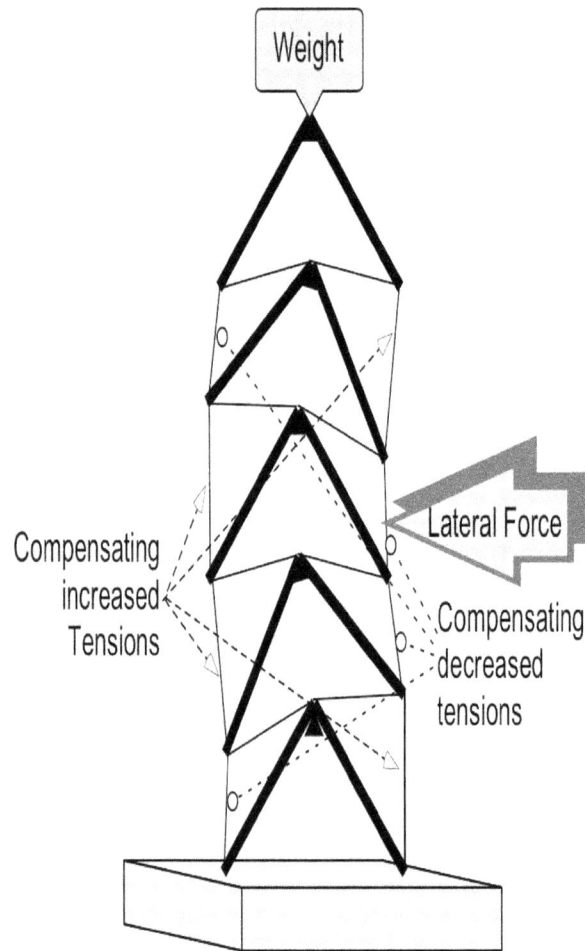

Figure I.8.3a A structure in a 2-D space that can support a weight and can bend without buckling either by changing the tensions on one or more lateral (vertical) wires or, as shown, when a lateral force is applied. The main stresses are taken up by the crossing (horizontal) wires, not by forcing the "bent rods" against one another. It is not a pure tensegrity structure, but shares some properties of one.

Figure I.8.3b A 12m tall tensegrity tower in the Science City, Kolkata, India. Each rod end connects to seven wires. (Photograph by Biswarup Ganguly, used under the Creative Commons Attribution Unported 3.0 Licence.

The considerable strength-to-weight ratio of a tensegrity structure might be useful for building, but currently is not widely deployed. At least, it is not widely deployed in houses, bridges, and skyscrapers, though its principles were much used in houses built hundreds or thousands of years ago in earthquake-prone zones. Such houses were built with a sturdy wood frame with the walls filled in by solid material. The wood could flex in the earthquake, and limited the propagation of cracks in the solid walls. After the quake, the house would be standing, even though some of the solid wall surface might have cracked off. Tensioning cables in concrete structures perform some of the same service, but do not block crack propagation. Other means are used for that.

All perception embodies information in the structure of a control unit. Sometimes the structure as perceived corresponds to something in the real world, and sometimes it doesn't (Part 4, Sections I.12.1-I.12.3). Either way, control requires action that moves the perceptual value closer to its reference value (Approach) or away from the reference value (Avoidance). Approach in a control system is analogous to a mechanical pull on the CEV toward a location reference value, whereas avoidance corresponds to pushing the CEV away from that location value. As we did for 'stiffness' in Section I.8.1, we follow this mechanical analogy to show how tensegrity is an emergent property of hierarchic control.

I.8.4 Approach-Avoidance and Control Tensegrity

An analogy is proposed which likens properties of mechanical tensegrity to properties of control systems. In this section and the next, we discuss the properties of mechanical tensegrity so that we can then consider how they may help us to understand analogous properties of the control hierarchy. For simplicity at the beginning, we ignore avoidance control for now. To represent something that can exert a pull, the mechanical representation uses a wire or a rubber band, and to represent something that can push it uses a rod that can be compressed end-to-end without buckling or bending. In a control hierarchy, the compressible 'rods' are differences between reference values, treated as 'fixed points' like the ends of rods because their values are independent of the control actions that provide the 'wire' tensions.

Consider a simple push-pull conflict. One mechanical analogue is a wire and rod connected at one end and both pushed and pulled from the same place at the other end, as in Figure I.8.1d panel a. Such a linkage is not very useful, but a nearly identical arrangement suggested in Figure I.8.1d panel b and panel c is found in all organisms with skeletons, and in many mechanical devices.[82] Figure I.8.1d panel c shows the 'stiffness' configuration described in the last section, since at any joint angle the counterposed tensions of the two muscles are variable between very tense and very relaxed.

---

82 Scarr (2014) points out that most, or perhaps all, joints do not act as simple hinges, but are themselves tensegrity structures that prevent the bones from applying force directly to each other. At this point in the discussion, however, a simple hinge serves the purpose.

Figure I.8.4a shows situations like those of the paired panels b and e, c and f of Figure I.8.1c, with the addition of a third component, a controller (Figure I.8.4a and b) or mechanical structural element (Figure I.8.4c or d) that opposes the lateral movement that would occur in the four right-hand panels (b-c-e-f) of Figure I.8.1c. Lateral movement in the controller diagrams is prevented because the reference value at the lower ends of the displayed arrows are fixed, and unresponsive to changes at the arrowheads of the arrows. In panel a and c of Figure I.8.4a, with three participants rather than the original two, the structure is stable even in situations that correspond to panels c and f of Figure I.8.1c. Panel a is a representation of the collective control of stiffness.

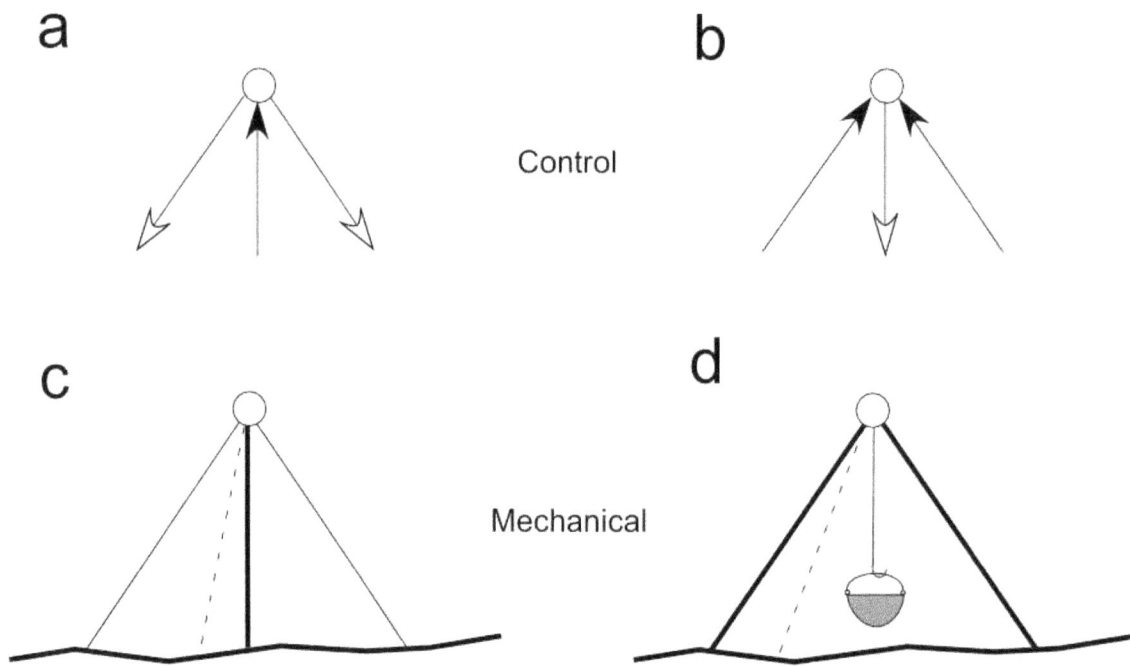

Figure I.8.4a Simple tensegrity structures. (a) Two approach control units and one avoidance unit, (b) two avoidance and one approach, (c) A mast held vertical by two guy wires, (d) a weight hanging on a wire supported by two rods. (In 3D, "two" becomes "three" in each example). The dotted lines in (c) and (d) represent possible changes in the fixed end of the wire or rod.

The mechanical structure is stable in two dimensions because of the tight linkage between the lengths of the solid wires and rods and the tensile or compressive forces on them. When disturbed, the system stores energy in the tensions and compressions of its members, and releases that energy in returning to its equilibrium position. If the entities are soft, such as rubber band guy-wires with a mast that was compressible lengthways but rigid against bending, it would still be stable, but the top of the mast would move further left and right when disturbed by a lateral force (such as wind), returning to its equilibrium position when the disturbance went away, like a ball pushed up the wall of a bowl.

In the absence of any external disturbance, the top of the mast in Figure I.8.3a and Figure I.8.4 panel c could be moved left and right by reducing the tension on one wire or the loop gain of one control loop while increasing it in the other. The guyed mast is a tensegrity structure that we see all over the place in our everyday world. Familiar but less frequently seen today is the cooking pot held up by rods (in our 3-D space, usually a tripod).

The mast-with-guywires physical structure is stable because the pegged locations are fixed in the ground. In the control analogue, those fixed points are reference values, which do not respond to changes in their controlled perceptions, though they might vary because of control actions at higher levels.

Does it matter that the so-called fixed points actually can vary? It does if the result of their variation leads to an unstable structure, such as would happen if the right-hand guy-wire anchor point (the reference value for approach) were to move to the left of the mast base. The mast could then fall leftward to a new stable configuration with the two guy wires still pulling together against the falling mast — the two approach controllers pulling together against the avoidance controller in one dimension, the conflict having been eliminated in the other direction.

There is, however, an intermediate position for the transition at which the system becomes unstable. This unstable region occurs when the moving anchor point (reference value) is too close to the mast base (avoidance reference value). How close is too close depends on the wire tensions (control loop gains). The higher the tensions and compressions (loop gains) the more energy is stored in the structure, and the more energy is released if the structure collapses. We will find this happening in many related conditions, of which the Bomb in the Hierarchy (Section I.6.5) is one.

As a practical matter, the compression members, 'rods', in a tensegrity structure need not be straight. They can usefully be bent, or have fixed 3-D shapes (which may be resolvable as smaller-scale rod-and-wire tensegrity structures, since there is always tension at the 'elbow' of a bent rod). Figure I.8.3a suggests a 2-D version of a weight-bearing structure in which the compression members never touch. The approximately horizontal light lines represent tension members, 'wires', bent because of the lateral force of gravity, and the lateral vertical wires are supposed to be somewhat elastic, so as to allow the column to bend smoothly but not so far as to destabilise the structure unless the applied force is very large.

In this structure, although the compression members are no longer straight disconnected rods, they still never touch one another. The weight-carrying is done by the horizontal wires, not by 'bricks' piled on top of one another. Scarr (2014) describes the vertebrate spine as just such a tensegrity structure, though appreciably more complex. In what follows, however, we treat the compression members as simple rods because we need not be concerned with bending stresses and such complications, which have no analogues in the control structure.

Figure I.8.3a shows the 2-D tower responding to a static lateral force, increasing the energy stored on the side of the tower opposite to the force while reducing it on the side where the force is applied. Energy is force times distance, so the force adds energy to the structure. This newly stored energy is distributed into the entire part of the structure that bends, and is not localised in any one specific component. If the bends straighten again, that stored energy is returned to the environment.

As Figure I.8.3a suggests, perhaps more clearly than the photos of 3-D tensegrity structures, the work of resisting pushes is done by the wires even though they can only pull. The compression members act mainly as structural impediments that prevent the wire ends from collapsing into themselves. In the control analogy, the rod ends may be reference values or they may be structural properties of the local environment which control cannot change without breaking the components apart, such as the configuration of the chair in Section I.5.3.

In two dimensions, the bending stresses can be eliminated by substituting for each bent compression member a parallelogram with opposite vertices connected, in which the cross-connections and boundary members of the parallelogram are of opposite type, one being compression, the other being tension (Figure I.8.4b). It doesn't matter which is which. So long as they are opposite, the net force at every vertex remains zero as changing external forces are applied to any of the nodes. Such 2-D parallelograms of wires and struts are often seen on masts, such as near the tops of the masts of sailboats.

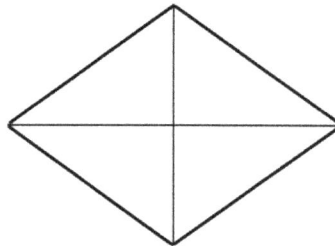

Figure I.8.4b A simple two-dimensional structure that could replace the bent compression members of Figure I.8.3a. Either the boundary members are compression and the crossing members tension or vice-versa.

A more interesting way to remove the bending stresses from the structure of Figure I.8.3a is to substitute for the bent member a tensegrity structure that is a miniature form of the whole, but with shorter wires on the lower side than on the upper side. Of course, this miniature replica has its own bent members, but they are much smaller, and can themselves be replaced by even smaller replicas of the original tower. Mathematically, such a sequence of successive reduction of scale to replace a bent member could be carried to infinity,[83] but in practice it has to end somewhere. After *n* down-scaled replacements of the bent members, the rigid

---

83   An affine self-similar fractal (Gouyet 1996). — Ed.

members become very small, but very numerous. At a sufficiently small, molecular scale the bending stresses become negligible, since the width of the arc would become comparable to its length, and no further down-scaling would be required.

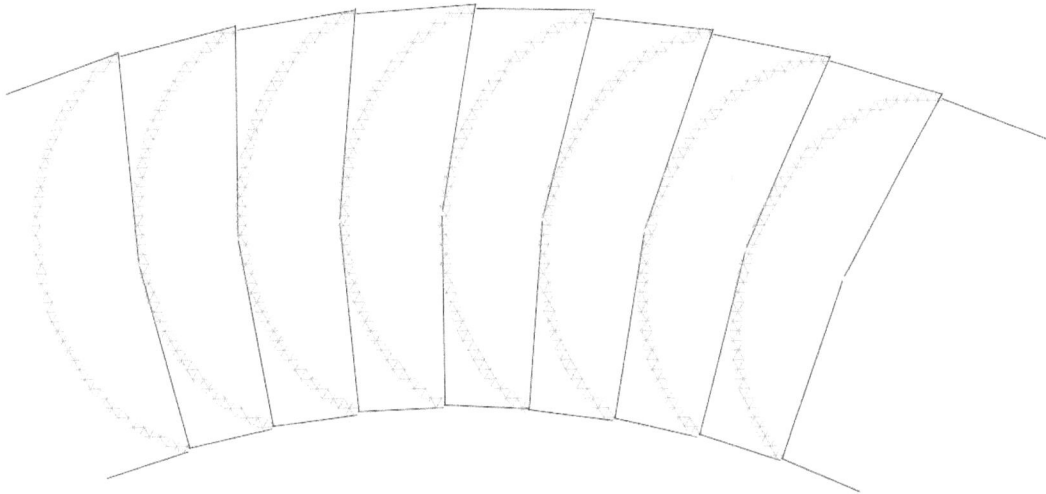

Figure I.8.4c. A bent tensegrity structure that might replace the rigid bent members of Figure I.8.3a. Each of the rigid curved members of this figure could be replaced by a miniature replica of the entire structure, and so on.

Scarr (2014) follows a much better trail of 3-D tensegrity support using as its base a 'tetrahelix' structure with no bent elements at any scale. It is necessarily more complex than our corresponding unit (a bent member, two cross wires, and two edge wires). Scarr carries this structure down to molecular scale, forming the different tissues of a living body along the way.

In contrast to the complexity of the tower in Figure I.8.3b, Figure I.8.4d shows a 'minimal' tensegrity structure, the smallest and simplest sort that can exist in a 3-D space. A minimal tensegrity structure is one in which no element can be removed without the structure collapsing.

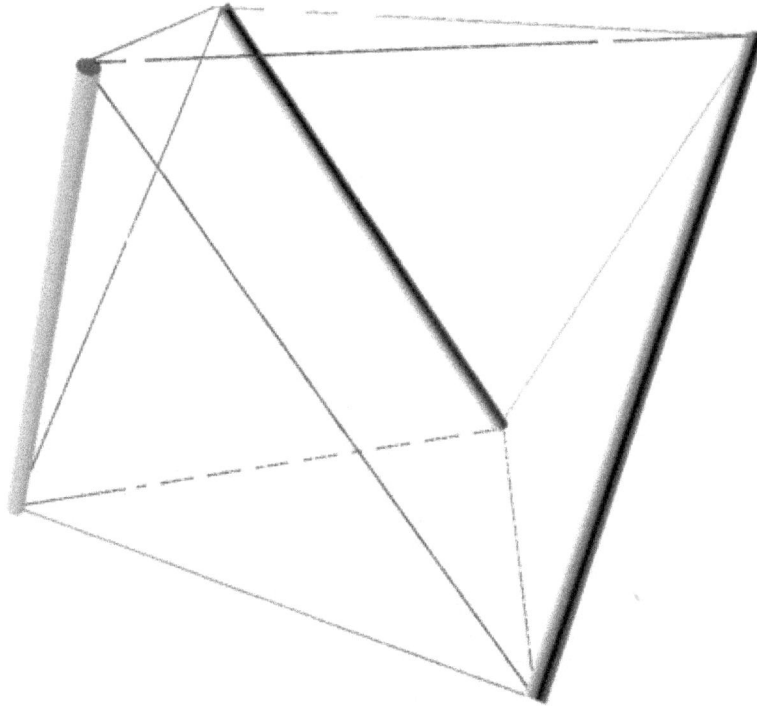

Figure I.8.4d A minimal 3-D tensegrity structure. Every rod is compressed by tension in the three wires connected to each end, and no rod end can move without increasing the tension on at least one of its connecting wires.

The essential function of tensegrity structures is typically to define a strong and resilient shape, which can be simple, as the octahedron in Figure I.8.4d, or quite complex, as the tower in Figure I.8.3b or the even more complex sculpture in Figure I.8.2. The smallest and simplest possible tensegrity structure in three dimensions, a minimal tensegrity structure, is made of three rods and nine wires, because it takes three rods to define the three dimensions of space, and because the end of each rod must be held from moving in all three dimensions. Examples are the structure in Figure I.8.4d and the topologically distinct structure in Figure I.8.10a. If there had been only two rods, the tensions of the wires would cause the rods to form a coplanar (2-D) structure such as a cross. Tensegrity is therefore an 'emergent' property of a structure in D dimensions, requiring at least D 'rods' and D^2 'wires' before the tensegrity property can emerge. The 'elbow' in Figure I.8.1d panel c is a one-dimensional structure requiring one 'rod' (the bone) and two 'wires' (the muscles).

Changing the lengths of the wires by changing their tensions changes the shape of a tensegrity structure. Wires with variable tensions are the 'muscles' of a tensegrity structure, and can enable an appropriately shaped structure to move. This capacity to change shape and move is related to the complexity of the structure. The form of a minimal tensegrity structure is highly constrained. The 'elbow' of Figure I.8.1d panel c moves only in one dimension, the angle of the forearm to the upper arm. Higher-dimensional tensegrity structures have more degrees of freedom to move. The tower of Figure I.8.3b could, if provided with motors to change the tensions, have all the flexibility and power of an elephant's trunk or the arm of an octopus, which is difficult to accomplish with a jointed skeletal structure.

Tensegrity structures intended for useful work will not be minimal. They will be redundant, usually with more than three rods and with more than three wires connected to each rod end.[84] The 12m tall Kolkata Tower of Figure I.8.3b seems to have seven wires connected to each rod end. If one of the rods or wires in a *minimal* tensegrity structure breaks, the load (stored energy) will not be redistributed among the remaining ones. Instead, the tensegrity property will disappear and the structure will entirely collapse, releasing its stored energy into the environment as the collapsing pieces first gain kinetic energy and then convert that energy into sound and heat and possibly other forms when they hit the ground.

Redundant tensegrity structures may be built in a modular fashion so that each module recovers locally from the loss of a member rather than distributing the resulting changes in tension and compression uniformly through the structure. In a redundant tensegrity structure, the loss of one wire may generate an avalanche breakdown of the structure in a way related to the 'Bomb in the Hierarchy' (Section I.6.5), but the size of that avalanche may be contained, as is also true of the 'Bomb'. In the video referenced in the footnote, the collapse is contained mainly to the region around the rod to which the cut wire was connected.

I.8.5 Dynamic Tensegrity

A tensegrity structure can be in static equilibrium, as are those illustrated in the previous photographs, or it can be dynamic, like the opposed muscle system of Figure I.8.1d panel c or the robots described by Piazza (2015). Either way, it stores energy in the tensions of the wires and compressions of the rods. The stored energy can be augmented by the application of force to a structure in equilibrium, or released as the structure moves autonomously. Piazza quotes a comment made by one of the creators of such a robot (Vytas SunSpiral of NASA):

> *Everything in biology is compliant. There are few very rigid structures, and yet that is how we have been building our robots to date. The awesome thing about tensegrity structures and their tensile networks is that when you apply a load to them, that load diffuses through the*

---

84 As noted, three wires are necessary in 3D space, so this is a more than sufficient condition for preventing collapse of the local structure, and such a structure as a whole may still have tensegrity properties after losing one or more of its elements, as is dramatically illustrated in the video of a bouncing tensegrity 'ball' at t.ly/b_w55 (Retrieved 2018.04.07)

structure and the whole structure adapts to its most efficient shape to manage that load.

I will argue later that the same is true of a control hierarchy. For now, however, we continue with mechanical tensegrity. Although a wire or a tensing muscle cannot push, a tensegrity structure powered by tensions can. Consider the opposed muscle system of Figure I.8.1d panel c. No matter which muscle is pulling the harder, the tip of the moving bone (the hand in a human arm) can both push and pull. SunSpiral also noted:

> *Living animals are never static! We are constantly breathing, moving, vibrating, and oscillating. We are constantly changing our orientation to gravity, and dealing with unexpected forces from every possible direction. These are all properties that tensegrity structures are well suited to deal with. So, my conclusion from all this is that tensegrity structures are an excellent design choice for a something that needs to move, but they are a poor design choice for static rigid structures (other than surprisingly beautiful art).*[85]

Excessive changes in the lengths and angles of the wires may cause a tensegrity structure to collapse if the angles among the wires at a rod end mean that the compression of a rod or the tension in a wire becomes negative. This is more likely to happen under an applied force if the tensions in the wires start low than if they are high, but the explosive energy release of the collapse is correspondingly greater when the tensions are high. The same is true of control tensions that have built up because of conflict. When something becomes intolerable, a person who is relaxed is a lesser danger to those around him than is a more 'highly strung' person.

Even static forces have an onset, which might be sharp or gradual. To get a feel for how the stress imposed by a force applied at a point in the structure comes to be distributed throughout the structure, it may be helpful to look at the propagation of the shock in the first instants after a sharp impact.

Imagine a 3-D tensegrity structure falling and sharply hitting the end of one of its rods on a hard floor (Figure I.8.5a). Just before impact, the forces on the rod end are in balance, the wires pulling just enough so that the pushing force of the rod is balanced in all three dimensions. Impact changes the force balance at the rod end, loosening the tension on the wires and sending a compression wave up the rod.

85   From SunSpiral's 'Being Human' blog: t.ly/Ug270, Retrieved 2024.02.02.

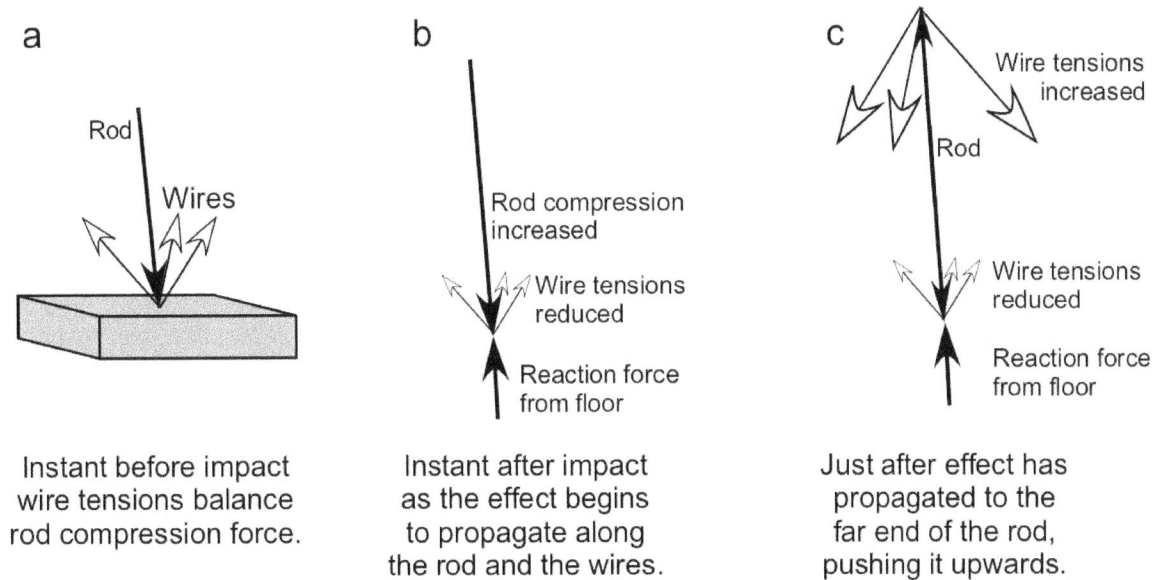

a

Rod

Wires

**Instant before impact
wire tensions balance
rod compression force.**

b

Rod compression
increased

Wire tensions
reduced

Reaction force
from floor

**Instant after impact
as the effect begins
to propagate along
the rod and the wires.**

c

Wire tensions
increased

Rod

Wire tensions
reduced

Reaction force
from floor

**Just after effect has
propagated to the
far end of the rod,
pushing it upwards.**

*Figure I.8.5a Tensions and compressions in a physical tensegrity structure are
altered by a sudden external force as a tensegrity structure falls onto a hard floor.
Propagation of the effect through the rod and the stays is not immediate, but
happens at the speed of the compression wave along the rod or the longitudinal
tension wave in the wires.*

When the compression wave reaches the other end of the rod, the rod regains
length and increases the tensions in the wires at its other end. The increased
tensions in those wires pull on the rods to which they are attached, but again
not instantaneously. There is a 'speed of sound' along a rod or a wire that
determines how fast a sharp change at one end is felt at the other. Eventually the
effect propagates through the whole structure, leaving few, if any, tensions and
compressions unchanged, but all are once more in balance and no single tension
or compression force has changed greatly from its value before the impact.

Scarr (2014) details many tensegrity structures at all scales from molecular to
body-part-sized in biological organisms. Each level of tensegrity structure is built
by complexes of smaller tensegrity structures down to the molecular scale, very
much as the control hierarchy of Powers is built up through control units of ever
increasing complexity and time-scale, and reducing speed of change. Following
either mechanical or perceptual control tensegrity down to ever smaller, simpler,
and faster-changing units, one can see how each level is supported by those next-
level smaller and faster units, down to the near-molecular scale of individual
muscle fibres or peripheral sensors. Scarr points out that individual cells are
tensegrity structures, and like SunSpiral, asserts that mechanically, tensegrity is
the fundamental structural fact of life, in the same way that PCT claims that
control is the fundamental function of life.

In practice, modularity is important in large tensegrity structures, however redundant their construction may be, just as it is in hierarchical perceptual control. Modularity, however, seems on the face of it to be inconsistent with the idea that changes in the loading on one part of the structure are distributed uniformly throughout the structure. To see that it isn't, we look at a two-level tensegrity structure.

Imagine on a table six copies of the minimal tensegrity structure of Figure I.8.4d, all with high tension in their wire, so that each individually is a rigid object similar to a crystal. Since they are all lying on the table, the structures as units can be moved independently. Now, in your imagination, think of these same minimal structures connected together pairwise by three rods and by nine wires, in such a way that each of the set of six is at a rod-end of a larger minimal tensegrity structure like that of Figure I.8.4d. We could call this larger tensegrity structure a 'second-level' tensegrity. At this level, it, too, is a minimal 3-D tensegrity structure, though as a whole, it is not, having no less than 21 rods and 42 wires.

The properties of this second-level structure are potentially quite different from those of the first-level structures that were lying on the table, even though all seven structures have exactly the same formal design. If a force is applied to a vertex of one of the original six, that force is distributed as changes in wire tensions and rod compressions throughout that minimal second-level tensegrity structure, resulting in more or less equal changes at all its vertices. In any one member, the second-level structure feels one-sixth of the force initially applied, and that one-sixth is distributed among its own vertices, transmitting 1/36 of the original force to the other five basic first-level structures. The applied force is indeed distributed throughout the complex two-level structure, but not uniformly. Nor is it uniformly distributed within any one of the minimal-structure modules.

The addition of the second-level structure does not eliminate the modularity of the first order structures, but it does allow some force-sharing among them, reducing the stress that any one of them must handle when an external force is applied to the whole. The effect is less pronounced in a minimal structure that has no redundancy to protect it against breaking, because the stresses will be roughly equally shared within a level of a minimal structure, which would not necessarily be true within a redundant structure.

One tensegrity system of ancient technology that depends for its stability on a single member is the bow and arrow. A lot of energy may be stored in the tensions and compressions of a bow and arrow (Figure I.8.5b). A bow is a nearly free-standing tensegrity structure, which we can diagram in 2D if we simulate the resistance of the bow to bending by two springs and fix the hand that holds the bow relative to those springs.[86] The three tensions — the springs and the

---

86   The springy resistance to bending is the effect of tiny tensegrity structures within the wood, which we ignore for simplicity, as we want to emphasise the analogies to approach and avoidance control in the large.

two halves of the bowstring — are in balance, and the compressions of the two halves of the bow balance the tensions in the bowstring and the springs. The arrow to be shot does not participate in the tensegrity structure, but when the hand-pull tension is removed much of the stored energy is transferred to the arrow as kinetic energy, the rest being dissipated as heat.

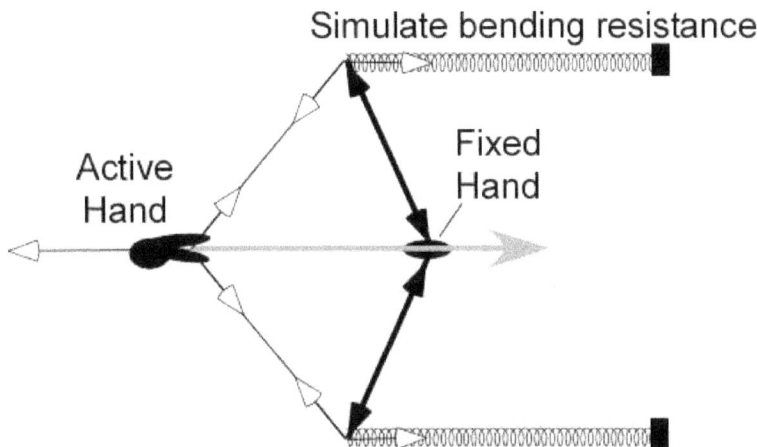

Figure I.8.5b A bow and arrow as a tensegrity structure in which the bending strength of the bow is simulated by springs in tension. The three tensions of the hand pulling the bowstring and the two halves of the bowstring are in a stable arrangement with the rigid rods and springs that together represent the wood of the bow. When the bow is drawn, it stores energy that is transferred into the kinetic energy of the arrow when the string is released.

Similar, but more subtle energy storage exists in Figure I.8.4a panel d which shows a structure holding a weight, such as a cooking pot over a fire or a bucket over a well. This structure is just as stable as the guyed mast. Unlike the guyed mast it is not necessarily unstable even if the right-hand support moves to the left of the centre wire, because although the outer (left-side) 'rod' would then be in tension, most rods can maintain their integrity with some tension. (Recall that the base or lower ends of the rods are fixed.) When the outer rod is in tension, the structure is equivalent to a guyed mast that leans like a simple hoisting crane. The rod, which in the depicted configuration was an avoidance control unit, would become an approach control unit. Eventually the configuration would become unstable if the approach and avoidance reference values came too close to each other.

The pot-hanger structure has a quite different instability mode that does not occur with the guyed mast. This occurs when the pegged base locations for the rods move too far apart, because then the circled meeting point of the three lines in the diagram could 'fall through', with the cooking pot falling into the fire, because at that point the two rods and the wires would be all working in the same direction.

In the basic perceptual control hierarchy, because lateral interconnections within a hierarchical level are disallowed, the issue of choice and switching to the chosen alternative has no clear solution. In Chapter I.9, we will extend the basic hierarchy by allowing lateral inhibitory connections that produce flip-flops and polyflops. If the lateral interconnections create a positive feedback loop of sufficient gain, the flip-flop produces two outputs, one at a high value, the other at a low value. These maintain their values fairly closely despite changes in the two corresponding input values, until the input corresponding to the low value sufficiently exceeds the input corresponding to the high value, after which they switch (Figure I.8.5c).

Figure I.8.5c suggests a tensegrity flip-flop, in which the key point (the grey circle) is stable some distance either left or right of centre, but not at intermediate points. Imagine the fixed points as being posts pointing out of the page mounted on a wall behind the plane of the picture. If downward force is applied to the rod ends in configuration a, the system could be driven into configuration c, at which point it would snap into configuration d. To do this requires energy, which is stored in the 'Stretchy' wires until the rods are just below horizontal and is then used to provide the kinetic energy that snaps the structure into its final configuration d, and is lost to the environment when the final configuration is reached.

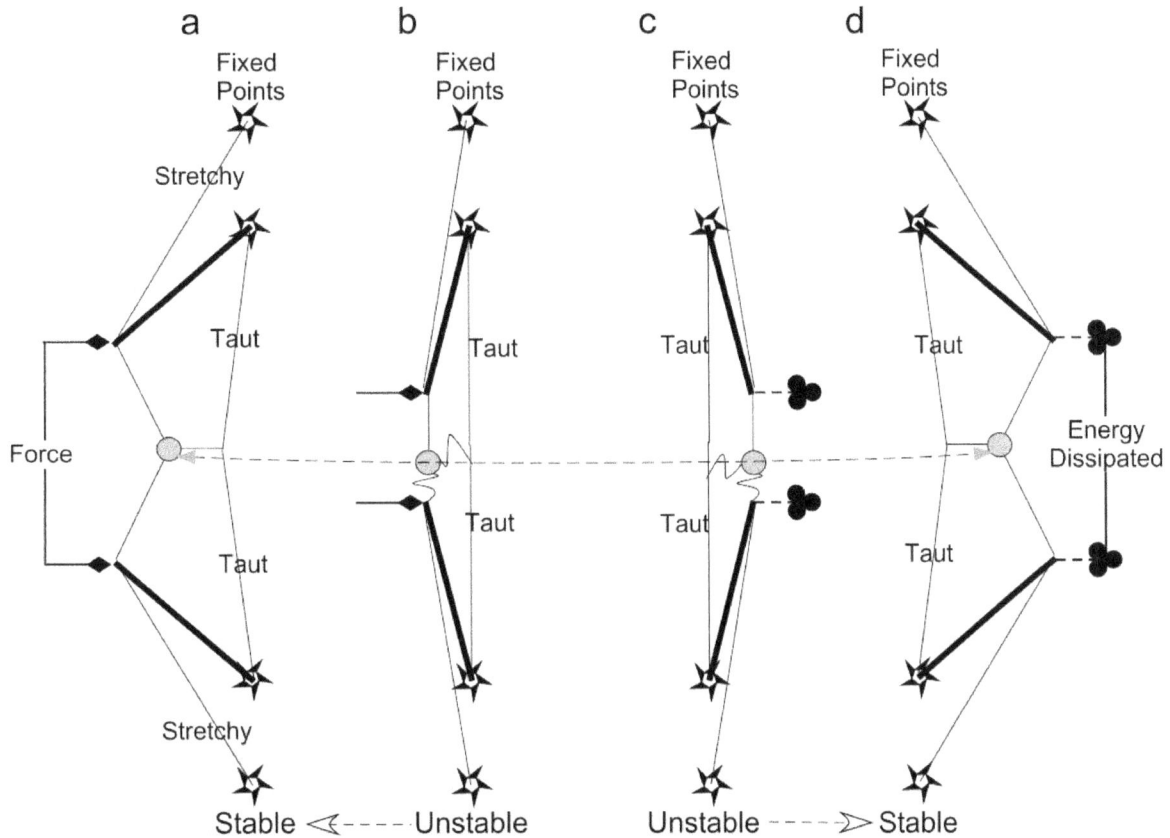

Figure I.8.5c A tensegrity structure based on the "Hanging Cooking Pot" of Figure I.8.4a-d. The grey circled point is stable in states a and d, but at some moment between them it will be in state b or c and is hanging free. From there it will snap back to a or d. Force applied to the rod ends can switch it from one stable configuration to the other. How much force is necessary depends on the stretchiness of the outer "wires". To switch back from configuration d to a, reverse the "Force" and "Energy" arrows.

In this tensegrity flip-flop, the grey circle that represents the item being switched is not connected to a rod, but is suspended by three wires, which lose their tension in the unstable stages during the switching. Energy was stored in these wires, so one might ask where this energy went. The two taut wires between the inner fixed points also lost some of their tension and thus their stored energy during the transition. That energy was transferred into the 'stretchy' outer wires along with the energy supplied by the external force on the rod ends, but this transferred energy is transferred back to where it came from when the system snaps into its final configuration. Unlike the energy gained from the external force that caused the switch, this transferred energy is not lost to the environment in the final 'snap'.

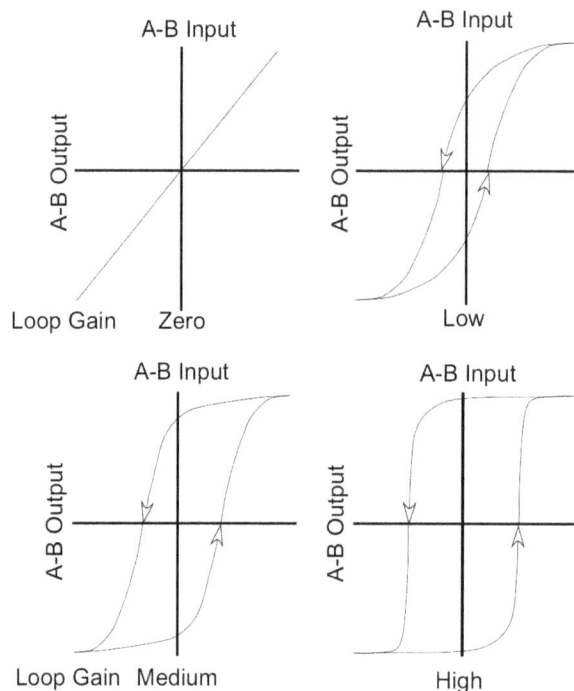

Figure I.8.5d Hysteresis loop of input-output changes of a flip-flop in which the output of each side of the flip-flop can vary between zero and a common maximum. As the cross-link loop gain increases, it takes a bigger change in the difference between the A and B inputs to cause the output to switch between A high B low to the reverse state, and the switch happens faster when it happens.

In Figure I.8.5d, the switch between the two states in one direction occurs at an input difference greater than the input difference required to switch in the opposite direction. The disparity is greater the higher the positive loop gain of the lateral interconnection of the flip-flop. The two switching trajectories enclose an area. This area represents energy used in completing a cycle between the two stable positions. It is a variable that is significantly independent of any of the other variables involved. The energy represented by the area can be seen in each of the arms of the cycle. When moving to the right and then up, it takes a certain amount of force to move the state only a little, which adds energy to the structure, after which that energy is released as the structure relaxes into a new configuration. The same happens on the way down and to the right. It represents energy corresponding to that supplied by the external force to the tensegrity flip-flop of Figure I.8.5c, energy that is lost to the environment as the switching movement is completed. Only when the loop gain of the lateral interconnection is zero is the energy gained and lost also zero.

All physical systems with hysteresis are the same. They absorb energy over part of each transition, and return it to their environment over the remainder of the transition, no matter what direction the transition is going. We will argue that the same is true of systems involving perceptual control and other less concrete systems.

I.8.6 Tensegrity and the Control Analogue

Stiffness can be a quantifiable global property of a tensegrity structure as well as of a single pair of opposed forces. Changing all the tensions by the same factor changes the stiffness of the structure without changing its shape. Low tensions allow the structure to flex, and redistribute the force by changing the angles at which the wires meet at the rod ends. The higher the tension, the less the structure can flex, although necessarily the tension at both ends of any wire must be equal, and the compressive stress at both ends of any rod must be the same.

To use the word 'compressive' is to attend to the effects on the rod. If, however, we think only of the forces at the ends of the rod, all that is required is that when the location of the end is stable, the forces balance in all directions. This balance does not require a rod. It can be achieved by a wire pulling against the combined forces of the other wires attached to the rod (Figure I.8.6). Another way of thinking about it is that a rod is stiff and does not buckle when compressed. Wires pulling in opposite directions have similar properties, and in addition the stiffness can be modified by varying the tension in the wires.

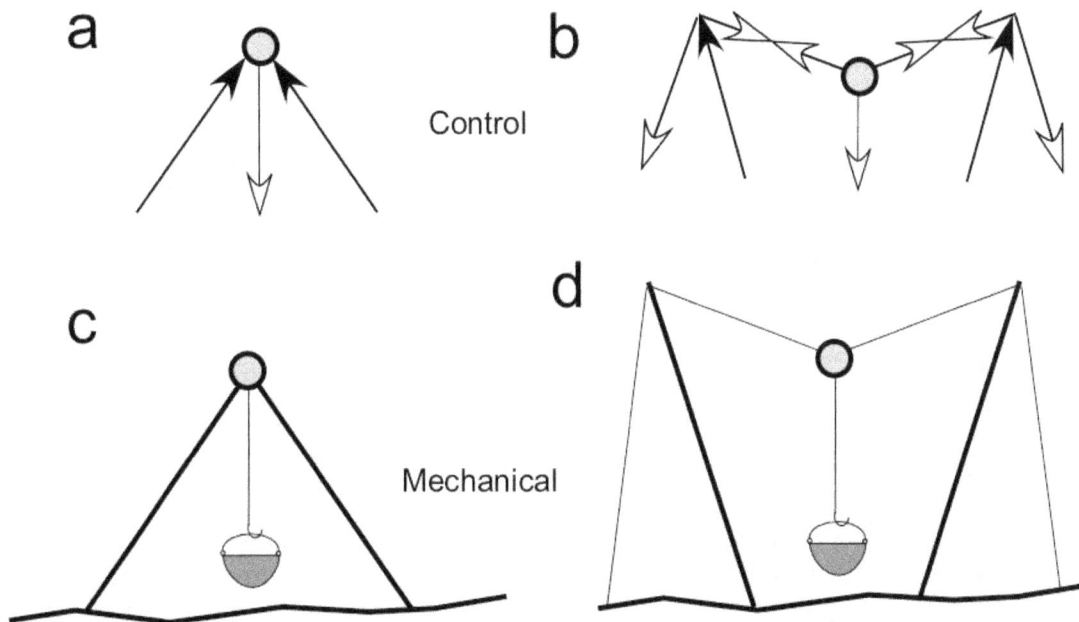

Figure I.8.6 The compressive forces on "rods" can be replaced by the stiffness created by tension forces in "wires" both mechanically and in control. In (a) and (c) the compression or avoidance forces around the grey disc are balanced by the tension. In (b) and (d) the forces are all tensile, and the conflict between the upper "wires" creates the stiffness property. that is characteristic of a rod.

When we talk of 'directions', we are taking the Analyst's viewpoint. If we think of control loops as 'wires' that 'pull' the perception toward its reference value we can see one possible analogy between the mechanical and control versions of tensegrity. Each of the elementary control units (ECUs) in question perceives only a value, and acts to influence that value and that value alone. All else to the ECU is a sideeffect. Sideeffects are, by definition, not effects on the perception being controlled, and are therefore in 'other directions'. The omniscient analyst, however, sees these otherly directed sideeffects, which occur in and through the external environment. Those side effects will become important later, when we are dealing with organisational structure development and a measure called 'rattling' (Chapter II.5).

The 'stiffness' effect is not a side-effect. Nor is it necessarily a controlled perception, though it could be. It is an emergent property of a particular class of structure. In Figure I.8.6 panel d, the cooking pot is held up by the conflict-induced stiffness of the upper pair of wires. Without the guy wires in tension from the ends of the posts to the ground, the pot would fall. If the guy wires were under more tension, the pot would be held higher, but in the control analogue, the controller responsible for the increased 'pull' would not have any direct connection with the controller that corresponds to the wire holding the pot. Their mutual influences would appear to be entirely by way of side-effects, although their total influence on the 'pot' would be the effect of the structure, not of any individual controller, not even that of the 'pot' controller. We will look much more closely at these 'organisational' (or perhaps better, 're-organisational') results of side-effects in Volume II starting in Chapter II.5.

It is natural to ask what the three controllers corresponding to the wires that meet to suspend the pot might be controlling in the control analogue of Figure I.8.6 panel d. The obvious answer is that since the tensions in the upper pair of wires in the physical structure determine the height of the suspension point for the pot, the corresponding controllers would be collectively controlling the lateral location of the triple connection point, creating the degree of stiffness that determines the reference location of the pot suspension for a given weight of pot.

In the analogue control tensegrity structure shown here, multiple higher-level controllers distribute their effect to one lower-level controller by way of the stiffness emergent. The structure shown later in Figure I.8.8 inverts the effect, so that multiple lower-level controllers distribute effects through one higher-level controller. Distribution of stress works both ways, up and down throughout the control hierarchy.

The 'omniscient analyst' could be a higher-level perceptual function inside the person doing the controlling. The perceptual inputs to this function would be the values of the lower level controlled tensions and compressions together with the angles between them (their correlations in the external environment). This higher-level perceptual function would develop and be sustained through reorganisation or evolution only if its control of those perceptions tended to maintain low error conditions in the intrinsic variables, e.g. if controlling this complex improved the stability of side-effects which otherwise could disturb intrinsic variables.

For example, having the cooking pot fall into the fire is not 'a good thing' if the person wants a good meal, which presumably affects some intrinsic variables. A higher-level perception of the configuration of sticks and the hanging pot might have a value that could be interpreted as the quality of the configuration as a stable way of holding the pot over the fire (and thereby of getting a meal).

Controlling perceptions of the individual legs in a structure such as Figure I.8.6 panel c is quite possible, but moving the legs away from the reference 'most-stable' configuration could be problematic from the viewpoint of a higher-level controlled perception. Perceptual control of the leg locations individually might be perfect, but some higher-level configuration control might have a non-zero error value, because the pot would be perceived to be more likely to fall if there was a small disturbance to one of the legs. The perception of the configuration's stability would then be outside its tolerance limit.

Non-zero error is analogous to the two ends of the 'wire' being separate, one end being the reference value and the other end being the perceptual value. Either the control force on the CEV in the environment or the perceptual error could be taken to be analogous to the tension in the wire. However, the analogy with control force is problematic, because the control loop may well include an integrator, which would operate as though the wire tension continued to increase while the separation of its ends did not, creating a hysteresis effect as the tension (error) fluctuated up and down. Physical wires do not have that property (though elastic bands may). Their tension is what it is, and changes almost immediately as the stress changes. It is better to equate the error in the controlled perception to the tension in a wire, as we have been doing, rather than the output force.

Environmental constraints embodied in mechanical tensegrity structures used by an individual can readily be converted into learned perceptions that we might label 'stability criteria'. Such perceptions can often be taught: "*If you build it this way, you will be able to do that reliably, but if this is too close to that it all might collapse under any significant strain.*" Some, however, are more difficult to teach, being based on simpler perceptions that must be taught, and that an expert may experience simply as "*that structure mostly looks good, but it depends too much on that bit, which might break unless it is reinforced — I can't tell why. I just feel it*", much as a high-level athlete may not be able to describe why a particular golf swing, tennis stroke, or running gait feels and looks graceful and effective while another does not. These statements represent conscious knowledge, and therefore belong to the Predictive Coding branch or track in the two-track form of the perceptual control hierarchy we introduced in Section I.1.6 and will expand on in Volume II, Section II.8.9.

For simple structures, the physical and control analogues are very close, but as we saw in Figure I.8.6, the analogy is imperfect in one important respect: control is active, while a tensegrity structure of physical rods and wires is passive. The compressive forces at each end of a rod or the tensile forces at each end of a wire are always equal, apart from dynamic variations that travel through the structure at the speed of sound in the rod or wire. This symmetry does not hold for a control tensegrity structure, in which a perception is 'pulled' toward its reference value but the reference value is not pulled toward the corresponding perceptual value. The 'rods' in the control tensegrity structure are partially defined by limits in the environment on possible values of controlled perceptions, such as that no two solid objects can occupy the same space.

In the physical tug-of-war, the rope has its own Elastic Modulus, but the side-force rod or wire does not, at least not as seen by the rope. It does, of course, in its own direction, whether it is like a physical rod in compression pushing on the tug of war rope or like a wire pulling on it. Either way, the force it applies is partially distributed to the rope, as is the energy used in moving the connection point at the middle of the rope. That is the principle behind the useful properties of any tensegrity structure, which distributes stresses throughout the structure, smoothing out the effects of concentrated disturbances.

Looking at the up-and-down transmission of effects in a control hierarchy, downward (output) effects are transmitted with nearly full strength, though that strength is usually distributed through multiple lower-level control loops (atenfels), while upward effects are attenuated at every stage by the countering influence of control. This directional asymmetry has no counterpart in a physical tensegrity structure. Consequently, when we want to use a specific physical tensegrity structure as a guide to a possible control tensegrity structure, we must be aware of the possibility that we might need to consider the upgoing perceptual paths separately from the downgoing action paths through the control hierarchy.

I.8.7 Conflict-induced Stabilities in Control

Every junction or node in a tensegrity network inherently involves opposing forces, and it is these forces that create the flexible strength of the structure. In the control analogy, these forces represent conflicts of the kind McClelland (1993) demonstrated, in which two controllers try to move a common CEV in opposite directions. In the stable configurations a and d of the switch in Figure I.8.5c, the key node (the grey disk) is tightly constrained by the tensions in its three wires. During the period in which the tensegrity configuration is unstable (represented by states b and c), the grey disk has considerable freedom to move within the configuration. Only during the approach to a stable configuration do the tensions in its wires build to create the stabilising 3-way conflict around it.

Later (in Volume II) we will show how a single conflict can provide a stability that improves control in a social context in which side-effects influence the ability to control by people who play different roles. We now illustrate a parallel possibility within the set of control units that constitute the control hierarchy of a single organism.

One control system may control an environmental variable that another can use as part of its environmental feedback loop. In the introduction to Chapter II.3 there is an extended quote by McClelland that discusses the importance of this possibility for control in a complex society. Figure I.8.7 illustrates a non-specific organisation of this kind. In it, a side-effect is shown by a dashed curve, with the location of its effect on a neighbouring control loop shown by a small circle at one end of it.

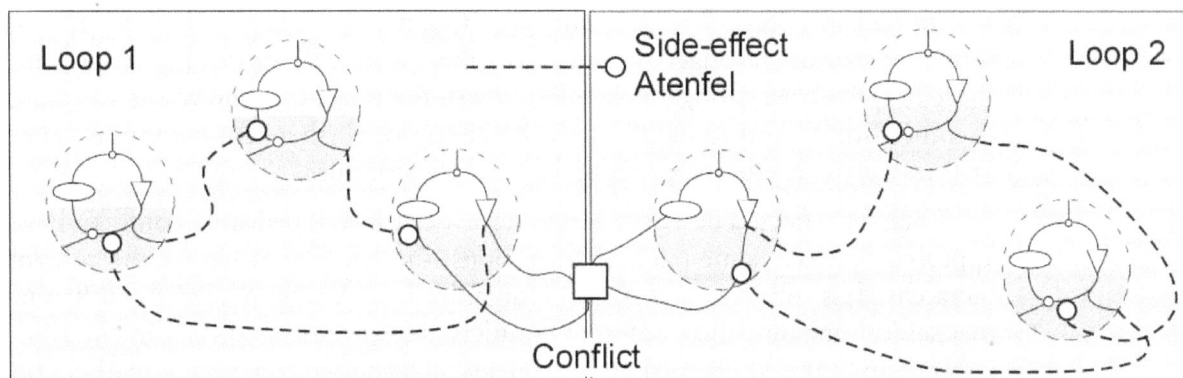

Figure I.8.7 Two loops of beneficial side-effects within the same control hierarchy. Each loop has a member that has a resource conflict with a member of the other loop. The box represents the CCEV over which there is a conflict. That CCEV is stabilised by the collective control created by the conflict.

All the side-effects, in this example, are beneficial, but the set of six control loops as a whole is stabilised by the existence of a conflict between two of them. Now, is it gratuitous to presume that such side-effect loops of enhanced control quality exist, just to make a point? No, their existence is almost inevitable. For example, if the probability is 1/1000 that a particular side-effect is beneficial to a randomly chosen control loop, the existence of at least one loop of beneficial side-effects is very close to 1.0 when there are as few as 400 interacting control loops.

Even if we use Powers's 'neural current' as an approximation to the effects of the firings of many correlated neurons, there are far more than 400 perceptual values available to be controlled in any one organism of reasonable complexity. If we treat instead the individual nerves, we are dealing in billions of possible interactions, which are combined into a relatively small number, perhaps thousands or even as few as hundreds, of states and dynamical changes in the environment that may be influenced by control.

The functional result of gradual reorganisation within the complexity of the neural system is to enhance beneficial effects and eliminate detrimental ones. Such loops of beneficial side effects are almost certain to be preferentially preserved because they improve quality of control by the individual systems, and are likely to be so many that interactions among them emerge. In Volume II we will examine how they may form loops of loops.

Now consider the arrangement shown in Figure I.8.7 and imagine other similar arrangements in which beneficial loops interact through a conflict between two of their members, one in each loop. As McClelland (1993) demonstrated, in such a conflict, the conflicted control units are likely to keep increasing their output up to a point, while the CEV over which there is a conflict has a reference value between their reference values, controlled with a gain equal to the sum of the gains of the conflicted units.[87] In other words, the property of the environment which is subject to the conflict is more stable than it would be if either control system had full control of its value. At the same time, the individual control units are creating stronger outputs than they would if controlling alone, and those outputs have stronger side-effects, with the result that the whole structure becomes 'stiffer' and less subject to the effects of further reorganisation. That 'stiffness' is the resilience aspect of a tensegrity structure.

When beneficial side-effect loops have been developed and conflicts have appeared between some pairs of such loops, the effect is to render the control hierarchy less subject to further reorganisation. Using a slightly different mechanical analogy, these internal conflicts that produce stiffness in the side-effect loops may be likened to the effects of alien carbon atoms in the iron matrix that result in the toughness of steel, or the rigid platelets that toughen some plastic materials.

Conflicts within a control hierarchy are usually assumed to be bad, and the psychotherapeutic Method of Levels (MOL; e.g. Carey 2006, 2008; Mansell, Cary, & Tai 2012; Carey, Mansell & Tai 2015) is based on removing them by reorganisation. But let's see what happened here. Two beneficial side-effect loops were created by reorganisation in a situation in which both were strengthened by conflict. It is not the existence of the conflict that matters, but the strength and localization of the conflict. In situations relieved by MOL, the conflict could lock parts of the control hierarchy in place, but in Figure I.8.5c the elastic energy temporarily stored in the conflict is what permits the switch to function. The effectiveness of some tensegrity structures, such as long bridges, may depend on high tensions that create rigidity, but it is unlikely to be so in the control hierarchy of a living organism.

Reorganisation preferentially leaves alone what works, while tending to eliminate structures that do not enhance the control of intrinsic variables. 'Preferentially' and 'tending' are probabilistic words, and when there is only a probability rather than a certainty that things will get better, sometimes they get worse.

The two side-effect loops of Figure I.8.7 probably each help sustain controlled perceptions that use particular atenfels in their environmental feedback path. If one of the controllers in the loop were to shift to a different mechanism, its side-effects would change, and the side-effect loop would be broken. This does not mean that the other controllers in the loop would fail to control. After all, the

---

87   To the Observer, this has the appearance that the CEV is controlled by a single control system, which the Analyst may refer to as a virtual control unit, but that is beside the point here.

side-effect loop persists only because they control better so long as it functions. Without the loop, they still control as they did before the loop came into being. By the definition of 'side-effect', none of their environmental feedback paths include side effects chained into loops, but nevertheless it is their side effects that enhance their Quality of Control.

Does a hierarchy so organised exhibit stereotyped behaviour or even obsessive compulsive disorder (OCD) at low levels, while at high levels it exhibits dogmatic high-gain control of a complex of opinions that are stabilised because they contain an internal contradiction? Or is a conflict of this sort helpful to keep the beneficial loops operating even in relatively difficult situations?

I think either can happen. Perhaps a person controls strongly against moving to a different environment, perhaps even refusing to leave the house, in order to sustain the stability of the side-effect loops that have developed to function in the existing environment. The rigidity is not all inside the head; the control loops involved go through the external environment. If the environment were different, the side-effects would be different, and maybe the stability would be lost.

From another viewpoint, syndromes of mutually supporting control systems that are not directly involved in each other's environmental feedback loops form stabilities that might act as 'rod-ends' in a tensegrity control module. The inability of the loop to change (much) without detriment to the control quality of its members depends on the particular structure of the environment. The greater the variety of environments in which the maturing hierarchy has been reorganised, the less dependent the side-effect loops are on any particular environment, and the more stable the 'rods' are in the tensegrity structure of the hierarchy.

Before, I remarked that a beneficial side-effect loop would be broken if one of its members changed its means of control. But as previously mentioned for a mechanical tensegrity structure, this is true only for a loop (or a tensegrity structure) that is minimal in the sense that each member's side-effects alter the control quality of only one following controller around the loop. It is extraordinarily unlikely that such would be the case. More probably the side-effects of each controller would influence many others' ability to control, some beneficially, but more often detrimentally — apart from the effect of reorganisation. We will have a lot more to say about homeostatic loops and their relation to tensegrity in Part 5, at the beginning of Volume II.

As Powers showed in his Arm 2 Demonstration (in LCS III), reorganisation is quite capable of quickly reducing detrimental cross-influences among control units. His cross-influences were mutual disturbances, which can only be detrimental, and reorganisation made them tend toward zero. If there are beneficial cross-influences, reorganisation would tend to enhance them while decreasing the detrimental ones. If that is the case, the result would not be isolated beneficial side-effect loops, but a mesh or network or even a whole ecology of them in which the loss of one member of a loop would not much affect the mutual support of the rest of the control units in the network.

Chapter II.2 on creativity deals with how such autocatalytic and homeostatic networks might come into being, and why they would tend to form overlapping clusters of mutual support. The effect of conflict between two control loops would still enhance the rigidity of a cluster, but more softly than would be the case if the beneficial structures were minimal side-effect loops. The mesh, however, would tend to be stronger, more resilient, than would any one minimal loop. These are tensegrity properties.

Tensegrity systems exhibit toughness, the ability to bend rather than break under external stresses. They exhibit resilience or elasticity, the ability to return dynamically to a prior equilibrium state after an external influence is removed. They store energy distributed throughout their structure, and energy from an external force applied at one place could be used at a different part of the structure.[88] A tensegrity structure that has too much tension can break, while one that has too little may collapse.

All of these characteristics apply equally to both mechanical and control tensegrity structures, but in control the 'tension' in a wire often (not always) corresponds to output forces that bring a controlled variable to its reference value, while the opposed tensions at a junction in a tensegrity structure correspond to conflict. The implication is that a functioning, healthy control hierarchy will be full of low-level conflict, and of perceptions that never quite match their reference values. If this claim is true, then reorganisation or evolution, which we consider in depth in Chapter II.4, is likely to provide for ways to change the tolerance range for most, if not all control loops.

I.8.8 Approach to Control Tensegrity I: One Using Three

We begin now an approach to the minimal control tensegrity structure that we will finally complete in Section I.8.10. We consider always a two-level hierarchy, adding units and then replacing some by linkages to pre-existing ones. In Section I.8.10 we will wind up with three controllers at each of two levels forming a minimal tensegrity structure.

If the analogy to physical tensegrity is to be useful in thinking about hierarchic control, control must have analogues of physical rods and wires. One end of a wire is probably attached to one end of a rod. The wire pulls its ends toward each other. In the control analog, a controller pulls its controlled perception toward its reference value. Now consider the rod, which is the component that prevents all the wire tensions from collapsing into an amorphous tangle.

Could a reference value be the analogue of the end of a rod? Yes, it could. The critical property of a rod is that it holds things apart. As we have seen above, if a higher-level perception corresponding to an environmental structure is being controlled by sending reference values to the lower level control units,

---

88   A fundamental principle in martial arts.

then the relationships among those lower-level perceptual values are largely determined by the structure implicit in the higher-level perceptual function. The reference values are fixed by the outputs from higher levels. Changes in their corresponding perceptions cannot change the reference values. They resist forces from their own controllers that would bring any pair of reference values closer together in any dimension they have in common in their descriptions.

When control is good, environmental values and reference values are in correspondence. If the environmental values for a structure do not relate to each other in the way specified by the perceptual function for perceiving the structure, the perception will not be well controlled, and reorganisation will tend to change the way output values are distributed to the lower level in the form of reference values. Crudely put, if a chair is to go in one corner of a room, a leg of the chair cannot go in a different corner of the room without breaking the chair, and a controller of a perception of the chair location will not provide to the leg location controller a reference location in a different corner of the room. For a well-reorganised higher-level perception, the reference values at the lower level will closely track the functional relationships which the higher-level perceptual function defines for those lower level perceptual values. So yes, reference values can hold things apart, one chair leg from another, just as can rods in a physical tensegrity structure, provided that these reference values are derived from the control of a higher-level perception which corresponds to an environmental structure.

Consider a controller of one variable: $x+y$, which has a reference value of 5. The perceptual function of this controller has only two components, x and y, which are the values of two lower-level perceptions. If '5', is a reference value for the upper-level controller's perception, and x changes to become 3 and is difficult to alter by action, then y must be controlled to become 2, and when some external disturbance then moves x to 2, the upper-level can control its perceptual value only by controlling y to become 3.

A slightly more complex perceptual function might be $x+y+z$. Suppose this perception, which we can call S has a reference value of 10, and its perceptual value is at its reference value with $x=1$, $y=6$ and $z=3$. Now a disturbance moves x to 4 in some way that the lower level x controller is powerless to oppose. How does the higher level controller now distribute its output as reference values to the lower levels that control y and z?

The answer is that the distribution is dynamic, and it is not the business of the upper-level S controller to do it. When x suddenly moved from 1 to 4, S jumped from 10 to 13, producing an error value of 3. The resulting output can be described as signifying 'Less, please' to both of the lower level y and z controllers. What this means to them is their business, and the y and z reference values change as a consequence of the change in the output from S. The upper-level output simply says to the lower level reference inputs — all of them — *do whatever you do to give me more* (or *less,* as the case may be), and the lower-level controllers act to change their perceptual values appropriately in respect to their environmental constraints.

Both y and z perceptions now deviate from their changed reference values, and act in ways that will reduce the value of S (or have in the past). As their effects take hold, the error in S is reduced, and its output changes further the reference values of the y and z controllers. Eventually, S is not in error, but this does not imply that y and z are in agreement with their reference values. They may be or they may not. If they are not, they will continue to act, which affects S and through S each other's reference values, until eventually all will be satisfied.

In a multi-level hierarchy, then, changes propagate through the hierarchy, up through perceptual values, down through reference values, and back up completing the loops through perceptions. But these loops are not control loops, they are reciprocal interactions between the levels of the hierarchy that are a consequence of the actions of control loops at the two levels.

In the preceding discussion, notice that S is a scalar — one-dimensional — variable, whereas x, y, and z are also scalar individually, but together describe values in three different dimensions of perceptual space. The control of the value of S is 'One through three'.

Figure I.8.8 illustrates a two-level structure slightly differently from the way such structures are usually shown in the PCT literature.

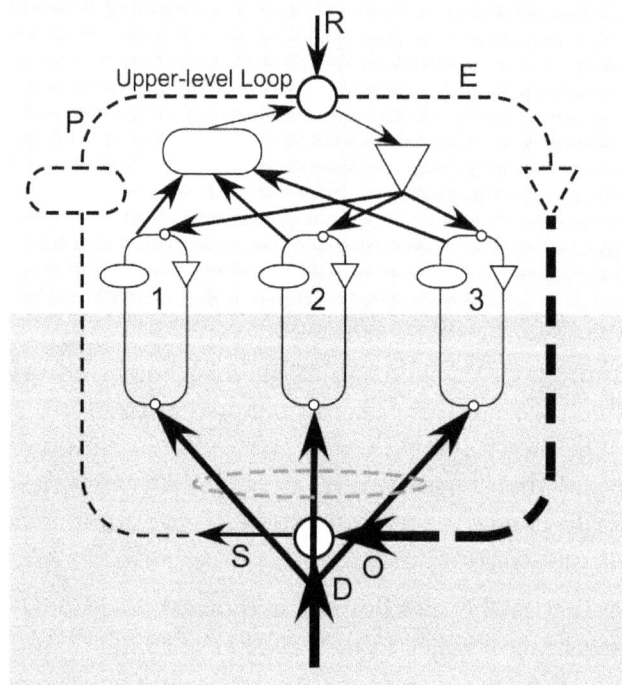

Figure I.8.8 A two-level control loop. Three simple control loops send their perceptual signals to the second-level perceptual input function and receive their reference values from the second-level output function. The CEV of the second-level loop is disturbed and the disturbance opposed by the combined outputs of the simple loops, shown here as the output of a virtual output function. Likewise, the second-level loop can be seen as having a virtual perceptual function that incorporates the actual perceptual functions at both levels. Line weights suggest uncertainty values and the distribution of tensegrity effects.

In Figure I.8.8, dashed lines represent a virtual loop controlling a scalar variable. Within this loop, solid lines represent two levels of control in the conventional manner. The virtual loop has a virtual perceptual function and a virtual output function, but its comparator is that of the second-level control system. This comparator and the CEV are not virtual, they are as real as are any of these hypothetical control structures. The real CEV is the variable that is important. If use the same example as in Section I.5.5, the CEV for the second level is the location of a chair, and the lower level CEVs (the small circles in the lower-level loops) are the locations of a leg, the seat, and the back of the chair.

To continue the analogy to a physical tensegrity structure subject to a varying external force at one point, in the two-level loop of Figure I.8.8 suppose an external influence disturbs the second-level structure CEV by only affecting lower-level CEV 1. The second-level controller sends error output to all three lower-level systems. All three systems now have some changes in their errors, the amount by which the perceptual inputs differ from their newly changed reference values. Variable error demands variable output to the corresponding CEV. For loop 1, however, the error uncertainty also includes that due to the locally applied disturbance.

What we wind up with is a situation in which the uncertainty due to the applied disturbance is being countered in four places, all three lower-level loops plus the higher-level loop, which acts independently despite being implemented through the actions of the lower-level ones. The relationship pattern among the reference levels for the lower-level perceptions is acting rather like rods in a physical tensegrity structure.

To see how this works, imagine a single control loop with a fixed reference value. Whatever way the disturbance 'pushes' the CEV, the control property of the loop 'pulls' the perception back toward the reference value, analogously to a tensioned elastic 'wire'. In our two-level example, the reference values of the lower-level loops are not fixed. When a disturbance 'pushes' the CEV of loop 1, it also 'pushes' the CEV of the higher-level loop. With a rod in a physical tensegrity structure, pushing on one end of a rod changes the tensions of the wires at the other end. Likewise with the control structure, 'pushing' on the CEV of the higher-level loop changes the reference values and hence the 'tension' (the error, multiplied roughly speaking by the loop gain) in all the 'wires' they represent.

One of the core attributes of tensegrity structures is the way they distribute forces applied at one point through the entire structure. Up to this point, what we see is that the same is true of a two-level control hierarchy that has one higher-level controlled perception, when a sudden change or a continuous variation occurs in the disturbance to one of its lower-level variables.

This example is neither an example of three-dimensional tensegrity of control nor an example of tensegrity in one dimension. It does exhibit some of the properties of a tensegrity structure, but whether it should be called a fully fledged tensegrity structure is dubious, since the effects depend largely on maintaining only one value stable by means of variations in three others.

Both the small size of the segment of the control hierarchy and the special nature of the stress lead to questions as to whether what happens in it is representative of what happens in general. We next extend the example by slightly increasing the size of the part of the hierarchy that we examine.

I.8.9 Approach to Control Tensegrity II: Two Using Five

We could extend the example system of Figure I.8.8 in at least two ways, by adding more levels to the hierarchy or by extending the hierarchy to include more perceptual functions at either of the two levels considered in the previous section. The more fruitful extension is to consider more perceptual functions at each level. To be specific, we will add one more higher-level perceptual structure to be controlled. The new 'upper-level loop 2' uses a set of five simple lower-level loops that partially overlaps the set used by upper-level loop 1 (Figure I.8.9).

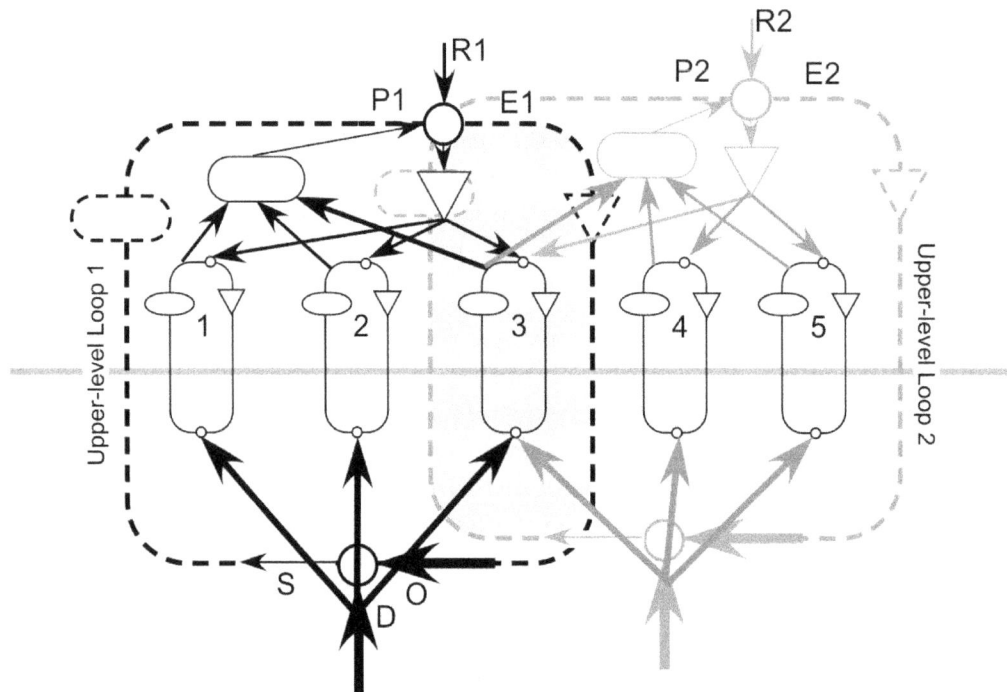

Figure I.8.9 A second upper level control loop may share some of the same lower level perceptions and/or outputs as part of its control of its structured perception. In the usual introductory diagram of the Powers perceptual control hierarchy, every lower-level loop is connected to every controller at the next level, exaggerating the lateral overlap effect illustrated here.

Now lower-level loop number 3 contributes its output to the perceptual input functions of both higher-level loops. Ignoring for now any questions about the form of the reference input function that combines the two output variables, we can ask what happens in this more complex portion of the hierarchy when the CEV of loop 1 is disturbed while all the other external variables are undisturbed?

The answer is not difficult to see. The original loop corrects its error (E1) by creating error in all three lower-level loops by changing their reference values. As part of this, it creates error in loop 3 by changing the loop 3 reference value. That has an effect on loop 3 similar to the effect that the disturbance has on loop 1, though the uncertainty of the variability of the loop 3 reference value is almost certainly much less than that of the original disturbance to loop 1 because of the fact that lower loop 1 controls its own perception.

Control by loop 3, after a change in the contribution to its reference value, changes the perceptual value that it contributes to upper-level loop 2 as well as to upper level loop 1. This induces error in the perception controlled by upper level loop 2, which generates output which changes the reference values of loops 4 and 5. The job of opposing the initial disturbance has been spread further through the larger chunk of the hierarchy, even though the perceptual values P1 and P2 in the two higher-level loops can take on their own reference values without mutual interference. This structure has a two-dimensional vector variable controlling through five dimensions of lower-level perception.

When everything has settled down after the onset of a disturbance to lower-level loop 3, it is unlikely that both upper-level loops will be asking lower-level loop 3 to take on the same reference value. They will be in conflict. This conflict looks functionally like the diagram of Figure I.8.1a that illustrated 'Stiffness' of the rope in a tug-of-war, and it would be the same, were it not that both upper-level controllers could control their perception through two other low-level controllers. This being so, there need be little or no conflict-based tension over the controlled value of low-level perception 3.

As McClelland (1993) demonstrated, the result is a collective controller for the resulting lower-level reference value created by the opposed tensions in the 'wires' pulling the collectively controlled variable toward both desired reference values. This creates conflict-driven 'stiffness', and the equivalent of a tensegrity 'rod' between the two upper-level controllers.

I.8.10 Control Tensegrity: A Minimal 3-D Structure

Spreading the 'stress' through the larger structure reduces the 'strain' at any particular place in the structure. Whereas by itself lower loop 1 has to control its perception using only its own private mechanisms, by participating in the larger structure of the higher loop that controls P1, its control problem is reduced. By participating, even slightly, in control by higher-level loop 2, low-level loop 1 can control even more easily. The bits and pieces of the structure might seem to invite conflict, but they need not, and are more likely to have reorganised so that they do not, even when higher-level controls share a lower-level controller such as loop 3 in Figure I.8.9.

Figure I.8.9 shows one way that the multi-level control structure can spread the tensegrity property among control loops that have only a tangential 'lateral' relationship. Figure I.8.10a illustrates a physical tensegrity structure in a 3-D world that has similar lateral relations cyclically among three rods. This model differs from the actual 3-D control tensegrity structure, but if you mentally follow what pushes and pulls on what, you should see how stresses imposed from outside are distributed through the structure.

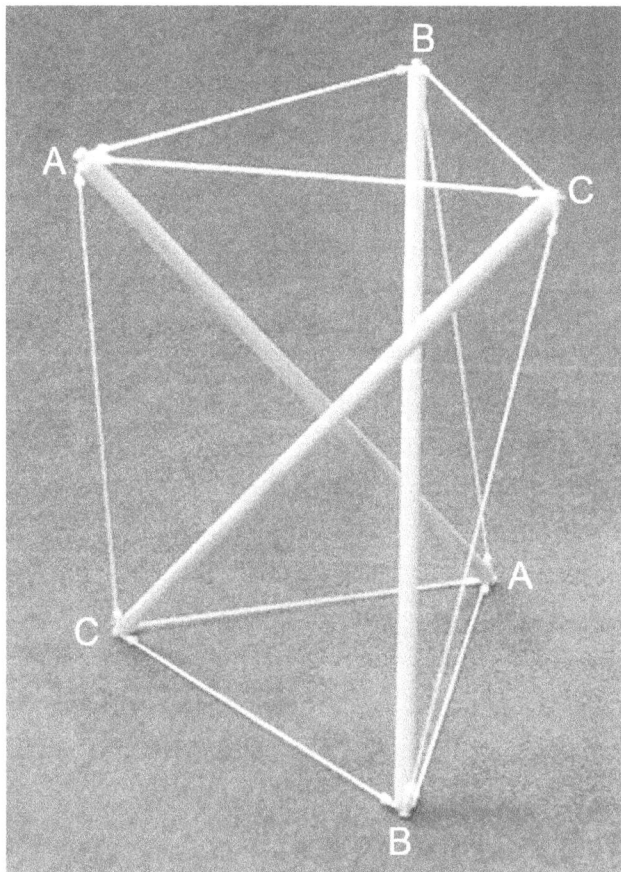

Figure I.8.10a A minimal 3-D tensegrity structure like that of Figure I.8.4d. Remove any wire, and the whole structure will collapse.

At first sight, this structure may look complicated, but look more closely. It is made entirely of triangles, two at the top and bottom that ring around the ends of all three rods, preventing those rod-ends from moving apart from one another, and six that connect one rod-end to both ends of another rod, such as the triangle that connects the upper end of rod A to both ends of rod C, while the other end of rod A connects to both ends of the remaining rod, B. Likewise, one end of rod B is wired to both ends of rod A while the other end of rod B connects to both ends of rod C, and one end of rod C is connected to both ends of rod B while the other end of rod C is connected to both ends of rod A.

In Euclidean geometry, the angles of a triangle are determined by the lengths of its sides. The same is true of the triangles in this minimal tensegrity structure. Does this mean that the structure is completely rigid? No, it does not, because wires will stretch under tension and rods will shorten under compression, changing the angles in ways illustrated earlier in this chapter (e.g. Figure I.8.3a). These changes in length also represent energy storage in both rods and wires, using the formula energy = force   distance. This energy is imposed from outside, whereas in a control system, the energy is supplied from a source independent of whatever forces are applied to the control structure.

It is therefore impossible for a control system to be exactly analogous to a physical structure. Nevertheless, we can make approximate analogies. In doing so, we must represent each rod or wire in the physical structure with two parallel signal paths in the control structure, one for a downgoing output signal, the other for an upgoing perceptual signal. With that caveat in mind, let us investigate the inexact mapping between the physical structure in Figure I.8.10a and the control structure in Figure I.8.10b. Both are tensegrity structures in the sense that they have the properties of resilience, flexibility depending on tension, and distribution of energy, but the structure of Figure I.8.10b has differently arranged virtual 'rods' which connect directly, a connection which is disallowed in a pure tensegrity structure.

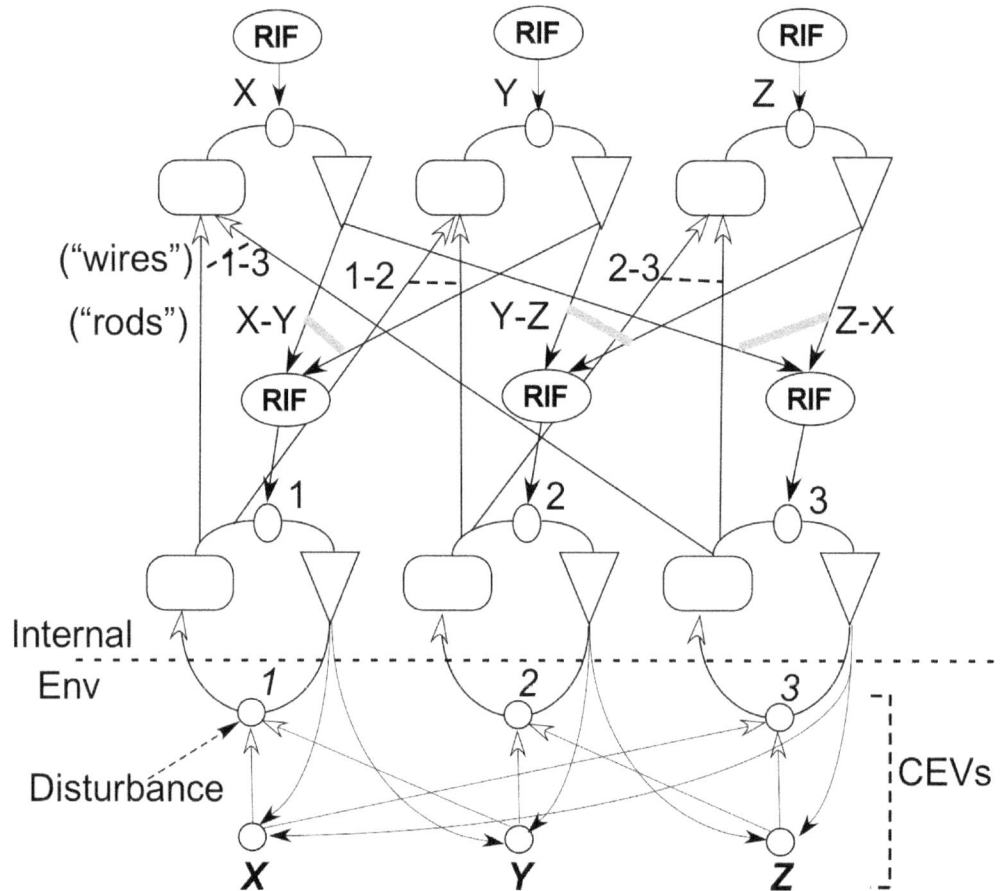

*Figure I.8.10b A control tensegrity structure. lines and arcs with white arrowheads represent upflowing perceptual information, those with black arrowheads downflowing action output information. Italicised symbols represent CEVs, Roman symbols represent controllers or their reference signals. It differs from a tangible tensegrity structure in that three **rods** link controllers X, Y and Z, and three rods may but need not also link the respective CEVs, X, Y, and Z.. "RIF" means "reference input function".*

The key observation about Figure I.8.10b is that it contains three overlapping and interacting copies of Figure I.8.9. Figure I.8.9 incorporated a conflict between the attempts of the output functions of the two higher-level controllers to set a reference value for lower-level controller 3. As noted in the discussion of Figure I.8.9, this conflict looks functionally like the diagram of Figure I.8.1a that illustrated 'Stiffness' of the rope in a tug-of-war, except that because both upper-level controllers can control through two other low-level controllers there need be little or no conflict-based tension.

In Figure I.8.10b, this conflict cannot be so readily avoided. The three upper-level controllers X, Y, and Z have reference values set from above, and reorganisation is likely to have built the structure so that these reference values do not correspond to unattainable relationships among the corresponding environmental variables

(CEVs) *X, Y,* and *Z*. The reference values X, Y, and Z cannot be identical, and there will be three 'tug-of-war' stiffening conflicts like the one in Figure I.8.9 — the three being X and Y conflict over reference 1, Y and Z over 2, and Z and X over 3. (These conflict 'rods' are informally represented by grey bars connecting the arrows that represent the conflicting outputs.)

These pairings set up a triangular tug-of-war conflict that cannot be resolved by using other low-level controllers to bring the higher-level perceptions to their reference values. Nor could they if extra 'wires' were to be added to form a complete many-to-many set of connections between the two control levels. The same conflicts would persist, and the same 'virtual rods' would separate X, Y, and Z or their respective CEVs. Environmental constraints might add independent 'rods' separating the variables in their environment. For example, two chairs cannot physically be in the same location, or nearer than the width of a chair. Adding either rods or wires to the structure in Figure I.8.10b would not eliminate the functional tensegrity properties of the structure in the figure, but they could stiffen the structure, perhaps even so far as to make it act like a solid block, which would seldom be useful for perceptual control.

Powers, in his 'Arm2' demo (CD in LCS III) reorganised a 14 perception-wide hierarchy of three levels to function with no interference between controllers at any level anywhere, no matter what the 'arm' was asked to do. Why should this smaller section of a completely linked hierarchy be different? The answer is in the environment and the ways the environmental variables *1, 2,* and *3* behave consistently together to produce the environmental variables *X, Y,* and *Z*. The environment of Arm2 consisted of the bones of the arm, wrist, and fingers, and the 14 variables to be controlled by coordinated muscular tensions were the angles formed by these rigid elements. These coordinated tensions were reorganised to be functionally independent, allowing smooth control of complicated actions of the Arm as a whole.

Powers could use reorganisation to adjust the interconnections between neighbouring levels, in the absence of any intrinsic variables other than Quality of Control (QoC). Here, however, we are assuming that to control each variable is of some advantage to real intrinsic variables, presumably biochemical.

The hierarchy reorganises over time to match the internal perceptual and reference input functions and the inter-level relationships to the environmental relationships among *X, Y,* and *Z*, so we should assume that this, in addition to the mutual non-interference within controllers at a level, determines the eventual parameter settings achieved by reorganisation. If the end result is sustained error and therefore sustained tension somewhere in the structure, so be it; the reorganisation process has arrived at a minimum for the energy stored in the tensions induced by the error.

Now it is true that the disturbance values of the higher level CEVs X, Y, and Z will vary dynamically as the whole structure settles toward mechanical equilibrium in the absence of disturbance, but control of higher-level variables is typically slower than control of lower-level variables. If it were not so, the high-low feedback processes with leaky integrator output functions at the higher level would very probably lead to an exponential runaway as the overall phase shift approached 180° at frequencies where the absolute gain was above 1.0. This would not happen if the output functions at the higher level were simple multipliers, but that is a detail for the producers of simulation models. Either way, controllers 1, 2, and 3 will not be significantly affected by the ongoing control of X, Y, and Z.

The sustained error tensions are thus concentrated in the three-way interaction among X, Y, and Z or the corresponding CEVs *X, Y,* and *Z.* As we noted above, any pair of them in isolation will show no sustained error. Only the introduction of the third prevents the first two from simultaneously bringing their perceptions to their reference values. What happens now when we note that most controllers have non-zero tolerance bounds?

Tolerance eases the restriction on how close a perception needs to be for it to be functionally equal to its reference value. If the 'third' upper-level controller, in this case Z, had a wide enough tolerance zone, the other two would be able to bring their perceptions to their reference values just as though the Z controller were not connected. But when all three have sufficiently wide tolerance zones, the whole structure is inert, with all perceptual values satisfactorily close to their reference values, despite normal levels of disturbances.

Such a structure is rather a waste of its components, since they do nothing most of the time. It would probably be reorganised out of existence. On the other hand, if the three-way conflict of errors and their correcting processes created an escalating conflict, as in McLelland's (1993) demonstration of a one-dimensional two-way conflict, the structure would be even less useful to the organism.

Finally, if three-way tension did not cause escalation of outputs, but could be held stable, then the structure would react to disturbances quickly and more powerfully than would the initially disturbed control loop were it controlling alone. It would be a true tensegrity structure, and (in 3-D) a minimal one from which no link can be removed without it losing its tensegrity properties.

How might such a boundary state be achieved? Let us look at McClelland's demo for a clue. Why does the conflict escalate at all? A positive feedback loop necessarily connects the two opponents through their individual negative feedback loops, since each negative feedback loop contains a sign reversal at the comparator, and two sign reversals cancel each other out. But this argument does not hold for a three-way conflict, since there are three sign reversals on the round trip. Nor can we appeal to the possibility that there would be an escalating conflict between any two of the three controllers, because when the circuit is complete, the third pairwise conflict opposes the first.

The three-way circuit is a negative feedback loop, tending to stability, and that is what we want for the tensegrity structure to hold its form without producing escalating error in any of the controllers. That circuit is not a control loop, it is a homeostatic loop, which we will discuss further in Chapter II.1 and Chapter II.2.

Chapter I.9. Lateral Inhibition

Powers's strictly hierarchical version of PCT disallowed lateral interconnections among elementary control units at the same level of the hierarchy. However, talking about sensory nuclei (Powers 2005:101), he did write:

*Ised*This chapter contains an attempt at such modifications of the model. In this section we argue firstly that lateral inhibition is commonplace and should be a part of the control hierarchy, and then that lateral inhibition does much more than cause edges to be enhanced. In a biologically natural way, it solves some open problems with the strict hierarchy. It supports tensegrity in the hierarchy.

Some of the functions that can automatically be produced by lateral inhibition include parsing complex input into informationally efficient forms (Taylor 1973a). This sharpens discrimination of details, creates category perceptions, creates conditions in which category perceptions and labels are associated so that the perception of either enhances the probability of the partner[3] being perceived, and enables associative memory more generally.

We will explore those other possibilities, some of which may not be intuitively obvious, a little at a time. But such effects presuppose the prior existence of lateral inhibition. Why did lateral inhibition evolve in the first place?

I.9.1 Why Lateral Inhibition?

One reason why lateral inhibition is not merely plausible, but necessary is that the brain needs a way to keep its energy usage as low as is compatible with effective operation. Every nerve firing dissipates some energy in the form of heat, and this heat must be dissipated outside the brain that holds the nerve that fired. In an organ as convoluted and tightly packed as is the human brain, dissipation of heat is a major problem.

Every firing of a neuron is sent down the nerve axon to many synapses that may activate or inhibit the recipient nerve. If all the connections are excitatory, then for each connected neuron the likelihood of its firing soon is increased. Each firing adds some heat that must be dissipated outside the brain and eventually outside the body. The temperature of the brain will rise until there is a balance between the overall rate of firing and the rate of dissipation of heat.

One implication of this is that the effect of a neuron firing cannot be simply activation of the firing of specific other neurons. It must be accompanied by an equal total inhibitory effect, averaged over local regions and moderate times, so as to maintain a fairly steady total firing rate across the local region. If few nearby neurons have fired recently, *this* neuron will be less inhibited than on average and will be more likely to fire. The opposite is true if several nearby neurons have fired recently. One could see nerve firings as an analog of infection, with lateral inhibition an analog of immunisation.

Lowering the energy requirement of the brain provides a potential increase in evolutionary fitness because it reduces caloric requirements and because it reduces the need for means of dissipating the heat of computation. Increasing its computational capacity increases evolutionary fitness by helping the organism to find food while avoiding becoming food, among other benefits. These two fitness enhancements are in conflict. A mechanism that lowers the energetic cost of computation allows more computational capacity while maintaining the balance with energy expenditure and heat dissipation.

The negative feedback provided by lateral inhibition seems likely to approximate a locally stable total firing rate over appreciable regions of the perceptual system, but only if a strongly firing neuron suppresses the firing rates of its neighbours. Although the result of that suppression is an enhancement of the firing rate of this particular neuron, the reduced firing rates of its neighbours compensates energetically for the enhancement. Hence, it is reasonable to suppose that a widespread use of diffuse lateral inhibition is evolutionarily beneficial, especially if there is a general mechanism for changing the level of inhibition over substantial regions as need arises.

These considerations deal with spatially distributed nerves. The energy-relevant inhibition requirement is the same whatever the function of the nearly co-located nerves. What lateral inhibition does functionally is a quite different matter, which we consider in the rest of this chapter. Before we do that, however, we should do two things.

The first is to note that the firing rate of a single neuron exposed to a constant pattern of input should not be expected to stay constant. If it were constant, energy that must be dissipated would continue to be used, even though that neuron soon ceases to provide any information to neurons further down the line. Changes provide information, steady states do not. From an evolutionary viewpoint, it would be parsimonious to signal a change when a steady state begins, reverting to the resting state while they remain steady. This pattern is the equivalent of inhibition across local time as well as of space, just like an on-centre-off-surround system (Figure I.9.2e) in the early visual system, but with one of its dimensions being time — as also is likely to be the case in the early visual system.

The second thing is to return to Powers's concept of the 'neural bundle' that was his elemental 'neural current' carrying link in a part of a control loop within the brain (Chapter I.4). A 'neural bundle' is a collection of nerves that tend to fire together when activated by a particular pattern of inputs at their synapses, but, as we discussed in Chapter I.7 (see especially Figure I.7.3c in Section I.7.3), the bundle's edges are not sharp, but fuzzy.

A neuron in the bundle may be tuned to the central pattern of the bundle, or to something similar but not identical to that central pattern. If every firing of a nerve in a bundle tends to inhibit its neighbours, the ones that are more strongly excited will survive the inhibition, while those that are marginally excited will tend to be more depressed, as they do not inhibit their neighbours so strongly. If there are two 'bundles' that would have a lot in common, each central group will tend to inhibit any intermediate peripheral members of both bundles, moving the overall best tuning of each bundle as a whole away from that of the other.

If only from the viewpoint of the survival of the organism against the possible heat death of its brain, we should expect lateral inhibition to be pervasive throughout the brain. Now let us examine a few of the particular effects that we should expect (and that are observed) in real brains and in experience and experiment.

I.9.2 Edge Enhancement and Displacement

The most obvious effect we should expect from a generalised lateral inhibition might be edge enhancement. If there is an edge between neurons responding to a region where some property is constant at one level and a region where it is constant at another level, the low-level nerves near the edge will be more inhibited by the nearby high-level neurons than by their compatriots. On the other side of the edge, the opposite is true. The high-level neurons will be less inhibited by their low-level neighbours than will neurons far from the edge, surrounded by other high-level ones. The ones in the central parts of the regions will, on both sides, tend toward the global average firing rate, giving the effect of edge enhancement (Figure I.9.2a).

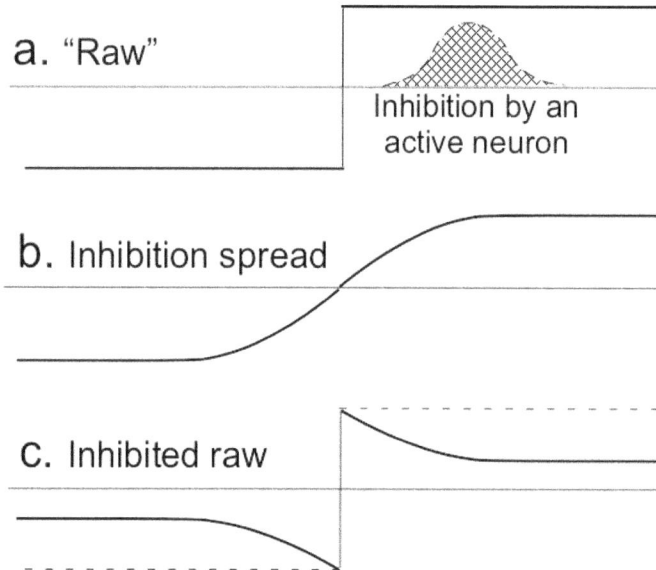

Figure I.9.2a Local inhibition leading to edge enhancement (horizontal midline in each panel represents local average value) (a, top) the relative amount of activation received by each of many neurons reporting the value of nearly the same variable. Shaded area represents the spread of inhibition from any one neuron to its neighbours. (b) The total spread of inhibition from the individual neurons. (c) The resulting contrast enhancement near the edge between the two regions, making the interiors of the regions more alike than the border regions. Eventually, all the neurons will tend toward firing at their average rate, if the boundary does not move.

If we look at what our senses actually provide to the rest of the perceiving apparatus, we see that they do not give a consistent output for a given physical intensity of input. The output of almost every sensor depends on the recent intensity of input as much as it depends on the present intensity. When you go from a sunlit area into a dimly lit room, at first you can't see anything at all. Everything looks black. After a while, contrasts begin to appear, and a few minutes later, the dim room may look quite bright. If you enter another room, you may sense a particular odour, but after a short while you may not sense that odour at all, even if it was initially quite offensive. If you have been in a noisy environment and move to a quiet one, it may be a little while before you hear anything at all (except the ringing in your ears, perhaps).

The same is true of local spatial differences of intensity in vision. Over time, all the neurons converge to their overall average rates of firing. If the boundary does not change or move, it will disappear, an effect dramatically realised in experiments on stabilised vision (for a review, see Martinez-Condé, Macknik, & Hubel 2004).[89] One's eyes normally cannot keep looking in a constant direction, because they are subject to a steady 'microtremor' that moves all boundaries back and forth across

---

89   PDF available at https://tinyurl.com/2caypvyw, retrieved 2022.11.30.

the retina. It is sometimes possible, however, for one to make an object vanish to vision if in a dim light one looks very steadily at one point on its boundary.

The contrast effect itself can be seen quite easily. A grey square among darker areas seems much lighter than a square of the same grey among yet lighter areas. In Figure I.9.2b, the right-hand pair of embedded rectangles may seem to be about the same lightness, but physical measurement or elimination of the surrounds shows that it is the outer pair that are the same and the middle one that is different.

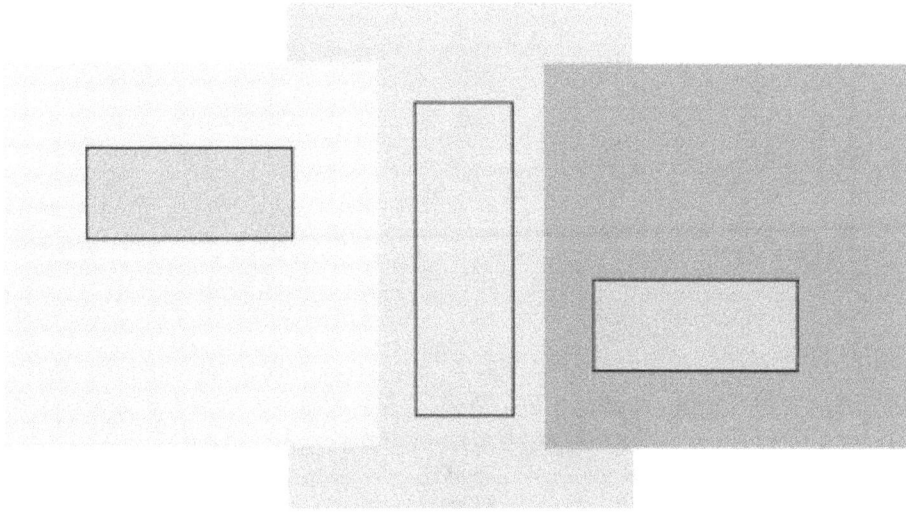

Figure I.9.2b The effect of local contrast on the perceived lightness of an area. Two of the embedded rectangles are the same shade of grey. Which two?

Similar effects occur in most areas of perception, including vision. For example, edges are instantly visible between regions of the visual field that have characteristics that are steady over the individual regions but that differ in some way between neighbouring regions, as in Figure I.9.2c and Figure I.9.2d, even though edges in those figures are nowhere explicitly marked. In Figure I.9.2c the regions mostly differ in the nature of the grouped objects, though the right-hand group and the next door group, and the upper and lower centre groups, contain the same objects, differing only in their spatial arrangements. Yet the boundaries between the regions are visually distinct, as though the regions were separate entities.

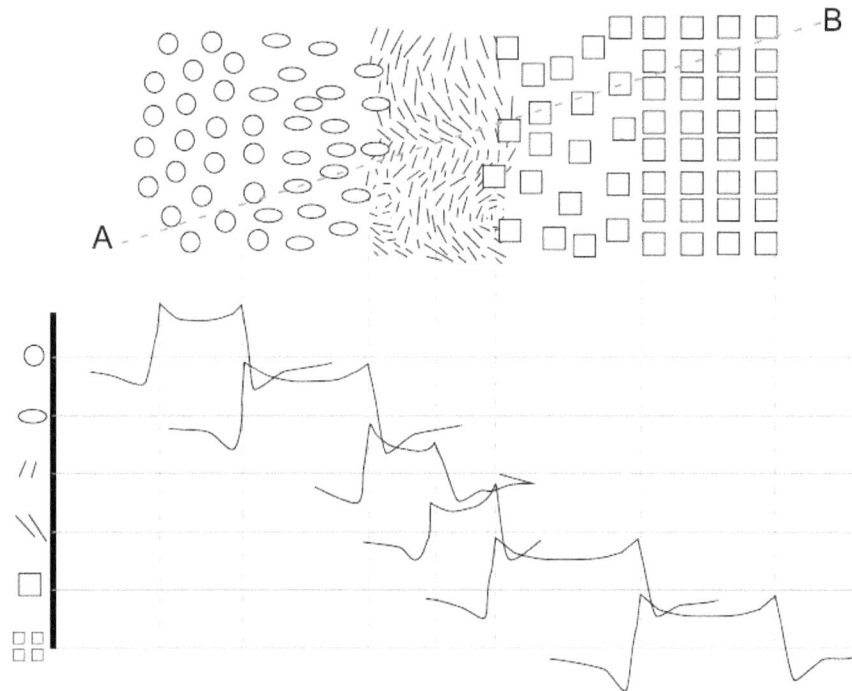

Figure I.9.2c (Upper) Regions of the visual field with easily perceived edges based on differences among the characteristics if the regions: shape of the elements, regularity of the arrangement of the elements, spacing of element groups (in the right-hand region) and orientation.texture (the two areas of short lines). The transect A-B represents the line along which the perceptual function outputs are shown in the lower half of the Figure. Note the asymmetry in the output graphs for the two scattered-line regions, between which the edge is less clear than in the other cases.

The sameness within most patches arbitrarily located contrasts with the lack of sameness within patches that straddle the invisible lines and curves that are region boundaries. So how might that function? Lateral inhibition is one possibility. We can suppose that at any level of the perceptual structure every perceptual output provides a low-level inhibitory input to neighbouring perceptual functions whose perceptual variables correlate highly with its own. When the neighbouring region is 'looking at the same thing', the mutual inhibitions do not extinguish each other, but they do depress each other's output. But when the neighbour is 'looking at something different', its output is either lower or higher. The mutual inhibition will then enhance the contrast between them, leading to outputs that might vary as suggested by the lower diagram of Figure I.9.2c, a cartoon of the possible outputs of the relevant detectors along the transect A-B.

Thinking about prediction, in most places in the upper half of Figure I.9.2c, if you look at a small patch, you will predict that nearby there will be more of that same kind of element, whether they be circles, short lines of a particular orientation, longer lines of a particular orientation, ovals or squares. But for some of the patches you can't do that, because the patch contains more than one kind of element, which means that taking only that patch into consideration, outside it there may be yet other kinds of element beyond the two in the patch. That change of predictability defines the edge. Edges can therefore be interpreted as places in the visual field across which prediction becomes unreliable.

These phenomena are equally if not more evident at much higher levels of perception. Social perceptual control is the topic of the latter volumes of this book, but we may observe here that the social environment in which you live is an unperceived background to ongoing events. It may not be ideal, but it is normal. An old question is "Does a fish know about water?" We do notice when something changes that had been steady. If we go to a different region or country, we may notice aspects of our previous social environment that we had not perceived consciously before, because of the contrast with the society to which we moved. If social policy changes after an election, we are likely to notice the changes more than the things that the new government leaves alone. We compare neighbour against neighbour at all levels of perception, from the visual system to the nature of the government between region and region (nation and nation, perhaps) and between one time and another. In Figure I.9.2c, it is easy to see the individual circles, ovals, lines and squares, but it is also easy to go beyond these and see six regions with more or less clear demarcation edges between them, and in the lower of the two swirly-line regions one might also see two smaller regions that merge smoothly into one other without an edge between them, around a pair of centres. The six regions do not need analysis in order to be distinguished; they are immediately evident to the eye.

In Figure I.9.2d we easily see three distinct regions. although nothing distinguishes them except the distributions of the angles relating the invisible lines that connect neighbouring circles. Even then, we see patches that are 'the same' even though their elements may be irregularly strewn, and we see never-drawn lines and curves across which 'things are different between here and there'. We don't look for the difference edges. They are just there, in the contrasts between the neighbouring areas of sameness in the statistics of neighbour orientations.

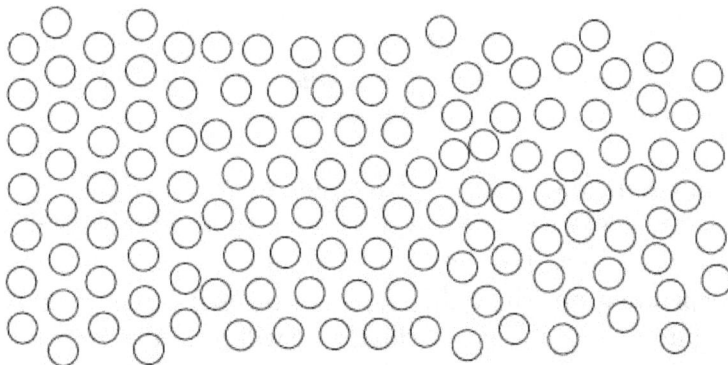

Figure I.9.2d Three regions distinguished only by the distributional statistics of their elements. The edges between them are perceived clearly.

A neuron with lateral inhibition has many of the characteristics of an on-centre-off-surround retinal process, as suggested in Figure I.9.2e, but occurring in N-dimensional feature space rather than 3-D normal space or the 2-D and 1-D spaces of the figure. It also has much in common with a category recogniser. Cossell et al. (2015) found that excitatory interconnections in the visual cortex have just this characteristic. They did not study the inhibitory connections, but some inhibition is necessary, to prevent all the neurons for all the features firing together, which would be a waste of energy and create heat that would need to be carried away to the cooler environment.

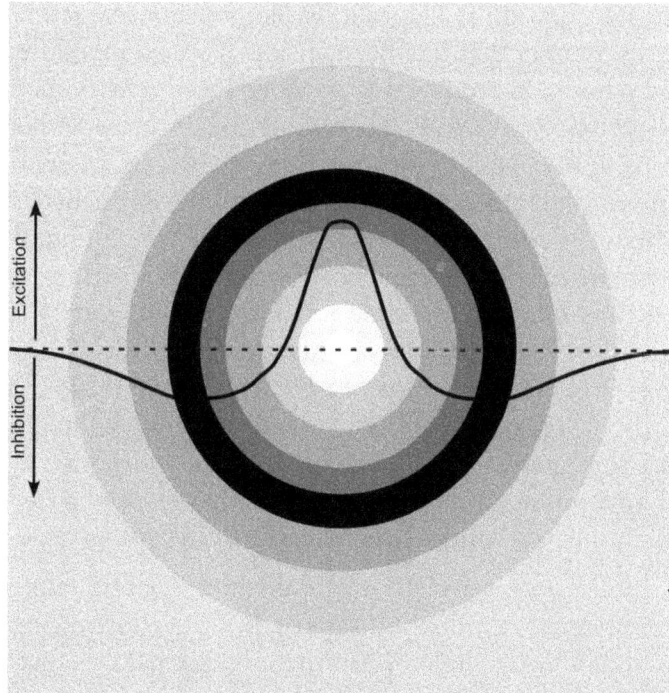

Figure I.9.2e. The relative degree of excitation or inhibition in a map of feature weights for other units around any one unit of a polyflop, graphed in one dimension and shown in grayscale in two dimensions. The same form can be extended to many dimensions. The unit of interest is at the centre, and the axes represent feature values for other units with lateral connections to it. Another unit with very similar feature weights will reinforce and be reinforced by the unit's output, whereas if the other unit has feature weights in the black ring, that the two units will mutually inhibit one another in flip-flop fashion.

If the environment includes patterns that would excite individual on-centre-off-surround units close to each other in the feature space of a level, the apparent locations of the patterns will be displaced either toward each other or away from each other, depending on where they lie in each other's ring of excitation and inhibition (Figure I.9.2e). If they are close, they will seem to attract one another, and might even merge to be perceived as one item, as in the upper panel of Figure I.9.2f, whereas at a greater separation, they will seem to repel each other. These effects will be important when we talk about illusions and after-effects.

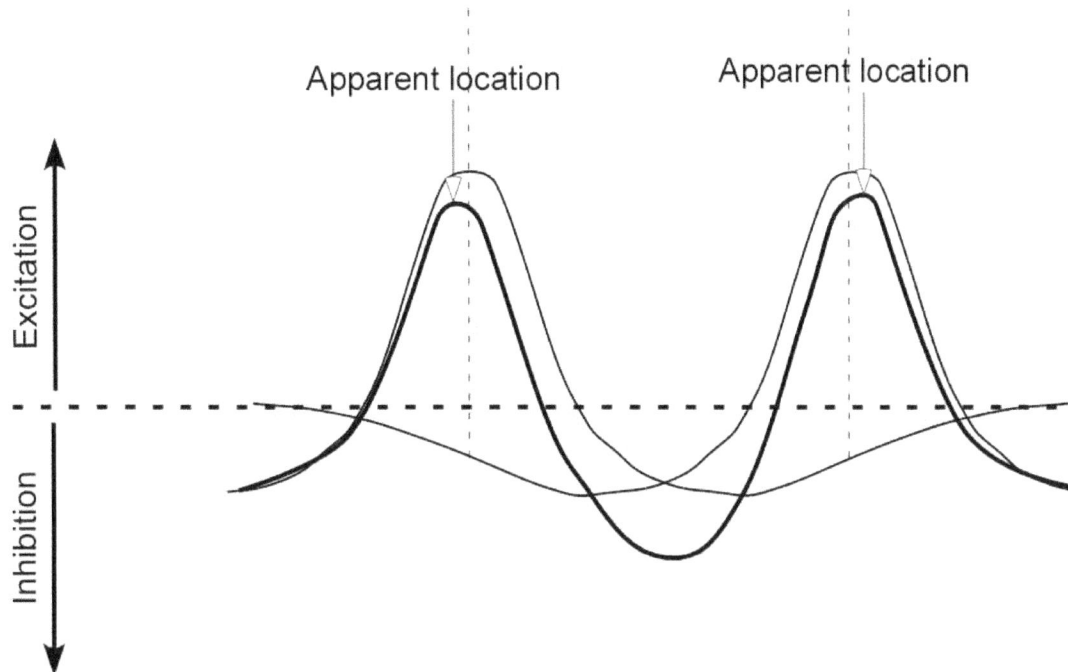

Figure I.9.2f Shifts of apparent location in a lateral inhibition field when two nearby units would have been individually excited by environmental patterns of features. If the locations are very close, only one is seen, at a location between the two, but if they are more separate, they repel each other.

An evolutionary reason for the existence of lateral inhibition and experimental evidence of it are necessary but not sufficient. It also requires a mechanism for its development. So now let us deal with one possible functional mechanism, after which we will consider various consequences of lateral inhibition beyond edge enhancement.

I.9.3 Hebbian-anti-Hebbian (HaH) Learning Process

'Hebbian learning' in the present context means the strengthening of a synapse that occurs when a presynaptic spike at that synapse is closely followed by a post-synaptic spike in the neuron; conversely, anti-Hebbian learning refers to the weakening of synaptic strength when the reverse timing pattern occurs, or when a presynaptic spike is not followed by a postsynaptic spike, or when the neuron fires with no closely prior spike having occurred at the synapse. In 'HaH' learning, both processes are operative.[90]

---

90   The detailed situation is considerably more complicated than this simple description, involving several different kinds of molecules, receptor channels, and feedback loops that affect the timing and even the momentary degree of plasticity of a synapse (e.g. Tigaret et al. 2016). However, the main points are reasonably valid.

In (Taylor 1973a) I proposed that HaH learning should exist, a proposal subsequently independently supported in many areas of the brain as well as in various species (e.g. Bar-Gad and Bergman 2001, Bell et al. 1993, Bell et al. 1997, Carlson 1990, Koch et al. 2013, Kullman and Lamsa2008, Lamsa et al. 2007, Markram et al. 1997, Roberts and Bell 2002, Roberts and Leen 2010, Tzounopoulos and Kraus 2009, Tzounopoulos et al. 2007). The process must be important in the ever-changing brain if it exists in so many brain areas and in different species.

In the 1973 paper I reasoned that the effect of the HaH process would be to create lateral inhibition, which would sharpen sensory discrimination and recode the sensory input into a more efficient representation, such that clusters of neurons would come to be tuned to mutually independent sets of salient features. In other words, it would produce new perceptual functions that were statistically more efficient in representing the sensory world than the corresponding vector of lower-level perceptions. In a continuous analogue world with partially correlated values (as with the perceptions away from the edges in Figure I.9.2c), they would approximate a 'principal components analysis' of the incoming data (Figure I.9.3a).

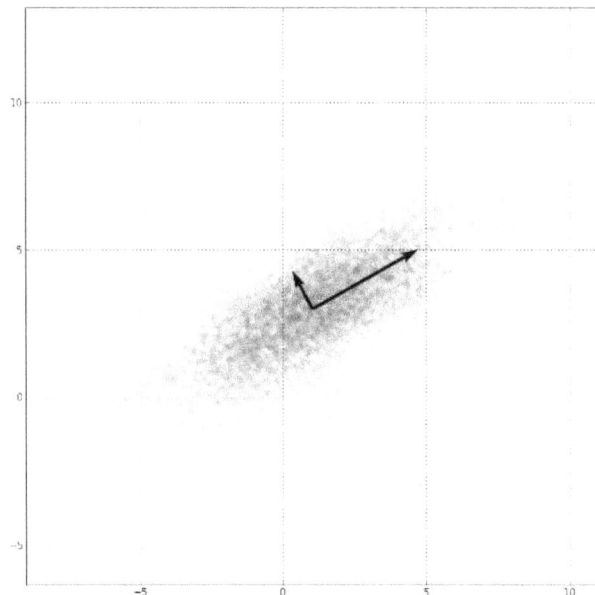

Figure I.9.3a. A scatter plot with correlated x and y values, showing the principal component directions in which the data would be more efficiently encoded. (from Wikipedia http://upload.wikimedia.org/wikipedia/ commons/1/15/GaussianScatterPCA.png, retrieved 2015.13.10)

In a principal components analysis of high-dimensional natural data, it usually turns out that a very small number of components carry almost all the relevant information about the source; the many remaining components largely represent minor statistical fluctuations. Redundancy in the patterns of perceptual input would therefore result in the production of a smaller number of higher-level perceptions that were less redundant (more independent, less mutually predictable) than the original set, as in the case of the location and orientation control units of Figure I.5.5c for which there were many fewer 'chair' units than 'leg', 'seat', 'back' units. What changes together goes together. Several researchers have since independently confirmed that aspect of my 1973 proposal (e.g. Falconbridge et al. 2006; Földiak 1990; Girolami & Fyfe 1997; Hyvärinen & Oja 1998; Plumbley 1993a, 1993b).

The importance of reducing redundancy can be illustrated by a case study in data analysis (which is fundamentally what perception is). As part of a major study of sleep deprivation (Pigeau et al. 1995), I asked the sleep-deprived subjects to perform a suite of tracking tasks of different kinds and of different difficulty levels. In all, several thousand tracks were recorded. To fit a PCT model to those tracks I used Powers's e-coli hill-climbing method of approaching the optimum, working in the space of the raw data. Some years later in a different sleep-deprivation study, to compare two PCT models each with five parameters against 1300 human tracks, I compared the e-coli fit based on the five raw parameters against a fit using genetic algorithms that included a parametric rotation of the data space.

The results of the comparison were reported to the CSG annual meeting in 2005 (Taylor 2005). The e-coli fit, which apparently satisfies many criteria for efficiency, was less successful than the genetic algorithm fit in consistently finding near-optimum sets of parameters that allowed a comparison between the models in their ability to produce tracks that matched those made by the human subjects. Why?

In the e-coli fit, very often a change in one parameter of a hypothesised control model could be offset by a compensating change in another. For example, changing the (quasi-logarithmic) power law in the comparator function (Figure I.4.6a, panel b) could be simulated by changing the gain rate of the integrator in the output function. There are many such correlations among the effects of the different variables. The genetic algorithm fit included parameters that rotated the axes to an optimal configuration akin to a principal components representation, and produced much better fits than were found using the e-coli method without rotation.

In producing a principal-components representation of the incoming data by lateral inhibition, the HaH process incidentally sharpens the perception of edges (Figure I.9.2c) and category differences (discussed later).

A principal-components analysis is well suited to data whose statistics are invariant over the different sensory environments encountered over time, but biological organisms are exposed to different kinds of statistics in one environment as opposed to another. Someone who worked in a city with many straight edges and right-angles, and who also took extended field trips to a jungle environment would not be well served by a single perceptual analytic structure that was a compromise between the two environments. Either the principal components basic structure would have a considerably enlarged number of axes that account for substantial variance, or the representations would be inefficient for either environment. One might well expect that not only would a generalised principal components representation be imprecise, but also that different competing sets of principal components-like analytic structures would be developed to suit the different statistics of the different environments.

Whereas the different representational dimensions (axes) of any one principal component structure are orthogonal, between two differently tuned analytic structures the axes of one would not be orthogonal to those of the other. Lateral inhibition between entire structures would be expected then to enhance the learned structure most relevant to the statistics of the current environment, and to suppress those more appropriate to different perceptual environments. In other words, what you would see for a given input would be context-sensitive, depending on the general category of the environment, such as 'urban' or 'vegetation'. For example, a sharp corner around an area darker than the surround might contribute to the perception of 'leaf-not-window' in a 'vegetation' context but to a perception of 'window-not-leaf' in an 'urban' context because of lateral inhibition of the currently irrelevant representational basis category (Figure I.9.3b).

Figure I.9.3b Context affects the most efficient representation of picture elements such as corners. (Photos by the author).

Although one is seldom in a city and a jungle at the same time (the 'concrete jungle' metaphor notwithstanding), the enhancement of one representational basis in favour of the other is unlikely to cover the whole visual field. As implied by Figure I.9.2b and Figure I.9.3b, lateral inhibition, though widespread, covers only a neighbourhood, and there are edges where the suppression fails because the 'neighbourhood' applies to the feature space as well as to physical space. The same should be expected to be true of the lateral inhibition of categories of representational bases. Some areas of the visual scene might be most efficiently represented in one basis, some in another (Figure I.9.3c).

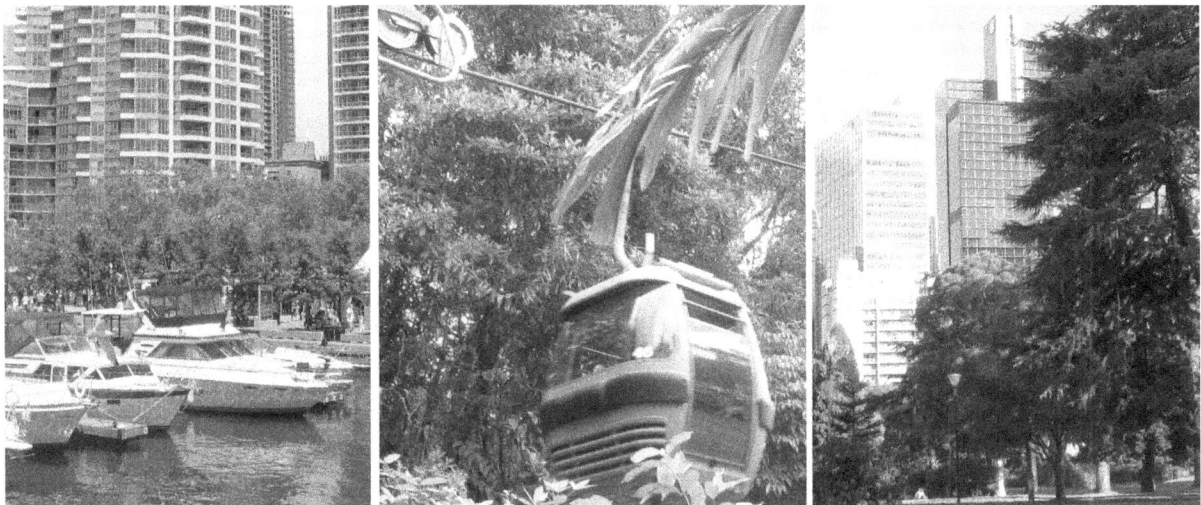

Figure I.9.3c Regions of different statistical characteristics require different perceptual basis structures across clearly visible edges (as in Figure I.9.2c). (Photos by the Author)

We are dealing here with each and every level of the Powers perceptual hierarchy separately. We have no justification for considering lateral inhibition to work across the different levels of the hierarchy. The photographs show scenes that we interpret as objects of different categories, but they also show changes in the local statistics of intensities, relationships, and so forth. Being static images, they do not, of course, show events and time-based sequences, but the buildings do illustrate what we could call space-based sequences, repetitions of much the same set of features.

Principal components representation is efficient when the data are continuously variable in a multidimensional space, with distributions that are reasonably like a multidimensional Gaussian. However, when one looks at a wind-ruffled water surface, the myriad distinct blobs of varied light and shade are more efficiently treated as a single perception that we might call the water's roughness, with possibly some modulating perceptions of how the roughness changes from region to region and moment to moment. The same thing applies when patterns are constructed from feature values that cluster around discrete locations in a feature space, as they do for objects and as suggested in Figure I.9.2d for an abstract space of two dimensions.[91]

---

91   Later, in Volume II, we will call such fixed locations in a descriptive space 'syncons'. We leave further explanation until then, but note the future use of the neologism.

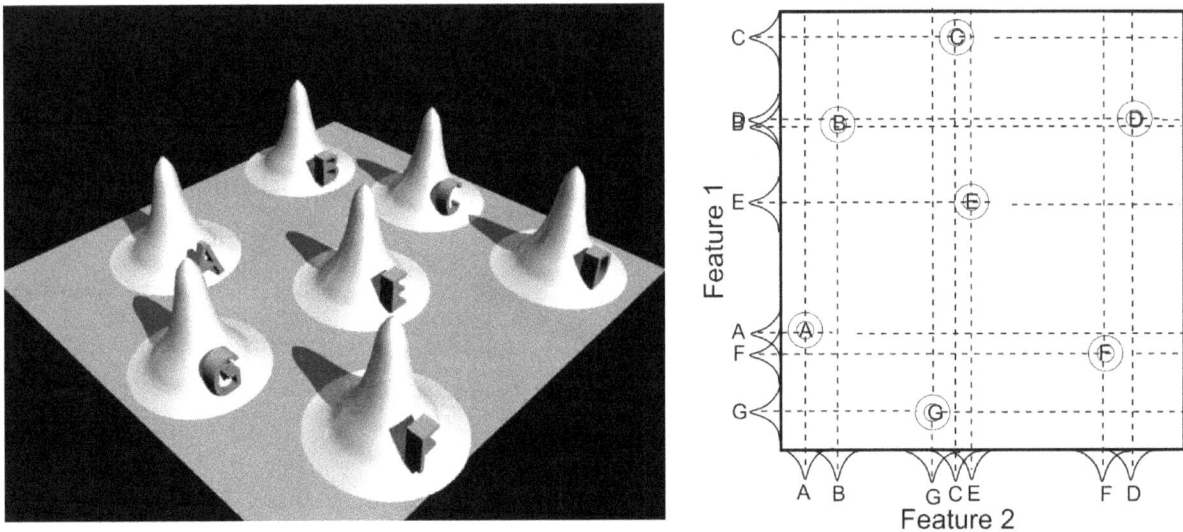

Figure I.9.3d. A 2-D space in which data cluster around just seven canonical values, seen as a 3D view and as a 2D plan view of the distribution.

In Figure I.9.3d, rather than every feature being continuously variable over its whole range independently of the values of the other features, the uncertainty of the feature values is much reduced by identifying a specific location, say location 'G' in the figure Instead of the continuously variable space of feature variation being recoded into a principal components (i.e mutually independent) derived feature set, a better recoding would result in something akin to a set of discrete entities, identifiably different, even if the distributions within individual features are much wider than the very tight cones shown in the figure.

As we shall shortly see, the difference between creating a principal components representation and creating a set of categories by lateral inhibition is not a difference of kind, but of degree — a difference in the magnitude of the lateral inhibition among related potential perceptions of a given complex sensory input.

The distribution suggested by the cones in Figure I.9.3d is a two-dimensional surrogate for distributions of features at a single perceptual level in what is usually a much higher dimensionality space. If such a distribution represents the 'sensory' world after some processing through different levels of the hierarchy, so that the processed data appears almost always near one of the locations marked in the figure by the cones, then to identify a point as being 'A' plus a vector of small values representing the deviation from an ideal 'A' would be more efficient than simply to create a perceptual function that was an analogue function of all the actual feature values. In practice, the distributions might be appreciably less sharp, and could overlap appreciably on any one feature dimension, though as suggested in Volume IV, Appendix 5, they would be unlikely to overlap much in the higher-dimensional feature space.

But what would be the "ideal 'A'"? Only the external analyst knows about 'A's and 'B's. The sensory system has no labels (yet we show later how it may develop that capability). It 'knows' only that feature patterns cluster near certain areas of the feature space, whereas the values taken by individual features do not. The 'ideal A' or 'ideal B' is simply a location in the feature space defined by a cluster of patterns.[92] One might call the evocation of the new perception 'one of those', as opposed to a random set of features.

I.9.4 Flip-flops and Polyflops

Figure I.9.4a shows the functional operation of a common electronic circuit that lies at the heart of digital computation, a 'flip-flop' and its extension to several categories as a 'polyflop'.[93]. However many units there are, the output of each provides an inhibitory signal to all the others, so that the one with the strongest output is also the least inhibited. If the inhibitory connections are strong enough, the resulting positive loop gain results in only one of the units having a significant output, the others all being thereby inhibited. The polyflop circuit thus has the effect of a set of on-centre-off-surround units such as is diagrammed in Figure I.9.2e. Lateral inhibition is at the heart of the operation of flip-flops and polyflops. If the HaH process does create lateral inhibition, it will necessarily create flip-flops and polyflops when the mutual inhibition loop gain is sufficient.

---

92 Platonic Ideals are discussed in Chapter II.7 of Volume II.

93 The right side of Figure I.9.4a shows only three categories, in an arrangement we call a 'tri-flop'. My colleagues and I developed a hardware system that used many triflops for running experiments in psychoacoustics, published in J. Acoustical Soc. Amer. under the generic title "MDCC" over the period 1969-75. These serve as a practical demonstration of their feasibility.

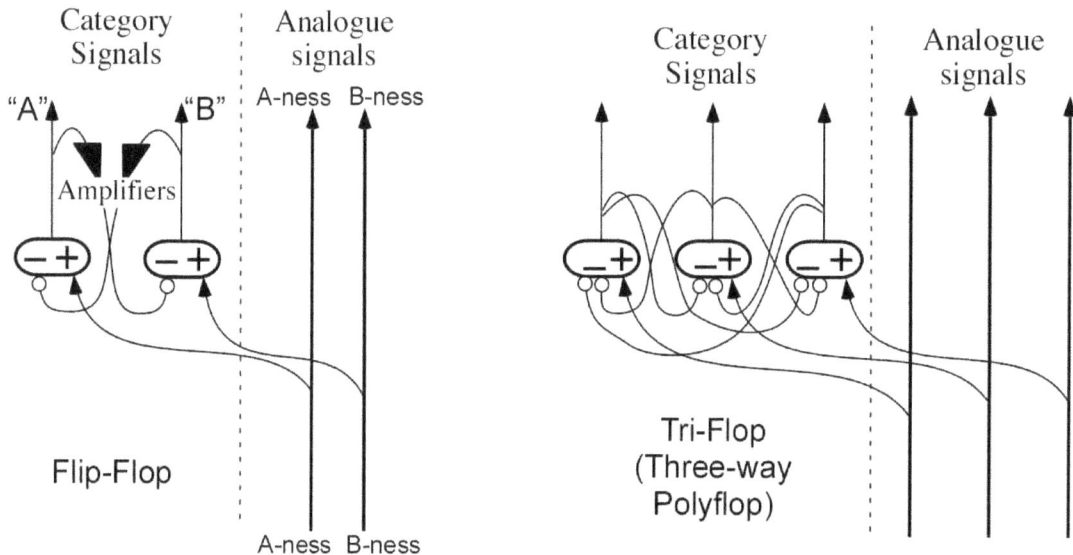

*Figure I.9.4a. (Left) a flip-flop. If there is more A-ness than B-ness in the
analogue signals, the left function produces more "A" output than the right
function produces of "B". Consequently the "A" output inhibits "B" (indicated by
the small circle as opposed to the arrowhead) more than the "B" output inhibits
"A". If loop gain is less than unity, "A" is enhanced and "B" reduced relative to the
values in the Analogue side of the diagram. If the positive loop gain is greater than
unity, the "A" output will go to a high value, and the "B" output will go to zero.
The outputs will stay that way until the analogue signal balance clearly changes
to an excess "B-ness" sufficiently strong to overcome the "A" inhibition, at which
point the "B" output goes high and the "A" output goes to zero. (Right) The same
effect can be created with multiple possibilities in a circuit called a "polyflop". The
diagram shows a circuit called either a "three-way polyflop" or simply a "tri-flop"
(lateral connection amplifiers omitted for clarity).*

The usual job of a flip-flop is to output a decision as to which 'one of those'
perceptions best represents the current vector of inputs, taking into account the
recent history of the vector values. In that role, it can be seen as a perceptual
function that reports identifiable categories. But the flip-flop or polyflop circuit
shown in Figure I.9.4a does not necessarily lead to an either-or -output. Whether
it does depends on the loop gain around the competing units (Figure I.9.4b). The
circuit may permit intermediate states that represent preferences for one possibility
or another, states that we further investigate in Chapter I.12, and in greater detail
in Volume II when we introduce the concept of 'crumpling' (Chapter II.6).

In Chapter II.10 we consider how the identifiable categories are consciously
perceptible, whereas the analogue versions are not. Flip-flops exhibit hysteresis,
which implies that their switching back and forth demands energy beyond
that inherently involved in either direction of switching considered by itself. A
hysteresis loop is involved only in conscious thought, because the non-conscious
processes make no choices other than those implied by the lateral inhibition
among the analogue versions of the perceptual categories.

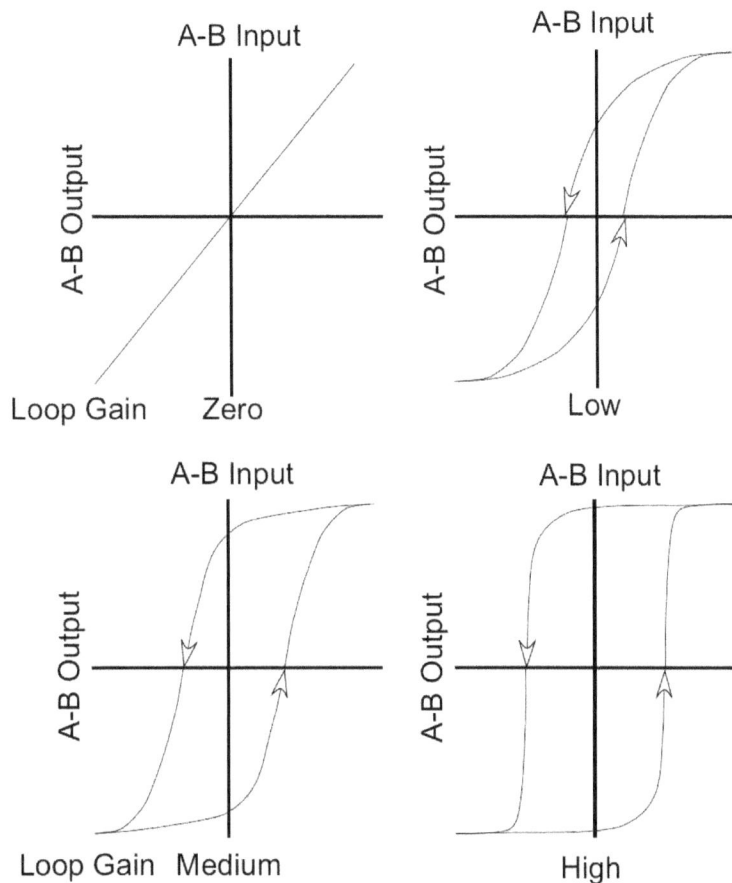

Figure I.9.4b Hysteresis loop of input-output changes of a flip-flop in which the output of each side of the flip-flop can vary between zero and a common maximum. The one that has produced the greater output continues to do so even as its input decreases to fall below that of the other. As the cross-link loop gain increases, it takes a bigger difference between A and B to make the switch, while the outputs of the "winner" and "loser" remain closer to the maximum and zero respectively.

The loop gain from, say, the 'A' side of the flipflop back to itself depends on the strength of the lateral inhibitions and the gains in the individual recognisers (the Perceptual Input Functions of any ECUs that might control these perceptions). If the loop gain is high or if the 'A' analogue input is much larger than the other analogue inputs, only the 'A' output will be active, but if the loop gain is low and the 'A' input is not much stronger than the others, then the result will only be an increase of the 'A-ness' and a reduction of the 'B-ness' of the set of outputs. In other words, with low loop gain the flip-flop or polyflop acts to enhance contrast, rather than generating a categorical perception.

Figure I.9.4c shows typical outputs from the two elements of the flip-flop at the left of Figure I.9.4a, as a joint function of the difference between their inputs and the loop gain between them. Three symbols are shown at points that have very different outputs even though the input is the same for all of them. The outputs are at 'o' if the loop gain is low. The A output is slightly enhanced and the B output slightly depressed by lateral inhibition compared to the two inputs, thus sharpening the perception of the difference. When the loop gain is high, the same A and B input values produce either the 'x' or the '+'state. All three symbols show the same values of A and B inputs, and moreover, 'x' and '+' both have the same value of loop gain. The only difference between 'x' and '+' is the history of how the outputs got to where they are.

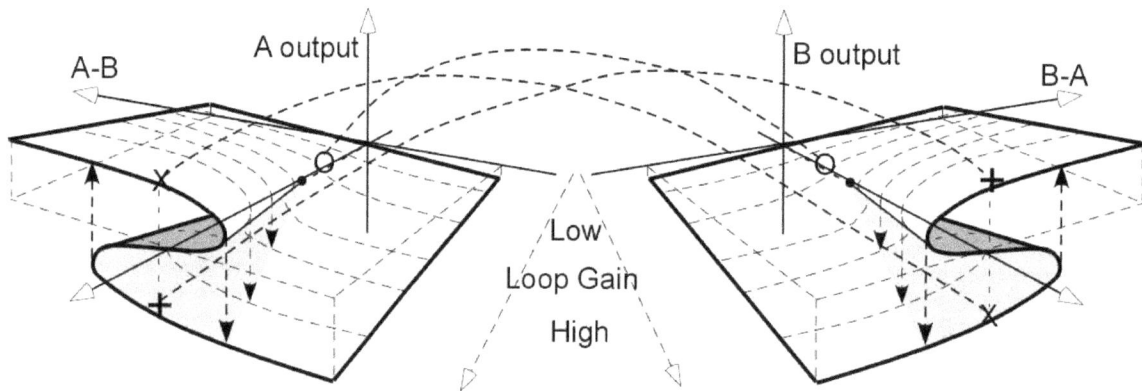

Figure I.9.4c. The possible output states of a flip-flop, as a function of the difference between the two inputs and the loop gain through the lateral inhibition connections. The figure depicts a "cusp catastrophe". At high gain one of the outputs is high and the other low, whereas at low gain both outputs can be moderate. The points marked X, o, and + all have the same value of A-B. O differs from the other two in loop gain; X and + differ only in their history, X having been in a state of greater A-B difference, + having moved from a value of A-B that favoured B.

Figure I.9.4d shows cuts through the surfaces of Figure I.9.4c at the common value of the A-B input difference shared by all three symbols in Figure I.9.4c. The lack of a path from the o symbol to the + illustrates that the x state is the one that will be found if the state is initially at o and the loop gain is then increased without change in the input.

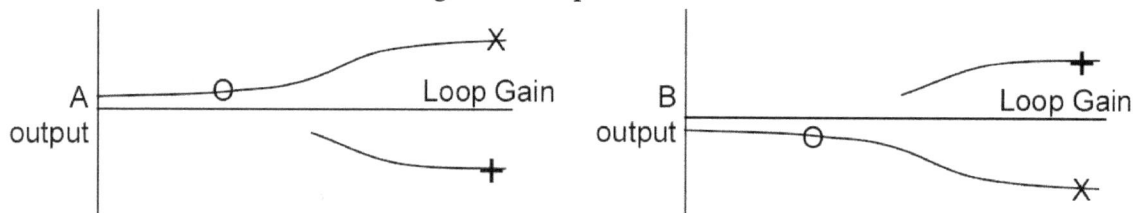

Figure I.9.4d. A slice through the diagram of Figure I.8.5a at the value of A-B corresponding to the three symbols. If the loop gain were to increase slowly from the value at o, the outputs would have to move to a state like the x, because there is no path from o to + at that value of A-B.

If the loop gain between the two elements of the flip-flop is low enough, both A and B may produce output (the o symbols), even though the input pattern cannot be both an A and a B. If both outputs are used at a higher perceptual level, the complex might be perceived perhaps as a 'B-ish A' or 'kind of A but with a bit of B', as in Figure I.9.4e, but would be used as one or the other. All cusp catastrophe surfaces have this same characteristic, though seldom are they employed in the context of a tutorial on perception.

Figure I.9.4e A "B-ish" letter "A".

The inconsistency might not be perceived unless at some yet higher level to have both A and B outputs creates a conflict between controlled perceptions. The input might, perhaps, be a sound pattern that could be /l/ or /r/ or just a random noise, but unless the context requires the pattern to represent a phoneme and a choice must be made as to whether a word was 'lug' or 'rug', all three possibilities might be represented as fairly low strength inputs to higher-level perceptual functions. As the circuit is shown in Figure I.9.4a, the raw analogue input values might in any case be available as non-categorical inputs to higher perceptual functions.

Figure I.9.4f shows the way that a single pattern of lines may be perceived as two quite different letters without changing the pattern in any way. The figure is intended to suggest a bunch of sticks seen on the forest floor. The central member of the group at the top of the figure might look like 'A' or 'H' or just a bunch of sticks overlaid on one another by happenstance, but at the bottom of the figure, where the context is clearly intended to be words, the same configuration of sticks is seen once as 'H' and then as 'A', both instances quite unambiguously. At a higher level, the word-constraint and implied situational contexts suggest why most people immediately read the bottom part of the diagram as

'THIS WAY ➡' rather than 'THIS WHY ➡' or 'TAIS WAY ➡'.

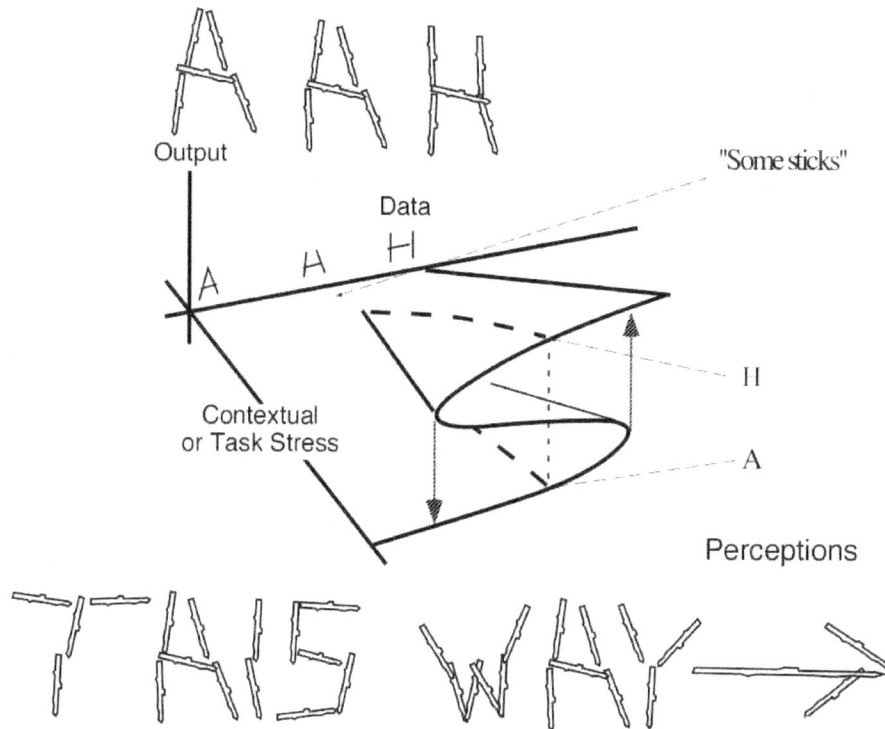

Figure I.9.4f The way the same visual pattern is seen differently in different contexts. Outside the context of what might be words, the group of five "sticks" in the middle of the upper set might be seen as a more or less random arrangement in space, but once it appears to belong to a word, the "contextual or task stress" increases the flip-flop loop gain, and the pattern is likely to be seen as a letter appropriate to the context. Why do you (probably) see the right hand word as "WAY" rather than "WHY"?

Flip-flops show hysteresis, at least when the loop gains ('task stress') are high enough to generate the fold shown in Figure I.9.4c and Figure I.9.4f. If they currently produce a particular output, they will continue to do so if the input returns to a neutral state, or even if the input becomes slightly biased toward a different output, as illustrated by the × and + symbols in Figure I.9.4d. We will see the cusp catastrophe and hysteresis in a wider scope when we talk about crumpling in Chapter II.6. Crumpling is a widely occurring and socially important phenomenon which will come to the fore again in later volumes. Within the individual, hysteresis can be seen at higher perceptual levels in a reluctance to change a reference value for a consciously controlled perception, which in an extreme version is sometimes called 'pig-headedness' or 'stubbornness'.

The effect is similar to what we see in many levels of perception, where something is seen one way until sufficient contrary evidence is amassed, at which point the perception switches, and the person says to him/herself "How could I not have seen that?" Yellen (1980) found hysteresis for as simple a task as brightness discrimination and explicitly identified it as the kind of 'cusp catastrophe' that is depicted in Figure I.9.4f. No hysteresis was evident in an easy version of the task, but hysteresis was clear in a difficult version. Hock et al. (1993) showed hysteresis for apparent motion direction, and Brady and Oliva (2013) showed it for facial recognition.

Hysteresis also occurs in reversing figure perception (Taylor and Aldridge 1974). In this case the data suggested that the cause was the existence of a small finite number (in the upper 30s) of independent units with analogue values jittering between two possible interpretations of the physical pattern, together being interpreted categorically at a higher level that required more than a simple change of majority vote to cross between one possibility and the other.

The end result of these processes is that if at some level the pattern of inputs is redundant, then it is probable that a set of perceptual functions will develop to take advantage of the redundancy in two ways. If the feature values vary continuously, the new perceptual functions should represent a principal components analysis of the feature pattern distribution, but if they cluster around discrete locations in the feature space, the new perceptual functions should have properties like those of category recognisers, and will be available for control with hysteresis.

I.9.5 The HaH Process, e-coli Reorganisation, and Novel Perceptual Functions

The HaH process directly implements at least one aspect of the e-coli-type reorganisation proposed by Powers. If we take the strength of connection between neurons A and B to be a value along one dimension in a space of a huge number of dimensions, and the vector of these values (representing the state of all the neural interconnections) to be a location in that space, then HaH tends to continue moving the location in a consistent direction in the space. However, HaH does not implement in any obvious way the other essential component of the e-coli process — the random change of direction when things begin to get worse.

In a polyflop, however, if the input corresponding to the currently strong category output becomes weak enough while none of the other outputs is strong compared to the others, the polyflop output switches in a quasi-random manner to some other category. (This is also suggested by the results of Taylor and Henning 1963, on changes of what is perceived during long-term presentation of ambiguous figures in various perceptual dimensions.. Such a change alters the properties of the HaH process after the switch, which could implement the random change of direction required by the e-coli proposal. None of this has been investigated either theoretically or by simulation, so the suggestion is little more than pure speculation.

On the other hand, as Powers's 'Arm 2' demonstration shows clearly,[94] e-coli reorganisation does tend to orthogonalise the output side of the control structure,

---

94   At http://www.livingcontrolsystems.com/demos/tutor_pct.html,
    'Arm with 14 degrees of freedom'.

which is the same as performing the Principal Components reorganisation that the HaH process would perform on continuous analogue input data. Since we argue that the HaH process should implement at least the continuing direction aspect of the e-coli process, and they both achieve the same final result, it seems reasonable to suspect that HaH might possibly be the mechanism underlying the whole e-coli reorganisation process.

The flip-flops and polyflops that occur with lateral inhibition do not arise magically out of nothing. As suggested by Taylor (1973a), the HaH process produces lateral inhibition because when two neurons are confronted with a change common to both their inputs, but one more strongly than the other, the neuron with stronger input tends to fire sooner than the neuron with weaker input. In the HaH process, synaptic connections that occur just before the receiving neuron fires tend to be strengthened, whereas those that occur just after the receiving neuron fires tend to be weakened.

My 1973 proposal assumed that the input change occurred at the sensors, but perhaps a more plausible mode of synchronisation is provided by quasi-regular firing rhythms such as those which are given Greek letter names in EEG records ('alpha waves', etc.). The original concept of Hebbian learning (Hebb 1949) was that neurons that fire together join together. Anti-Hebbian learning occurs when an excitatory synapse becomes weaker, as happens when an incoming pulse at an excitatory synapse follows an outgoing pulse by an appropriate small interval, or when an inhibitory synapse becomes stronger, which is mathematically almost equivalent if widespread regional inhibition maintains an overall average firing rate in an ensemble such as the mutually incompatible members of a polyflop category cluster like duck, goose, turkey, swan, etc.

In a large number of incoming fibres that have branching lateral connections to each other, those that tend to fire together are likely to form excitatory connections to the same downstream ones, as Hebb proposed, while the connections from members of that pattern to members of other patterns to which they do not contribute would become more inhibitory. Rather than the single 'wires' suggested in the figures in this section, the flip-flop and polyflop units would represent whole patterns that could acquire labels, as illustrated in Figure I.9.4f. The results of Taylor and Aldridge (1974) hint that the numbers of inputs for this might be in the low tens rather than the thousands that usually seem to be implicated in neural operations.

The HaH process thus offers a possible route to the production of new perceptual functions that respond to frequently encountered associations of perceptions. Everyday experience suggests that some such mechanism exists, as we clearly are able to identify patterns as individual quasi-objects. For example, we see a certain kind of cloud pattern in a particular area of sky and we think "That's rain", or we see a particular set of types of furniture in a room in a house we have never visited before, and are able to say that the room is a dining room, a child's playroom, a kitchen, or a living room.

In the rest of this work, we assume that some process exists for producing recognisers for patterns, and possibly labels for them, whether or not it is the HaH mechanism.

I.9.6 Labels and Association

With HaH, we might expect to see the development of neurons that show 'association', meaning that if a particular pattern of inputs recurs, those synapses which were strengthened the first time will be further strengthened, and those that were weakened before will be further weakened. Association is thus almost a description of what each single neuron does, at least in the computational approximations which are commonly used. A neuron receives input from other neurons through many synapses. Of these, the excitatory synapses add to the potential that causes a firing spike when a threshold is exceeded, and the inhibitory synapses subtract from that potential. The neuron fires when enough of an excitatory pattern of inputs is encountered to overcome whatever inhibitory input there may be. If this happens, the postsynaptic neuron requires only a subset of the 'associated' inputs to fire in order for its likelihood of firing to increase. The firing means 'I saw my pattern'. In this way, it can provide a perception of the entire pattern even though some of its inputs might be missing or deviant. A single-neuron form of association is thus inherent in the basic neural structure of the brain.

The polyflop structure suggests a different kind of association that might supplement the single-neuron form in a way that fits neatly into the Powers hierarchical control structure on both the input and the output sides of the hierarchy. For this, the Powers hierarchy must be extended to permit lateral connections.

Suppose a cluster of visual features that resulted in a perception of 'A' also occurred frequently in a context that included the sound 'eh'. The HaH mechanism would tend to strengthen synapses in neurons 'reporting A' as well as in those 'reporting eh'. Positive feedback between the letter form and the sound pattern perceptions would tend to create or enhance a perception of 'A' when 'eh' was heard, and vice-versa (Figure I.9.6a). When the positive loop gain is less than unity, the visual pattern for 'A' acts as a 'prime' or sensitiser for hearing 'eh' and vice-versa as in the 'Stroop Effect' (Stroop1935), but if the gain is greater than unity, it acts as a selector that causes A to be perceived when 'eh' is heard and vice-versa. 'A' and 'eh' have each become a label for the other. In this way, the polyflop structure creates both association and the kind of context sensitivity discussed above using the examples of 'urban' and 'vegetation' contexts (Figure I.9.3b).

Category   Analogue   Category   Analogue
Signals   signals   Signals   signals

"A"   "B"   A-ness B-ness   "eh"   "bee"   eh-ness Bee-ness

Amplifiers   C,D   Amplifiers   "cee"
              etc   "dee"
                                                   etc.

                                                                     "cee-"
                                                                     "dee-
                                                                     ness"
                                                                     etc.

A-B-C-D-...   C, D-ness   eh-bee-cee-dee...   eh- bee-
polyflop   etc.   polyflop   ness ness

                  A-ness B-ness

      Shape perceptions   Sound perceptions

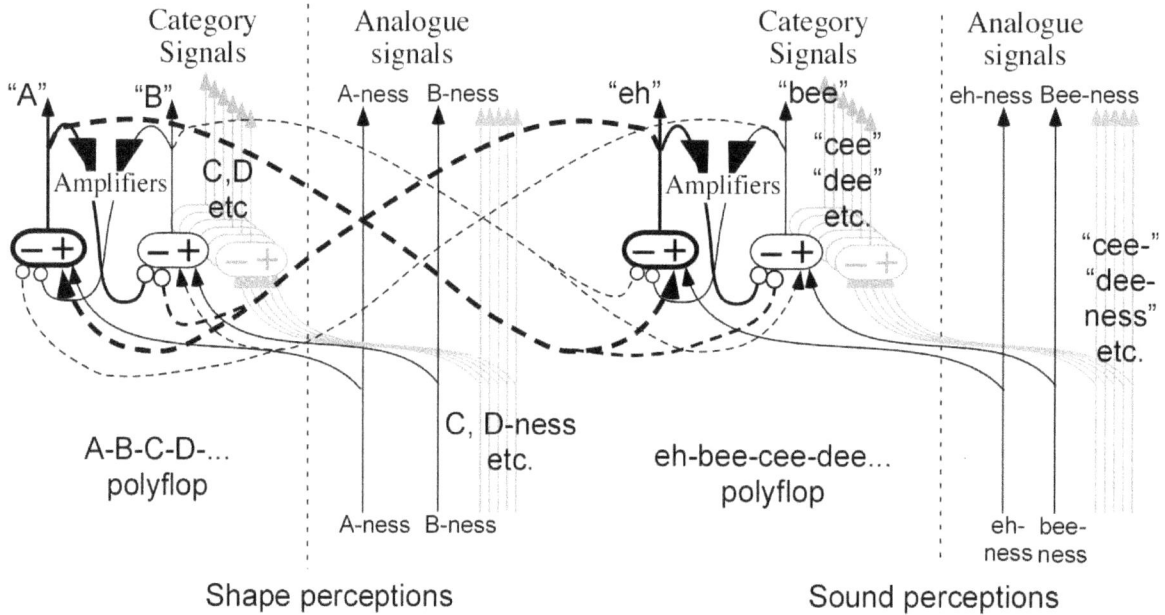

Figure I.9.6a. Cross-coupled polyflops can implement labelling. If, say, "eh" is heard, it increases the input strength to the "A" recogniser, and if "A" is seen, it increases the input strength to the "eh" recogniser, reducing the input strengths to the "B" and "bee" recognisers. The sound and the image are labels for each otherIf the analogue inputs change so as to provide either a strong "B" or a strong "bee", the category signal outputs may switch to "B" and "bee". The same applies to perceptions of "C", "D" or their labels "cee", "dee", and so forth. If "eh" is heard while "B" is seen, the consequence is a classic "Stroop effect", more usually demonstrated by showing the letters of a colour name (e.g. RED) in a colour other than the one named (e.g. green).

Any tendency toward positive feedback between 'A' and 'eh' would be enhanced if at the same time the polyflop structure within each perceptual type suppressed the outputs of the 'B', 'C', 'D', ... recognisers and of the 'bee', 'cee', 'dee'... recognisers. Suppression on both sides implies that both sides have developed at least part way toward becoming category recognisers. If no development toward category recognition had occurred on, say, the acoustic side, then 'eh' might be heard not as 'eh' but as a nondescript waveform, providing nothing that would consistently affect the 'A' synapses preferentially to those of the 'H' synapses, since the voiced sound of the vowel has much in common with the voiceless vowel which is perceived as 'aitch', and both have quite different waveforms when spoken on different occasions by different people. Only if enough of the category perception had developed for the listener to feel "I've heard one of those before" would there be much tendency for occurrences of 'eh' to establish an excitatory connection to an 'A' recogniser.

If an 'A' recogniser already exists, the occurrence of 'A' in conjunction with waveforms appropriate to 'eh' (but not 'bee' or 'aitch') should facilitate the development of an 'eh' category recogniser. 'A' becomes a 'label' for 'eh' and vice-versa. At high perceptual levels, the result is sometimes called 'reification' — if a word for something exists, then the thing referenced by the word must be a true property of the environment, and its properties must be open to exploration. Much philosophical confusion can occur; how many angels can indeed dance on the head of a pin? There's a word for angels (and for pins), so there must be opportunities for real angels to dance on pin-heads, must there not?

If 'A' is a label for 'eh', when 'A' is perceived, the inputs to the 'eh', 'bee', 'cee' detectors are biased toward 'eh'. Sometimes the bias might be sufficient for the other form actually to be perceived. A label is just a form of association, having no necessary relationship with linguistic forms, though in everyday speech we ordinarily think of a 'label' as a linguistic tag for a class of perceptual configuration such as a 'chair'.

The 'label' effect can occur without language. For example, suppose we go for a walk on a dark, gloomy day, and after a while we feel a few drops of water on the face. We probably would 'feel rain' without using the words. But if we experienced the same sensations on the face when the sky is blue and the sun shines brightly, we probably would not 'feel rain' and might look around for an artificial spray of water such as a fountain or a lawn watering device.

Another way of thinking about the concept of 'label' justifies the phrase "to perceive as". The 'A' is perceived as an 'eh', and 'bee' is perceived as a 'B'. The water on the face on a gloomy day is perceived as rain, but on a sunny day it is perceived as spray from some artificial source. As noted above, such association allows the perception of a complete complex when part is missing. The missing part may be perceived as being present even though it is not sensed. Of course this filling-in does not always happen. Whether it happens in any particular situation depends on (among other factors) the amount of context that is perceived directly as compared to the amount that is induced by the polyflop process or derived from imagination and memory. We return to the perception of 'missingness' and 'wrongness' in the next section.

In both Figure I.9.4a and Figure I.9.6a, the analogue signals are shown as continuing upward to provide potential inputs to higher level perceptual inputs, in parallel to the category signals, though if the cross-coupling links and the context are appropriate, only one of the category outputs may have a positive value. In other contexts, and with weaker inhibitory cross-coupling, several of the category outputs may have non-zero values, and the conscious result (if there is one) might be uncertainty about the identity of the environmental pattern than created the analogue values. *"Is that a dark patch of wet concrete or just a shadow?"* Sometimes we will use the word 'syndrome' for a set of analogue values that usually lead to a decisive category output.

In Figure I.9.6a, only one positive feedback loop is shown between 'A' and something else, in this case 'eh'. As suggested in Figure I.9.6b, the diagram could as easily have illustrated a link between 'A' and 'a', 'B' and 'b', 'C' and 'c', 'A' and 'First (1st) quality grade', 'B' and 'reasonably good grade', 'C' and 'acceptable grade', between 'A' and ' ', 'B' and ' ', and so on. All these possibilities are conventionally called 'associates', and the loop gain of each positive feedback loop is an index of the 'associative strength' of that connection.

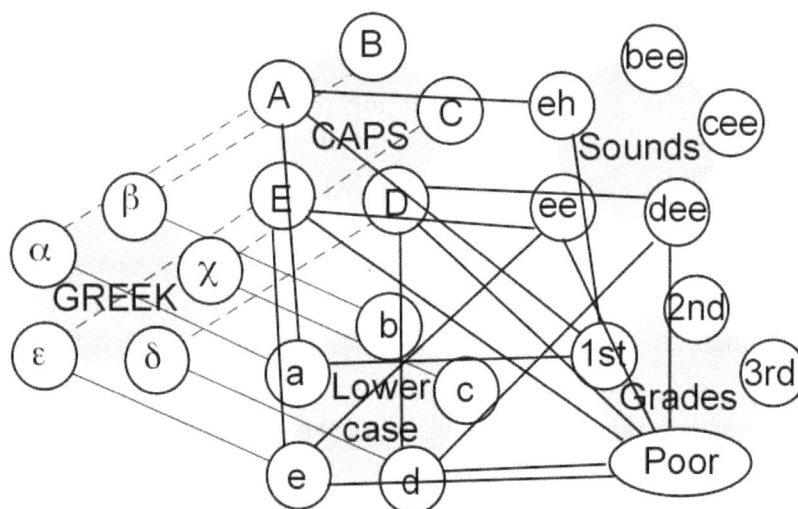

Figure I.9.6b Modules of mutual lateral inhibition (polyflops) cross-connected by excitatory links between items encountered in similar contexts or together. Gray regions indicate mutual inhibitory connections; lines indicate mutually excitatory connections. "D" and "E" are both considered to be "Poor quality grades" and are not perceptually distinguished between 4th quality grade and 5th quality grade. Many links are omitted, for clarity.

The gain of any individual loop is hard to assess, because the value of every output is influenced by many inhibitory connections among the different signals in each domain. Only in a pathological case (we may see 'Obsessive Compulsive Disorder' as a high-level example) will the overall loop gain of the set of positive feedback loops induced by inhibition of inhibition exceed unity. However, as suggested in Figure I.9.4a and Figure I.9.6a, situational context might alter the loop gain sufficiently to produce a categorical output from an unsensed part of an associative complex. This might be especially true if other associated members of a complex such as that of Figure I.9.6b were directly excited at the same time.

Labelling, or 'seeing as' is a kind of imagination, but it is not the imagination in Powers's 'imagination loop'. In B:CP, Powers (2005:219, 227ff.) describes a phenomenon like labelling, without proposing a mechanism. He considers an imagined component of a complex perceptual pattern, most of which is derived from the sensory input. We have been considering a complex perceptual pattern of which much is derived from sensory input, but some is induced by the polyflop labelling process. The difference between the two processes is that Powers obtains the imaginary element as a result of producing an addressed reference value that would ordinarily apply to a lower-level ECU, whereas the polyflop labelling process is entirely within the perceptual input system. The two processes are not in conflict; if they both exist, they would complement one another in producing perceptions relevant to ongoing perceptual control.

I.9.7 Analogue and Categorical Hierarchies in Parallel.

As if the anticipated connections in a single level of both analogue and categorical perceptions are not complex enough, consider how two such levels of perception might interact. A two-dimensional diagram cannot do justice to this added complexity, since not only do we have to consider the analogue levels interconnecting as Powers described, but also must consider how the category stages are connected and how the 'labelling' cross-connections might function in control. When we have done this, we may find ourselves asking about conscious as opposed to non-conscious perception and control.

Refer back to Figure I.9.6a, which showed how analogue perceptual functions for sound patterns might be connected in positive feedback loops with category perceptual functions for letter identities, and then refer to Figure I.9.6b, which suggests how category functions for different ways of signifying the letters might be interconnected. Figure I.9.6b could be thought of as a slice through the category side of Figure I.9.6a, cutting through the page, and each of the sets of 'Greek', 'Caps', 'sounds', 'lower-case' and 'grades' (plus others) would have the same kind of connection with the analogue-side equivalents of squiggles on a page or concepts in the context of school.

Figure I.9.7a merges these two diagrams . The 'category interface' consists at each level of myriad polyflops that receive biasing data from the analogue hierarchy in the manner of Figure I.9.6a. Those polyflops are created by mutually inhibitory links among the incompatible possibilities and excitatory links among compatible possibilities supported by the same data patterns, such as 'A', 'a', 'a' and perhaps ' ' (Greek) or 'eh' (sound).

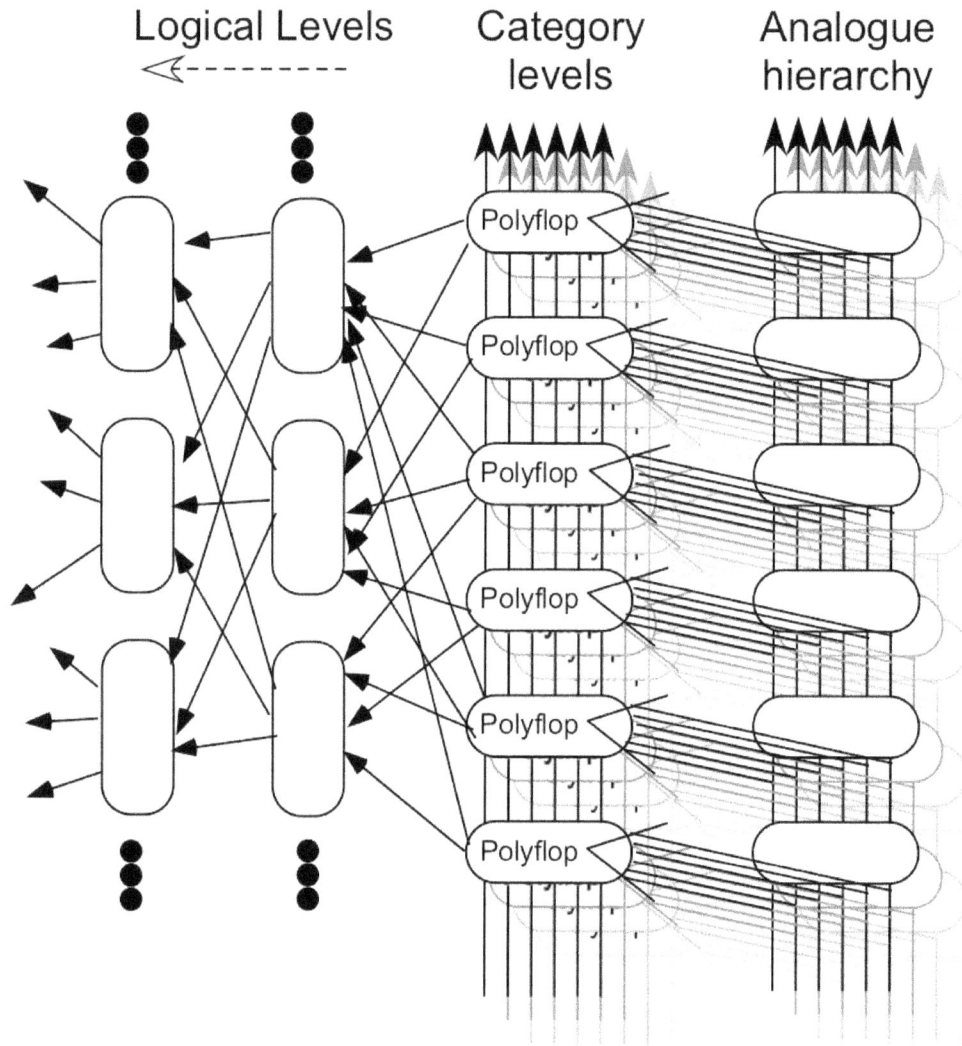

Figure I.9.7a The same interface structure as Figure I.9.4a emphasizing the multiplicity of analogue perceptions at each level that contribute to a polyflop at that level. On both sides, you should imagine that each analogue and categorical perceptions is distributed onward to many perceptual functions at the next higher level, analogue to analogue, categorical to categorical.

Context is important, because if, say, the neighbouring 'squiggles' on a page had excited a category perception for 'Greek', that would have fed back to all the categories of 'Greek' letters and the other possible categories of Figure I.9.6b would then be inhibited by the cross-connections in the usual manner of a polyflop. The shape 'A' would then be more likely to excite the perception of the sound label 'alpha' than of 'eh'.

'Greek' is a higher-level category than is 'alpha', in that many different letter labels form inputs into a perceptual function that would deliver the output 'Greek'. But here we have a feedback loop between 'alpha' and 'Greek' and back again. The Powers hierarchy admits no such inter-level feedback loops, but when we come to category perceptions, the very idea of 'level' seems somehow irrelevant. A red hue is a rather low level perception but a 'Red' category is so tightly linked to a high-level perception of Stalinist communism that the label 'Red' in the USA could at one time be directly substituted for 'Communist'. The two labels each strongly excited the other in a tight positive feedback loop.

In the analogue hierarchy, 'Communist' would be a very high-level perception, probably at Powers's 'system' level, whereas 'red' would be near the bottom, very close to the sensory input. If the polyflop mechanism for category perception is correct, does this mean that we should ignore the concept of levels of categories? Well, yes, and no.

Each category that is developed through polyflop feedback loops is based on a distinct analogue perceptual level. The categories developed by this mechanism have the same level structure as do the analogue perceptions on which they depend. Similarly, categories such as 'Greek' or 'Cyrillic' or 'Roman' must be at a level higher than categories of letter names. The lower-level category perceptions provide inputs to the higher level polyflops just as do the same-level analogue perceptual functions. So the answer to the question of whether categories should be treated as being at different levels is that we cannot ignore category levels.

How, then, should we deal with the Communist⇔ Red feedback loop and others, where the interconnections cross level boundaries. How should we deal with logical perceptions such as the Powers 'Program' level, in which the organism controls for perceiving the execution of a logical program such as '*if colour of traffic light is red then stop, else if colour of traffic light is green then go, else if safe then go, else stop*'. The answer is that we should consider the categories, which from the analogue point of view are at different levels, as being the base of a different hierarchy, a 'logical' hierarchy. The categories are an interface, much as sense-organs and muscles are an interface between the organism and its external environment (Figure I.9.7b).

Logical Levels   Category   Analogue
levels   hierarchy

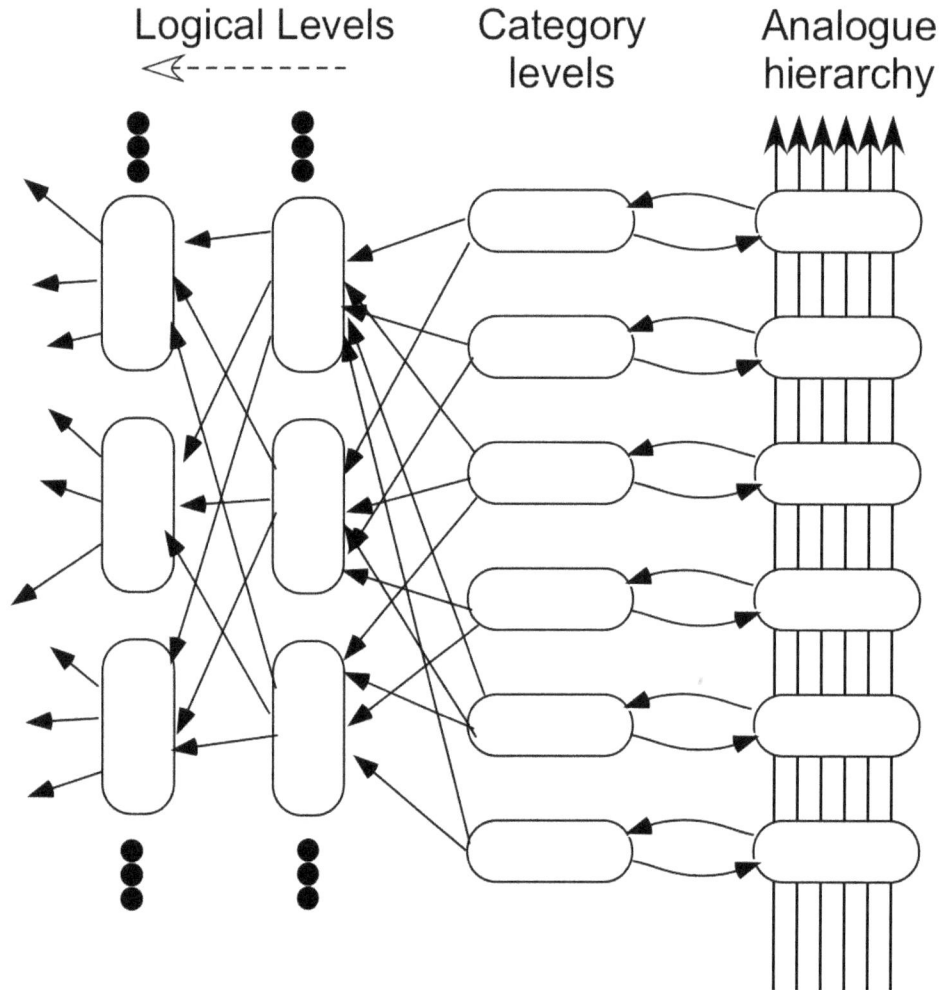

Figure I.9.7b Schematic showing how categories can be seen both as being organised in levels that correspond to the perceptual side of the analogue control hierarchy, or as the base level of a different, logical hierarchy. Each lozenge in the analogue structure represents several perceptual functions, all of the same kind and mutually incompatible, such as different colours, different kind of furniture, different triangle shapes, different political systems, etc.

Something new has been added in Figure I.9.7b. The analogue inputs at level *n* are shown as providing inputs to the category polyflops, as they do in previous figures, but now the category polyflops are shown as returning inputs to the corresponding analogue and perceptual functions. These connections from the polyflops to the corresponding analogue perceptual function inputs serve to emphasise the analogue value of the perception that a polyflop selects as the preferred category. The degree of emphasis depends both on the internal loop gain of the polyflop, which may or may not produce a clear winning response, depending on context or task stress (Figure I.9.4f), and on the weight given to the category output by the corresponding analogue perceptual functions.

Within the logical hierarchy that builds sideways (leftward) in Figure I.9.7a and Figure I.9.7b, everything is exactly as it would be in the levels above the category level in the Powers hierarchy. The problem, if one exists, is in the interface to the analogue hierarchy. These two figures suggest only the perceptual side of that interface. It differs from the Powers hierarchy in that the inputs to the category perceptual functions — the polyflops — come from a range of levels of the analogue hierarchy, not all from the same level. Is that important? And what about the output side of the interface, which is ignored?

The fact that the category interface connects to analogue units at many levels rather than one constitutes what Powers called 'level jumping'. He argued that in general level-jumping on the output side of the hierarchy would create interference and conflict. This is because a single control unit at level n would set, or at least contribute to, simultaneous reference values for units at levels N-1 and N-2, where N-2 also has its reference value influenced by the output of the level N-1 unit. Imagine that a general orders a colonel to execute a manoeuvre and at the same time orders the colonel's subordinate to do something that interferes with what the colonel wants him to do in order to perform the manoeuvre. The problem is that both of the units are controlling their perceptions in support of control of the same level n perception.

This problem is more apparent than real. The output side of any control loop must be analogue at the lower levels. In the Powers hierarchy with a 'category level' to separate the logical levels from the analogue levels, the shift between categorical and analogue occurs at the category level. Something similar occurs with the category interface shown in Figure I.9.7b as the category interface feeds back into the analogue hierarchy. In both cases, the Powers hierarchy and the proposed category interface, a category is not directly controlled, the analogue variables that contribute to a category are controlled.[95]

I.9.8 Similarity and Dissimilarity

At this point I want to expand on that interface interconnection between logical and analogue sides of the perceptual control hierarchy, and introduce a model I developed before I knew anything about PCT: the "Bilateral Cooperative Model of Reading" (the BLC Model) (Taylor and Taylor 1983; Taylor 1984; Taylor 1988a,b). The BLC Model was based on psychological and neurological studies known up to that time, in ignorance of Perceptual Control Theory. That the two theories produce essentially the same model structure is, I think, suggestive that the structure may have some validity and some value.

---

95 In controlling a category, the 'General' does not command that subordinates do anything in particular, only that enough of what they do maintains his status as General. To control 'Red' many hues will do; at the analog level we control 'that red colour'. — Ed.

The BLC Model was built upon what was known at that time (about 1980) about the differences in the language processing functions of the two hemispheres in the brains of right-handed people. Those studies depended largely on the effects on reading behaviour of accidental trauma and of surgical interventions. References can be found in Taylor & Taylor (1983). Though the neurological sciences have moved on in the last four decades, the principles of the BLC model remain valid. They can be summarised more or less as follows.

The Model postulates two 'Tracks' of processing, a 'Left Track' (LT) and a 'Right Track' (RT). The LT performed analytic functions such as sequencing, using categorically identified units and making distinctions among similar possibilities for the identity of a unit. The LT was concerned with a question of the kind "This looks like X, but can I see any reason that it might not be X?" In right handed people and most left-handed people, the LT processing was almost entirely done in the left hemisphere of the brain. RT processing was less committed to a particular hemisphere, but was largely (far from exclusively) done in the Right Hemisphere. The RT was concerned with similarities — was this sufficiently like X that it could possibly be X? LT looks for reasons to exclude, RT looks for reasons to include a unit in a category.

The two tracks feed each other at every level of perceptual complexity. When the RT produces a perceptual signal, it biases the equivalent analytic LT polyflop; when the LT provides a clear selection of X, its signal becomes a perception in the RT complex. Usually the LT is slower than the RT to produce clear distinctions among the RT possibilities, but if the writing is in an unusual script, such as Cyrillic for an English-speaking reader, or if a word is unfamiliar or placed in an usual manner within the text, the RT may not produce any clear winner but the LT analytic processing will eventually do so, enabling higher levels of the RT structure to proceed.

The BLC Model need not be described here in any greater detail, but it is implicit in much of the rest of the book. When a choice is made at any level, the LT might be involved. When an explicit difference is mentioned the LT is involved. When, however, a similarity is mentioned we can be sure that the RT is concerned. Which predominates at any point, if either, is usually unimportant to the argument, but is often easy to determine from the context.

In the previous section we have anything 'logical' (to the left of the 'category interface' in Figure I.9.7b) producing dissimilarity by inhibiting incorrect possibilities, while the analogue track to the right of the category interface produces similarity from the outputs of perceptual functions for data patterns close to what the perceptual function is tuned to report. The match between these 'sides' of the category interface and the two 'tracks' of the BLC Model is too close to be simple coincidence. The hypothesis that these separable parts of the control hierarchy should exist seems to be justified by the psychological experiments that led to the formulation of the BLC Model 40 years ago.

The emergent property we call control requires that the gain on the output side of the control loop should substantially exceed that on the input side. Might this be the case at the category interface, or do analogue perceptions influence categories as much as categories influence analogue perceptions? Even when an analogue perception is clearly specified by the sensory input as, say, a 'b', one may perceive it as an 'd' in the context of a familiar word in an ordinary context such as "My whole boby aches." When that happens, did the category for which a reference value was set by the higher logical levels override the 'disturbance' of the sensory input?

In a case such as this, it usually does. The second 'b' either is perceived as a 'd' or the meaning 'body'is perceived despite the letter string 'b-o-b-y' being clearly perceived, there being no Perceptual Function tuned to that letter sequence in English. As another example, if someone says something that sounds something like 'aigwánnudwit' the listener might hear them say "I'm going to do it."

The level-agnostic connections of the discrete category hierarchy may explain this phenomenon. The outputs of the hierarchy contribute to analogue perceptions at any level. If the logical hierarchy contains a polyflop that produces an output corresponding to having heard "I'm going to do it", the logical implication is that these were the words that would have been needed in order to produce that decision as a perception. (In Chapter II.7 of Volume II, we will treat this as the perception of a trajectory.) Those values will be returned as inputs to analogue levels lower than the level that perceived the sentence, creating a positive feedback loop between the inputs from the analogue levels and the discrete decisions produced by the polyflops.

How often have you said to your friend or partner "I'm sure I heard you say ..." when the other has told you they said something quite different. It is impossible to say who is correct, the sounds having passed into history, but at least sometimes is it not clear you might have misheard? When reading, do you never backtrack to look again at a word you read clearly but that makes no sense in the context of what you read later? Such experiences are to be expected if the foregoing argument is correct.

The same is true of the infilling that occurs when fluent speech is interrupted by an unexpected sharp noise (e.g., Warren 1970). The speech may seem to continue uninterrupted through the noise burst. This postulated positive feedback loop could infill the interruption. Having received enough information, the polyflops use redundancy to create a clear perception through the gap.

Can a clear sensory input flip the polyflop that selects the category from its set of similar categories? Indeed it can, for this possibility is the essential function of a polyflop. So it seems that there is no obvious direction of greater influence between the logical and the analogue hierarchies. As the BLC Model describes, each influences the other more or less strongly, depending on the circumstances. If the analogue inputs from lower levels tend to excite more than one perceptual function more or less equally, the the category will tend to select one to predominate.

To identify something as belonging to a category such as 'alpha' or 'Greek' is to say "It is this, and it is not those other things." An analogue perceptual function, even with localised lateral inhibition such as is suggested in Figure I.9.4a, in contrast, says "It is like this to such and such a degree." For a particular data set, several perceptual functions may simultaneously say that the data resemble it to some degree, the pattern for which they are tuned. A category recogniser implemented by a polyflop is the opposite. It says that the data are NOT like those.

When one is dreaming, the analogue values may be biased by momentary changes in neural noise, but the logical levels maintain a kind of coherence in what is perceived in the dream. It makes syntactic structural sense, even though the semantic relationships in the dream might seem to the waking memory as being quite incompatible with any possible reality. Yet the categories that drive the analogue perceptions in this case are themselves affected by reference values output from higher logical levels, reference values that have some connection to the values they may have had before sleep intervened. If we have been worried about some problem before sleeping, the random semantic nature of the dream may embody some of the elements of the problem.

To speculate further about the psychology of dreaming is far beyond the scope of this book, but it is interesting that the 'Powers of Perceptual Control' together with Hebbian and anti-Hebbian learning seem to point the way toward both Freudian and non-Freudian utilitarian approaches to dream analysis.

Part 4: Uncertainty, Novelty, and Trust

We next introduce the perception of uncertainty and belief, and spend some time quantifying the concepts, using the mathematics of Information Theory introduced by Shannon (1949) and to some extent Kolmogorov (1965). I try to avoid the actual mathematics so far as possible, instead introducing some important concepts by means of examples, most of them in a very simple toy world of checkerboard squares with markers on some squares.

We start, however, by looking at trust, belief, uncertainty, illusions, and consciousness, these being some specific ways in which uncertainty manifests itself. This we do in Chapter I.12, but first we begin the more technical discussion of uncertainty, information and structure in Chapter I.10. By the end of Chapter I.10, I hope that the reader will have a sufficient grasp of the principles involved that the further development of tensegrity principles in Volume II will make sense.

Using information-theoretic concepts, Chapter II.1 illustrates in a different way how and why the Powers control hierarchy should be expected to have tensegrity properties. The key property is the way unpredictable stresses imposed from outside (disturbances) are diffused beyond single control loops at a level, so that apparently unrelated control loops can assist in resisting the disturbance, much as the stresses imposed on a physical tensegrity structure are spread through the network, making a tensegrity structure tougher than one would anticipate from the strengths of its constituent components. The collective is stronger than its parts in control, as well as in mechanical tensegrity structures.

Chapter I.10. Uncertainty and Structure

We must now bite the bullet and talk about uncertainty, information, and structure, both formally and informally, rather than keeping them around the edges of our discussions of control. The perceptual complexes, misperceptions, trust, and surprise that we have discussed and will discuss further all relate in their different ways to these core concepts.

'Information' is a slippery idea. In everyday speech, one thinks of "my information" as "what I know", and in a way that meaning does conform to the technical definition provided by Claude Shannon (Shannon and Weaver 1949), but to see how that is true, we must look at what Shannon actually wrote, rather than at what many writers have said about it in the ensuing decades.[96]

Information and uncertainty are complementary concepts. At Bell Labs, where Shannon was working, the business was all about communication. Shannon used the concept of uncertainty and its complement, information, to analyse telecommunication systems. His communication interest has led many to believe that information theory is about communication, but Shannon's mathematics is no more limited to communication than is Fourier analysis, which also applies to communication but was originally developed for the analysis of heat flow (Fourier, 1822). Shannon titled his book The Mathematical Theory of Communication, and in this book we will sometimes use it in treating communication, but the theory is much more widely applicable and we will often use it in other domains.

For a few years after Shannon's publication, many writers (including me) used his 'Communication Theory' in a wide variety of other domains. At first, it seemed the heaven-sent solution to all problems, but eventually that promise became generally considered to be a useless mirage. The problem was akin to what might happen if someone invented a super-screwdriver, and people who did not understand its domain of application and its limitations tried to use it to pound nails, to drill holes, and to cut timber. A chorus of disappointment leads to a newly accepted wisdom that the super-screwdriver is completely useless, despite having seemed at first to show such promise. But as a screwdriver, it nevertheless remained super.

Information theory is a mathematical tool among many others, and is no more the answer to all problems than is Fourier Analysis or the general solution of a quadratic equation. When it is appropriate, it should be used.

Unpredictable disturbances increase the uncertainty of perceptions (reduce their information). Once you know the reference values for controlled perceptions, control can be understood as reducing the uncertainty that unpredicted disturbances introduce to perceptions and their environmental correlates. When control systems and hierarchies are analysed from this point of view, information theory is an appropriate and natural tool to use, among others.

---

96   References in this work to 'Shannon' refer to Shannon's mathematical work in the first section of this book. The second section contains Weaver's interpretation of the work to explain it to non-specialists.

I.10.1 What is 'Information'?

Shannon defined 'information' as the reduction of uncertainty. I follow Garner (1962) in using 'uncertainty' for what Shannon called 'entropy' because I think the term 'uncertainty' more clearly conveys the essence of the concept, and because the familiar meaning of 'uncertainty' about something is more relevant to the domains in which we shall use the concept. There is no familiar meaning of 'entropy'. Though 'uncertainty' was his first preference, Shannon was persuaded to call it 'entropy' because the underlying mathematical basis is the same as for physical entropy in statistical mechanics.[97]

Uncertainty in this sense is a measure, a statistical property of a probability distribution, the same kind of statistical measure as variance, mean, kurtosis, etc. Uncertainty as a concept, however, is always *about* something, because probability is always the probability of something in some context of prior assumptions. The statistical measure we call 'uncertainty' is about that 'something' *in the context of what is already known or assumed*. In the communication context, the receiver always has some background knowledge, even if it is no more than a knowledge of the coding or the language in which the message is couched.

As an example, if you know before you pull a ball out of a bag that it holds only a black ball or a white ball, but don't know which, you are uncertain about the colour of your yet-to-be-pulled ball. The colour of the ball means to you that you will or will not be accepted into a club you want to join. After you take your hand out of the bag, you see the colour, and no longer have any uncertainty about it, nor about whether you will actually become a member of the club. You have gained one bit of information (a binary choice) both about the ball colour and about your membership chances. If you don't look at the ball, and someone tells you "You're in", then because you have that information you do not need to look at the ball to have lost all uncertainty about its colour. Meaning and information are inseparable.

Your uncertainty about a thing and your information about the same thing are complements; but your uncertainty is quantifiable, and your information in the sense of your subjective understanding is not. Quantitatively, information is measured by the difference between uncertainty at one time and uncertainty about the same thing at another time, or more generally the difference between two uncertainties about the same thing. If you knew only that there were balls of a variety of different colours in the bag, you would have been more uncertain about the colour, and would have got more than one bit of information when you saw that the ball you took was white. You would have learned that you would be allowed full membership in the club rather than being 'blackballed' or allowed some kind of provisional membership.

---

97   John von Neumann persuaded Shannon to use the word 'entropy' instead of 'uncertainty' because the mathematics of Boltzmann-Gibbs entropy are the same as for uncertainty, "and more important, no one knows what entropy really is, so in a debate you will always have the advantage" (Tribus & McIrvine 1971:180).

Once you say "I am uncertain", the obvious question is "What are you uncertain about?". The word *'about'* is crucial. The ball colour and your club membership are very different concepts which in this case have the same uncertainty. Without the 'about', there is no way of knowing what information you will get from the ball you pull. Maybe you will learn about its weight, the purity of its colour, the material of which it is made, or even about what time is dinner, about your friend's birthday, about the number of words written by Emile Zola on the Dreyfus affair. There is no such thing as 'information in the ball' unless you describe what the complementary uncertainty was *about*.

It all depends on what the ball might mean to you. Maybe you are participating in a quiz show, and you believe that someone has written a number on the ball which will be the correct answer to the Zola-Dreyfus question. *What* you learn is the meaning of the ball in that context; *how much* you learn is your reduction in your uncertainty *about* something — the information you have received.

When we talk about a person being uncertain about something, we may be dealing either with variability in some perceptual quantity, or with the way something the person may observe relates to what he or she already believes. One perceives oneself to be quite sure about something, or mildly uncertain, or perhaps pretty much at a loss about it. *"What's for dinner?" "Was that animal a dog or a coyote? I didn't see it well enough to be sure." "Is he telling the truth about his unbelievable exploits? Probably not, but his manner suggests he may be."*

Many writers have wrongly said that Shannon showed that information is technically unrelated to meaning.[98] This is a complete misunderstanding, as I have attempted to demonstrate above. The numerical quantity of information from an observation, '3.2 bits', is related to the meaning of that information in the same way that the number '123 cm' is related to the height of a person or the width of a piece of furniture just measured.

The meaning of '123 cm' is very different when you are measuring a child and when you are measuring a piece of furniture, and the meaning may include your uncertainty about whether a child can get a reduced fare on the bus or your uncertainty about whether a piece of furniture will fit in the available wall-space. Measures of information and uncertainty quantify how much you were uncertain about something before as compared to after you have observed or been told it. The nature of that 'something' is as important as its uncertainty, as it is for any other measurement. If someone tells you out of the blue '123.456 cm', does this tell you about anything other than that the person probably believes you already know what they were measuring?

---

98 Including contributors to the 2017.05.22 version of the Wikipedia article on 'Entropy' (subsequently corrected). It said that 'Entropy' is a property of a signal, whereas it is actually a joint property of a signal and the process that acts on (perceives) the signal. For example, a sequence AAAAAAA… appears to be of lower entropy (uncertainty) than ABXIUVD… but is not when seen by a process that lacks memory for preceding characters but has +knowledge of the set of characters available to the sender. This is true whether one is using Shannon (1949) or Kolmogorov (1965) uncertainty measures (see Section I.10.2 Basic Concepts).

Shannon was considering how much information a communication channel could deliver in a given time. Necessarily, the information was about something, but what that something might be is irrelevant to the capacity of the channel to deliver it. What he did *not* say was that the messages conveyed through the channel would be meaningless. Messages do convey meaning, and that meaning is quantifiable as reduction in uncertainty — information that the recipient did not have about something before receiving the message.

To measure uncertainty, Shannon started with a realm of possibility, the realm about which there is uncertainty. He based his examples on the alphabet in which a message might be written, in which the realm of possibility for material written in English in Roman upper-case letters is 26 possible letters plus spaces and a few punctuation marks. The channel that interested him was phone wires, but to psychologists, the channels of interest are equally related to our sensory apparatus, our history of experiences, and our internal coding. In that context, a 'message' is an observation or, in more precise PCT terms, the magnitude of a perceptual signal.

The realm of possibility for any message or observation could be very large or very small, but there always is such a realm, such as the possible properties of a ball in a bag. The meaning of what you will observe when you draw the ball out of the bag begins with technical information about the ball, such as the colour or weight of the ball. You did not know beforehand what colour the ball might have, and the colour you observe means to you your future as an honoured member of a society. Meaning is technical information because it is the reduction of quantifiable uncertainty about something specific, which could be at any level of perceptual complexity.

The mathematical formula for the uncertainty of a probability distribution has some key properties that are intuitively required of it. Quoting Shannon:[99]

> *Suppose we have a set of possible events whose probabilities are $p_1, p_2, ..., p_n$.*
> *These probabilities are known but that is all we know concerning which event*
> *will occur. Can we find a measure of how much 'choice' is involved in the*
> *selection of the event or of how uncertain we are of the outcome?*
>
> 1. *If there is such a measure, say $U(p_1, p_2, ..., p_n)$, it is reasonable to require of*
>   *it the following properties: U should be continuous in the p_i.*
>
> 2. *If all the p_i are equal, $p_i = 1/n$, then U should be a monotonic increasing*
>   *function of n. With equally likely events there is more choice, or uncertainty,*
>   *when there are more possible events.*
>
> 3. *If a choice be broken down into two successive choices, the original U should*
>   *be the weighted sum of the individual values of U. ...*

Shannon showed that there is only one function that satisfies these criteria, $U = -\sum p_i \log(p_i)$. Using 2 as the base of the logarithms gives a measure of U in 'bits'. For a single equiprobable 'yes-no' observation, the uncertainty before the observation is 1 bit, which is the most information that any observation could

---

99   I substitute U for his H because elsewhere I use the U symbol for 'uncertainty'.

provide about that particular set of possibilities — for example, *"Am I going to be accepted as a club member?"* If there are N equiprobable possibilities, the uncertainty is $U = -\log_2(N)$. The numbers in themselves do not relate to the meaning of the uncertainty or information. The meaning is what the uncertainty is *about*, while the numbers indicate *how much* uncertainty there is about it.

Shannon recognised that the mathematics of uncertainty are exactly the same as the mathematics of physical statistical entropy, as described by Boltzmann (1877) and extended by Gibbs (1902). For our purposes, the important concept to take from their work is a distinction between 'macrostate' and 'microstate'. A macrostate is a collection of microstates that can for some purposes be treated as 'the same' state of the world — a category. A macrostate might be 'a dog', within which a set of different microstates might include Labrador, Cairn Terrier, Basset Hound, Dachshund, etc. At the same time, this 'dog' macrostate contains a quite different set of possible microstates, such as brown dogs, dogs with a white chest, black dogs, and so forth. The number of ways a macrostate might be divided into different microstates is essentially boundless. A macrostate is a perceptual category, within which a microstate is a sub-category.

A microstate, a sub-category, could be a macrostate with sub-sub categories. For example, imagine an alien confronted by a moving fluffy object. Through a translator device, the alien asks "What is that?" and gets the answer "Trevor." "What is a Trevor?" "Trevor is a dog." "What is a dog?", "A dog is an animal." When considering animals, "dog" is a microstate because there are many different kinds of animal, but when considering "Trevor" and "Rover" and "Fido", "dog" is a macrostate within which Trevor, Rover, and Fido are microstates. At all of these levels, the category is a microstate when included in a wider category and a macrostate when seen as including more refined categories.

A microstate represents a specific case, whereas a macrostate represents a set of specific cases, all of which are the same for some purpose or from some viewpoint. Since Boltzmann's conceptual Universe was a box containing a pure gas, his set of possible microstates identified the positions of each atom separately, but a macrostate would treat the atoms as interchangeable, and a larger macrostate in which this macrostate was a microstate according to Gibbs would be one in which the temperature of the gas had a given value, no matter where the atoms might be in the box.

What is 'the same' depends entirely on the receiver of the information, whether mechanical or living—possibly an observer of some physical event, or, as in the Boltzmann-Gibbs case, a device which produces a perceptible measure of some property. In the case of pulling a ball out of a bag, you might not have known what sorts of objects were in that bag, but you knew one bag contained golf balls, another wooden alphabet blocks, and a third had billiard balls. When you put your hand in and took out a white billiard ball, the colour might have been irrelevant. The only information you might have received was 'this was the bag with billiard balls'. Or, as before, the information might have been 'white' if the possibilities of interest to the observer were colours.

The universe of possibility might have been defined by possible types of object considered 'different' by the observer (recipient of a message), or the colour might have defined it. On another occasion the observed macrostate might be different, perhaps defined by the surface smoothness of the ball, even though the physical observation of a white billiard ball was identical. It would all depend on what perceptions were being controlled by the perceiver, living or mechanical.

I.10.2 Basic Concepts

Remember, Shannon defined 'information' as the reduction of uncertainty about something due to some event such as having made an observation. He was interested in the message rate capacity of a channel such as a telephone wire, regardless of what the message was about, so his analysis was quite general. He defined uncertainty as $U = -\sum p \log_2(p)$ bits, where p is the probability of some condition being true, and the summation is over all the possibilities. For example, if you believed that the next person you would meet would either be wearing a tie or not with equal probability, your uncertainty before meeting another person (U) would be one bit. But if it mattered what colour the tie would be if one was worn, and it might equally likely be red, blue, green or yellow, each tie possibility constitutes 1/8 of the possible conditions. The summation then is

$$U = -(1/2) \times \log_2(1/2) - (4 \times 1/8) \times \log_2(1/8) = 1/2 + 3/2 = 2 \text{ bits.}$$

One can arrive at this value of two bits in a different way. Uncertainties are additive provided they are uncertainties of independent variables, in the same way as the variances of independent variables are additive when the variables are summed. Indeed, if the variables have a Gaussian (normal) distribution, their uncertainty is proportional to their variance. To emphasise this point, it is worth noting that Garner and McGill (1956) showed that it is possible to describe an Analysis of Uncertainty that is exactly parallel to an Analysis of Variance. The main difference is that the Analysis of Uncertainty remains valid even when the data distributions are very far from Gaussian. The interaction components of Analysis of Variance all have their counterparts in the Analysis of Uncertainty.

In what follows, we are not concerned with the intricacies of Analysis of Uncertainty, but with the partitioning of uncertainty in a complex situation that involves many sources of 'information'. If there are several data sources, as is the case for every perceptual input function in the perceptual control hierarchy, each may provide some evidence that reduces uncertainty, but the sources may not be independent. and lack of independence requires us to deal with conditional rather than absolute probabilities. In fact, we always should treat a probability as depending on some specified background condition.

What, for example, is the probability that the 165th symbol in my text after this '@' sign is *ʄ*? You can't tell without providing a precondition. If your condition is that you will treat this peculiar symbol as 'h', which is what you would do if you saw it in a handwritten '*fiʄing*', it is simply one of 26 possibilities, or 52 if you remember that some letters are capitalised, or 204 if you consider there is a difference between roman, italic, bold, and bold-italic letters. Or maybe you recognise ƕ as a technical symbol, and think of all the other technical symbols that might occur if this one is possible, or maybe you realise that the probability that it is the 165th symbol is close to zero since this is a text in ordinary English, albeit with a peculiar character introduced as an example. The probability that the letter in question is ƕ depends entirely on what the basic set of possibilities contains. The probability 'p(165th letter is ƕ)' should be written 'p(165th letter is ƕ | (italic is same as roman) & (technical symbols are possible) & (ƕ differs from *h*) & (syntax is ignored) &...)', where '|' means 'given that' and '&' has its usual meaning, 'and'.

If you don't know whether something is A or B, but a precondition is that it must be one or the other, the probabilities of A and B are pA and pB, which sum to unity, and the uncertainty is $-pA \times \log_2(pA) - pB \times \log_2(pB)$ bits. But knowing that it is A may not be enough for you, since if the observation yields only a simple A, you may care about the microstate and be uncertain about which version of A it is among A1, A2, A3, ..., An, as in the case of tie colour above, and similarly if it turns out to be B. However, you can still find your initial uncertainty about the macrostate by calculating $pA \times U(A) + pB \times U(B)$, where U(x) means the uncertainty of x after the observation that distinguishes A from B. The basic summation, of which $-\sum p\log_2(p)$ is a special case, is $U = \sum p_i \times U(i)$ where the *i* are the various independent possibilities, each of which has its own uncertainty.

Often, however, the possibilities are not independent. Uncertainty is still additive, but instead of just dealing with, say, *pA* and *pB*, you ask about *A* and then about *B* given that you already know *A*.

$$U(A, B, C, ...) = U(A) + U(B|A) + U(C|A,B) + ...$$

If $U(B) = U(B|A)$, then knowing *A* tells you nothing about *B*; they are independent. $U(B)-U(B|A)$ is the information you get about *B* by observing *A*. Conditional uncertainties, or rather their informational complements such as the aforementioned $U(B)-U(B|A)$, are related to correlations and to interaction effects in Analysis of Variance (Garner and McGill 1956).

In everyday parlance but not in information theory, if *X* is 'redundant', it means that *X* is irrelevant, and can be ignored without consequence. The word is less severe in information theory. It simply signifies that if *X* is redundant with *Y*, then the uncertainty of X and Y together is less than the uncertainty of either by itself: $U(X|Y) < U(X)$, or equivalently $U(Y|X) < U(Y)$. You can get some information about *X* by looking at *Y* and vice-versa. 'Redundancy' just means you can get some information about one variable by learning about the other. 'Mutual Uncertainty' is what is left over, or what you don't know about one if you know the other, and Mutual Information is what is gained by combining them, the difference between the Mutual Uncertainty and the sum of the Uncertainties of each if you know nothing of the other.

From the above, the total uncertainty of the XY complex is

U(X, Y) = U(X) + U(Y|X) = U(Y) + U(X|Y)

Swap the sides of the latter equality, and we have

U(X) - U(X|Y) = U(Y) - U(Y|X)

Each side of this equation represents the uncertainty that is left in one variable after you subtract what can be learned about it from observation of the other. The amount is the same for both variables. Another corollary is that observation of Y cannot give you more information about X than the original uncertainty of Y, and vice-versa. A natural measure of quality of control (QoC) of some variable is the amount by which bringing it under control reduces the uncertainty of its variability over time. This has significant implications for analysis of the QoC in a control system. This concept will become important when we deal with 'rattling' in Chapter II.5.

The ordering of which probability is conditional on knowledge of which other variable(s) is completely arbitrary, since if you observe A and B, you have gained the same amount of information no matter which you observe first, although that information is represented as U(A)+U(B|A) if you observed A first or U(B)+UA|B) if you observed B first. However, for a particular purpose quite often one ordering is more convenient than another. This is not the place to go into these niceties of uncertainty analysis, which can be found in Garner and McGill (1956) or Garner (1962). We describe some specific features of uncertainty and information in Volume IV, Appendix 5.

An important quantity is 'mutual information' I(X,Y), a measure of redundancy in the X, Y system. It is the difference between the system uncertainty U(X|Y) and the uncertainty that the X, Y system would have if X and Y were independent and the system uncertainty were the sum of the two individual uncertainties.

I(X,Y) = U(X) + U(Y) - U(X|Y)

'Mutual information' I(X,Y) is the numerical equivalent of covariance, apart from a multiplicative factor when the variables have Gaussian distributions. If X and Y are perfectly correlated (informationally, if you know the value of X you can deduce the value of Y and vice-versa) then

U(X|Y)= U(X) = U(Y) = I(X,Y)

I is a measure of redundancy in the X, Y system, and sometimes it is useful to normalise it by dividing it by U(X|Y), in which case it might be called 'relative redundancy' R(X|Y) = I(X,Y)/U(X|Y). However, since uncertainties, like variances, are additive, R is more useful as a final descriptive measure than in further computation.

Macrostate and microstate with resolution-limited observation: The total uncertainty over a wide range observed with detailed resolution can be partitioned into two components, the uncertainty of the original observation and, independently, the uncertainty within a 'box' delimited by the resolution of the observation. The box is a 'macrostate' (as is the set of boxes) and the value within the box a 'microstate' (as, within the set of boxes, is the specific box, as illustrated in Figure I.10.2). The axes that define the dimensions within which a box can be specified are perceptual categories of properties of any definable kind such as brightness, hardness, kindness, density, and so forth.

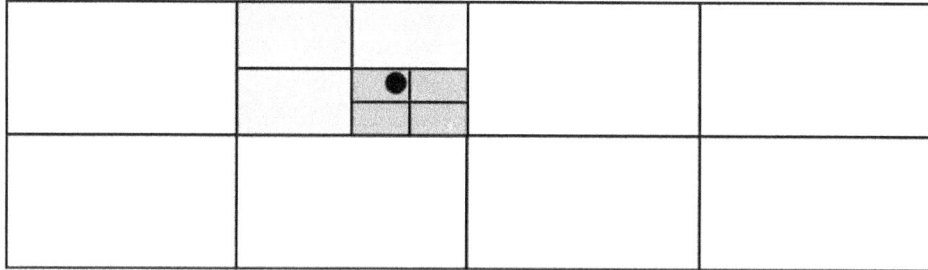

Figure I.10.2. You want to tell a friend where the black dot is within the outer big rectangle. Since the space is continuous, his total uncertainty could be infinite, but it isn't. Your friend only wants to know where it is to some level of precision. If you identify the lighter grey rectangle as the macrostate that contains the dot, you have provided three bits of information (there are 8 possible macrostates). The remaining uncertainty is microstate uncertainty within the selected macrostate. Identifying the dark grey nested macrostate provides two more bits of information. Identifying the top-left quadrant of the dark grey rectangle provides two more, or 7 bits total starting from the initial big rectangle. This process of nesting macrostates can be continued ad infinitum, or until the resolution limit of the relevant sensor has been reached.

Uncertainty partitioning in nested macro-microstates: A new observation using a higher-resolution means of observation (e.g. a telescope or microscope) can define smaller microstates nested within a macrostate that was originally a microstate. So can the introduction of a new dimension. When considering the uncertainty of macrostates, the i of p_i in the probability formula represents only the identity of the macrostate, regardless of the microstate of variables within the macrostate. Taking someone's age as an example, the total uncertainty depends on whether the age is measured to the decade, the year, the month, the minute, or the millisecond. A macrostate might be taken as the decade: "He's in his fifties", within which "He is fifty-three" might be a microstate. "Fifty-three" then might be a nested macrostate within which "Fifty three years, four months and twenty-two days" could be a microstate.

Uncertainty of a continuous variable is relative to a unit of measurement: When a variable is conceptually continuous, as for example is the age of a person, the nesting of microstates within macrostates can in principle be continued indefinitely. The total uncertainty depends on the resolution of the finest microstate of interest, which is partitioned into independent parts, for example the uncertainty of the decadal macrostate within the range of possible ages, the uncertainty of the year within the decade, and the uncertainty of the day within the year. The choice of resolution, and hence of total uncertainty, is arbitrary. There is more uncertainty when time is measured in days than when it is measured in months, and more yet when the unit is milliseconds or nanoseconds. Change of unit is analogous to changing a nesting level of the finest macrostate. The total uncertainty is infinite, but there is no problem with this, since it is always possible to base uncertainty and information calculations on a microstate unit much smaller than any variation of interest, provided all measures to be compared use the same microstate unit. (Shannon put this in a much more elegant mathematical language, but we need not.)

If the total range of a continuous variable is 1 m, and over that range it is impossible to tell two values apart that differ by less than 1 cm, there are 100 possible values. The same is true if the total range is 10,000 km and it is impossible to tell apart two values that differ by 100 km. Magnification of the variable together with its unit of measure does not change its uncertainty. However, if a variable is continuous, and the magnified version is observed using the instrument that can distinguish values that differ by 1 cm over a 10,000 km range, there are now a billion distinct possibilities rather than 100. Before an observation, the uncertainty is correspondingly higher (about 13 bits higher, as the precision ratio is near 2^{13}), and more information is gained about the distance by making the measure.

If a variable x is discrete, uncertainty about its value cannot go lower than zero, as zero means absolute certainty about which of the n possibilities for x happens to be true. For a continuous variable there is no such zero, but changes and differences in uncertainty can be computed so long as the unit of measure, which provides a working zero point, is the same in all the calculations.

A summation such as $-\sum p(x)\log(p(x))$ in a discrete calculation becomes an integral in the continuous calculation, and the probabilities become probability densities. Often, we finesse this issue by arbitrarily defining as the unit of measure a smallest microstate that has a size appreciably less than any variations of interest to the immediate question. For example, if the finest possible resolution is 1 cm, we might use 1 mm as the unit of measure, and accept that our choice entails just over 3 bits of uncertainty within a macrostate of 1 cm size.

I.10.3 Uncertainty and Perceptual Information

For most perceptions, the uncertainty of the perception itself is low, as distinct from the uncertain relationship the perception has with the state of a variable in an uncertain external world. One perceives what one perceives, and that is that. However, all sensory data has a resolution limit. The environmental variable may have a value 'c' which is, in principle, exact, and the perception of it has a value 'p' which is also in principle exact. But the precise value of p given a particular value of c has some uncertainty, U(p|c), about what that perception represents, if anything, in the external world, as well as about how much of it there is.

Perceptual uncertainty about what something is in the external world is associated with category perception. Mackay (1953) called this 'logon' uncertainty, whereas the uncertainty associated with sensory resolution limits would be 'metron' uncertainty.[100] One may perceive a lake in the desert, though the omniscient analyst can see that the 'lake' is just an effect of atmospheric refraction over hot sand. One clearly perceives the lake, but depending on one's background knowledge of the current temperature and the properties of light propagation, one might have as much as one bit of logon uncertainty about whether the perceived lake contains water that could be splashed or tasted. In the city, one may see a person in the distance on a misty day, but be uncertain whether it is your friend or a stranger. We consider this further in Chapter III.7 when we consider the public and private natures of 'truth', and before that when we discuss protocols in Chapter II.14.

Sometimes, the perception of uncertainty is itself a perception to be controlled, usually but not always with a reference value near zero. A race handicapper wants to adjust the handicaps on the entrants so as to maximise the uncertainty about which horse will win. But usually one is controlling to reduce uncertainty to a sufficiently low level for some purpose. In our introductory example of Oliver weighing a rock, Oliver is controlling to reduce his uncertainty about the rock weight to as near zero as he can manage with the apparatus available.

For another example, Robert might want to know from Terence whether dinner will be at 6 or 7 because he has forgotten the invitation, and thinks one or other is certainly correct, and considers either to be equally likely. Robert has only 1 bit of uncertainty about the message Terence will send in reply. Robert may not perceive that his uncertainty is quantitatively one bit, but he surely perceives that he is less uncertain than he would be if he did not know whether it would be at 5:30, 6, 6:30, 7, 7:30, or 8.

Terence could reduce Robert's perceived uncertainty to near zero in many different ways, all of them conveying the same single bit of information about when dinner was planned.[101] For example, after Robert has asked: "*Did you say dinner will be at 6 or at 7?*", Terence might say any of the following, among many other possibilities:

> *I don't remember what I said, but it will be at 6.*
>
> *Six.*
>
> *I said 6.*
>
> *Dinner will be at 6, but come earlier if you want.*
>
> *8 on the dot.*

---

100 Logons (Gabor 1946) are a structural measure representing the logical dimensionality of a signal, and Mackay's metrons are an index of the weight of evidence for a probability (Mackay 1969:4). 'Logon' information concerns what is perceived and 'metron' information is much of that is being perceived.

101 Near, but not exactly zero, because there is always a possibility, however slight, that Robert mishears Terence, or that Robert believes Terence may not be completely sure.

Six heures, mon ami.

By 7, we should all be much less hungry.

OK, now think. It's a barbecue, and what time is sunset these days?

… and so forth.

No matter how Terence phrases the message, he cannot pass more than one bit of information to Robert about the time of dinner. So what should we say about all the information that seems obviously to be 'in' the different forms of message, all of which Robert can easily understand to mean that dinner will be at 6 and not 7? It seems obvious that the communication channel (voice to ear) carries a lot more than one bit of information, and that is true. The extra information is just not *about* dinnertime. It is about a lot of things, including Terence's perception of his relationship with Robert, Robert's mood at the moment, his seriousness or whimsicality, and much else. But they are not about dinnertime.

In this example, Robert's macrostates distinguish only two possible dinner times: '6' and '7'.[102] Some microstates within this macrostate might be the nature of the dinner (barbecue), Terence's tendency to indirection ('we should all be less hungry'), Terence's pedanticism ('on the dot'), and so forth, as well as unstated things about which Robert would remain uncertain (Ribs or chicken or both? Other guests? Whether to bring wine? …). These possibilities are all within the macrostate of '6' and within the macrostate of '7'. They are orthogonal to (informationally independent of) the time of dinner.

The receiver (or recipient) of a message, Robert in our example, defines the macrostates that might have their probabilities changed by the message that is received. The intention of the transmitter (Terence) is irrelevant. If Robert understands Terence's '6' to mean between 5:45 and 6:15, the fact that Terence intended to mean 'between 6 and 6:01' is irrelevant and unknowable without a further message from Terence. The fact that Terence hesitated and pronounced '*six*' in a drawn-out way may have been imperceptible to Terence, but Robert might perceive it as Terence's own uncertainty, and on perceiving it, he might not reduce to zero his own uncertainty about the dinner time, and might call Marge, Terence's wife.

Robert can do no better than observe Terence's output, though his perception may well also incorporate imagined inputs that are based on his prior observation of Terence. The reception of a message is an observation; Robert's interpretation depends on his internal state — his perceptual functions, his memories, and so forth — which are the prior conditions on which both Shannon and Kolmogorov uncertainties depend.

---

102 There is almost always a third possibility, generically labelled 'other', but if Robert is almost certain that dinner will be at 6 or at 7, the low probability of 'other' means that it contributes almost nothing to the overall uncertainty. Terence could say something like: "I'm glad you reminded me. Marge called to say she was not feeling well and we have to postpone dinner."

The overt messages contained in Terence's words are not the only things Robert observes. He might observe not only hesitations, but also tensioning of Terence's face muscles, slight movements of his head as he speaks, and so forth. Maybe he perceives that Terence is (or is not) keen on hosting this dinner. Every observation changes perceptions, and every perceptual change may change the perceiver's uncertainties about some things.

Observations provide information, even in the metaphorical sense of observation, as in: "I observe that every time Evelyn says something she thinks is clever, she tosses her hair." But they do so only if the observer has a perceptual function that allows the observation to occur. At any level of refinement, the information transmitted by a message or observation is the selection of a macrostate from among the possibilities entertained by the receiver-observer, not from among the possibilities considered by the sender.

No third-party observer can accurately identify macrostates and microstates for either the sender or the receiver of a message. For example, in the example of a printed message, the main macrostate for both writer and reader might be whether the font is one with or without serifs, and the actual text may matter only insofar as it has a wide variety of symbols — *let sleeping quick brown foxes lie!*. Within this macrostate, the most important refinement might be the name of the font, and only after further refinement might the selection of letters be considered. The definition of the macrostates and microstates at different orderings of the levels of refinement are completely up to the observer, whether that observer be the recipient of the message or a third party. There is no guarantee that they are the same for any two observers. This is something about which we have much to say later in this book.

This might seem odd, since it would allow for, say, the specification of a perceptual macrostate such as '*occasions when the temperature is above 35C and I can see red rocks and a distant rainshower*'. But such a macrostate seems odd only because, in the experience of the one who perceives its oddness, to control that perception has not proved useful for the maintenance of intrinsic variables or for the evolutionary survival of genes. However, it might be very useful for a nature photographer, to whom it would not be odd at all to use the heat shimmer to enhance the effect of the rocks and the distant shower (and perhaps produce a picture that earns more money than would a picture taken on a more temperate day).

We perceive as 'natural' those structures and patterns (macrostates) for which control of their perception has worked usefully in the past of the individual or of the species. Such structures will seem natural only if they have recurred and proved useful on more than one occasion. Those structures, and perhaps only those, define the perceptual functions that have been built by, and that have survived, billions of years of evolution and a lifetime of reorganisation.

I.10.4 Channel Capacity and Perceptual Speed

In Section I.1.4, Oliver's weighing algorithm, measuring the rock weight, is a perceptual process. Every perceptual process has a resolution limit, and a rate of gaining information about the thing being perceived. That rate is a perceptual channel capacity. In Oliver's case, we prescribed a perceptual channel capacity of 1 bit/second, because he could change the scale pan weight only once per second, and each change would halve the uncertainty range of the weight on the rock pan. It took 5 seconds for Oliver to obtain the 5 bits of information needed for him to determine the rock weight within a range of one brick.

Oliver's measurement is an artificial gedanken experiment designed to illustrate a point. But does actual perception behave the same way? Is there a limit to the channel capacity of normal human visual perception? Yes there is. We can't say that the perception of a magnitude is a control process like Oliver's but we can say that the perceptual channel has limited capacity, at least in vision, and one bit of information gained about the value of what is perceived halves the uncertainty of that value.

Any physical communication channel has a limited capacity, and the neural channels are no exception. The communication channel between the CEV and the perceptual signal passes through the environment and lower-level neural structures. Information about the CEV at the perceptual signal is gained at a uniform rate that is the channel capacity of that communication channel. Ideally, one might expect continued observation to produce increased precision, and indeed this is the way astronomers deploy their telescopes when seeking to discover or to learn about very faint sources; they use hours or days fixedly observing one part of space.

Most stellar processes that are not explosive, however, take years or millennia rather than hours to change observably, not the hours available for a single telescope observation, whereas in the world in which we live, perceptions are disturbed by environmental changes on time-scales ranging from parts of seconds upward. A disturbance to a CEV acts as a source of uncertainty that leaks away the information about the CEV that is available at the perception. The asymptotic information about the CEV available at the perceptual signal — the precision of the perception — is limited by the interplay between these two rates, g, the channel capacity of the path from CEV to perception and r, the steady rate of information loss because of changes in the disturbance. The ratio $G = g/r$ is the asymptotic information gain of the process.

Schouten and Bekker (1967) performed an experiment that directly addressed the problem of how rapidly people gain information about one particularly easy discrimination in which the disturbance was a simple step, the lighting of one or other of two lights. Their results provided two measures, the channel capacity for that discrimination and the path delay of the communication. Subjects were asked to push one of two buttons simultaneously with the third of three closely spaced auditory 'pips'. The choice of button was determined by which of two easily discriminated lights was lit. The three pips might occur at any time, so the third pip might happen long after the lights were lit, shortly after, or even before the lights came on. In the last case the subject could do no better than guess. The less time between light onset and the third pip, the less time the subject had available to gain information about which light was lit before the button had to be pushed.

The probability of a correct response in a forced-choice experiment defines a commonly used discrimination measure, d, which can be turned into an information measure I, using the conversion formula $I = (d')^2/2\ln2$ bits (Taylor, Lindsay, and Forbes1967). This is not just the rate at which the one-bit macrostate choice of lit light can be perceived, but the rate at which the lit light might provide any and all information about its properties, including not only the one-bit fact that it is or is not lit, but also about the light's colour, brightness, position, variability, etc., which were not of interest in the Schouten and Becker study (analogous to Terence's mood or what the dinner menu is to be, in our earlier example of Robert's invitation to dinner at 6 or 7). In other words, the macrostate that allows the subject to choose a button (Robert to plan his arrival time) is determined by its one-bit property of 'litness', but as time goes on, uncertainty also is reduced about microstates which are not required for controlling the perception of the relation between the light and the response button.

When the $I = (d')^2/2\ln2$ transformation was applied to the d data of Schouten and Bekker (1967) by Taylor, Lindsay, and Forbes, the result was as shown in Figure I.10.4. After an initial well defined lag time, the perceptual information available about the light increases extremely linearly over a span of at least 100 msec. In this particular case the average subject gained information from the lights at a rate of about 40 bits/sec, while the one subject 'B' who provided a lot of data gained information faster but had a longer lag time than average. In other experiments with deliberately difficult stimuli, Taylor, Lindsay, and Forbes found information rates ranging from 3 to 25 bits/second.

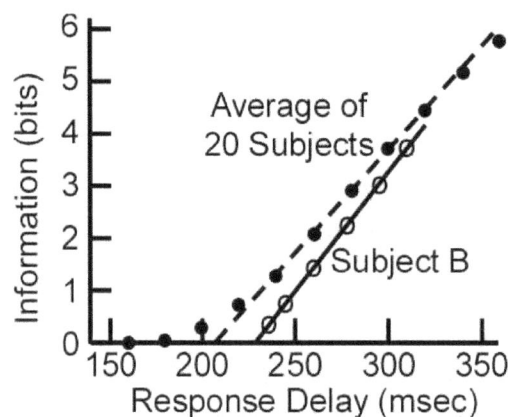

Figure I.10.4 Perceptual information gain as a function of time since the onset of one of two easily discriminated lights (Redrawn from Taylor, Lindsay and Forbes, 1967).

The slight curvature at the low end of the 'Average of 20 subjects' line in Figure I.10.4 is presumably an artefact of the different initial delays that different subjects have before they can use any information from the light. There is no curvature evident in the data from Subject B, for whom the information gained is extremely linear as a function of time.

As another example, in what now would be seen in the PCT community as a classical pursuit tracking study, Standing, Dodwell and Lang (1968) studied the Pulfrich effect (Pulfrich 1922) in which a laterally moving object is seen in stereo vision at a different depth if one eye is covered by a neutral density filter. The filter slows perception by way of that eye, leading to an apparent parallax and hence a changed depth perception. Since the study used a small artificial pupil, their results cannot be attributed to changes in the biological pupil of the eye.

The authors were able to use the change in perceived depth to estimate the effective perceptual delay as a function both of dark adaptation and of filter density, which had independent effects. The perceptual delay for each of two subjects was linearly proportional to log filter density, at around 15 msec per log unit immediately after the filter was introduced and 9 msec per log unit after the eye was adapted to the filtered darkness. The linear relation between the log filter density and the time lag of the perception is highly suggestive that the perceiving system is getting information (reducing uncertainty) about the location of edges in the visual field at a constant rate proportional to the log of the photon rate received at the eye.

Many years ago, I was able to use my knowledge of the reduced speed of visual perception in low light levels to good advantage at a party. Late in the evening, we set up an impromptu table-tennis table and played by the combined light of the full moon and an outdoor deck light. People found it hard to hit the ball because they tended to swing the paddle after the ball had passed them. Instead, knowing of the Pulfrich effect, I quickly adapted to swing at the ball when it appeared visually to be just over the net, often hitting it properly, to the discomfiture of my opponent. I would hear the sound of the paddle hitting the ball at about the moment when the ball was visually over the net on the way to me. That much sight of the ball's trajectory allowed me to predict its continued flight well enough to hit it with reasonable regularity and success. After the sound of the hit, I would visually perceive the ball continuing its path to my swinging paddle, but by then the ball was actually well on its way back to my opponent.

In biological perception, do these same effects come into play? Yes, they do. The information available about the state of the CEV increases with observation time at a steady rate determined by the sensor precision and other internal limitations — the integration gain rate 'g' — but is lost at a rate determined by the information rate of the disturbance — the leak rate 'r'. The result is a perceptual uncertainty that reaches an asymptote exponentially.

I.10.5 Good Form and the Reality of Structure

The relation between structure and macrostates may be easier to understand if we consider the distribution of cards in a deal of bridge. Four players are each dealt 13 cards from the 52-card deck. In Figure I.10.5a, the dots indicate which player received each of the 52 cards in two separate deals. One might be tempted to say that the upper distribution has more structure than the lower. But why would we do so when we know that every individual distribution of cards is equally probable *a priori*?

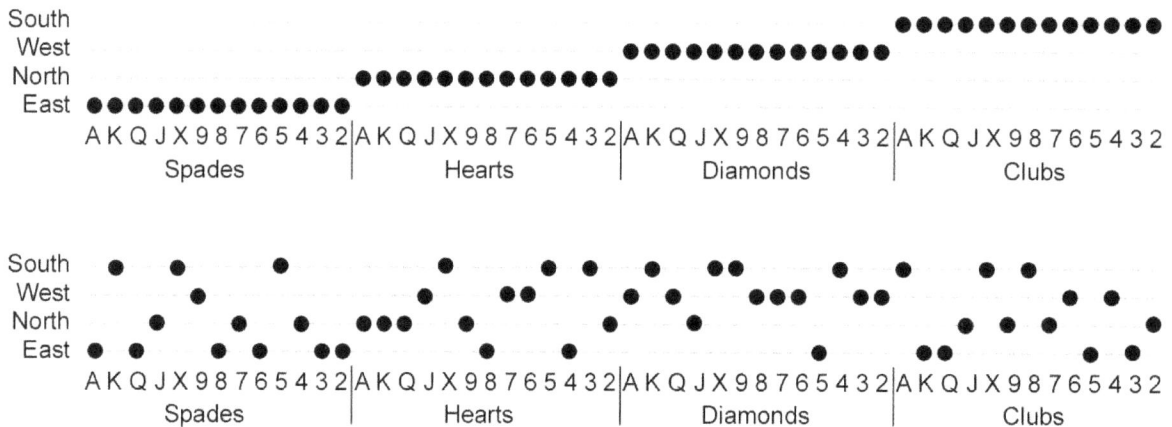

Figure I.10.5a Two possible deals of a hand in bridge. Both are equally probable, but do they look as though they are?

We can say that it is highly improbable that each player gets all 13 cards of the same suit. If we go further and ask about a deal in which specifically South gets all the spades, West gets all the hearts, East gets all the diamonds and North gets all the clubs, such a deal is even less probable, by a factor of 4! or 24. These are statements about specific deals that we can reference by label, in the same way that we can reference only a few numbers such as π by label.[103]

A deal in which each player gets *any* specific set of cards is just as improbable as one of these. Yet in duplicate bridge tournaments, each specific distribution of cards among the players sitting South, West, East, and North occurs at exactly two tables if it happens at all. If these deals are random, this would be an extraordinarily improbable event, but it happens many times over in every tournament. How is that possible? The answer is control.

However the first card distribution is arrived at, someone or some machine has a reference value for perceiving the duplicate deal to have the same distribution, and changes the cards around until the perceived distribution matches the reference

---

103 All the rational numbers can be referenced by the names of their integer numerators and denominators, but rational numbers are infinitely few among the real numbers, of which only an occasional few, such as π, e, √2, or Feigenbaum's Number, can be referenced by label. Even these acquire their labels only because they are part of a structural relationship such as the ratio between the diameter and circumference of a circle.

distribution. The duplicator control system does not see 'structure', because the location of each card is set independently. But naively we see structure in the upper deal of Figure I.10.5a, and we don't in the lower deal. Why?

Part of the answer may come from some old studies by W. R. Garner and his students (e.g., Garner1962; Garner and Clement1963; Handel and Garner1966). They studied 'good form', linking the 'goodness of form' to the size of the class of patterns their subjects considered to be alike. The patterns used by Garner and Clement (1963) were arrangements of five black dots in a 3x3 array of squares, like a tic-tac-toe board. There are 90 different such patterns in which each row and column contain at least one dot. Three of these patterns are shown in Figure I.10.5b.

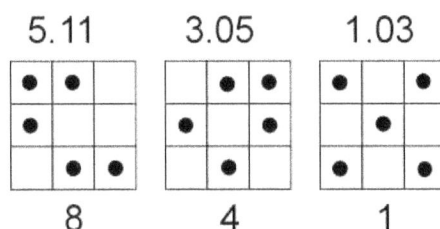

Figure I.10.5b Patterns of five dots in a 3x3 square array, as used by Garner and Clement (1963). The numbers above the patterns are the average rated goodness of form (scale of 1 good – 7 bad), while the numbers below the patterns are the number of patterns that can be transformed into that one by reflection and rotation.

Figure I.10.5b shows above each pattern the average ratings given by one set of subjects on a seven point scale of figure goodness (lower numbers are better), and below each pattern the number of patterns that can be transformed into that pattern by rotation and reflection ('1' means that the pattern rotates and reflects only into itself).

A different set of subjects were asked to sort the 90 patterns into 'about eight' groups of 'similar' patterns. The size of a similarity set for a pattern correlated 0.84 with the rating, the patterns found in smaller similarity sets being seen as better forms. Clearly, the patterns that tend to be found in smaller similarity sets are more 'special' than those in larger sets. For a subject who agreed with the group average sorting pattern, each of the eight similarity sets would be a separate macrostate of the set of 90 patterns. Within each macrostate, the particular pattern would be a microstate.

Garner's 'good form', a term used by the Gestalt theorists of the 1930s (see Wagemans et al. 2012a, 2012b for a history of the Gestalt theory) is equivalent to our 'structure'. Most of us probably would agree that of the three patterns in Figure I.10.5b, the 'goodness' of the forms increases from left to right, though 'goodness' being a perception, we would probably not all agree perfectly. For people who would agree, we would say that the amount of structure increases from left to right, as the size of the corresponding macrostate diminishes within a common Universe of possibility.

Think of a larger possibility space, say a chessboard of 8x8 squares, on any of which may be placed a marker dot. Rather than Garner and Clement's 90 possible patterns on the 3x3 board, there are now 8! = 40,320 patterns that have exactly one dot in each row and column, and in all there are 2^{64} or nearly 18 quintillion possible patterns of markers, almost none of which could be said to be a 'good form'.

Now suppose that on this board there happens to be a straight line of five markers while the rest of the board is empty. There are several possible placements of the line, each of which is as likely as any other of the 2^{64} possible patterns, but something about the line patterns makes them seem rather more special than most of the other possibilities. They have 'good form'. Let us see if the numbers bear out this assertion.

How many such line patterns are possible on this board? There are sixteen (2^4) squares that could be the central square of a line in any of the four possible orientations (north-south, east-west and two diagonals) on the board. There are also 32 squares that could be the central square of a line parallel to the nearest edge of the board. Of the 2^{64} possible patterns, there are only 2^6 (64)+32 = 96 possible patterns that are a five-marker line on an otherwise empty board. This is $1/(3 \times 2^{56})$ of all the patterns there might be. In other words, there is roughly a one in 200 quadrillion chance that such a five-marker line would appear on an otherwise empty board if every square had an equal probability of being occupied by a marker. It is a much better form than would be a randomly chosen pattern of dots.

The point is that if you see an 8x8 board with exactly five dots on it and those dots form a straight line, you can be pretty sure either that, as in the coin game (Section I.2.5), someone controlled a perception with that macrostate (any five-dot line) or microstate (this particular five-dot line) as a reference pattern, or that there is some natural phenomenon such as magnetic attraction and repulsion to connect these dots and repel all other dots from the board. Such patterns can occur by chance, but almost never do. The structure almost certainly was not created arbitrarily by your perceiving system, but exists in the external environment.

But what is it that 'exists in the external environment'? It is not the board containing the dots, because whether the board does or does not exist in the environment, the answer would be the same no matter where on the board the dots were placed. What is special is the set of relationships among the dots, whether the 'dots' are marks on paper, checkers pieces on a standard chess board, or soldiers on a parade square. The structure is the set of relationships. The checkers pieces are analogues of the soldiers on the parade square because the structural relationships among them are the same.

So it is with metaphor and simile, the metaphor works because the structure of the relationships among the content components is the same for both the explicit and the implicit set of elements. In the metaphor with which I opened the

Preface, the 'mental illumination' of PCT as applied to many apparently different phenomena was made explicit as the ever-increasing visual illumination provided by the rising sun. Physically, the two concepts are completely incompatible, but structurally, the relationships among the components of the metaphor are the same. Those relationships determine the structure. The contents do not. The structure is abstract, the contents may be either abstract or concrete.

If you are talking on the phone to someone who has a chessboard and you want to convey a message that will allow your listener to set up a pattern or structure identical to the one on your board, it is much simpler to say "A five dot line north-south centred on 4, 5 and otherwise empty" than to say for each individual square whether or not a marker is present.[104] The macrostate of the entire set of all possible patterns is 192 quadrillion times greater than is the smaller macrostate of possible five-marker lines, and that smaller macrostate is 96 times bigger than the microstate of the actual line pattern intended by the talker.

When you say "A five-dot line and nothing else" you have provided about 2^{58} bits of information, though it doesn't sound like it. The reason is that the listener already has such a structure as an available perception, and of all the patterns there could be, very few have corresponding perceptual functions. You just have to select that one out of a group that might include names, such as 'a cross', 'an arc', 'a tee', 'an Ell', 'a line bent in the middle', and perhaps a few dozen more. To make that selection takes perhaps between three and five bits.

Saying "and otherwise empty" specifies for each specific line one single microstate, one pattern of emptiness for the entire board, so the macrostate that includes all the 'otherwise empty' microstates is exactly the same size as the macrostate that defines the pattern of dots. The negative of a photograph on film conveys the same information as the printed positive, but only if you have the tools to extract it.

The listener on the other end of the telephone line can control a perception of the pattern 'five marker line on an empty board' or 'Vertical Cross' or a number of other named patterns, simply because there are extraordinarily fewer such patterns that have recurred in the game, or indeed in life. Lines and crosses and circles are pattern categories of relationships that have been experienced often enough to be communicated by the use of labels, with a reasonable possibility that the listener will perceive something like what the talker controls for having them perceive. Almost all of us have built perceptual functions for them. We might ask "Why?"

---

104 The lengths of the two descriptions could represent their relative Kolmogorov uncertainties. To say 'line' presumes that the listener knows what a line is. If the listener does not, then the talker must explain its properties before using the word, thus increasing the length of the verbal description of the pattern, though still keeping it shorter than would be the case if each square were described independently as 'empty' or 'with dot'.

I.10.6 Structure and Objects in Perception

When we observe the natural world, we do not see myriad isolated dots of different colours. We see objects. But what is a perceptual object, and why do we perceive it when we detect one? We approached this question from one direction when we considered perceptual control of the orientation and location of the 'chair-object' in Section I.5.5. Looking at it from an information-theoretic viewpoint leads to another approach to the concept of the control hierarchy. We will use a larger chessboard as a toy universe, a 15x15 space of squares that might be occupied ('dots') or unoccupied ('empty' or 'blank'). Much of the time we will have prior knowledge that there are exactly 25 dots in the 15x15 space, as in Figure I.10.6a. These dots may sometimes seem to depict objects.

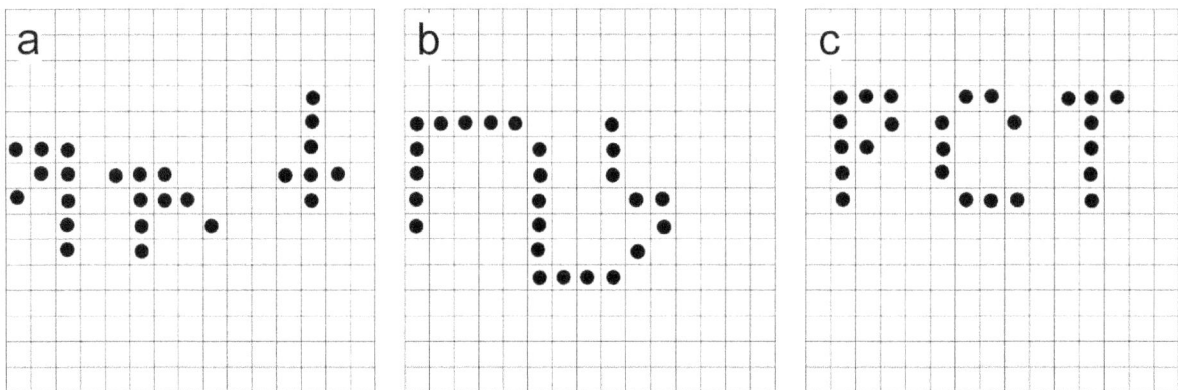

Figure I.10.6a How many objects are in these diagrams? There seem to be three in (a), one, or you may see two or even three in (b), but how many are there in (c)? Are there three or one? All three diagrams are made from the same seven groups of three to five dots, always in the same orientations. Does each diagram therefore have seven objects?

Each panel of Figure I.10.6a consists of the same set of seven dot-pattern members: two vertical 5-dot lines, a vertical 4-dot line, three 3-dot horizontals, the 'loop' of the 'P' in panel c, and the three-dot bent top of the 'C' in panel c. The three panels differ only in the locations of the seven fixed patterns. When we first look at them, however, we perceive the panels as containing objects, very different objects, but no panel seems to contain seven. The first is perceived as having three objects, the second as one sinuous object or perhaps as two or even three objects that touch one another, but the third is more ambiguous. Does it contain the familiar acronym 'PCT' as a single object or three objects 'P', 'C', and 'T'? The answer depends on the perceiver. If you perceive three objects, that is how many there are.

We can ask the same question about the three panels of Figure I.10.6b, which are supposed to represent a time sequence in the dot patterns. Panel (a) is a copy of panel (a) of Figure I.10.6a. Panels (b) and (c) represent the same dot array at two later times.

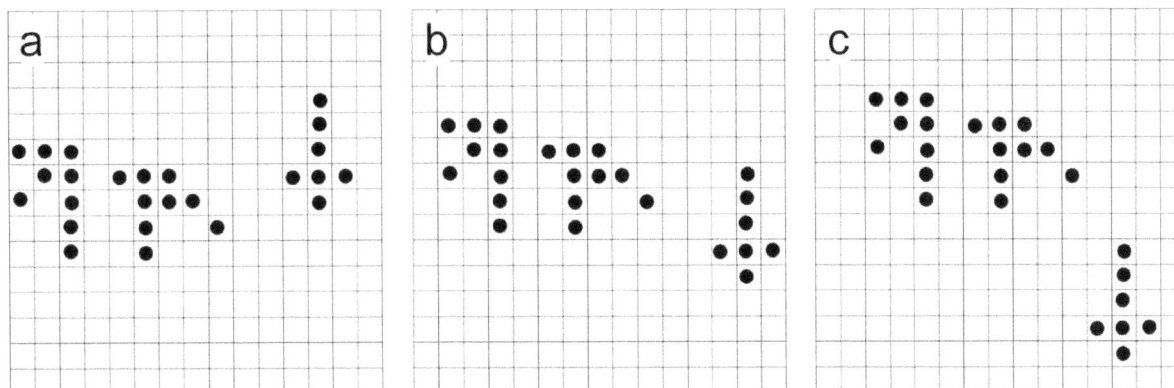

Figure I.10.6b The dot patterns of panel (a) above, at three different times.
How many objects are there? Three as in Figure I.10.5b-a, or two that move
independently, one upward and rightward and one that moves downward with
a jog to the right as though it was avoiding the other?

Do we now perceive three distinct objects, or two that are distinguished by a
difference in their 'common fate', as the Gestalt school would have called it?[105]
'Common fate' is a surface description, based in 1930's Gestalt Psychology, of
the way an external observer would perceive the motions of the elements of
the sensed environment. That description is analogous to the description of
overt actions as the behaviour of a person, such as "the finger is pushing the
doorbell button", as opposed to the PCT statement "the person is controlling
for perceiving the doorbell ringing by controlling for perceiving the finger to be
pushing the button" (Figure I.5.6b).

The 'common fate' description is convenient, but from a PCT viewpoint, it
is somewhat misleading. Earlier, we ascribed the development of multiple levels
of hierarchic perception to the way that ECUs control independent perceptions
of different aspects of the environment that would otherwise interfere with each
other. We would do the same thing here, but there is a problem, since these
dots, as dots, are not coherent objects, but are independent. Their locations were
controlled separately by me, and no placement of them on 25 different squares
would create a conflict, unless I tried to place two on the same square.

It would be very easy to ascribe agency to the 'arrow' object in Figure I.10.6b,
dodging rightward to avoid the slow linear drift of a bipartite object. But these
are just dots on an array. They have no perceptions to control, such as avoiding
being hit by other dots. But I, who configured the dots, did control perceptions,
perceptions in imagination of how the viewers of these patterns might perceive
them. When we deal with 'protocols' in Volume II of this work, we will often
be talking about such control in imagination of what other people will perceive
when we control our own perceptions in particular ways.

The concept of 'object' here can be extended over time, as 'cause and effect',
which is captured in Powers's 'sequence' level and in his 'event' level. For example,
if a noise happens immediately after a visual flash, one usually perceives there to

---

105 For a general review of a century of Gestalt psychology, see Wagemans et al 2012a, 2012b.

have been an event that caused both the flash and the noise. The time sequence depicted in Figure I.10.6b probably is not perceived as 'cause and effect', unless one sees the 'motion' of the small 'arrow' object as being a controlled avoidance of the large two-component object that would hit it if it stayed where it was. That perception would be a perception not only of the coherence of the small arrow object, but also a perception that the pattern represents a perceptual control system — a live organism.

Why would we perceive the coherent 'objects' that we do perceive in these figures? The answer, as is often the case, comes down to the reorganisation of perceptual systems to control perceptual variables in ways that tend to maintain intrinsic variables and enhance survival of the organism. One of the generic ways reorganisation does this is to create perceptions of relationships, and in particular the relationships of relationships that constitute object perceptions (such as the chair we discussed under Perceptual Complexes).

Not all sets of individual perceptions that might belong together as an object can be controlled as a higher level 'object' perception, but many can, and if we return to the uncertainty analysis of the dot patterns, we can see why evolution built us to perceive 'object' by default rather than to try to control the multiple contributory perceptions in ways that might conflict if the coherence happens to have been a chance coincidence.

Let us consider the size of the macrostate of a 15x15 array that is defined by the fact that it contains, say, at least one straight row of 5 dots at any angle, horizontal, vertical, or diagonal among its 25 dots. Further calculations are much easier if we imagine the 15x15 array as having a toroidal topology, continuing from top to bottom and left to right, as in Figure I.10.6c, in order to avoid special calculations involving distance from an edge.

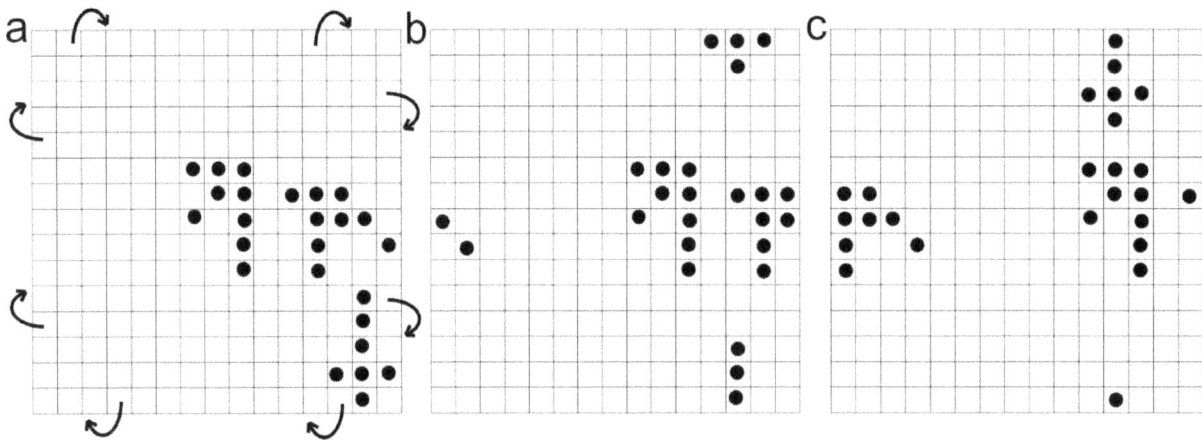

Figure I.10.6c Continuing the "object" movements of Figure I.10.6b, but now treating the 15x15 array as a toroidal space, in which the top is continuous with the bottom and the left with the right. The two-component object moves off the displayed square rightward and reappears on the left, while the "arrow" moves down, reappearing at the top.

In a 15x15 square array with edges, no square within two spaces of an edge can be the centre of a 5-dot line, leaving only 121 squares that could. But in the toroidal space with no edges, every one of the 225 spaces has an equal chance of being the middle of a 5-dot line. How many such lines are there? There are four orientations for a line with a specific mid-point, so there are 900 possible locations for a 5-dot line. What is the probability that a pattern of exactly 25 randomly placed dots will include at least one 5-dot line? We find the probability of creating a specifically located line, which we then multiply by 900 to find the probability that at least one such line has been formed.

We consider the probability in turn that each space of the potential line is occupied. The first square has a 25/225 chance that a dot will land on it. If that chance pans out, the chance is 24/225 that the next square will be filled, and so forth. The probability that the whole line will be completed is the product of these probabilities, and the probability that there will be at least one such line is 900 times as large, or almost exactly 0.01.

Next we ask about the probability that a second such line exists in the space, and that it intersects the first line. There are 25 possible relative locations of intersection, 5 for each line intersecting at each place on the other line, but only three possible orientations for the second line, since lying alongside the first line is not an intersection. So there are 75 possible relative locations and orientations for the new line. If the new line intersects the original, it has only four new points. We use the same arithmetic to get .0037 for the probability the new line exists given that the first one does, and multiply the two probabilities to determine the probability that the space will contain two intersecting 5-point lines. The answer is that the macrostate with two intersecting 5-dot lines is about 15.7 bits smaller than the 105-bit macrostate with 25 dots.

If the orientation and meeting point of the new line is specified, so that we are talking about a 9-dot configuration in some location in the space, that macrostate is nearly 23 bits smaller than the 25-dot pattern universe. The same calculations apply to any prespecified asymmetric configuration of 9 dots, placed freely in the space, such as a letter 'P' of Figure I.10.6a 'c' set in a random location and orientation. In all these 9-dot configurations, the remaining uncertainty concerns the locations of the other 16 dots in the 1515 array. To find the size of the macrospace containing the letters 'PCT' spaced exactly as in Figure I.10.6a 'c', oriented horizontally, and placed in an arbitrary location, we do similar calculations. The relative placements of all 25 dots are specified, but the placement of the ensemble is not, so the first dot can go anywhere. This location uncertainty is the only remaining uncertainty, about 9 bits. The same would be true for any figure in which the relative placements of all the dots was prespecified.

Now we are in a position to do another macrostate refinement like that of Figure I.10.6d, but with a wider range than before, as a precursor to the one we will do for language perception and control. This time we introduce information obtained from memory (analogous to what Nevin calls 'objective information' about language, in his chapter of LCS IV) which adds to information obtained through the senses.

There is a problem, however. The problem is the determination of what constitutes a perceptual object. Since, like beauty, the 'object-nature' of a pattern is in the eye of the beholder, we cannot do the same kind of probability analysis we have been doing for other aspects of the macrostate analysis. The best we can do at this point is a zero-order approximation, recognising that if the dots are randomly placed on the array, any 'objects' would be purely fortuitous and that object perception must depend on some perceptible deviation from the probabilities derived from the random distribution.

There are many possible deviations from randomness, most of which are unrelated to the perceptual existence of objects in the everyday world. We usually expect some perceptual quality to be more consistent within an object than between the object and its surroundings. Such differential consistency is the central design feature of much camouflage, dating back at least to the zebra-painting of ships in the First World War, and much longer in evolutionary terms.[106] We expect lines and curves that we perceive to be co-aligned to belong to a single object, because the probability that they are so related by chance is much lower. Hence we perceive partially hidden objects as objects despite parts of their component edges being absent from direct view. Likewise, we do not perceive distant parts of the visual field to belong to the same object unless they move similarly in relation to other parts of the visual field.

In light of all these possibilities, we cannot properly assign probabilities to the numbers of objects that a particular perceiver will perceive in a specific microstate of the 25 dots. So we do not attempt it, leaving blanks instead of numbers in the diagram. But we can nevertheless make some assessments of remaining uncertainties at the later stages of refining the macrostates (Figure I.10.6d). For that purpose, we assume that there are 64 symbols in the known script (six bits per symbol) and that the perceiver remembers 32 different 'TLAs' (Three-Letter-Acronyms). Other values might change the estimates by one or even two bits, but not much more.

---

106 For examples, see t.ly/6b--n or the article "Dazzle camouflage" in Wikipedia (both retrieved 2011.03.24).

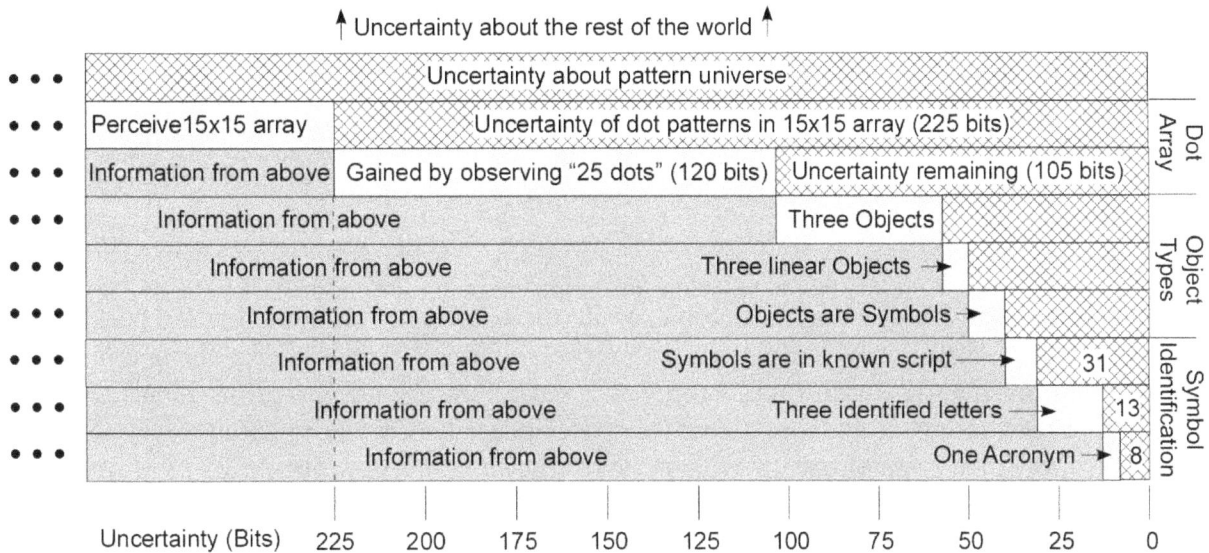

Figure I.10.6d A possible set of macrostate refinements of the pattern of Figure I.10.5b-c, by adding more information from observations and/or memory (e.g. linguistic memory is needed to identify the symbols as being in a known script and perhaps to recognise the "PCT" acronym). Missing bit numbers depend on the perceiver.

In diagrams such as Figure I.10.6d (and the similar Figure I.10.8d below), the ordering of the information used to illustrate the refinement of macrostates by acquisition of information about different relationships is to some extent arbitrary. For example, it would be quite reasonable to suppose that the perceiver would perceive very early that the pattern contained entirely 'linear' elements (continuous sequences one dot wide), which were then perceptually combined into three individual objects, or even that the 'object' stage was omitted entirely, in favour of the linear sequences being recognised as symbols. Each of these orderings would lead to a diagram similar in principle to Figure I.10.6d but different in detail. All of them would be equally correct as possibilities, but would describe different processes in the perceiver.

One might also partition the diagram according to the Powers levels of perceptual control. We do not do that here because the diagrams are intended to ease the later interpretation of a similar diagram (Figure I.10.10) indicating the information available from different structural aspects of a text. The main point instead is to illustrate how much more structured a small macrostate is than an even slightly larger one. 'PCT', for example, is much more structured than 'P'-and-'C'-and-'T'; and a pattern of symbols, even in an unknown script, is much more structured than a pattern of irregularly arranged lines and arcs.

The Powers perceptual levels have exactly the same property, as we discussed in 'Perceptual Complexes' (Section I.5.5), and it seems intuitive to suggest that as we go up the levels, the perceptual signals represent ever more structured aspects of the perceptual environment.

I.10.7 The *a priori* Improbability of Structure

Let us do a little more arithmetic to show the *a priori* unlikelihood of structures that we perceive as existing in the external world. We will now use the 'chair' example of Section I.5.5 rather than abstract patterns of dots.

Our simple 'chair' of Section I.5.5 was composed of six even simpler objects, two front legs, two back legs, one seat, and one back. We can presume that the two front legs are interchangeable, as are the two back legs, but front and back legs are different. When we talked about controlling the 'chair' object perception rather than the individual components separately, the advantage of using the structure was that it sent coordinated reference values to the controllers for the individual components to avoid conflicts that might arise if they tried to move their components in unrelated ways. Now we will ask about the likelihood of the components being perceptually arranged to form a chair if they do not form a structure in the outer world.

Of course, if we start with four legs, a seat, and a chair back floating randomly about in infinite space, it is infinitely unlikely that they will ever come together in the form of a chair. So let's make the situation a little less artificial, and assume that we have a *very* stupid apprentice to a furniture maker. We will call him Ted. Ted's boss gives him the six components and tells Ted to fasten them together to make a chair, but Ted doesn't know what a chair looks like. To make his job easier, we assume he knows that legs go on seat corners and backs go on seat edges. How many ways could he fit the components together, only four of which are correct?[107]

Ted starts with the seat, because he knows that all the components fit individually onto it. The seat has a top and a bottom, and a front and back, but Ted doesn't know anything about that. First, he has to worry about where the back fits. There are four edges, each of which has a top-bottom choice, and the back itself has a top and bottom. Ted can see the difference between sides and the top and bottom of the back, but not where on the seat it fits so there are 16 ways to fit the back, fifteen of which are wrong. The only correct way to fit the back onto the seat is to mate the bottom of the back to the top-back of the seat, but Ted doesn't know this and has to guess.

Having placed the back, Ted has to install the legs, the first of which has six possible places to which one or other end of the leg must fit, or 12 fittings. The second leg has five, or 10 fittings, the third has 8 fittings and the last leg has 6 possible fittings. In all, there are 92,160 ways Ted could put these six components together, only four (or 24) of which make a chair that his boss would recognise as being correctly made. Even this simple structure is extremely unlikely to be produced by chance connection of mutually attracting components. To be exact, the chances are 23,040 to one that Ted will fail.

---

107 Four, because the two front legs are interchangeable as are the two back legs, making a four microstate 'chair' macrostate. If all four legs are identical, there are 24 microstates in the 'chair' macrostate.

Because of its improbability, when we have seen one such configuration, it just might have been a random rearrangement of the parts that we happen to see at a propitious moment, but if we see the same arrangement again, the odds become much longer that it is a configuration that might be useful to control as a unit. So when we see a configuration that appears to be a 'chair', it is almost certain to be a chair in the environment unless some control system has arranged the parts so that some other configuration will give the same sensory data as would a chair.

Deliberate illusions and advertising displays have been created to do this, as the example in Figure I.10.7 illustrates. The same is true of the Ames Room illusion, in which a room appears from one specific viewpoint to be a perfectly normal rectangular room, but a room in which people and objects change size as they move about. From any other viewpoint, the walls, floor and ceiling are easily seen to be far from rectangular, and one back corner to be much further from the prescribed viewpoint than the other. It took careful control to achieve either illusion, which is caused by the fact that the perceptual configuration is extremely unlikely to happen by chance and is the same as a configuration of perceptual angles where flat surfaces meet in a room we have previously learned to perceive as a unit, which makes it even less likely to have happened by chance.

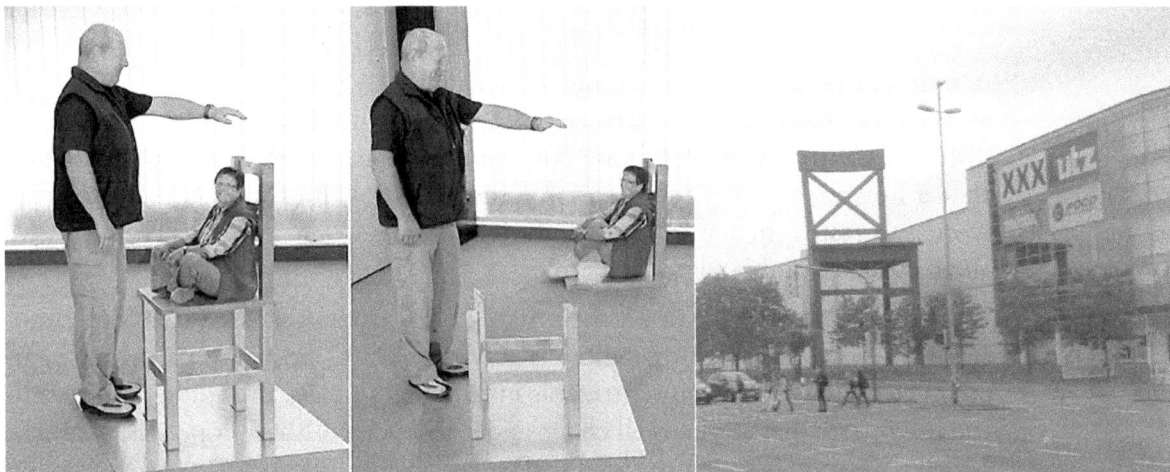

Figure I.10.7 (Left and Middle) What seem to be components of a chair on which a miniature person sits are not, when viewed from a different angle. The appearance is created by the action of a control system (a person who set up the picture). (Picture from <https://www.weirdoptics.com/dwarfism-visual-optical-illusion/>). (Right) Even though the components are correctly arranged, this is not a chair on which one can easily sit. Its actual size is seen by comparing it with the cars and people in front of it (Picture by the Author.)

These illusions are amusing and perhaps surprising, but they illustrate the point that we see what our perceptual functions have reorganised to produce. What they produce is the degree to which the data at their inputs resembles structures that recur and that have proved stable when used for control. Visual structures like those of Figure I.10.7 are the same as those that produce the real 'chair' that has occurred many times in the viewer's experience because 'chairs' are produced by controllers more competent than our apprentice chair-maker 'Ted'. The illusory 'chairs' are also constructed by controllers, but do not remain 'chair' if, in the left example, the viewpoint is changed or, in the right example, if one tried to use the chair as an atenfel for perceiving oneself to be seated against the back with feet on the ground.

The chair is a simple example of a general property. Even when we are talking about structures much more complex than lines or arcs of pixels and structures made from only a few highly structured components, yet the structures we perceive as structures are extraordinarily unlikely to have occurred by a random arrangement of the parts. They are 'Good Forms', very small macrostates within much larger macrostates of unorganised structural possibility. They are therefore unlikely to produce the requisite patterns to our sensors unless the actual structure exists in the environment being sensed, or unless some control system has produced an illusion such as the Ames Room or the illusory 'chairs' in Figure I.10.7.

The category 'chair' represents a set of perceptual control atenfels, prime among which is that it can be used to perceive oneself to be sitting. There are many other configurations of the natural and manufactured world on which one can sit, but which are not usually considered to belong to the category 'chair'. To be a 'chair' demands that it appear to have been made to an intentional design and perhaps that it should be located in a suitable place. The design, as for most intentional structures, allows the chair to serve as an atenfel for different controlled perceptions. For these CVs, the action hierarchy includes that a human person sit and stay for more than a brief moment — in other words, to rest from standing or walking, perhaps while 'doing' something else such as eating from a plate on a table.

On the edge of a cliff or on the top of a chest of drawers are places one could sit, but few would agree that those places are 'chairs'. So being perceived as an atenfel for the perception of 'sitting' is insufficient to place the configuration into the macrostate that contains all chairs and only chairs. Other properties, such as a lateral separation between the level surface suitable for sitting and the surrounding area, distinguish 'chair' from a 'non-chair atenfel for sitting' such as 'bench', 'couch', 'tatami mat' and other things on which people commonly sit. Perhaps 'manufactured' or 'moveable' are also properties required for the core concept, though some natural configurations might also be members of the category.

I.10.8 Fuzzy Nested Macrostates

Macrostate boundaries are generally fuzzy (a technical concept in Volume II and Appendix 2, as well as an everyday one), and there will be cases of 'chair' that lie between the clear non-membership of a cliff edge and the clear membership of a dining room chair. An example might be a moss-covered rock ledge that is depressed between two slightly higher ledges about 60 cm above a firm surface. Sometimes, one might think of this place as a chair, especially if it overlooks a pleasant vista, but usually one would not. Technically, it would have a fuzzy membership midway between zero and unity, perhaps 0.4, in the macrostate 'chair'. Is a throne designed for the monarch to sit on when executing formal royal functions a 'chair'? It is a manufactured object intended for someone to sit on, which usually defines a 'chair', but if a tourist tried to sit on it, a guard would probably act quickly to remove the trespasser. Maybe the royal throne has a membership of around 0.8 in the 'chair' macrostate.

When macrostates become large and the fuzzy edges wide compared to the core of membership values near unity, we are more likely to call the concept a 'syndrome' than a 'category'. The category 'red' is not likely to be called a 'syndrome', but the parcel of perceptions ('symptoms') related to an illness such as flu is. Not everyone with the flu has all the symptoms in the syndrome, and not everyone with most of the symptoms has the flu. Nevertheless, to say that the person "has the flu" is usually to say that the person reports perceiving most of the symptoms in the syndrome, not that the person has been tested for the virus that is prevalent in this year's epidemic.

These characteristics suggest that 'category' and 'syndrome' should perhaps not be thought of as near synonyms in which 'category' has sharper boundaries than 'syndrome', but instead 'syndrome' might represent a collection of property values that are often encountered together, whereas 'category' is a distinct identity into which a perception might belong, as we suggested from a different viewpoint in Section I.9.5. In Figure I.9.4a and Figure I.9.6a, 'syndrome' and 'category' would apply to the same set of signals, 'syndrome' to the analogue signals, and 'category' to the output of the polyflop that has the syndrome as its basic inputs. Another way of looking at the distinction is that a category perceptual function has a syndrome as its input variables.

Both the syndrome and the category would be perceptions, available as inputs to higher perceptual control levels, either together or separately. If both are available to a particular higher-level perception, a clear category output is likely to suppress the syndrome, leading to differences in sensitivity to analogue variation within the category as opposed to across category boundaries, a common psychophysical finding. Habets, Bruns, and Röder (2017) even produced a synthetic audio-visual syndrome that resulted in a similar category effect on the perceptual discrimination of simultaneity.

'Good form' is clearly not an all-or-none property of a pattern. We may think some forms are excellent and precise, while others are pretty good but flawed, yet others are passable but a bit messy, and some are not good at all. To put it another way, macro- and micro-states often do not have crisp clear boundaries, and it is impossible to state that *this* microstate has good form while *that* microstate, which differs by the placement of only one dot, has not.

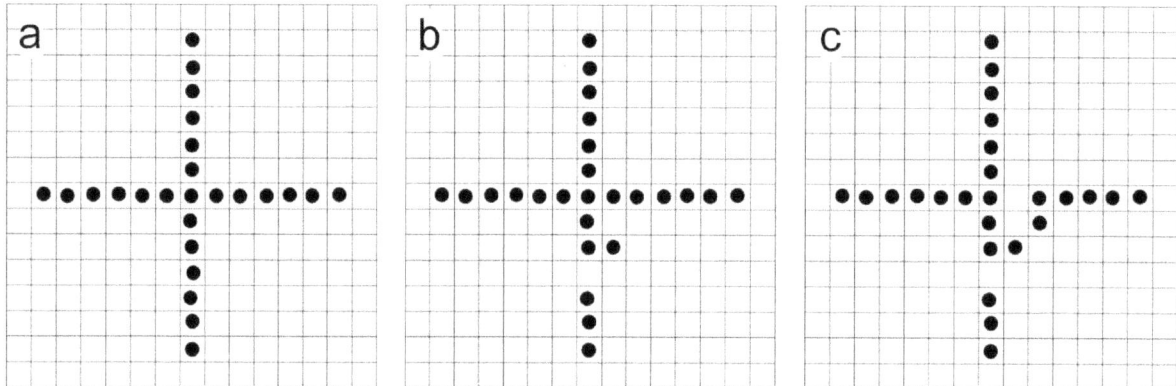

Figure I.10.8a Moving a random dot moves the microstate (b) out of a small macrostate (a). Moving a random dot again (c) is unlikely to return the pattern to the original macrostate.

It may be questionable whether Figure I.10.8a's right panel is as good a form as the middle panel, but it is certainly better than any of the panels of Figure I.10.8c (below). The answer to the question is that though the boundaries of many macrostates are fuzzy, yet the sizes of the macrostates, and hence the information gained by observing that a microstate is in a particular macrostate, need not be imprecise.

Instead of asserting that a microstate is or is not in a particular macrostate, we assign to each microstate a 'fuzzy membership' in the macrostate. The size of the macrostate is then given not by the number of microstates it contains, but the normalised sum of the memberships of the microstates in its class. This being the case, for now ignoring the problem that for most purposes the macrostates have diffuse boundaries, we can make two assertions: Firstly, that the macrostate size in bits of microstate uncertainty may be well defined even if its boundaries are not, and secondly, that the information gained by identifying the macrostate to which a microstate belongs is weighted by the microstate's fuzzy membership in that macrostate.[108] We will seldom need to use the second assertion, but it should be remembered.

Figure I.10.8b illustrates how fuzzy membership relates to the logon-metron partitioning of information (logical dimensionality vs. weight of evidence for a

---

108 Since the membership values are normalised to make them sum to unity, they can be treated mathematically as though they were probability weightings in the computation of uncertainty.

probability). In this Figure, each panel represents 12 bits of information, and each column in a panel represents one macrostate. The macrostates are, say, inputs to a perceptual input function at some level of a control hierarchy. The horizontal top panel represents a pattern in which twelve different macrostates are crisp, not fuzzy, because they have only two possible values. In this panel, each individual macrostate either exists as an input present in the data or it does not. The other panels might be used to describe the fuzzy membership of the data in (respectively) 6, 4, 3, and 2 different macrostates, and (on the right) 1 macrostate. In each case, 12 bits are used to describe the complex structure that is input to the perceptual function, with more precision for each membership, the fewer macrostates there are to consider.

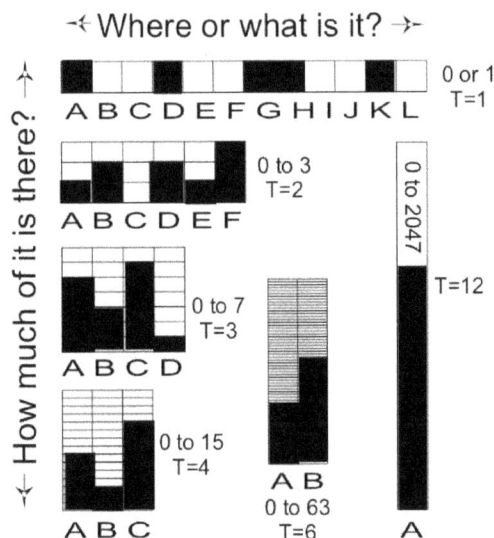

Figure I.10.8b The metron-logon trade-off of information. Each outer rectangle shows a way that 12 bits of information might be used to transmit how much of each of n *different entities exist in some data, from a decision between "some" and "none" for each of 12 entities, transmitted in one burst of 2047 pulses that each could be zero or 1 (top horizontal panel), to a slow precise measure of a single entity (right-side vertical panel), and anything between. The fewer the entities, the greater the precision of each and the longer the total transmission takes. "T" is the number of 12-bit transmission bursts to transmit the specified precision for all the variables.*

These same diagrams could be used for a variety of situations, such as to describe the inputs from higher levels to lower-level reference input functions, with varying levels of precision about how much of each lower-level perceptual variable to produce. Later, we will consider them in a rather different context, the channel capacity of parallel channels, where a wide set of channels could dump the 12 bits in unit time with very low precision for each channel, while a single narrow channel would take a long time to provide a precise value for one variable.

Nested macrostates yield a successive reduction of uncertainty. We start with macrostates defined by the numbers of dots in the rows and in the columns of the patterns in Figure I.10.8c. The lower two rows show possible outputs of perceivers of those numbers. They might serve as inputs to a perceiver of the location of the body of the pattern, but they would provide little information about the nature of the pattern. They would certainly provide some, because just such projected patterns, taken from different angles, are used in clinical and industrial settings for computerised tomography.

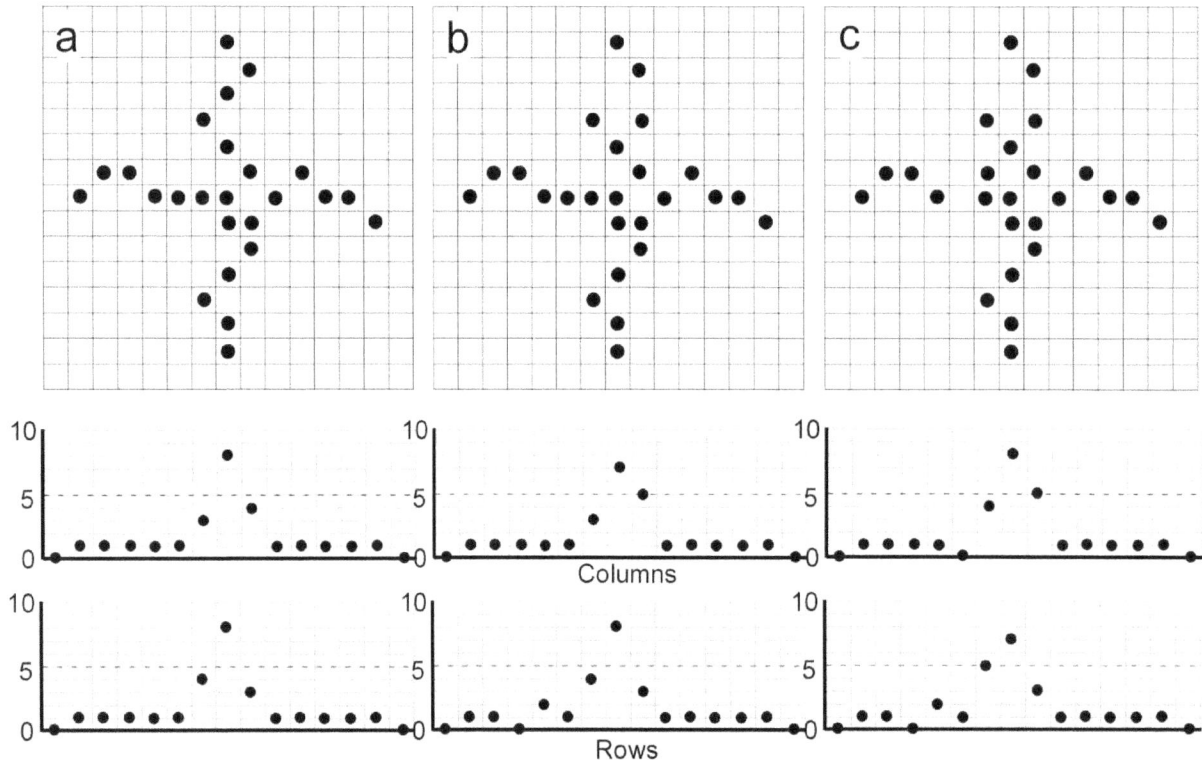

Figure I.10.8c Moving a random dot does not move the microstate out of a slightly larger macrostate, and nor does moving a random dot again. The lower panels show (upper) the number of dots in each column and (lower) the number of dots in each row.

It is not obvious how much information the one-dimensional distribution patterns could give about the 25-dot pattern, but the uncertainties of these patterns is rather low, since most of the rows or columns have only one dot whereas three have several dots, so the information they could provide is rather high. It could be between 20 and 30 bits each, say 50 bits in all, out of the 105 bits of uncertainty of the set of possible 25-bit patterns. These particular one-dimensional patterns all suggest that the corresponding two-dimensional pattern is a cross, which would not have been the case if the views had been from diagonal directions. Diagonal views would have given distributions with higher uncertainties and thus less information. Viewpoint matters here, in exactly the same way as it does for the chair illusion of Figure I.10.7.

In a tomography application, scans from other directions would provide further information, the remaining unpredictable amount reducing as the number of non-orthogonal directions increases. Here, and in much of what follows, the uncertainty estimates are mostly 'ball-park' estimates, quite possibly in error by 30% or more, but they should make the point that macrostates can be indefinitely refined by further observation, and that the order of observation determines how fast the uncertainty is reduced.

The two one-dimensional distributions still leave, by our ball-park estimate, about 55 bits of uncertainty for patterns with similar distributions of dot numbers in their rows and columns. If we assume that a 'wavy cross' is a good cross in which no dot is more than one place out of position, the size of the wavy cross macrostate is about 30-35 bits. If we do not consider a 'good form' or a 'slightly damaged' cross (Figure I.10.8a) to be 'wavy crosses' we should remove their smaller macrostates from the 'wavy cross' macrostate. Doing so would take from the core of the 'wavy cross' macrostate roughly 15 bits, 3 for the good form cross and say 12 for the damaged, leaving something like 15-20 bits. There remain 20-25 bits for other 25-dot patterns having similar distributions of dot number in their rows and columns, 50 bits for miscellaneous 25-dot microstates, and 120 bits for patterns having other numbers of dots.

Successive reduction of the 225 bit uncertainty as a consequence of observing different features of the display in the sequential way just described can be diagrammed as we did in Figure I.10.6d to show the way each observation reduces the residual uncertainty. Figure I.10.8d shows nested macrostates (cross-hatched), the information gained by each successive observation (light grey), and the information already obtained by earlier observation or prior knowledge (dark grey). We shall use a similar diagram later (Figure I.10.10) to illustrate a possible partition of the information available from different sources when understanding some text or speech.

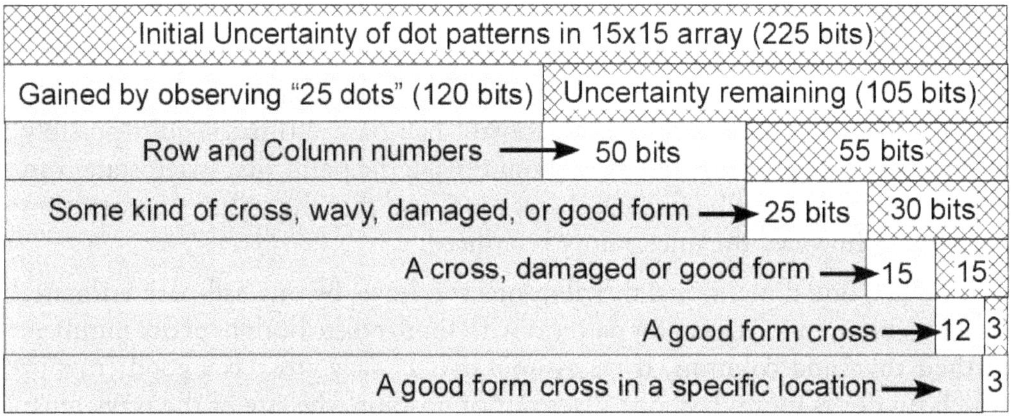

Initial Uncertainty of dot patterns in 15x15 array (225 bits)		
Gained by observing "25 dots" (120 bits)	Uncertainty remaining (105 bits)	
Row and Column numbers ➔ 50 bits	55 bits	
Some kind of cross, wavy, damaged, or good form ➔ 25 bits	30 bits	
A cross, damaged or good form ➔ 15	15	
A good form cross ➔ 12	3	
A good form cross in a specific location ➔	3	

Figure I.10.8d Successive reduction of the residual uncertainty (the size of the macrostate) by observing different features of the display. The numbers of bits are reasonable guesses intended to illustrate the concept, not accurately computed values.

I.10.9 The Expanding Universe of Possibility

Throughout, we have been talking about the same 'Universe of possibility', namely, the locations of dotted and empty squares in a 25x25 array. But Universes of possibility come in different sizes. When we extend the idea to the Universe of manufacturable objects, we see that each invention increases the possibility space open for more inventions. The invention of insulated wire made possible the invention of all kinds of electrical equipment. The invention of railways allowed the invention of centralised markets and distribution hubs. The point can be illustrated by some more 'dotty crosses' in spaces of different sizes, as in Figure I.10.9.

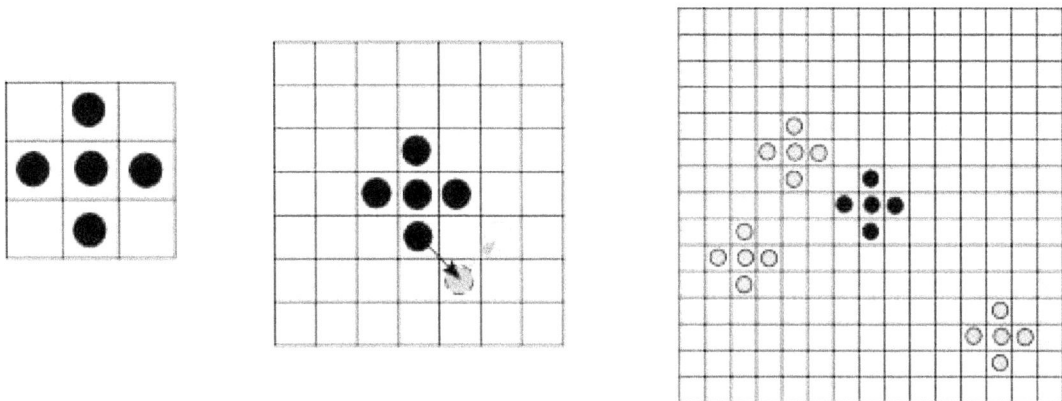

Figure I.10.9 An expanding Universe of possibilities. The relative structure of the 'same" pattern differs, depending on the size of the Universe of possibility.

The three panels of Figure I.10.9 show a 3x3 cross in the centre of an array whose size increases from 3x3 to 7x7 to 15x15. One might think that the structure of the cross was numerically similar in each case, and in absolute terms, it is. Given that there is a dot in the centre of the array, one asks what the probability is that there is a dot above, below, to the left and to the right of it, but none in a diagonal direction, if the probabilities of dot or no-dot are *a priori* equal in each square.

The answer is easy to calculate for each of these small Universes of possibility. Each location has eight neighbours, so, given that the centre of the array has a dot and if the probability that each neighbour is occupied is 0.5, the probability that the form around the centre is a 3x3 cross must be 2^{-8}, or 1/256. Once one knows that the centre location of the array has a dot, to say that it is the middle of a 3x3 cross provides 8 bits of information. This is true whatever the size of the array.

If we know not that there is a cross or that the probability is 0.5 that a randomly chosen square has a dot, but instead that there are exactly four additional dots besides the centre one in the Universe, the calculation is different. In a 3x3 Universe with a dot already in the middle, the first dot to be placed has a probability 0.5 of filling one of the locations forming the cross, the next has a 3/7 probability, the third has a 1/3 probability and the last has a 1/4 probability, so the probability that these four dots form a cross around the already known centre one is 3/168 or 1/56. A probability of 1/56 is worth about 5.8 bits, so the observation that the five dots form a cross provides 5.8 bits, rather than 8. Knowing that there are exactly 4 dots to be placed, as opposed to knowing beforehand that each location has a 50-50 chance of having a dot, provides 2.2 bits of information in this 3x3 Universe of possibility.

In a 7x7 Universe and when the location of the centre dot is already known, the probabilities are much lower. There are 48 unoccupied squares, of which four are to be occupied. There is a 4/48, or 1/12, probability that the first dot will be in a position to contribute to the cross, a 3/47 probability for the second, a 2/46 probability for the third, and a 1/45 probability for the fourth. Overall, the probability that these four dots, randomly placed, will create a cross with the first one as its centre is 5.1×10^{-6}.

The resilience of the cross form against disturbance also changes in Universes of different sizes. In the 3x3 Universe shown in the left panel of Figure I.10.9, an event that moves any of the dots one square can move it only onto a corner square. A subsequent event may move the same dot or a different one, so there is a 1/5 chance that it moves the same dot, and if it does, the only place the dot can move is back where it came from. In the other two panels, the dot that is moved by the first event can go to any of four or five places (as suggested by the grey dot in the middle panel). Even if the next event moves the same dot again, there is only a 1/8 chance that it will be moved back to its 'good form' location, rather than to somewhere else, for example the square suggested by the faint ring in the middle panel.

Even for a 3x3 cross, if the Universe of possibilities is large enough that the edge is at least two squares from the tips of the arms, as it is in a 7x7 array, the probability that two successive 'hits' return the cross to its good form is only 1/40. For larger structures, it becomes increasingly improbable that a second hit will move the structure back into its original form, but less so if other configurations such as the 'wavy crosses' of Figure I.10.8c are perceived as belonging to the same macrostate as the original.

The right panel of Figure I.10.9 illustrates a different possibility. How big is the macrostate represented by the 3x3 cross? As we have seen, there is no unique answer. Does the macrostate consist only of the cross at the centre of the array, would any of the shaded crosses belong to the same macrostate, or would the macrostate contain only those crosses at least two spaces from the edge, which are less stable against external events than are crosses closer to the edge? It depends entirely on the perceiver.

If the perceiver controls not only the shape of the cross but its location, this matters, because these uncertainties influence the rate of entropic decay that must be repaired by perceptual control. If location is controlled and the cross is actually a rigid object in the environment, the entropic decay is only in the eight possibilities for a one-position movement (3 bits per second, if one event occurs every second). If the cross is a shape the perceiver is controlling by replacing moved dots individually, entropic decay adds just over two extra bits (four or five positions for one dot to move) to the control problem.

The size of the array determines how much information could be available for structure, as we noted above. The flat 15x15 element array has about 225 billion possible 5-dot patterns, only 169 of which are 3x3 crosses. The 7x7 array has nearly 229 million possibilities, of which 25 are 3x3 crosses, while the 3x3 Universe has 15,120 possible patterns of 5 dots, of which only one is our cross form. The relative uncertainty of the 3x3 cross structure compared to the possible five-dot patterns changes in proportion to the binary (base 2) logarithms of these numbers, namely 14.9 bits for the 3x3 array, 26.8 for the 7x7, and 31.2 for the 15x15 array. To observe that the array is empty apart from a 3x3 cross therefore provides 14.9 bits of information in a 3x3 array, 22.2 bits in the 7x7 array, and 24.8 bits in the 15x15 array. This is the relative amount of structure provided by observing the lonely 3x3 cross in arrays of those sizes.

The lesson to be taken from this perhaps bewildering set of numbers is that the more elements you have, the numbers of ways they can be combined grows very much faster than is intuitively obvious, and that any particular structure represents the possibility of getting commensurately more and more information from an observation of the space as the varieties of possible combinations increases. As a corollary, if the structure has appeared in the sensory data more than once, it probably is not a random array, but represents something that actually exists in the real world. Repetition is a good way of making something seem real, especially if the same structure is seen in different contexts. We will see examples of this when we discuss politics in Volume IV of the book.

Here is a small-scale example of the increasing information available from a structure as the world of possibility grows. In the 1940s, if you had a telephone, it was almost certainly one with a rotary dial, though phones that connected you with an operator when you picked them up still existed.[109] If in 1940 you had told someone that Angela, whom you both knew to have a phone, had one with a rotary dial, the telling would provide almost no information. After a while, touch-tone phones began to replace rotary dial ones, and at some time to tell your hearer what kind of a phone Angela uses might have provided as much as one bit of information. Now, supposing rotary phones still could be in use when the world is full of a great variety of mobile phones, to tell your hearer that Angela uses a rotary phone might provide quite a few bits of information. The phone did not change its structure over the intervening decades, but the Universe of possible phone types expanded.

A perceptual Universe is enlarged by the invention of possible components or by the creation of novel perceptual functions. The bigger the Universe, the more opportunities exist for creating new kinds of higher-level structures, not just chairs from chair-parts, but living rooms, dining rooms, office spaces, auditoriums, and arenas created from different relationships among chairs and related furniture structures. The ability to create these different kinds of furnished spaces where people may assemble provides opportunities for different kinds of interpersonal social structures, those social structures can have different modes of interaction, and so on to ever more complex possibilities, some of which will be discussed later in this book.

One of the essential points about structures is that the 'none of the above' macrostate is almost always vastly bigger than the individual small macrostates defined by perceptual functions at any layer of the hierarchy. We saw this with the simple line, but it was reinforced by consideration of the movement pattern of the line. It is the a priori unlikelihood of encountering those patterns more than once in space and in movement that makes them structures worth perceiving when they do occur. Even a ten-bit difference between the sizes of a small macrostate and the corresponding 'none of the above' macrostate represents for a random arrangement a thousand-to-one better chance that it will be in the bigger macrostate. Even in the small 8x8 space we talked about 30 and 40 bit differences (billion and trillion to one against), and in the vastly larger spaces of possibility we encounter in everyday life, these numbers become truly astronomical.

When you see something that has an a priori one in a billion or quintillion odds against occurring by chance, and it stays around while all about it is changing, it is probably something worth keeping track of by perceiving it as a unitary entity, something real in the world. This is just an information-theoretic way of restating Hebb's "Nerves that fire together link together" axiom, as a way to create perceptual functions such as a chair from perceptions of its component legs, seat, and back (Section I.5.5).

---

109 My parents in a rural community had such a phone as late as about 1978. If someone called you, it rang with a particular code, such as one long ring followed by two short. Another subscriber might be called by two long rings, and a third by three short rings, for example. There was no privacy, since anyone on the same line could listen in to any conversation using the line.

I.10.10 Uncertainty Constraints in Language

As a conceptual example of microstate-macrostate refinements in the case of language use, we will successively refine ever smaller macrostates relevant to the meanings of messages within a universe arbitrarily defined as the set of letters and punctuation marks available for constructing the messages. The example represents only one of many possible ways different perceivers might construct a hierarchic set of macrostates from the same microstates. You can think of going up the perceptual hierarchy, with each perceptual function defining a new macrostate from a combination of ones at a lower level.

The symbols in this example are analogous to the individual squares in the 15x15 array of 25 dots. They certainly will not include all the symbols used to write all the languages of the world.[110] If a writer happens to use a symbol outside the set we can interpret, either it will be seen as similar to one in the set that we can interpret or it will be perceived as an irrelevant scribble. The initial macrostate, the entire Universe of our consideration, consists of all possible sequences of symbols from our chosen set.

Each time we define a new refinement of macrostates, their sizes will depend on the constraints among the microstates of which they are composed. For example, not all sequences of letters form words that would be recognised by a reader of English. The macrostate that contains all the words and only the words of English is much smaller than the macrostate that contains all possible letter sequences without spaces up to the length of the longest English word.

Next, we suggest a possible structural description of text, in the spirit of Figure I.10.6d and Figure I.10.10. We choose our smallest microstates to be collections of letters and marks that might appear in an English text. We could have chosen the strokes from which the letters were formed, or we could have incorporated the font used in producing the text, or we could have chosen words or phrases. We could use any arbitrary starting point, down to the positions and momenta of the atoms in the paper and the ink, but we arbitrarily choose the 128 symbols that can be coded in ASCII, representing the letters and punctuation marks used in English.

Ignoring all structure, which means taking these different symbol microstates to be the smallest macrostates, one microstate per macrostate, the uncertainty is given by $U = -\sum p_i \log_2(p_i)$. Since, however, we start by ignoring the different probabilities of the individual letter types, this becomes $U = \log_2(N)$. As our symbol set is restricted to the characters available in ASCII code, $N=128$, so U is 7 bits per symbol. If the message has a length of L symbols, the uncertainty of the message at this level of refinement is $7L$ bits. That is the maximum amount of information that a recipient could get about what was intended by the creator of the symbol string. We will continue to use bits per symbol as the ever-reducing uncertainty measure in our hierarchy of macrostates.

---

110   We could have allowed all extant languages, or we could have used phonemes and talked about speech. This would change the numbers but not the concept.

The successive levels of macrostates that we will create by combining smaller macrostates are distributed over ever larger chunks of text, based on the way the probabilities are inter-related among the smaller macrostates. We sometimes call those relationships 'constraints'. For example, if in an English text one encounters a lower-case 'q', the uncertainty of the following letter is very low, because the next symbol is almost always 'u' (except for words from other languages which are seen occasionally in English text, such as 'coq-au-vin' and 'Iraqi'). Most of the constraints we consider are nowhere near as tight, but the reduction of uncertainty because of the successive constraints allows us to give names to structures of different sizes, such as syllables, words, phrases, sentences, paragraphs, topics, essays, and books.

One possible use of constraint in the reduction of uncertainty is to ask about the next letter to come in a message that is part of a text that has been scanned up to a certain point.[111] All the letters up to that point constitute prior observations, but before the reader even started reading, the uncertainty was less than 7 bits because of what she already knew about the language being used and the situation in which it was being used. The uncertainty is less than 7 bits per symbol by an amount related to what Nevin in his chapter in LCS IV calls 'Linguistic Information'. It will be further reduced by information the reader has observed about the pragmatic situation and other elements of the context such as whether the text is likely to be a description of scenery, a theological sermon, or an answer to a question.

Figure I.10.10 illustrates one possible ordering of the information added by considering one constraint after another that might affect a listener's uncertainty about the next letter. Other orderings and different kinds of constraint might be equally valid, but however the breakdown is done, the extra information gained (reduction in uncertainty) at each stage is conditional on the preceding constraints having already been taken into account.

---

111 See Harris (1955) for the first computational demonstration identifying word and morpheme boundaries as points at which the set of possible next successors rises to or near the maximum, i.e. points of freer combinability.

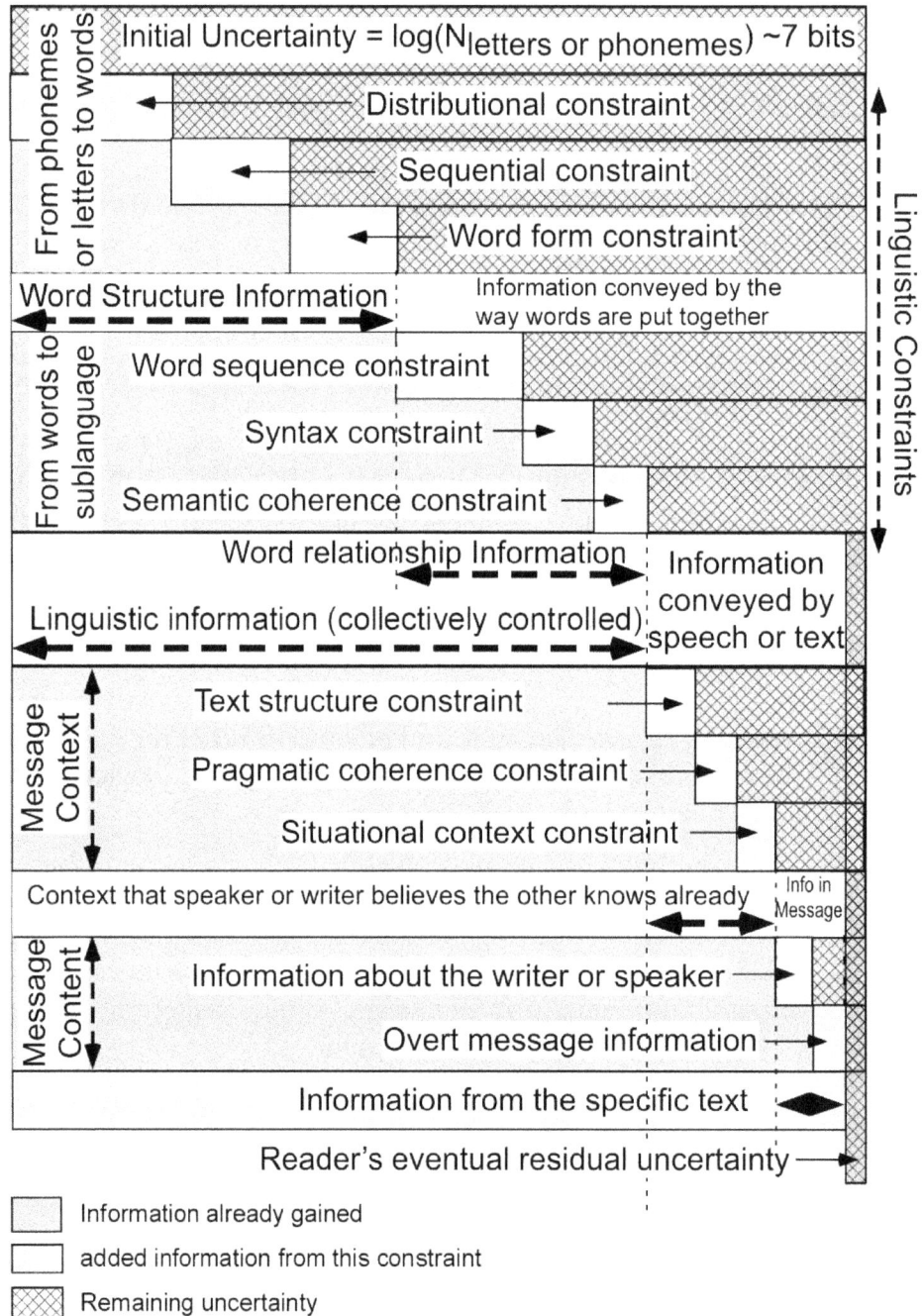

Figure I.10.10. One possible breakdown of the average uncertainty a reader or listener might have about the next letter or phoneme to occur in a text or utterance. "Message" could mean as little as a letter or phoneme, or as much as an entire election campaign. This breakdown could equally have been done in four stages: Information gained from intra-word constraint, language constraint, pragmatic and dialogue context, and the text or spoken message. The first two stages could be considered together as "Linguistic Information", and the first three specify what the speaker or writer perceives the listener or reader to know already at that point in the speech.

The values in Figure I.10.10 represent averages, not the results of particular observations, which can vary wildly. For example, as indicated before, if the reader perceives this to be an English text, and the preceding letter was 'q', the probability that the next letter will be 'u' is very near 1.0, but if the preceding letter was 'W' many other letters such as 'r', 'h' or any vowel might reasonably follow and the uncertainty is relatively high. On average, however, the uncertainty change represented by each step in Figure I.10.10 will be intermediate. Shannon estimated that just the word-level constraints reduced the letter-level uncertainty in written English by about 50%.[112]

Going eight levels down in Figure I.10.10, we come to pragmatic coherence. This constraint might be illustrated by imagining a couple who have just walked onto a viewpoint over beautiful scenery. One might turn to the other, wave a hand and sigh happily, to which the other says "Yes", the first having effectively communicated "I think that is a gorgeous view. Do you agree?". The sentences needed no words at all, but they were accurately received.

Suppose the sequence '[space]W' was observed after a material that had all been about US federal politics. With quite high probability, the next few letters are likely to be 'a', 's', 'h', 'i' and so forth. After the 'a' and the 's' appear, the rest of the letters in 'Washington' provide almost no information, because of all the earlier constraints, to which is now added the 'Situational context constraint', nine levels down in the arbitrary ordering of constraints used in Figure I.10.10.

Additional examples of syntactic and all the other constraints are not needed to make the point. Observation of specific instances can provide much or little information, but considered over all material of a given type (a particular writing style, for instance) the averages converge to reasonable numbers, each of which builds on what is already known from other constraints. An unusual and unexpected word can provide a lot of information, but such words, by definition, occur only rarely, and contribute little to the average.

Indeed, a particular observation at a specific point in the text might even increase uncertainty, providing negative information, meaning that after the observation what earlier seemed highly probable now becomes just one among a range of other possibilities whose probability has been increased. For example, after 'Wash...' in the political discussion, the next letter is highly likely to be 'i', which then would make 'Washington' highly probable, but if the next letter turns out to be 'b' the following letters become much less certain than they had been.

Outside the language domain, if a soccer team is well in the lead at a late stage of the game, it is highly likely to win, and the uncertainty is low; people start leaving the stands. But then the other team scores a couple of quick goals to tie the score. The uncertainty about the eventual winner is increased by each

---

112 Which, he pointed out, enables the construction of crosswords.

of those goals. But again, such negative information observations are rare, and on average an observation reduces uncertainty.[113]

Returning to language in use, the speaker or writer usually has an audience in mind. It might be one person, a professional group, a team, or the general public. Whoever is the audience, the speaker or writer is likely to assume that the audience already has all the information marked as 'Linguistic information' in the area of Figure I.10.10. Someone outside the target group reading the text might not have the same 'Linguistic information'. For example, a physics professor and a lawyer might not be able to make much sense out of texts each other wrote for professional colleagues. In a face to face interaction between close partners, the situational context might provide a lot of information, but in broadcast language over mass media, the speaker might not be able to trust the listener to be able to apply much if any situational context constraint, because the listener at the time might be anywhere, doing anything, and not know what the speaker was seeing or doing.

From an information-theoretic viewpoint, the basic question in analysing language use is why what is included is included, not why certain items are omitted. Whether at the phoneme level, the word level, or at a rarified intellectual level, in a face-to-face interaction it is generally a waste of time and energy for the speaker or writer to provide data that would not ordinarily reduce the listener's uncertainty. However, from a grammatical analysis point of view, the omissions are the focus of interest. Later, particularly in Chapter II.6, we will see how control processes can lead to reductions and apparent omissions at many levels of language.

Here we turn around the grammarian's idea of items being omitted from speech, and ask again about the items that the speaker explicitly includes. The listener may perceive these items as being included in order to provide information that the speaker believes the listener not to have. If the listener had believed the speaker to know that the listener did already have the information, the listener is likely to think that the speaker intends to say something new, and may try to understand just what the new thing might be. Such a misperception could easily lead to confusion and a cycle of misunderstandings that might escalate into conflict.

The old-fashioned retort "*Go teach your grandmother to suck eggs*"[114] illustrates the point. Not only is it inefficient to offer unnecessary information, it can also add to the listener's uncertainty about the situation as a whole.

---

113   In the soccer example, observing these goals does provide positive information, but not about the outcome of that particular contest. They provide information about, for example, the competence and mental strength of the teams and the probability structure of the sport, among many other possibilities.

114   Which can loosely be translated as " You must have known that I knew that."

These considerations either do not apply or apply with much less force in broadcast language, which is treated in Chapter II.6.3 since a listener who already has the information is likely to perceive that the speaker believes there may also be listeners who do not have it. In broadcast language to an indefinitely large audience, a speaker is likely to take advantage of the available phonetic syntactic constraints and use them explicitly in audible speech, speaking clearly and 'grammatically', whereas in face-to-face interaction the speaker relies more upon constraints among the types of information that the speaker assumes to be known by the listener. In broadcast language there is usually little or no feedback, so the dynamic reasons for ensuring minimal redundancy do not apply. What is 'structure'? Another word for it is 'organisation'. Yet another is 'predictability'. And here are two more: 'redundancy' and 'low relative entropy'. These words have different connotations to people with different backgrounds, but all have the same core meaning, that learning something about one part of the structure can reduce your uncertainty about other parts that you have not observed. We have discussed this in different ways in various places, and now we do it from yet another viewpoint, the information-theoretic approach to the tensegrity properties of the control hierarchy.

Structure decays. Sometimes, as with a mountain, it takes millions of years; sometimes, as with the spherical shape of the pressurised air in a bursting balloon, it takes milliseconds or less, but eventually all structure vanishes, its components widely distributed throughout the local region of the Universe.

Structure decays, but not all at once, nor all of its parts together. Palaeontologists infer a lot about an extinct animal simply by observing one tooth, but they learn more if they have a jawbone, and yet more if they have a whole skull. They can believe their inferences because the structure of a skeleton influences the behaviour of the animal, its eating habits affect the tooth, the tooth shape works better for a herbivore than for a meat-eater predator, and so forth. In other words, the palaeontologist perceives a structure that includes not only the animal to whom the tooth belonged, but also its behaviour and eating habits, not only for it, but also for a multitude of related animals, living or dead. The animal's structure may be less coherent than it was when the animal was alive, but the structure of the palaeontologist's perception of it and similar animals becomes increasingly coherent with increasing experience, as more fossils are seen, analysed, and compared with living animals.

Chapter I.11. Boxes, Objects, and Objects

We are conscious of a world full of objects,
but control perceptions only of their properties.

Complex organisms, especially mobile ones such as animals, birds, and fish, probably are not born with their control hierarchies prebuilt or even pre-designed, any more than they are born with all their material parts in good functioning order. Newly hatched birds cannot leave the nest under their own volition without some days or weeks of maturation. Even newborn antelopes that can run within a few minutes of struggling to their feet at birth are not as competent as they will be as adults.

What most complex organisms are probably born with is a genetic predisposition to develop their control hierarchies in certain ways, just like their bodies. These genetic plans, however, do not completely specify what the adult body will be in its physical shape. Presumably the same is true of its internal functioning. The adult form will depend in detail on the environment in which the growing body matures.

We should expect it to be so with the control hierarchy as well. Some species appear to specify most of their control hierarchy in the genes, and others are more adaptable to different environments. Among the most adaptable are humans, both in body and in what they learn to perceive and how they learn to control it. How the perceptual control structure changes over time and with experience in a particular environment is called 'reorganisation' in PCT. We will reconsider and refine our ideas of reorganisation as we learn more about PCT through the course of this book, but this chapter introduces some preliminary ideas.

I.11.1 Reorganisation: Changing Hierarchy Parameters

Somehow or other, the output of an ECU must act on the real environment in a way that the input from the sensors affected by the real environment influences the ECU's perception, and influences it in the direction that reduces the difference between the perception and its reference. Moreover, even though all the influences happen through the real, rather than the perceived, environment, controlling *this* perception must serve to keep the important life functions operating smoothly better than controlling that one does.

Powers called the physiological variables representing the state of these life functions 'intrinsic variables'. Later, we will deviate from Powers, in that we will consider 'intrinsic variables' to be members of homeostatic loops rather than variables with genetically pre-set reference values. For now, however, it suffices simply to call them variables that are important to the physiological survival of the organism. Their maintenance is the maintenance of relationships that help the organism survive, which, according to the Analyst, is in retrospect the evolutionary rationale of control.

If controlling *this* perception helps maintain the intrinsic variables, it is likely to be integrated into the organism's repertoire of controlled perceptions, and its quality of control becomes as important as the maintenance of the physiological intrinsic variables. Ignoring for the moment the maintenance of the relationships and dynamic interactions among the intrinsic variables, we will first address reorganisation as a way to develop good control of some perception or other.

We first consider the so-called 'e-coli' method of reorganisation based on the 'intrinsic variables' proposed by Powers (following Ashby 1952/1960). Later we describe two other possibilities, one based on a particular understanding of consciousness, the other on interactions among numbers of controllers. All or none of these forms of reorganisation might be eventually found to be used in live organisms, but that is for future researchers to discover.

If one is trying and failing to control a perceptual variable, madly flailing about like the proverbial 'bull in a china shop' will probably influence the perception in question, but almost certainly not in a way that is likely to improve the chances for long-term survival of the organism or the propagation of its genes. Nor is a general rampage any more likely to reduce the error value in the ECU than it is to increase the error. Indeed, the side effects of the rampage will inevitably disturb the values of perceptions controlled by other ECUs, usually increasing their error and their countervailing action. In short, rampaging is usually counter-productive. Yet many children and some adults do have episodes we call 'temper tantrums'. Why would this be? To answer this question we ask how the functioning of a partially constructed control hierarchy can be altered by reorganisation.

We hinted at reorganisation in Section I.5.5 when we were adding a higher level of 6 controllers above the 36 that control the orientations and locations of parts of the chair. The six controllers are not very useful if their connections to the lower-level controllers are random, nor if their connections differ much from corresponding ones in Real Reality (RR). Reorganisation changes the influence of any one upper-level output on each of the lower-level reference values, and likewise for the upgoing perceptual signals used as inputs to the higher Perceptual Input Functions. Reorganisation adapts these inter-level connection influences so as to improve the upper level control quality.

Reorganisation is the equivalent of 'trying something else' when what you are doing isn't working very well. Randomly 'trying something else' would take a very long time to produce a useful result in a complex system, but 'trying something else' need not be random; Powers (2008; also Marken & Powers 1989b) described an effective 'hill-climbing' reorganisation algorithm known conversationally as the 'e-coli' method because it was conceptually based on the movement of the e-coli bacterium.

Like all hill-climbing algorithms, an e-coli hill-climber in three-dimensional space can easily get trapped in a local optimum. but this becomes less likely as the dimensionality (number of independent variables) increases. In Section I.11.6

we discuss Kauffman's (1995) finding that optimization in his toy Universe became worse (more difficult) if the dimensionality of his reorganisation modules exceeded about six, and we expect that having four or fewer dimensions in a module is likely to result in the hill-climber getting trapped in local optima. It seems that Kauffman and we, from quite different viewpoints, converge.

The e-coli bacterium (at least when taken as a model of the reorganisation process) moves more or less in a straight line through a solution of a chemical it favours until it comes to a place where the concentration of the desired chemical begins to decrease, at which point it 'tumbles' and starts moving in a new randomly chosen direction. If that direction turns out to be down-gradient, it immediately tumbles again. It continues tumbling until it finds itself again climbing the chemical gradient, after which it continues in a straight line until it once more reaches a place where the concentration again begins to decline.

Similarly with reorganisation. The pattern of connections among the ECUs at different levels is taken to be a location in a high-dimensional space, so the direction of 'movement' is represented by a vector of weight changes. So long as control continues to improve, the same direction of weight changes is retained, but when control begins to get poorer, a new random 'direction' of weight changes is chosen.

Powers demonstrated the effectiveness of the e-coli technique in a space of 14 higher-level controllers in a demonstration called 'Arm 2' that is included with LCS III (Powers 2008).[115] If the apparent convergence with Kauffman's finding is real, we should not apply this method over such large spaces, but instead should work with modules of around 5 or 6 parameters at a time, arranged perhaps as a hierarchy, but more probably overlapped to avoid edge effects.

The actual neural mechanism of reorganisation in living control systems is unknown and the detail of it is not very relevant to most discussions in the rest of this book, though we do discuss other possible mechanisms in this chapter. One may, however, presume that reorganisation involves synaptic modification, and in Chapter I.9 we hazarded a guess at some possibilities as to how Hebbian and anti-Hebbian synaptic modification might implement at least some of the e-coli reorganisation process described by Powers.

E-coli reorganisation changes control connection patterns that do not work more quickly than those that do work, but does not leave totally untouched even controllers that are working well. The result is what is sometimes called a 'winter leaf' effect. Dry fallen autumn leaves get blown around by gusty winds until they pile up in some relatively calm place under a hedge or in a corner. A reorganised control structure contains control units which have worked and continue to work together in the environments in which the organism (person) has learned to control. If reorganisation changes them so that they work less well, they are likely soon to change back again, or at least change to a state

---

115 Also at http://www.livingcontrolsystems.com/demos/tutor_pct.html,
       'Arm with 14 degrees of freedom'.

where they work better. We will follow the winter-leaf effect much further in discussing reorganisation and social self-organisation in Chapter III.8. Crudely, reorganisation approximates the adage "*If it ain't broke, don't fix it*", although it occasionally does 'fix' something that ain't broke.

This context touches on an aspect of the stimulus-response notion of 'carrots and sticks'. In the reorganisation process, the 'carrot' is not a reinforcement, it is simply an indication that control is improving or working well, and the 'e-coli' principle simply says that if there had been ongoing changes, to continue to change in the same direction.

A 'carrot', according to PCT, is something that has positive value because it improves control of some perception, not necessarily or even probably the one for which the carrot is offered. "If you mow the lawn, I'll give you $10." Your perception of the state of the lawn comes to have reduced error, as does the mower's perception of the amount of money available to her. The $10 'carrot' is to the mower something that disturbs her perception of your state of mind, for which the compensatory action is for her to act as you wish, and mow the lawn. We deal with this kind of transaction when we talk about barter and trade in Volume III of this book.

'Carrots' do not affect the rate of reorganisation. 'Sticks' do. A 'stick' is a disturbance to a perception that is not currently experiencing much, if any, error. A 'big stick' is such a disturbance applied in a manner that cannot be corrected by the controller using the means at hand. It is usually called 'punishment', and because it produces a state of sustained error in the control of some perception, it is likely to be accompanied or followed by an increased rate of reorganisation.

The problem for the one using the big stick is that reorganisation can have quite unpredictable results. The one being punished may have been controlling for a wide variety of different higher level perceptual results, using atenfels that involved the 'punished' actions. The ways that those higher levels can be controlled are usually numerous — '*many means to the same end*' — and not all of them would fail to disturb some variable the punisher might be controlling. In plain language, the punishment intended to make the evildoer see the straight and narrow might instead turn him into a rebel. It can achieve the punisher's immediate intention, but the probability that the 'evildoer' will find an acceptable way to achieve the higher-level goal is rather low, although the punisher may in addition guide them to acceptable means. We saw a similar issue when we discussed the perception of 'not' and the problem of avoiding having an unwanted value of a perception in Section I.6.6.

I.11.2 Reorganisation: Growing the Hierarchy

Reorganisation has a second aspect of at least equal importance, on the perceptual side of the hierarchy. As the chair example illustrated, reorganisation of perceptual inputs uses the principle of "*If controlling this doesn't help, maybe you can see things differently*" to go along with the output-side axiom "If it ain't broke, don't fix it." Perceptual reorganisation has seldom been addressed in PCT discussions, but it is obviously important. The organism should be able to generate novel perceptual functions when the environment changes. It must also be able to recycle perceptual functions into new forms when control of the perceptions that they generate uses energy without benefiting the intrinsic variables.

As a control hierarchy matures, whether it be in a human, a tree, a bird or a fish, the higher levels cannot develop effectively unless the levels below control their perceptions at least moderately well. If the 36 lower-level units acting on the parts of the chair of Section I.5.5 did not control their perceptions very well, control by the six higher-level ones would also be impaired. Indeed, the higher-level units probably would never be formed, since the patterns of perceptual values and effects of outputs to the lower-level references would be quite inconsistent. Nothing would then be controlling 'chair-object perceptions' as opposed to 'chair-part perceptions'.

The ability of a control system to oppose the effects of disturbances and have its controlled perception reflect changes in its reference value is limited by the stability of the effect it has on its CEV. If its influence on, or its perceptions of, the environmentally constrained collection of properties constituting the CEV of a well controlled perception keeps changing character, it will control badly. The effect of a higher-level controller on its CEV is actually implemented by the actions of lower-level control systems controlling their perceptions to reference values responsive to the higher-level systems. If those lower-level systems are unreliable, the higher-level systems will be unstable or worse. In the end, reorganisation can develop control systems only to as many levels as will allow the effects of changes in organisation to improve both control itself and the effects of control on the intrinsic variables that keep the organism alive.

The apparent consequence of this is that early control will be best in a stable environment, which allows new, higher levels of control to be built on top of stable lower-levels, as suggested below in Figure I.11.2. Later, the stability on which new control systems are built is no longer required to be an aspect of the environment, but is the stability created by well-functioning already-built control units. Newly built higher levels will continue to operate for as long as the lower levels maintain good control, and at the same time new perceptions at existing levels may be built, forming new 'top-level' control sub-hierarchies, as in Figure I.11.2 panels c and d.

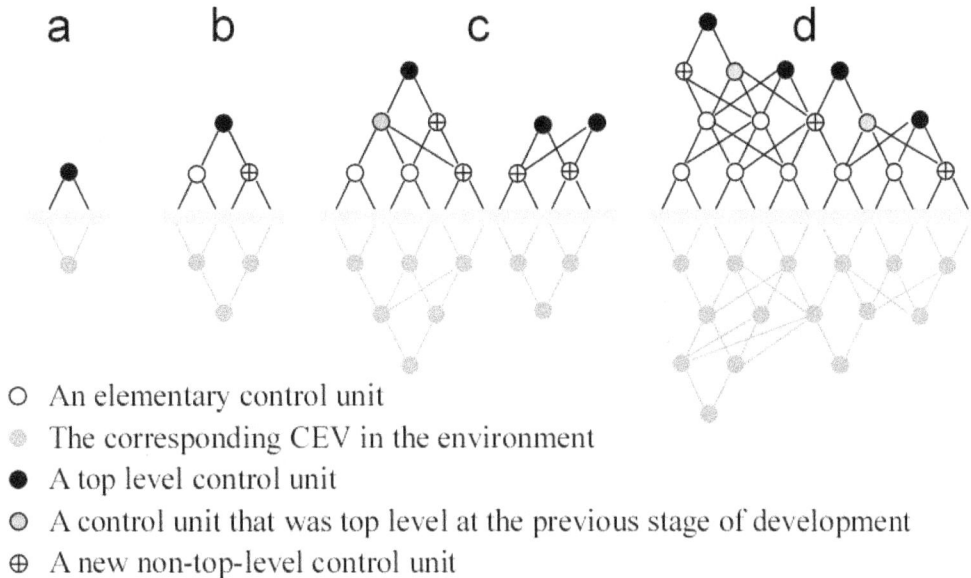

O   An elementary control unit

●   The corresponding CEV in the environment

●   A top level control unit

◎   A control unit that was top level at the previous stage of development

⊕   A new non-top-level control unit

Figure I.1.2 A developing control hierarchy builds control of ever more complex perceptions (with correspondingly complex environmental variables) onto previously reorganised control units. A "top-level" unit is one that receives no reference input from any higher-level unit. For example, in Panel d, there is a top-level unit at level 2, two top-level units at level 3, and one top-level unit at level 4. The grey mirrored structures below the line are the "Mirror World" created by the developing hierarchy. The Mirror World is Perceived Reality, which can be stably controlled only insofar as the controlled perceptions match corresponding properties of Real Reality.

In the figure, panels a to d show successive stages in the growth of a simple control hierarchy. Investigation of the development of successive levels of perceptual control in the growing child showed that well-defined changes of behaviour can be observed when each new level is achieved (Rijt-Plooij and Plooij 1992/2019, Heimann 2003). The figure illustrates that 'top-level' control units do not always have to be at the same level.

The observations of Rijt-Plooij and Plooij suggest that perhaps there is a mechanism that facilitates the development of new instances of a control unit at a level once the first instance has been created. In Volume II (Section II.2.6, illustrated in Figures II.2.6a & b) we suggest how this might happen in a toy evolutionary proposal for the descendants of a trivial primaeval 'e-coli-like' bacterium. The toy descendant bacterium contains in what I call a 'Template Store' the instructions for creating a duplicate of itself, together with a mechanism that interprets the template and builds replica instances of the entity described, whether it be a new instance of a class of perceptual control loop or an entire child bacterium.

An analogy might be the easy creation of new instances of an object class in Object-Oriented-Programming (OOP) once the class itself has been programmed, though of course the mechanism of creating a new instance of a type of ECU cannot be anything like the method of creating a new programming object!

I.11.3 Object-Oriented Programming and Its Objects

It has long been obvious (though often forgotten) that the brain does not work like a digital computer, but there are aspects that form helpful analogies to processes that are hypothesised to operate in the physiological brain. From a PCT standpoint, the conscious perceptions of 'objects' such as tables and chairs, not to mention more complex structures such as dining table arrangements and less complex parts of tables and chairs, bears a very close structural relationship with the organisation of 'Objects' in an Object-Oriented-Programming (OOP) environment. Why? Let us first offer a simplified description of OOP.

In any OOP environment, an Object is a kind of package of functions with some allied parametric data. The package has input terminals and output terminals. The Object is completely specified by the functional relationships among these terminals. For example, we might specify a 'Trivial Arithmetic' Object that has three input terminals and one output terminal. One of the input terminals accepts discrete values from one to four, which we could label 'Add', 'Subtract', 'Multiply', and 'Divide'. The other two terminals might be labelled 'first number' and 'second number'. Both accept only analogue data, a magnitude that could range anywhere from a large positive to a large negative value. The output terminal is labelled 'Result'.

This 'Trivial Arithmetic' Object does what the labels suggest. Depending on the discrete value sent to the first input terminal, the Result terminal will output an analogue value that represents the sum, difference, product, or quotient, according to what the first input was. Nothing about the object description tells how the Object does what it does, but it is not impossible to guess something about its functioning. For example, if the two number inputs are held constant, changing the value sent to the category input wildly changes the value that appears at the Result terminal. The category value input must act as some kind of a switch operator, but how it does this is quite unknown. The switch must somehow select a function to apply to the two number inputs, but again, how that is done is also unknown.

All Objects in OOP are like this. They have certain input and output specifications that can be discovered since the relevant terminals are open to external observation and testing. Only the programmer knows how the functions that relate the input and output terminals work as the specifications say they do, but it is sometimes possible to determine some of the functional linkages internal to the object without knowing how the functions they link actually are programmed.

There are two kinds of OOP Objects, the 'Class object' and the 'instance' of a class. In the Class objects, internal parameters are defined but not provided with a value. The instances of the class differ in what values these parameters take. For example, a Class might be labelled 'Garden' with an internal parameter 'formality'. One of its instances might be a garden that had been left to its own devices for many decades, in which the flowers and grasses and even trees had spread irregularly since the garden had been tended. In this instance, the 'formality' parameter would have a value near zero.

One of the properties of the class 'Garden' might be the number of well-defined routes specified by defined paths between entrances and exits. Our untended garden would have lost all these paths, so it would have that parameter set to zero. A visitor would nevertheless perceive that it was an instance of a 'Garden', a complicated object in her perceptual world (unless her personal perceptual category definitions included that a garden must show evidence of currently being tended, but our proposed Garden Class Object does not).

A perceptual function in the visitor produces a value, say of 'formality' when it is supplied with input data, but the property of 'formality' exists in many other Class Objects. The whole set of such properties, however, determines whether some perceived object is or is not a garden rather than a parking lot. A 'Garden' has no property such as 'number of parking spaces', while a 'Parking Lot' has no property such as 'ratio of flower-bed area to grass area'. OOP Objects are defined by the functional properties that link input terminals to output terminals; equivalently, perceived objects are defined by the set of properties that can take on different values for different occurrences of that kind of object.

In OOP, given the specifications, a competent programmer should be able to program a Class Object so that her version behaves exactly like the original, insofar as any third-party observer or tester could determine. The replica Objects would have the same internal functional connections as the original, insofar as their existence could be discovered by observation and test, but might be programmed in an entirely different language using quite different algorithms and even different physical substrates. For an extreme example, the internal functioning of the Object might be performed by a human who was informed vocally through earphones what to do with two numbers that appeared on a screen, and then performed the desired function and used a keyboard to output the results.

The corresponding observation for objects in Perceptual Reality is that Real Reality could create the observed functional properties of any perceived object in an unfathomable number of ways, provided that the functional relationships between our actions on the environment and our perceptions of what happens when we act in those ways is exactly what we observe (including all the possibilities of hallucinations and illusions). Nothing we do or observe can constrain *how* Real Reality does what it does, but our experiments and their results can constrain *what* Real Reality does. In this sense, Real Reality is inscrutable at base, but not in the inter-relationships of its discoverable functionalities.

Humans pretending to be the mechanical functioning of an inscrutable object are by no means a new idea. Some touring illusionists in 'The Age of Reason' exhibited marvellous automata such as 'The Turk' chess -playing 'automaton' from 1780 to the mid-19th century.[116] The Turk actually hid a Chess Master

---

116 Wikipedia article "The Turk", retrieved 2020.12.03.

within its base, where the audience was supposed to assume that the machinery had been placed. Modern robots are not at this stage of ability to mimic a human in face-to-face interaction with living humans, but their abilities in many realms show rapid advancement, to the stage where it is easy to believe that in a century or two, they might pass a face-to-face Turing Test in which the tester cannot tell whether the purported human is living or mechanical.

How could The Turk and similar chess-playing simulated automata fool the public, including chess-playing antagonists for as long as they did — several decades — without being revealed as a hoax? The answer is that The Turk, an ornamented but opaque box, was similar to an OOP Object in that nothing about its working was accessible from outside. It was able to play pretty high-level chess. How it was able to play could be discovered only by looking inside the box that hid the human chess master.

At the time, intricate automata of various levels of complexity were popular objects of amazement, so it would not have been implausible to add chess-playing to the list of things an automaton could do. Since it was advertised as an automaton, and all that could be observed, as with most of the other automata, was its functionality, the substitution of a human for the promised automation would be easily accepted.

The Turk vividly illustrates the point that OOP Objects can be constructed in many different ways to do specified functions, even playing good chess. The fact that you can produce a theory of the internals of the Object that works in simulation very precisely just as the Object does in no way argues that you have a correct theory of how the Object works. But as noted above, sometimes there are relationships among the variables that require there be certain internal connections among them if the observed functions are to be produced.

There is more to the Objects of OOP. The simple specification of an Object does nothing. To execute the described functions requires an 'Instance' of one to exist. The specification is of a class of Instances, of which many might be constructed, possibly using different material substrates such as the chess-playing human and the chess-playing automaton, or different programming inside a computer. What all these Instances must do is perform all the functions included in the specifications. Importantly, this requirement includes any internal connections that are required by observations of the variable values at the instance terminals.

One particular category of interconnections can be observed if two input terminals show observable correlations that are manifest in their effects on the output terminal or terminals. The observer would know that inside the Object, the input terminals are both connected to a common process of some kind that identifies the existence of the correlation and produces an output related to its magnitude.

An observer who is also an experimenter can determine what signals are applied to the terminals of an Instance and get a much better idea of necessary linkages within the Object, always without being able to discover how the Object does what it does. In the next section, we will be talking about Black Boxes and

White Boxes; a Black Box has all the properties of an Instance of an Object class, and a White Box is an openly accessible counterpart, an Object that has been constructed to function exactly like the Black Box, but whose workings are accessible to public view.

Before we get there, however, we must add two more characteristics to the idea of OOP. The first is the idea of 'class parameters', numbers or packages that are intrinsic to the class and the same for all Instances of the class. The second is the idea of a 'subclass' and inheritance. For example, we might have a class of 'mammal', for which a subclass might be 'dog', of which a subclass might be 'hunting dog', and so forth until we arrive at an instance 'my dog Rover'.

Each level of subclass is distinguished from the 'superclass' at the level above by what the subclass inherits and what it does not, together with new properties that the superclass does not have. A dog, for instance, has four legs, but the unfortunate Rover was once in an accident and had one leg amputated. Rover, as a subclass of 'dog' through many intermediate levels of subclass, should be expected to have four legs, but as an Instance of 'dog' Rover overrides that class attribute and substitutes 'three legs'. The number of legs is a class parameter of the class 'dog', inherited from a superclass of 'quadruped', but Rover's leg-number parameter is not inherited, as it is a property of Rover, not of most dogs.

Another property Rover has, and other dogs do not, is that he has an owner who happens to be me (in real life, I don't have a dog, but for the sake of the example, Rover is 'my dog'). Many dogs have owners, so 'owner identity' is a class parameter of 'dog', which would have a 'null' value in the case of a feral dog.

You may have noticed that I slipped the word 'property' in when discussing Rover's number of legs. It was in reference to a parameter intrinsic to Rover, an instance of 'dog'. Rover has many properties, many of them variable, such as his location in space, his fur colour, his aggressiveness, the floppiness of his ears, and so on and on. To our perception, the world is full of objects (not OOP Objects), all of which have many properties. Some are animate, some inanimate, but all have many properties.

This fact highlights a distinction between conscious perception and the perceptions that are controlled in the reorganised perceptual control hierarchy. We consciously perceive objects, not properties except when our attention is drawn to a property such as Rover's fur colour, or where Rover has got to at this moment, or how hot is the tea in my cup. But the Powers control hierarchy controls only scalar values of perceptions — perceptions of properties. It does not incorporate the concept of an object entire, and even consciously we control object properties, never entire objects.

We control properties of the objects by acting on the object; in OOP terms, we use one of the Object's functions to change an Object parameter of the individual Instance. We warm up the tea in the cup, or wait until it cools to our taste. To warm up the tea that we have allowed to grow cold, we influence other properties of the object, such as its location, which we might change to 'in the microwave oven'.

In doing so, we take advantage of functional linkages within the object. Its location and its temperature are observable at its 'output terminals' (considered as an Object), but one would not expect *a priori* that those two properties would be functionally interlinked. We learn it by reorganisation, by learning some internal functional linkages of the object that we can use when we want to control one of its properties by influencing another.

I.11.4 Black Boxes and White Boxes

We continue this line of thought by following Norbert Wiener's (1948/1961) discussion of 'Black Boxes' that are opaque to an observer and 'White Boxes' that perform the same functions but for which the workings are open to view. Wiener wrote about Black and White Boxes before OOP was invented, so the language he used was different, even though the ideas were the same at heart. Wiener's main concern was the ability of a White Box builder to discover the internal functional linkages of a Black Box by experimenting on the signals applied to, and received from, the terminals accessible to an engineer building the White Box. The procedure is very much like what in PCT we call 'Reorganisation'.

We continue looking at the environment of control more generally. In place of the elephant partially perceived by the 'Hindoos' of Section I.2.2, we use a 'Black Box'. The box is called black because we have no access to what is inside except by way of two sets of terminals, input and output. We can apply signals to one set, the input terminals, and can receive signals from the other set, the output terminals. Those signals give Wiener all he can ever know directly about what is in the Black Box, just as the 'Hindoos' touch different parts of the elephant and report different findings about what it is.

Wiener's engineer can, however, look at any relationships that might exist among signals emitted from sets of two or more output terminals, or between signals at output terminals and signals we provide to one or more input terminals. Using these relationships, he may not be able to find out what is inside the black box, but he might be able to determine *what the contents of the black box do*, their functions and how the functions are interconnected. Never, however, can he discover how the Black Box implements those functions.

The signals at the input terminals of the Black Box are metaphors for the ways we, as actors, can directly influence our Real Reality (RR) environment, mechanically by using our muscles, chemically though our waste products, or even electromagnetically through our internal electrical effects that are captured in EEG and EMG examinations. Some organisms such as electric eels use electricity as the action output of perceptual control loops to stun potential prey or predators.

The signals at the output terminals of the Black Box are metaphors for whatever RR does that influences our sensors, the sensors standing for the terminals themselves. From our own point of view, our outputs are inputs to our RR environment, and our inputs are outputs from RR. We do thus and so, and we can sense such and such that happens more often than it would if we had not acted that way.

Nothing has been said so far about Perceptual Reality (PR), the environment we consciously perceive, as opposed to the Real Reality (RR) environment to which our muscles and sensors are exposed. PR is where we *experience* controlling. PR is where we see objects like chairs, landscapes, stars and living things. The 'Hindoos' used subsets of the elephant's 'terminals', their fingers providing inputs to and outputs from the elephantine Real Reality Black Box, and making their deductions about what the elephant was from the few properties that they perceived from these limited subsets individually. PR is where we control, but we can control only by acting on RR.

If the whole elephant was a Black Box (or an 'object' or 'Object'), the parts examined by the 'Hindoos' were its interacting components. A component of an Object is internal to the Object, and is quite different from an instance of a subclass of that class of Object. When a 'Hindoo' touched the elephant's ear, the elephant might have moved its head, and therefore the tusks another 'Hindoo' was examining. If the 'Hindoos' communicated with each other, they might have deduced that the ear and the tusks were part of one object that in some way were functionally mechanically connected. Similar connections might be discovered among other parts of the elephant that could be examined by the individual 'Hindoos'.

The 'Hindoos' function very much as analogues of the different sensor types with which we are endowed. We may see one object hitting another, and simultaneously hear a sound, for example. We consciously perceive the sight and the sound as belonging to the same event, not as discrete events. We eat a tasty morsel from a plate, and only by experimental analysis did scientists determine that what we perceive as a unitary 'taste' is actually composed from sensors in the mouth that physically touch the morsel and sensors in the nasal system that react to gases and vapours emitted by the same morsel.

The processes of evolution and 'reorganisation' that build our perceptual systems are in part based on the survival value of noting these inter-sensory consistencies. Even the visual appearance of a simple object depends on the fact that different individual rod and cone sensors in our visual system produce correlated patterns that change in coordinated ways. The specialised detectors in the early visual system discovered many years ago by Hubel and Wiesel (1962), such as on-centre-off-surround, directed edges and lines, moving edge, and their like all are White Boxes boxes pre-built into visual systems akin to ours and presumably to other species that use vision to direct their actions in similar environments.

Such genetically developed devices save the newborn individual from taking time to build them when first exposed to the world, and allow the newborn to develop White Boxes that emulate the properties of the part of Real Reality Black Box — city, jungle, desert, or whatever — in which they will probably need to act when they control whatever perceptions they will individually develop, some like others of their species, some unique to themselves. Evolutionarily developed White Boxes reduce, perhaps substantially, the length of extreme vulnerability experienced by every newborn, from bacterium to forest tree to lion cub.

Those consistencies include relationships among different kinds of data sources such as vision and hearing, for example the sight and sound of a stone falling on a hard surface, which have been correlated ever since our ancestors had both hearing and vision, as candidates for evolution to link into a single perceptual event. Such correlations of events in different senses are so genetically built into our perceptual apparatus as to make us unaware that the separation of the sources of the data even exists, unless we think specifically about the fact that we have different sense organs for sight and sound, taste and smell, and so forth. The perception is unitary, and it is only consciously thinking analysts who consider this to be a problem that should be addressed.

Let us move on, to consider why our perceptions seem to be of what is 'really there'. Since our actions on RR and the related sensory data we get (e.g. the sound and sight of some event) might on some particular occasion be perceived as a chair moving to a place we wanted it, we have to assume that PR is fairly closely related in some way to RR. Does RR actually contain a 'chair' entity like the one we perceive? Possibly, but we can never know for sure. What we can say is that when we pull on one leg of what we perceive as a chair, either the rest of the perceived chair comes along or we perceive the chair to fall apart with a leg torn off. Or maybe the leg was real, but the rest of the chair was a hallucination. The question then is how we develop the apparently close relationship between what we consciously perceive in PR and the great unknown of RR?

That is where Weiner's 'White Boxes' come in. The builder of the White Box must create something that functions like a chair that either comes along when we move what seems to be a leg, or falls apart with the leg torn off. Our perceptual structures and the functional ways they interact are our biological White Boxes. We, however, are not their designers (though in babies we have been provided with the means of constructing new ones), so if we want to understand RR we have to build models, such as Powers's hierarchical model of perceptual control, to model both what happens in RR and what we perceive to happen in PR. Powers built his theoretical hierarchy, his White Box that should explain the behaviour of a living thing, by a consciously imagined process called reorganisation, building simple small perceptual consistencies first, and building more complex ones on top of the consistent perceptual functions created earlier.

To recapitulate, a White Box, as described by Wiener (1948/1961), is a construction completely open to observation with two sets of terminals, input and output, which has been constructed so that the relationships between signals at its input terminals and the signals at its output terminals emulate the measured inputs and outputs of a corresponding Black Box. Such a Box (Black or White) corresponds in function to an OOP Object. A user of an OOP Object is told how specified patterns of inputs to the input terminals will result in specified patterns of outputs to the output terminals. The programmer making a replica

Object can then use an instance of that Class of Object to perform particular forms of data manipulation without knowing anything about how the interior of the Object is constructed.

We go about our daily business neither knowing, nor most of us caring, how we do what we do. The Real Reality world does what it does with our actions, and we experience changes in our perceptions, which are, for us, our Perceptual Reality model of Real Reality. Much of the time, we just *use* the world as a Black Box without trying to build a White Box to explain why what we do works as it does. Some people are scientists, interested in trying to build little White Boxes that provide possible explanations for small segments of the world in which we act, but most are not and simply accept the Black Box as it is.

Into all this comes the Psychological Theorist, to whom the workings of the organism (perhaps another person or a mouse) is a Black Box. The psychologist builds a theory — a White Box — of how the organism seen as a Black Box is constructed. At the same time, the person, as organism, might be developing a White Box that describes how the Black Box that is Real Reality is constructed. The psychologist is also an organism who is building her own White Box of how Real Reality functions, but the organism under study is part of her Real Reality, that part for which she builds her theory. But that part cannot be divorced from the rest of RR, especially the part with which the organism interacts. This complex set of Black and White Box relationships is partly sketched out in Figure I.11.4a.

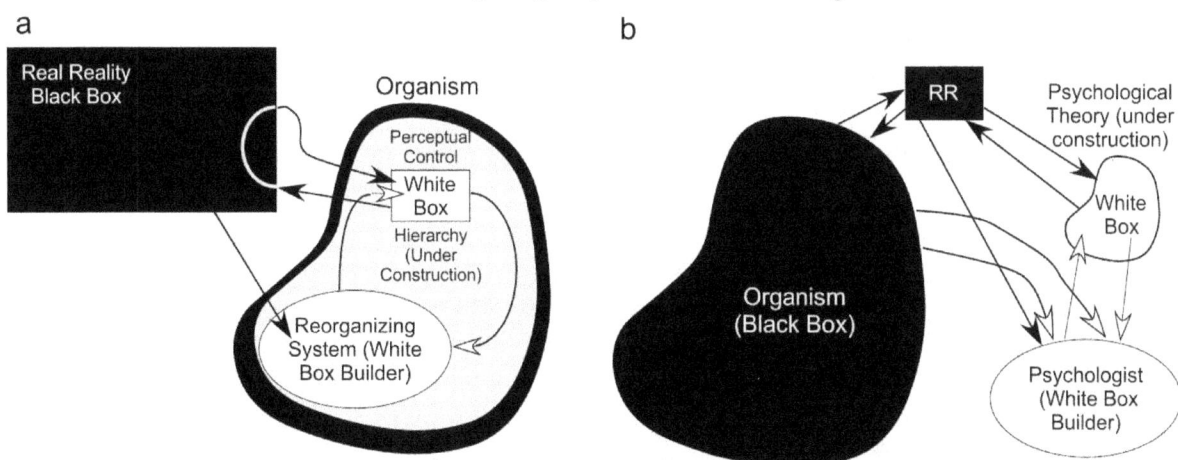

Figure I.11.4a An outline sketch of the problem facing a psychological theorist. (a) An organism creates by reorganization a White Box emulator of a part of Real Reality that acts to improve the relationship between the White and Black Box. (b) a Psychological Theorist does the same for the Real Reality Black Box that includes the organism.

W. T. Powers was a Psychological Theorist in the sense intended here. He observed that an organism needed feedback in order to stabilise the local environment and maintain its internal stability, and that this feedback had to stabilise internal variables representing states of the outer Real Reality that potentially might lead to damage to the welfare of the organism. He identified these internal representations as 'perceptions' and called the feedback process control of perception, as in the title *Behavior: The Control of Perception.* (Kent McClelland introduced the term Perceptual Control Theory, PCT, to distinguish it from the different aims and methods of control theory in engineering applications.)

Powers hypothesised a hierarchical series of perceptual input processes and action output processes that could partake in the feedback loops, and hypothesised how the organism, to him a Black Box, might generate such a hierarchical organisation of control by what he called 'reorganisation'. All of these hypotheses were implemented in a White Box that emulates the Black Box organism by incorporating two interacting White Boxes he called the Perceptual Control Hierarchy and the Reorganising System. The Reorganising System in its turn had the role of building the Organism's White Box emulation of the part of the Real Reality Black Box which the actions of the organism could influence and which could reciprocally influence internal states of the organism.

Figure I.11.4a depicts this recursion in Powers of his White Boxes emulating the operations of an organism as a builder of White Boxes (perceptual functions and their associated control structures) produced a theory which has had both practical and philosophical consequences.

If we now reframe this in terms of the analogy of a White Box to a newly programmed OOP Object that conforms to the specifications of another OOP Object with unknown internal mechanisms, the Psychological Theorist's problem becomes one of finding out what the 'specifications' of the Black Box object are. Given the skeleton White Box of PCT, the organism's reorganising system's task is to discover the 'specifications' of Real Reality.

Neither task can, even in principle, be performed with infinite precision, but with sufficient experiment and observation, both the reorganising system and the experimenting theorist may approach their goals indefinitely closely so long as the relevant external environment remains constant. However, evolutionary changes ensure that it doesn't. Instead, while both kinds of White Box builder (reorganising system and theorist) are in process of constructing a White Box emulation of it, the Real Reality environment changes. So, as with an ordinary perceptual control loop in a changing environment, what the reorganising system builds for the organism, and what the theorist builds, must accommodate the changes in the Black Box being modelled. We will content ourselves here with a broad-brush description in the language of OOP.

In most object-oriented programming (OOP) languages, the inputs and outputs can be Objects or other data structures, but when we consider White Box Objects as corresponding to perceptual functions in a Powers hierarchy, we can use only scalar variables as the inputs and outputs to each White Box. If a White Box is to emit a structured object, the values of the variables in the structure will be separately emitted as scalar variables, their structural relationship being lost in the process. The functions constituting an Object may operate independently of each other, or may use shared scalar inputs to produce their independent scalar outputs (Figure I.11.4b).

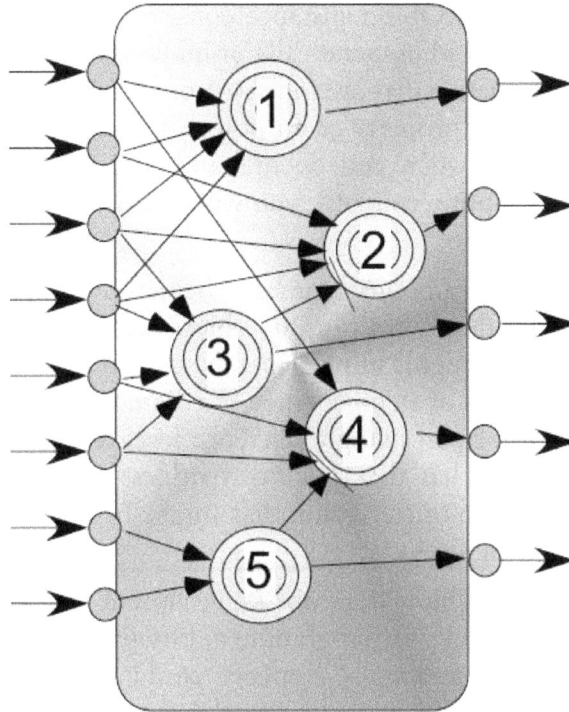

Figure I.11.4b Some functions within an OOP class object. Function 5 is independent of the other four, which all have some shared inputs. Function 5 does, however, provide its output as an input to Function 4, as Function3 does to Function 2. Each function is analogous to a perceptual function in the Powers perceptual control hierarchy or, equivalently, to a property of a consciously perceived object.

Objects individually are Instances of some Class. A Class Object is like a category, in that the class description specifies the functions and parameter values that are common to all instances of that object class, even though each instance may have its own values for variables which are not specified in the class description, and may even override some of the specifications of the class description. For example, the class 'bird' specifies some things a bird may do, one of which is 'fly', but an instance that is a penguin will override that ability and add the ability to 'swim underwater'. Since there are many penguins, the many instances that are penguins may be treated as members of a subclass of 'bird', called 'penguin'. We discuss the architecture and control implications of perceptual categories in Chapter II.6.

The Object is not only a member of a category, its Class, but also identifies by its specification a category of input patterns. A change in the value input to any of the input terminals is likely, but not guaranteed, to produce change in one or more of the values sent to an output terminal. Apart from the fact that Powers's perceptual input functions output a single scalar value rather than a vector of values (five in Figure I.11.4b), the Object would have the structure of a perceptual input function, as does each one of the five internal Objects represented in the figure.

A full description of an Object instance consists of descriptions of all the functions it contains either idiosyncratically or inherited as a member of a class, together with any parameters that are inherited or intrinsic to the object. Each function corresponds to a property of an object in the perceptual world. The description specifies that when you do this and that to the object (sending signals to the Object's input terminals), you will perceive thus and so (from the Object's output terminals).

In PCT language, then, an object comes to be perceived as such because the collection of properties that constitute the object recur as a group on various occasions. Each property might, in principle, be controllable. One may not be able to control the hardness of an object such as a diamond or a lump of wet clay, but one could control for perceiving a reference level of hardness by choosing or making an object that has that value of hardness as a property. In OOP language, hardness is a parameter value that might be inherited or might be intrinsic to the individual object.

The user of an OOP Object does not know how it was programmed, nor even the language in which it was programmed, but given Object specifications and the appropriate tools and skills, someone could program an Object that performed the same functions according to the same specifications. Without access to the programming, another user could never tell which Object was Black and which was White, but the programmer of the White Box Object would know how the White Box performed its functions without knowing anything about how the Black Box performed the same functions.

Functionally the Black and White boxes would be identical, while internally they might be constructed very differently, as they are in Figure I.11.4a — the organism, the psychological theorist's concept of the organism, and the Real Reality for which the reorganising system creates a White Box emulation we call 'Perceptual Reality' (PR) in the organism.

Wiener's problem was how an engineer/programmer might find the specifications of a Black Box so that they could create a White Box to replicate the functioning of his Black Box when the signals at the input terminals of the Black Box could be freely varied and the resulting outputs freely observed. His solution was to assume that he would be able to replicate the entire functioning of the Black Box as closely as he wished by a process of successive approximation.

Since, unlike Wiener, we are dealing with a Real Reality (RR) environment, we must assume that the Black Box has signal sources unknown to us as well as the ones we can freely influence at the known input terminals. We might as well call these sources 'hidden input terminals' of the Black Box that is Real Reality. Since we know nothing of these hidden terminals nor of the signals they receive, the parsimonious assumption is that (from our PR point of view) they provide noise uncorrelated with any other signal we can apply to the input terminals of RR by our actions.

In spite of the possibility of these hidden terminals, nevertheless our reorganising system can employ Wiener's method of ever more closely emulating that part of Real Reality whose effects we can sense. Powers's approach to the reorganisation done by a maturing and learning organism was the same as Wiener's, emulating the production of simple (low-level in the perceptual control hierarchy) controllable perceptions, and level by level building further perceptual functions upon them in a hierarchy rather than directly building them all on the sensory input. If PCT is in its essentials a correct theory, we will encounter this successive approximation process in the various places throughout this book where we discuss 'reorganisation' in living, learning, organisms.[117]

So what is Wiener's successive approximation method? At heart, though not in detail, it is the old Gestalt idea that our perceptual field separates itself into distinct areas according to 'common fate'. Parts of the visual field that change together in similar ways are likely to do so because something, perhaps their belonging to an object in the environment, makes them cohere. The same basic idea of common fate is inherent in the Hebbian mantra "Nerves that fire together grow together" (Hebb 1949). We will use this concept to look at a different approximation process based on categories or Class Objects in Chapter II.6.

Wiener assumes that the Black Box contains internal functioning structures that produce statistically detectable effects on the relationships among the output terminals, and tries to make White Box structures that emulate the functions performed by these mini-Black Boxes. Then he builds slightly bigger mini-White Boxes that internally use the first level of mini-White Boxes, to emulate structures in the Black Box that perform functions that use functions performed by the first-level mini-Black Boxes emulated by the initial White Boxes. And so forth, producing a hierarchy of White Boxes of ever greater complexity that emulate the functions performed by ever larger portions of the greater Black Box.

White Boxes can emulate the effects produced by the Black Box, but can never provide any information to Wiener's engineer about how the Black Box produces those effects, beyond discovery of linkages among sub-parts of the Black Box that are themselves smaller Black Boxes in the same way that OOP Objects can use other OOP Objects as part of their functioning. These interior functioning Objects may be of a class that is used also by more complex Objects.

For example, there may be an Object class that averages the last n values presented to an input terminal and presents the running average at its output terminal.Instances of that class may be used inside any Objects that require that functionality, either as part of the enclosing Object class or as part of a specific instance of another class. Wiener's engineer may be able to discover the linkages

---

117 In Volume II we will introduce another approach complementary to this one — identifying in a set of frequently observed relationships smaller motifs that recur as components in various parts of that set. This is the approach taken by much of science, physics being a prime example.

involved, but still does not know how the function of averaging is actually performed. The engineer can, however, build a White Box that performs the same functions transparently.

Wiener's engineer or the processes of reorganisation can build White Boxes that use functions available to other White Boxes, and use the White Boxes already constructed to build functions for yet more complex White Boxes. When the Black Box of interest is Real Reality (RR) and the 'engineer' is an organism's reorganising system, PCT calls the White Boxes 'Perceptual Functions'. If a White Box at any level of this construction hierarchy emulates effects produced by RR with sufficient precision, then for all practical purposes, RR is likely to incorporate internal structures linked in the same way, though their working mechanisms are unknowable. The network of White Boxes is what constitutes Perceptual Reality (PR), whether conscious or as a component of the non-conscious perceptual control hierarchy.

As an organism matures, it learns more and more about what seems to be there in RR by trying to control what it perceives to be there. Perceptual Control Theory (PCT) thus can be understood as a theory of learning to see the world more and more accurately by applying an ever greater variety of patterns of influence on the actual environment and thereby learning to control perceptions in ways that keep the organism alive and healthy as it matures.

This procedure will work perfectly only if the Black Box actually incorporates structures that do perform functions that can be emulated in this way — something that is forever unknowable, but it can always work approximately, depending on the desired accuracy of the match between the observed Black Box behaviour and the behaviour of the synthesised White Box. The theorist (or reorganising system) can always attribute mismatches to mysterious forces, such as the actions of omnipotent invisible Gods and Demons who feed signals into unknown hidden input terminals of the Black Box, creating effects that the White Box builder cannot, in principle, reproduce in the hierarchy of emulation. Some of these effects have been called 'miracles'.

In the same vein, PCT itself can be seen as a consequence of evolution. We, and all living organisms, can only control our perceptions if they model something that functions like them in Real Reality. Those organisms that do it well are, in the evolutionary sense, 'fitter' than those that don't.

In Volume III, we will consider mainly situations in which Real Reality incorporates structures we may call 'Socially Constructed Reality' — a form of PR created by many people which enables someone that interacts with those people to control better if she believed than if she disbelieved what they believe to be true. But for now, the Real Reality of non-living things together with living things that we can use without controlling them is quite enough to be getting on with.

I.11.5 Reorganisation: Idealism and Rigidity

Living in an environment that is too stable does not lead reorganisation to create versatile systems that can control in a variety of changing environments. In a stable environment, the system need not use 'many means to the same end' because 'one means' is enough. The system becomes rigid, always using that one means. If the system is a human, they might be perceived by an external observer as being bound by habit and ritual. The more connections a unit has to the levels below, the more flexibility it probably has to control its perception in different contexts — the more atenfels it has available for different purposes.

The child learns to walk to a friend's house, but later learns how to ride a bicycle, and can use either technique to control for perceiving herself to be at the friend's house. The overprotected and coddled child has less opportunity to build new control loops that can serve as atenfels when the child encounters a substantially changed environment. In a child who has been allowed more freedom to play dangerously reorganisation is likely to interpolate new control loops within the existing hierarchy rather than constructing them at the top in the way suggested in Figure I.1.2a. Reorganisation may produce new perceptual functions whenever a pattern of inputs is encountered repetitively, in the way a partial construct such as 'th', 'sub', or 'tion', or a pattern such as 'e[consonant] e[space] appears in many written words in English.

As the system grows new levels, each new high-level perception must initially depend on only those lower-level perceptions that first are connected as inputs into its perceptual function, and can act only through those lower-level controlled perceptions to whose reference function inputs it is first connected. Hence, every new high-level controller is rigid, in that it has just one way to control its perception, even though it acts by sending reference values to lower systems that may have developed multiple ways to control their own perceptions.

As they add levels to their control hierarchy, children often have phases in which they insist on doing things by the rule-book. It is just 'the right way to do it'. If an adult 'does it' a different way, the child may object. "Mummy always hangs her coat up on the door. Why are you putting yours on the chair? You shouldn't do that. You have to hang it on the door." These are ideals, reference profiles for different levels of controlled perceptions below the level at which the overt intent (putting the coat in a convenient place) exists. The child is apparently conscious of what the visitor did and is able to compare it with what Mummy does.

'Idealism' implies the person has a reference profile for 'the way the world *should* work' (in OOP language, a reference Object to be compared with a corresponding perceptual profile or Object). Idealism is a concept that can be applied at several levels of the Powers hierarchy. If 2+2 is the question, then ideally, the answer should be 4. When a child begins to learn arithmetic, perceptions of such problems provide clear reference values for providing the answers. But as the child learns more, 2+2 often does not mean 4. In mathematics, if your addition is modulo 3, then 2+2 = 1. Geometrically, if in a curved space you go

2 units and then another two units in the same direction, you may well not be 4 units from your starting point. You might even be back where you started, if the curvature is positive like the surface of a sphere and the unit is ¼ of the circumference of the sphere. Think of the old riddle:

"You walk ten miles due South, then ten miles due East and ten miles due North, arriving back where you started. You meet a bear. What colour is the bear?"[118]

But if the only geometry you know is in Euclidean space, and the only arithmetic you know is what you learned in grade school, then 2+2 must equal 4 both numerically and as a distance from your starting point. Furthermore, no triangle can have three equal sides with every corner angle being 90°, although that is a perfectly reasonable possibility for a triangle on a sphere. In the riddle, the path you walked formed just such an apparently impossible triangle, making the riddle insoluble based on Euclidean space. Only by realising that the triangle was not on a Euclidean surface but on a spherical one could the riddle be solved. Idealism at a perceptual level tends to evaporate when one has reorganised to be able to control perceptions in a variety of contexts.

If this 'one way first and then become flexible' sequence is a general property of growing levels in the hierarchy, rigidity at a level is likely to last longer the higher up the hierarchy we go. When we come to system-level perceptions such as political or religious systems, flexibility may develop very late, if ever. Even though a prophet may have had many ways to control certain perceptions and been very flexible in his personal means of control, later followers often are very rigid in their requirements for formal rituals, behaviours, or clothing, and consider them, rather than understanding the prophet, to define their religion.

Rules, independent of context, may persist for a lifetime, but more are likely to be evident and rigid for some time in nearly all people as they approach adulthood, because the higher levels are likely to develop more slowly than the much-used lower-levels. These rules form the 'ideal' way to achieve the (fixed) reference value for a top-level structure. If the 'ideal' consists of, say, obedience to authority in the Confucian sense, then criticism of authority, or failure to obey, might result in appreciable error in the controlled system-level perception. The same would be true if the 'ideal' included a requirement for 'fairness' and the person perceived the behaviour of others (or herself) to be unfair. If the perception is being actively controlled, error in a control unit leads to action.

Error in a unit that has only a rule-based (single-means) structure available for output may sometimes prove uncorrectable despite violent activity directed at correcting it. Such would be the case if a dominant authority is perceived to be unfair and the person has no atenfels for influencing that perception. That kind of perceptual error is hard to correct, and any action taken to correct it (by a parent or public figure or institution) is likely to be seen by the authority as an unfocused temper tantrum or a directed rebellion to be suppressed by force.[119]

---

118 White. It would have been a Polar Bear, because you started at the North Pole.

119 We treat the possibility of institutions acting as control systems in Volume IV.

Reorganisation, however it functions, fits the maturing organism, usually helping it to survive and, with luck, prosper in the environment in which it lives. If that environment consists of mechanical and biological servants that attend to its every wish, it will learn to function by ordering its servants. If it lives alone in a jungle, it will learn how to identify ripe fruit and avoid those that make it nauseous, as well as learning how to avoid or outfight predators. In a city, it may learn how to navigate traffic, techniques of shopping or stealing, and so forth. Every environment demands different sets of skills, and any species that has few descendants per parent must either be found in a restricted environmental niche, or be capable of wide-ranging adaptation — with sufficient reorganisation in a lifetime to control its perceptions effectively for the maintenance of its intrinsic variables in many environments. Humans are the adaptable species *par excellence*, and may reorganise in a wild variety of ways.

Two properties of reorganisation are important. Firstly, perceptions must be controllable, and secondly, the perceptions to be controlled should be those for which control serves to enhance the organism's survival and/or to propagate its genes. As we shall see, this latter requirement leads usually to socially adapted behaviour, in which members of a culture are more likely to try to help than to hurt one another.

Next, we speculate about how reorganisation can become effective and efficient, in part by working not on the entire hierarchy of perceptual control as a unit, but by reorganising small modules which are then treated as units in reorganising higher-level modules. We start by examining the e-coli reorganisation process a little more closely.

I.11.6 Modularity of Reorganisation

Reorganisation using the e-coli process does work, but if there are a lot of parameters to be altered it may work very slowly. Mathematically, the issue is that each parameter in the control hierarchy may represent an independent dimension in a space of extremely high dimensionality. When a 'tumble' occurs in an e-coli procedure, the new direction of change can be described by the rate at which each of the parameters is changing. These rates stay constant until the next tumble. Some directions result in improved intrinsic variable function, some in reduced function. The chance is about 50-50 which way it will go.

The problem is that in a high-dimensional space almost all random directions are nearly orthogonal to any pre-specified direction, such as the direction toward the optimum set of parameter values in the space. If there is any random variation, such as an external disturbance, changes due to that variation will act as 'noise' that makes it difficult to tell whether the underlying change in parameters improves the situation or makes it worse. One way to resolve this problem is to increase the rate of tumble if the rate of change in the intrinsic variable function is too close to zero. Of many tumbles in quick succession, maybe one of them will result in appreciably better function of the intrinsics variables.

But what if the current parameter set is very close to optimum? If that is the case, almost every tumble will make things worse, or at least not detectably better, and very often tumble would have not much beneficial effect. Another solution must be found. That solution is modularization. Modify only a few parameters together, and then modify the parameters in that group together, treating each of a small number of modules as individual elements that have their own set of parameters. Kauffman (1995) found that his optimum module had five or six parameters, no more. Does this sound like what happens with control of perceptual complexes in Section I.11.1? It should, because exactly the same principles are in play, coordination of change using modularity of effect.

In Figure I.11.6c (below), I show how a trivial example of reorganisation can take the form of an explicit control loop. In the example, Quality of Control by a simple 'subject' control loop is the controlled perception, and variation by Powers's e-coli process of the parameters of the subject loop are the output. Thinking of control in the abstract, as manipulation of the environment to structure it in a way most congenial to 'happy survival' of the organism, the whole perceptual control hierarchy can be thought of as the environment of a different control hierarchy, the reorganisation system.

In Section I.5.5, the consistent real-world effects of moving a chair created correlations among the perceptions of its legs, seat, and back. The environmentally created consistencies become reflected in the construction of modular components of the perceptual hierarchy, namely the 'chair' perceptions together with their component 'chair-part' perceptions. Might we not expect something similar of a reorganisation system, whether its components are perceptual control units or something else entirely?

There is a problem with this question. For the perceptual control hierarchy, the consistencies are imposed by Real Reality (RR), not just the environment we perceive (PR). We may not notice that this 'chair-back' moves in a way coordinated with that 'chair-leg', in that we may not construct a perceptual complex that combines them. Even if we do not, that consistent relation nevertheless exists in the real world. We do not perceive gamma radiation, but if we are exposed to too much of it, we die. The 'real world' is 'boss' as Powers often noted in on-line discussions.

For the reorganisation system, its environment is the eminently malleable perceptual control hierarchy. Indeed its very job is to change the perceptual control hierarchy, so its local environment is not its boss. What is, if anything? What else but the system of intrinsic variables that determine our well-being and our very survival? These are intricately interconnected in ways that have been determined by evolution through our ancestors who lived in the 'boss world' as it was structured in their lifetimes. Intrinsic variables do not maintain themselves simply by internal homeostatic mechanisms, but are subject to the 'slings and arrows of outrageous fortune' — external effects — just as much as are any other parts of our structure. To reduce or avoid these 'slings and arrows' is the reason for perceptual control.

Our controlled perceptual variables generally correspond to environmental variables in RR which we have called 'Corresponding Environmental Variables' (CEVs). To set a CEV to a particular perceived value, however, is not the same as to adjust an intrinsic variable. Perceptual control affects intrinsic variables only through side-effects of control, which include the effects of any changes in the CEVs. But in these side-effects we have the way that 'boss reality' provides the reliable structure within which all organisms have evolved from the earliest days. Error in the intrinsic variables leads to reorganisation of the perceptual control hierarchy, the operation of the perceptual control hierarchy affects the environment, and effects in the environment influence the intrinsic variables in a 'Grand Loop' that is reminiscent of a control loop (Figure I.11.6a).

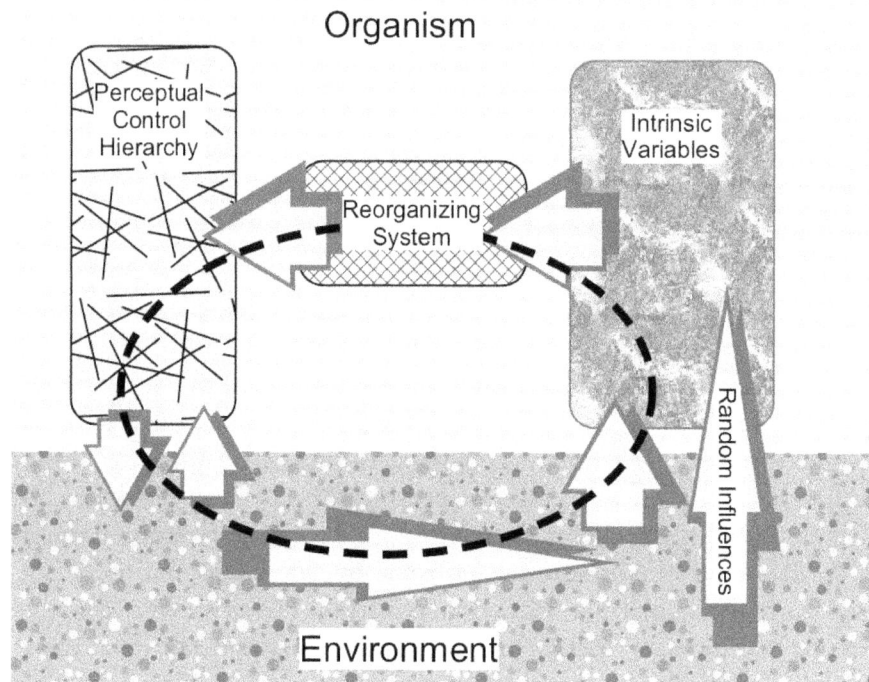

Figure I.11.6a The "Grand Loop" that allows an organism to survive in a complex environment. Organisms without a reorganizing system would either be short-lived or would be well armoured and live in a stable environment. Otherwise the species would survive by providing each entity with large numbers of descendants, very few of which would survive to propagate further generations. The place of reorganization in the Grand Loop would be taken by evolution.

At a very basic level, RR contains the physical constants which we presume have remained stable throughout the life of the Universe. According to present-day Physics, these determine what forms of matter and energy can exist, what atoms will be stable, what chemical molecules can form and how they interact, and so forth. On the shorter time-scale of the life of the Earth, the mass of the Earth and thus the force of gravity on an organism of given mass has not changed appreciably so long as there have been land-living organisms such as plants and animals that needed to counter it. As we consider shorter and shorter time-scales, more and more aspects of the environment seem to have changed hardly at all. Continental drift has affected the evolution of species, but not the forms of cultures, whether human, animal, vegetal, or microbial.

But some aspects of the environment do change on shorter time-scales, even within the lifetime of an individual, and these changes influence the way the side-effects of perceptual control affect an organism's intrinsic variables. Either an organism must live in a stable environment and produce many offspring, with few surviving to propagate further, or it must be able to reorganise during its lifetime to alter the side-effects of perceptual control in ways that continue to keep its intrinsic variables in good shape. 'Good shape', now, also refers to the way they were kept by our recent ancestors a few tens or hundreds of generations ago, in an environment that may not have been ours. Reorganisation must compensate for differences between our environment and theirs, as well as for changes in the environment during a lifetime.

Ignoring the species that do not reorganise during their lifetime, we can see two places in the 'Grand Loop' that could have complex stabilities that potentially might influence the reorganisation process. One is in the environment itself, while the other is in the genetically determined 'reference' structure of the intrinsic variables that might be affected by influences from the environment. Later, we will see these structures of intrinsic variables as describing homeostatic loops, but for now it suffices to treat the reference values of the intrinsic variables as genetically fixed.

To consider the implications of this, Figure I.11.6b simplifies Figure I.11.6a by merging the perceptual control hierarchy with the variables it controls in the external environment, since alteration of a reference value in the perceptual hierarchy is tantamount to changing similarly the value of a corresponding environmental variable. The environment of the reorganising system includes both the control hierarchy and the external environment.

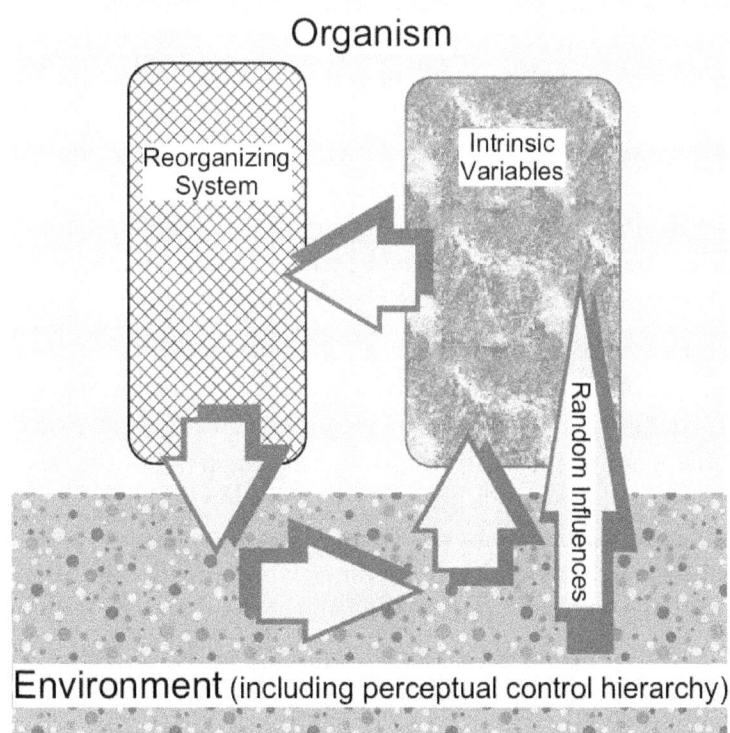

Figure I.11.6b The Grand Control Loop simplified. Seen this way, the whole structure has the form of a complex multi-variable control loop. The reorganizing system serves as an action component and the intrinsic variables as a perceiving component in which reference values for the variables have been provided by evolution.

In this simplified Grand Control Loop, the reorganising system acts on its environment through an interface that consists of altering the parameters and connections of the perceptual control hierarchy. These changes alter the perceptions that are controlled and the actions that are used to control them. If reorganisation is effective, actions performed by the outputs of the perceptual control hierarchy come to better influence some intrinsic variables toward their genetically determined reference values.

For example, an adult will work at a job to make money that can be used to buy food that when consumed will reduce perceived 'hunger'. As a side-effect, energy becomes available in chemical form for the operations of the cellular system, such as the brain, the muscles, and all the rest of the body, little or none of which is directly available as a perception in the perceptual control hierarchy.

In a system reorganised in a different environment, an adult will go out into the bush with a weapon to kill an animal that he will take home and eat, or he may go into the forest and pick fruit to eat. The office worker who has never seen bush or forest might starve or be poisoned in that environment until the error in his intrinsic variables induced some reorganisation.

If we accept the idea that the Grand Loop has the function of a control loop, then perhaps we can pick it apart in much the way that Powers did for the control hierarchy, treating each perceptual variable not as a complex but as a simple scalar variable represented in the brain by a neural firing rate. In Section I.5.5, we used the 'Chair and parts' example to illustrate how this approach combines with structural aspects of 'Boss Reality' to generate modular perceptions. We might well expect the same of the reorganisation system. It may be up to evolution to define the intrinsic variables for each organism, but reorganisation defines the perceptions and actions in the perceptual control hierarchy, and they are modularised by the structure of the environment, physical and social.

The environment is not alone in having internal structural and dynamic relationships. The many intrinsic variables do, too. Physiologists have found many interactions among them and discover more every year. These structures will be 'rigidities' (analogous to the relationships among the parts of the chair of Section I.5.5) which determine what kinds of reorganisation of the perceptual control structure will improve the states of the intrinsic variables.[120] Now we can call this improvement 'reduction of error', just as we use reduction of error to assess improvements in Quality of Control (QoC) in the perceptual control hierarchy.

The reorganising system thus seems to be 'squeezed' between the structural properties of the environment and the structural properties of the intrinsic variables. The reorganised perceptual control system would be most effective if it took advantage of both, which means the reorganising system must create perceptions which are useful to control and actions to control them which have side-effects that tend to reduce the errors in the intrinsic variable system.

---

120   Later, such as in Section I.12.1 and scattered elsewhere, we will treat both kinds of rigidities as analogous to 'rods', the compression components of physical tensegrity structures.

Because of the structural regularities or rigidities on both sides of the reorganising system, we should expect it to have structural levels analogous to the levels in the perceptual control hierarchy. In such a structure, reorganisation of the local neighbourhoods of single control units might be analogous to perceptual control influences on the smallest perceptual elements. Together with any parameters relating to a particular loop's connections with higher and lower level control units, a second-level reorganisation loop might act analogously to the 'chair-level' perceptual control to improve coordinated performance. If this idea is extended, one might conceive of a reorganisation hierarchy devoted to improving QoC throughout the perceptual control hierarchy, regardless of what perceptions are actually being controlled.

One thing we might expect of the reorganising system is that its smallest modules enhance the QoC of control units in the perceptual control hierarchy. A perceptual control unit that does not control well has inconsistent side effects, quite apart from any differences that might result from changes in its mode of control, such as choosing to walk to work instead of taking the car.

Figure I.11.6c suggests one possible structure of a reorganisation control unit — a possible reorganisation control loop which reorganises the parameters of a minimal perceptual control loop so as to optimise its QoC. The subject control loop has three parameters: integrator gain rate, integrator leak rate, and width of the tolerance zone. This problem is a simple hill-climbing optimization in three dimensions that might well have an analytic solution, but we here use it as an illustration of a general process, e-coli reorganisation.

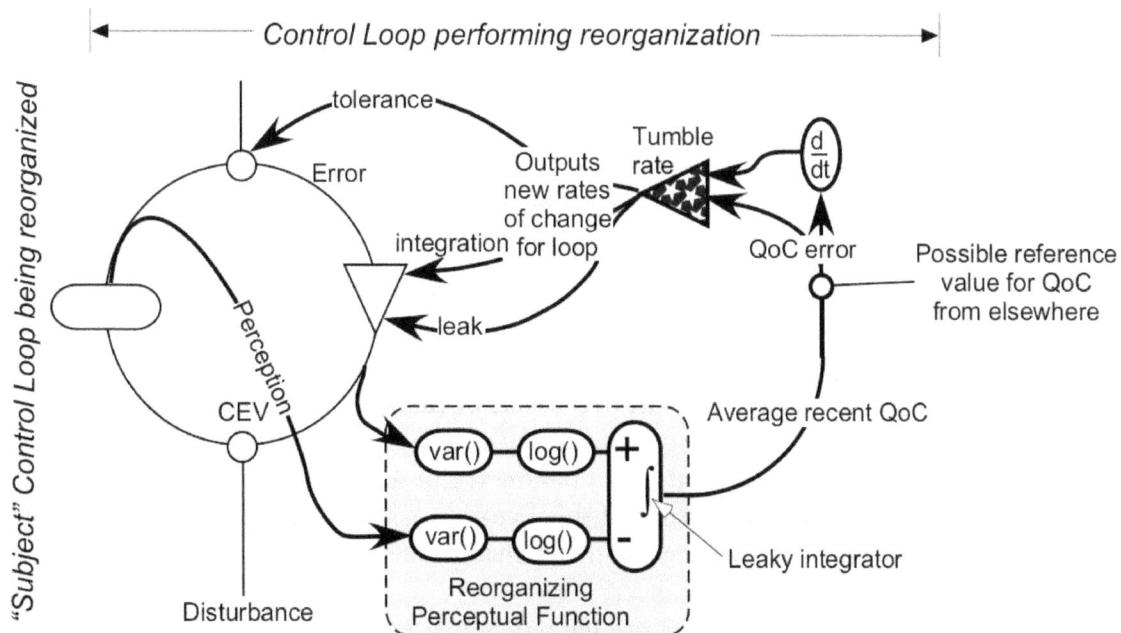

Figure I.11.6c A speculative reorganizing control system that has as its perceptual variable quality of control in the "subject" loop being reorganised and as its output rates of change of the three parameters of that loop. The output function produces these rates of change by "tumbling" to a random direction at a rate determined by the current Quality of Control, (QoC) and its rate of improvement or deterioration.

In this figure, the three parameters of the subject loop must have rates of change that are very slow compared to both the rate of change of the disturbance and the time it would take the loop to counter most of the change in the CEV induced by a steep change in the disturbance. The output function at the heart of the reorganisation loop determines what these three rates of change must be and when to change them — when to 'tumble'.

The perceptual function and the output function are the critical elements of any control loop, and the reorganising control loop is no different. In Figure I.11.6c, the perceptual function of the reorganisation loop is shown as the difference of the logarithm of the variances of the output and the perceptual value of the subject loop.[121] Since Quality of Control (QoC) is usually defined as the ratio of the variation of the CEV with and without control and the CEV is in the environment, those measures are not accessible to this reorganisation loop, which must work with internal variables in the loop.

There are, however, acceptable surrogates. The perceptual value is a surrogate for the value of the CEV, and the output value is a surrogate for the disturbance value. The ratio of their two variances is the QoC, but we do not use that ratio directly. Instead, the difference between their logarithms is integrated, since it is independent of the actual scale of the variances. The perceptual variance without control is what it would be if the link between the Output Function and the CEV were severed, namely the variance of the disturbance. If control were perfect, the output would exactly match the disturbance, and hence would have the same variance. The worse the control, the worse that match, but we still have no better simple surrogate for the disturbance variance than the output variance.[122]

Unlike the control loops of the perceptual control hierarchy, the leaky integrator in this control loop is in the perceptual input function to slow the computation of the average QoC. The output of the perceptual function is a measure of the average recent QoC, a variable we want to bring to some level, not necessarily perfection, within the tolerance of the reorganising system by varying the parameters of the subject control loop.

The output function has to produce an output vector consisting of the rates of change of the loop parameters. This vector has a constant direction in the parameter space until a tumble occurs, after which a new direction is taken. The vector has a magnitude that determines the overall rate of change of the parameters, which the output produces in the normal manner of a control loop by using the QoC error as the input to a leaky integrator that has the vector magnitude as its output. When a tumble occurs, the magnitude of the vector does not change, but its components are redistributed among the parameter change rates.

---

121 In Chapter I.9 we learned that this process measures the uncertainties of the two variables.

122 A more exact match would be the sum of the perceptual and output variances, but the wiring in the figure would be even more complicated, so that second-order improvement is ignored.

The rate of tumbling is a function not only of the QoC, but also of its rate of change, being low when QoC is improving and high when it is worsening. If it is unchanging, the tumbling rate is above zero even if the QoC is excellent, which allows the loop parameters to escape from a local minimum. Just what the function should be remains to be determined. Only qualitatively can we assert that the better the QoC, the less likely a tumble is to occur within the next small time interval. Put another way, the survival probability of a direction decreases with time no matter what happens to the QoC rate of change, but it decreases faster if the QoC is deteriorating than if it is improving.

This would not produce the best results for the intrinsic variables, because it would work independently of what perceptions might be controlled other than that they take advantage of structural complexity to control well. Only the question of whether controlling these perceptions is likely to reduce error in the intrinsic variables will affect their reorganisation. Hence, we might expect reorganisation of perceptual functions and the development of new perceptions to depend more on structural regularities in the intrinsic variables, and the reorganisation of the action outputs of the perceptual control hierarchy to depend more on structural regularities in the environment.

Much of this is purely speculation, and none of it is provable. But the speculation is based on general principles, so we might call it an 'envelope assessment' in the same way that limits on energy flow provide an 'envelope assessment' of how fast and long a runner might run or a stevedore shift loads on a dockside. Details may be wildly wrong, but the general idea may nevertheless prove valid no matter how the details change as the result of future experiments on real and simulated organisms.

I.11.7 Reorganisation and Evolution

The organism has another advantage over Wiener's engineer — time, evolutionary time. No present day organism must invent its Level Zero perceptual functions anew. Every simple or complex organism constructed from distinct parts is descended from ancestors that survived in an environment that had 'Laws of Nature' very like or perhaps identical to those under which we now live. The strength of gravity on this Earth has probably changed very little over the last few billion years. Most of the changes that matter are from things that move about, whether they be the movements of tectonic plates over many million years or the fall of a rock from a cliff face.

'Things that move about' include especially living control systems, even most of those that are rooted to solid surfaces. They control their perceptions and are equally the result of evolution over the same long time, no matter what the organism.

Whatever the truth of RR might be, creatures in our ancestry have been constructed and have acted in ways that allowed them to survive long enough to produce descendants, all the way back to the hypothetical 'soup' of Chapter II.2. There we suggest how homeostasis created structures that were more resistant to being changed by external influences than were arrangements of like components that did not belong to homeostatic loops and networks. We do not mention it there, but the influences that might destroy structure were influences in RR, not in an internal perceptual world. The homeostatic loop could be homeostatic only because its functioning in RR allowed it to be. If any structure such as a homeostatic loop functions at all, the way it functions is an aspect of RR, however hypothetical the analysis billions of years after the fact might be.

Figure I.1.2a, the second figure in this book (repeated here with a slight amendment as Figure I.11.7a), shows a control loop in which the output influences several variables in the environment, that coalesce in an environmental variable labelled CEV (Corresponding Environmental Variable), which then influences several sensors. The problem with this in its original form is that it asserts that the CEV is an entity in RR, which is not true. It may correspond closely with something in RR, but it is not a part of RR. It is a projection of a perception that was created by a perceptual function whose inputs were determined by RR. In Figure I.11.7a, this is represented as a hypothetical variable in RR called 'RREV' (Real Reality Environmental Variable), while the CEV is created by the perceptual value being back-projected through the perceptual function as part of the perceived environment or Perceived Reality (PR). (See Section I.12.1.)

Since the CEV is in the Perceptual World and is a creation there of the relevant perceptual function, for the most part it is irrelevant to actual control. When control affects the perceptual value it *ipso facto* affects the CEV, though perhaps not the RREV. The better the CEV corresponds to the RREV, the better control may be, so one function of reorganisation must be to improve the match between

CEV and RREV — between Perceived Reality and the functional connections (though not the mechanisms) of Real Reality. Because of the identity between CEV and perception, we may often treat one as being the other,[123] and deal with the perception, which has inputs from Real Reality.

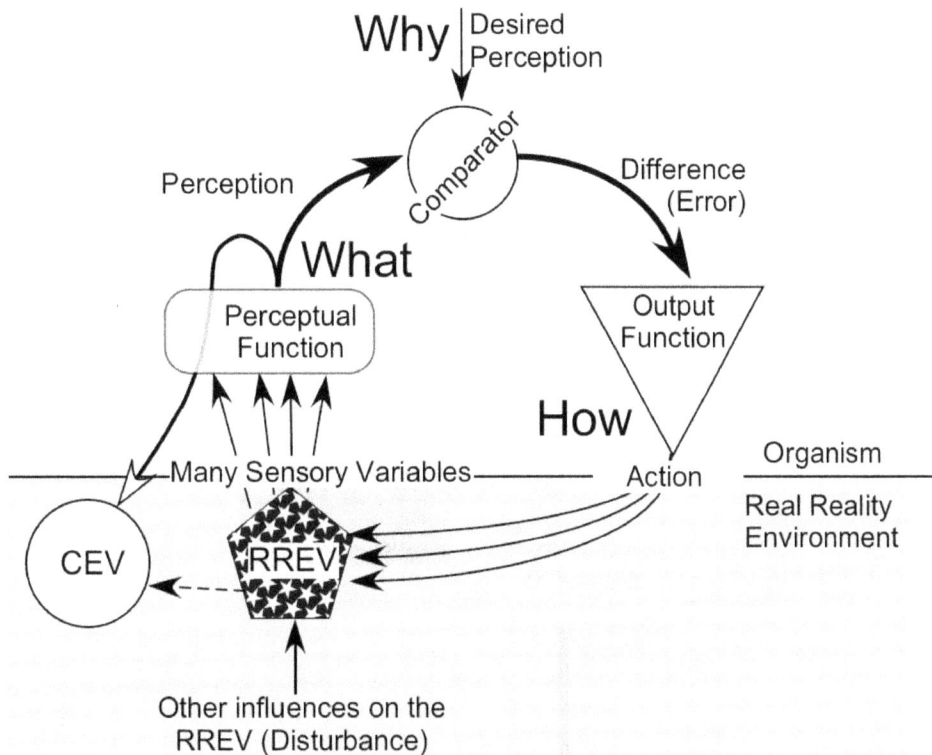

Figure I.11.7a (Figure I.1.2a amended). A simple control loop showing multiple paths from the output back to the perceptual input function. The dashed arrow implies that when a real disturbance or the Action Output influences the RREV, the change in the RREV must cause a related change in the CEV, which is the perception mediated through the channel from the senses through the Perceptual Function.

There may or may not be an entity in Real Reality that warrants being called an RREV. What matters to control is that the output actions that influence RR produce the desired effect on the perceptual input and therefore on the CEV. When we deal with abstract perceptions such as ideas, meanings of words, the powers allowed to an authority, the political style of a candidate for office, or even the virtual perceptions of a Collective Controller (Volume III), always what we are talking about is a perception produced by a perceptual function from its inputs, some of which may be in RR, while some, such as vivid memories, may exist only in imagination.

---

123   What Figure I.11.7a calls the 'CEV' is sometimes labelled 'CV' (Controlled Variable) by
      Powers and others, treating one as being the other, or not distinguishing them.

No matter whether the CEV corresponds to an actual RREV, the important question for reorganisation and evolution is whether controlling the perception that produces the CEV enhances or reduces the well-being and survival chances of the organism. If it does, on an evolutionary time scale, the ability to control that CEV is likely to depend on it corresponding closely to a true RREV in the real world environment.

Figure I.11.7b suggests an analogy to this situation. The perceptual function is represented as a cut-out silhouette in a slide of a 19th century 'magic lantern'. When lit by a projector bulb whose brightness corresponds to the perceptual value, the projector shines an image on Plato's cave wall. The image is cast over one of Plato's 'shadows'. Reorganisation or evolution of the perceptual function corresponds to the operator of the magic lantern refining the cut-out shape of the silhouette in the slide — the perceptual function — to create a better match with the shadow.

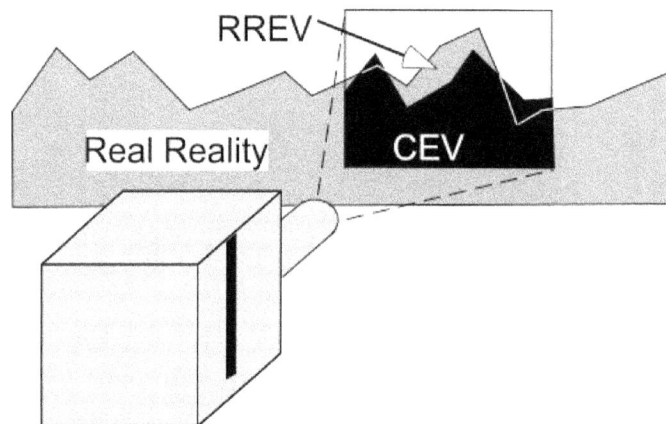

Figure I.11.7b Shadows on the Wall. Some part of Real Reality is accessible to perception. A small part of RR is perceived through the filter of the perceptual function, and projected back into the perceived environment, as a replica of unknown accuracy. The mismatch between the CEV replica and Real Reality is reduced in reorganization or evolution by changing the perceptual function.

The profile of the CEV is what the perceptual function produces from the profile of RR within the metaphorical 'frame' that is projected back into the environment. Within the control hierarchy, the individual points in the profile are perceptual values produced by lower-level perceptual functions. The points are 'logon' information (what is being perceived), while their heights are 'metron' information (how much of that is being perceived). For each point, a 'Shadow on the Wall' diagram could be produced, very like that of Figure I.11.7b.

Mismatch between the RREV profile and the CEV profile limits the precision of control of the perception, because in the environment the control loop's output acts on RREV (with its disturbances), while at the comparator it is the CEV that is compared with its reference value. If there were no relation between the outer reality and the inner CEV which is taken for reality, control would be impossible because the output would influence the CEV more or less randomly. If such a perception happened to be irrelevant to the welfare or survival of the organism, it would not matter, but to control it would be a waste of energy, and that would tend to be detrimental to the welfare of the organism. As we discussed in Section I.4.1, both J.G. Taylor (1963) and Powers (1973/2005) argued that perceptual functions persist only to the extent that they participate in feedback loops which the organism actually uses and which are not overwritten by loops that do the same job better.

Reorganisation using the e-coli process over a long enough period would allow the CEV profile to approach the RREV profile arbitrarily closely, if their relationship persisted steadily for long enough. For this to be the case, the CEVs of the lower-level perceptual functions would have to have had a stable relationship with their corresponding RREV profile over the duration of the reorganisation process. If we suppose that the organism was born completely naïve to the properties of RR, this stability would exist for the second level in the control hierarchy only after reorganisation of the first level had been substantially completed. The same would be true of the third level with respect to the second, and so on up the hierarchy. This predicts observed punctuation of infant development (Rijt-Plooij and Plooij 1992/2019, Heimann 2003).

We presume that the basic Laws of Nature, whether we know them or not, are stable over many lifetimes of living organisms. If any properties of RR that potentially might form an RREV remained stable in the environments of several generations of an organism type (species member) and the form of a corresponding perceptual function could be encoded in the genome, then that form could be matched to produce a CEV with an arbitrarily stable relationship to the RREV. If that happened, that particular perceptual function might not need to be reorganised within the lifetime of the individual. We may not know what it is about the RR Laws of Nature that allows us to describe with great precision what we perceive as gravitational force, but the effect that gravity has had on falling bodies on this Earth has probably changed very little since the time of the first single-celled organisms. Accordingly, all land-living species are built so that their bodies can sustain gravitational stresses, while some species are born into the world with control systems already built that allow the newborn to stand and walk, and perhaps even to run. These usually are species for which the environment has another stability that affects their likelihood of survival to an age where they can produce descendants — they are highly likely to be eaten if they can neither move to a hiding place nor run away when their mother does.

Another feature of the RR environment that produces effects that are stable over long periods of time is the effect of certain changes. If a sensor produces a change in its output from one moment to the next, probably something has happened in the environment (though it might have happened in the sensor). If a very similar change also occurred in a neighbouring sensor, what happened was probably in the environment, and if several sensors in a cluster usually produce the same change at the same time, the happening was almost certainly in RR.

The same argument applies to oriented differences between a sensor and its neighbours. If several neighbouring pairs produce the same relative difference in one direction, there probably is an edge in RR; carrying the argument up one level to where several such edge-clusters converge to form a connected shape, there is probably an object (which we may now also call an 'Object') in the environment. These relationships have been consistent since multiple sensor systems have existed, and if they were to be encoded by evolution into the genes of a species, that species would be able to reorganise much more quickly than if each individual needed to reorganise them from scratch.

One would think this argument could be carried up the hierarchy indefinitely, but it cannot. We already see a hint of why it cannot in the last example above. The configuration of 'edges' that signifies the presence of an object can be produced by patterns, in the perceptual world at least, that are not objects, such as a patch of sunlight through the trees falling on a rock face. The higher in the hierarchy we go, the greater the number of different RR configurations that could produce similar effects, and the more complex might be the ways they change over time. If they change slowly with respect to the time taken for substantial reorganisation at any level, then the perceptual functions in the control hierarchy can be produced by reorganisation. If they change too quickly, reorganisation of the control hierarchy cannot keep up, and here we probably come to one reason for the existence of conscious perception and thought, a question we will discuss more deeply in Chapter II.10.

I.11.8 So What is an Object That We Perceive?

We have said this before, and we will say it again. The fundamental facts on which all life depends are those of Real Reality (RR), of which we can know nothing other than how RR affects our sensors and our physiological functioning more generally. Even of that little, we can know only the perceptual effects that are produced by the RR operations of our bodies and by external processes independent of our outputs. The rest we know by our perception of the effects of RR on our instruments.

If hierarchic perceptual control is a reasonably correct analysis of some of these operations of our bodies, the limits are determined by the variety of perceptual functions that have been developed in our bodies by evolution and reorganisation. These perceptual functions seem to produce perceptions of things that exist outside of ourselves, but as philosophers at least as far back as Plato observed, we can never know whether what we perceive to be real is any more than "a shadow on a cave wall".

In Sections I.2.4, I.8.2, and I.11.4, we used Wiener's construct of 'White Boxes' which emulate relationships observed to exist in the signals sent to and received from a 'Black Box'. There we observed distinct parallels between Wiener's and Powers's development of hierarchies of relationships used in the building of emulators. Powers's 'perceptual functions' in the hierarchy are a direct analogue of functions performed by Wiener's 'White Boxes'. We also used the language and concepts of Object-Oriented Programming (Section I.11.3), arguing that an OOP 'Object' is directly analogous to a White Box and, more importantly, to an object that we perceive.

In Powers's hierarchy of controllable perceptions, each controllable perception is a scalar variable produced by some function applied to a set of inputs which are individually scalar variables, each one the output of a lower-level perceptual function. Each perception is of the state of some property of something otherwise unknown in RR.

We don't consciously perceive properties in isolation. They are properties of some entity and cannot be divorced from the entity. The abstract concept of the property, such as elasticity, temperature, volume, hardness, and so forth, can be divorced from any specific entity, but this divorce is the same as the difference between an OOP class and an instance of the class. The class has these abstract properties, but each instance has its own values for them. There are no 'hardnesses' without entities (objects) that are just that hard.

Not only can properties not be perceived except as attributes of entities, entities cannot be perceived without at least some properties (more than one). Perceptible objects, then, can be seen as a way our system has learned to bundle sets of properties that are more often than by chance found together in much the same bundle, the 'bundle' description being the class specification of an OOP object, a type of Black or White Box, an RR object, or a consciously perceptible object in PR. An object is an extension of the idea of a 'Gestalt', in which 'common fate' determines whether similar perceptions tend to change in synchrony. In other words, every perceptible entity is a potential atenex (Section I.2.4), capable of providing or influencing a means of controlling some conceivable perception.

I included RR in that list of possible objects, even though we are by this time clear that we cannot ever know how RR works. As Wiener argued, what we can know is some of the functional linkages among the mysterious working parts of whatever is in RR. We also have been arguing here that bundles of controllable perceptions tend to be found together in the same Gestalt-like ensembles in PR. Such a coherence would not occur if the effects of our action on RR did not produce these same coherences among properties that we perceive and control.

We might call such bundles or ensembles of perceptual functions 'syndromes'. A syndrome corresponds to an abstract object class in OOP, to an abstract collection of functions of a White Box, and to a perceptible object, whether concrete or abstract. We will use the word later to represent non-tangible (abstract) ensembles of perceptions, stable networks of perceptions whose control results in a feedback loop of mutual support. An example of an abstract object, or syndrome might be the personality of another person, or the style of a piece of music. The word 'style' fits the concept not only when applied to individuals, but also to social structures that cohere similarly.

On this basis, we can legitimately argue that the same kinds of objects as we perceive to be coherent objects in PR, defined by the packages of functional properties they include, very probably exist as coherent bundles, or objects, in RR. Just as 'very probably', RR objects have additional properties that do not exist in PR, some of which our senses unassisted by technology do not allow us to perceive, such as their opacity to gamma radiation or the isotopic distribution in their elemental composition. We know of them only by using instrumentation unavailable to anyone a few decades ago. RR objects presumably also have other properties we cannot yet imagine. Our PR objects are but partial approximations to such RR objects as may exist, if the PR objects are indeed fair approximations and not the results of fortuitous coincidental coordinations of functional properties, supplemented by imagination.

A White Box emulates the functioning of some processing in the corresponding Black Box without our being able to observe in any way how that processing is done in the Black Box, or even whether the Black Box actually exists in RR. In this, as we pointed out, the White and Black Boxes resemble the 'Objects' of Object-Oriented Programming. The engineer building a White Box may be able to determine that something in the Black Box consistently produces some pattern of outputs when the input terminals of the Black Box are fed some particular temporally varying pattern.

The engineer then can use the observed relationships to build a White Box that produces the same output pattern when fed the same input pattern. This relationship is a property of the Object —and the perceived object — that is the White Box.

To make the White Box, the engineer may attempt to influence the output pattern she observes in a way she desires by varying the input pattern she applies. The more kinds of variation the White Box correctly emulates, the more likely it is that the White Box truly mimics some processing going on in the Black Box.

Notice what we did here. We allowed the White Box to be more than the representation of a perceptual function, and instead accepted that one perceived object has many properties, each of which corresponds to a perceptual function within a White Box. A White Box corresponds to a perceived object, not to a single perceptual function, just as an OOP programming Object can include entire less complex Objects that perform some of its functions. Each of these smaller Objects (as 'classes') may be used within several quite different larger Objects, leading to a hierarchic structure of Objects identical in form to the Powers hierarchy of perceptual functions to be controlled.

Is it the same hierarchy? Not exactly, but close. It is a version of the freely connected Powers hierarchy, but with restricted freedom of interconnection because of the packaging of properties into defined Objects that may represent the everyday objects we perceive. These Objects not only have their particular sets of properties that act as perceptual functions, but also have defined values for those parameters as parts of a definition of the Object. For example, a teacup cannot be made of a material that melts at a temperature below the boiling point of water. Its hardness must be sufficient such that it does not change shape under the weight of the tea while one is drinking from it. A wooden chair cannot have the location property values of one leg changed without changing the values of location properties of its other components. And so forth.

Now we argue that any perception with which we interact is the output of a function in exactly the same kind of Object as an Object in Object-Oriented Programming. The Object itself is a White Box, which takes potentially many independent inputs and produces a set of scalar output variables. All we can know of a RR object which we perceive in the environment, such as the keyboard on which I type this, is that it has several possibly inter-related properties. Those properties are what we see and feel it to do when we or some other forces act upon it.

I know nothing of how a key on my keyboard works to magically produce a pattern on my screen that I call a 'letter'. But it does work (most of the time) and produces that letter on the screen. I might be able to learn what is in the keyboard if I were interested, because some engineer designed it and someone else produced it according to that design.

The physical keyboard, like any perceived object, can serve as an atenfel for control of many different perceptions. These atenfels are bunched, as properties of the keyboard. The keyboard has many properties that have nothing to do with the identities of the letters that can be produced by its keys. It has the properties of mass and of weight, for example. It has shape and colour and volume. It has a myriad of properties, all of which are bundled into the object we perceive as a 'keyboard'. They determine a whole suite of perceptions you could control by using the keyboard in different ways.

Suppose I did happen to be interested in the keyboard design. What might I learn? Basically I would learn that the keyboard has certain parts, each of which has a designed function. It does something when influenced by something outside itself. What influences it, and what its product influences, I could learn from a circuit diagram and a list of specifications. Each component, like the keyboard, is an Object in both the OOP sense and the perceptible sense. You do X to it and it responds by doing Y. How it does what it does internally is still hidden from my view.

What I would have learned about the keyboard Object is that the function of taking one of my key presses and having a pattern appear on my screen is performed by a bunch of miniature objects functionally inside the keyboard object. Treating it as a Wiener Black Box, Wiener's engineer might discover how these miniature Objects, as White Boxes, might be connected (the circuit diagram), but he would not discover how they worked internally. To do that, he would need to discover even smaller White Box Objects within the White Box Object, his emulations of components of the keyboard.

This regression of ever smaller OOP Objects to perform the functions of the larger object when appropriately connected together could, in principle, continue indefinitely. At the present state of scientific understanding, however, once you have arrived at the Objects that are the quarks and gluons whose interactions are the connecting channels of the circuit diagram for the components of the nucleus, you are able to go no further. You would still not know what is 'really there'. You might know all the functional influences among the components that are 'really there', but no more. For all we know, what is 'really there' might ultimately consist only of the interactions of forces. But for me using my keyboard to type this, the keyboard is the Object I perceive to be 'there' in my Perceptual Reality, and I need to perceive no more than is sufficient for me to control my perception of the shape appearing on my screen.

An Object, then, is a bunch of atenfels, an atenex. Do we need to consider it as an entity with a Real Reality existence? Perhaps not, but as with everything else in PCT, we must ask what perception the perceiver might be controlling, using the keyboard's Real Reality existence (or non-existence) as an atenex in itself. A philosopher, for example, might use it as a way of controlling their self-image as seen by self or by others: *"Aren't I clever to be able to make such a beautiful argument for (or against) the existence of the keyboard in Real Reality!"* .

Most people might not care at all, so long as they can use their keyboard to produce words that other people can read (or can use its mass and hardness to tack a circuit diagram to a wall). For all practical purposes (a loaded term), an Object is what it does when something is done to it. What it really is, what it is made of, is usually, perhaps always, irrelevant. As we shall see more clearly when we come to social structures, many of the Objects we use do not have atoms as components at all. They are pure structures of interacting influences, such as languages, polite forms of address, political concepts, and so on. They are bunches of properties, atenfels that in some way cohere together, combined into atenexes.

From the viewpoint of PCT, these intangible Objects, no less than those whose functionality depends in part on controlling perceptions of mechanical forces, are just atenexes, relatively stable bundles of the atenfels (properties) which we use for controlling our perceptions. The bundle exists in our Perceptual Reality as an entity — a tea-cup, a language, a law, a table — but what it IS in Real Reality will forever lie beyond our ken.

We can, in principle at least, discover at least part of the network of influences within any Object. That network of influences is likely to be a pretty good map of a corresponding network of influences in Real Reality, but whether there truly are Objects in Real Reality is a question bordering on the religious, not a question on which science can throw any light.

Here we have a conundrum. In most of the above, we have been taking the Object as something perceived, whereas the White Boxes are taken as analogues of the organism's Perceptual Functions, whose outputs are the perceptions. The implication is that not only are there Objects within Objects, but the internal Objects themselves may form inputs to other Objects within the same enclosing Object. In programming languages such as C (and C is not even an Object-Oriented language), this kind of relationship poses no problem, since the programmer can define a structure that is treated as a single variable. Hence, it should not be a problem in an organism.

The PCT analogue to the engineer's procedure is the development of the ability to control perceptions through reorganisation, the construction of new Objects such as perceptual functions and the patterns and parameters of their interconnections. It matters not how reorganisation happens — Powers's e-coli method (Section I.6.4, passim) is just one possibility, the control loop of Figure I.11.6c another.

What does matter is the number of degrees of freedom involved in the reorganisation process, which affects the time it takes to reach a useful approximation to an optimum match to Real Reality. In Powers's method, the degrees of freedom are manifest in the 'tumble' that occurs when the straight-line progression through the abstract design space stops improving performance and begins to lead to performance declines. The degrees of freedom are the number of independent directions in the space into which the tumble chooses a new random direction. It is important for the on-line operation of living species that the one-dimensional output of the perceptual functions in the Powers control hierarchy leads to the fastest optimisation.

Weiner's engineer faces the same problem of time and degrees of freedom. How many Black Box input terminals and output terminals should be manipulated and observed when trying to approach a consistent pattern of relationships to be emulated in the White Box being built? The number of possible changes to the White Box at any moment grows exponentially with the number of terminals.

Even if each terminal allows only two levels of signal, On and Off, then with just one input terminal and two output terminals that react instantaneously to input changes, at least eight possible relationships might be observed consistently as the input signal changes. One or other of the output terminals might remain unchanged while the other varies in phase or out of phase with the input terminal, both might vary together in or out of phase with the input, or both might always have opposite values with either output terminal varying in phase with the input.

Add another input terminal and the possibilities are not doubled, they are multiplied fourfold, to 32. And that is without considering the possibilities introduced by including the possible effect of any time delay between a change in the input and its effect on the output.

So Wiener indicated that his engineer would not try to make a comprehensive White Box, but would create small White Boxes that reliably produced certain relationship patterns among a small number of output terminals when the engineer fed one or a few particular input terminals with variable signals. These relationship patterns would have been observed when the engineer applied those same signals to the Black-Box inputs. The end result would be a set of what we might call 'level zero' miniature White Boxes.

The signals Wiener's engineer was asked to apply to the input terminals were 'white noise', perfectly uncorrelated with each other. Any persistent relationship among the signals observed to come from the output terminals would thus signify that some process within the Black Box was creating the persistent relationship. Powers theorised that the organism would start building the control hierarchy the same way, by constructing 'Level Zero' perceptual functions as the first stage of 'Reorganisation' (though since he starts from nothing, calling it 'organisation' might at first be more accurate).

But why should the perceptual functions, which we have said are analogous to White Boxes, mimic any processing that happens in Real Reality? The answer is that they wouldn't and at first need not. They do, however, produce signals we call 'perceptions'. We call them that because when we treat them as if they came from RR, most of the time things work out pretty well. Indeed, the signals input to Level Zero perceptual functions do come directly from RR, by definition.

Perceptions come from, or rather, are imposed on, our Perceptual Reality, almost tautologically. PR is not RR, but if controlling them works for us as it would if PR successfully mimics RR, then something in PR (an analogue of a White Box) must produce them. That White Box produces exactly the perceptual signal, with no mimicry involved. That White Box is in PR, and *is,* rather than '*is equivalent* to', the perceptual function. It is an 'Object'.

Now we can say that the perceptual function mimics the effects of some processing that happens in Real Reality. The perceptual function and the resulting perception are projected into the private Perceptual Reality of the organism. The same argument would apply equally to perceptions and perceptual functions at any level of the control hierarchy. The perceptual function is equivalent to a White Box which mimics some processing in RR, and the perception is what that White Box produces, which appears to the perceiver to be a state or property of the external environment — Perceptual Reality.

Real Reality differs in one crucial respect from Wiener's Black Box. RR is open-ended. No organism can know all the sources of effects on what its sensors tell it; its sensors cannot tell it about all the influences from RR that could affect its healthy survival. No human, for example, can without artificial assistance determine whether they are currently in a region of intense radioactivity which could kill them within hours or days. We do not have the appropriate sensors.

Nevertheless, the open-ended character of RR gives an organism which is learning its functioning one advantage over Wiener's engineer who is learning about the Black Box. RR produces effects on the organism's sensors whether or not the organism acts. The developing organism can be a passive observer while it builds its 'Level Zero' perceptual functions from relationships among the reports of the sensors.

Some ancient peoples called the unknown sources of these effects Gods and Goddesses or Demons. It doesn't matter what they are called, so long as they are unknown to the individual organism being (re)organised. From that organism's viewpoint, they are equivalent to the noise which Wiener's engineer deliberately introduces at the Black Box input terminals. If a relationship is observed at the output terminals of RR (the organisms's sensory apparatus), something in RR must be causing that relationship, and that something might as well be a Demon as anything else. The engineer's White Boxes or the organism's evolutionary and reorganisational structure development will eventually suggest internal connections among functional components of that something, even if it is a Demon.

Chapter I.12. Novelty, Belief, and Illusion

Illusions could not occur if our perceptions always correctly represented what they seem to represent in the real reality (RR) environment. Perceptions can be wrong, sometimes disastrously so. If a perception can be wrong about what is in the environment, it follows that there is always some uncertainty about the relationship between a perception and the reality it represents, some of which is testable, and some of which is not. 'Testable' means that if a perception indicates that something is true of the environment, then some other perception would be thus and so, or that the perception in question can be successfully controlled, or in some other way the accuracy of the perception fits with other perceptions we have of the way the world is or the way the world works.

We start by looking at ways in which perceptions normally do not correspond with the real world, in the form of illusions and after-effects. For some of these, we can say fairly precisely how the perception will fail. From this base of misinterpretation, we go on to consider various ramifications and concepts such as 'trust', 'belief', and uncertainty. Following this, we consider surprise, which leads us to think about the role of conscious perception in perceptual control.

I.12.1 Real Reality, Perceived Reality, and the Observer

The PCT loop passes through an unknowable region which, following Powers, I call 'Real Reality' (RR), which is influenced by the actions of a control loop's Elementary Control Unit (ECU), and which provides inputs to the perceptual function of that same ECU. We don't, however, perceive the external environment as an unknowable haze. We perceive it as being filled with a great complex of entities large and small which we can see, feel, smell, hear, and touch — and which we can influence by our actions.

It is important to distinguish between what we perceive and the reality that gives rise to our perceptions. Because we have no access to real reality other than our perceptions, we may be subject to undetected illusions. In particular, the 'CEV' of the PCT loop is as unknowable as is any other aspect of RR. The acronym means 'Corresponding Environmental Variable', which is accurate, whether the variable is in a PR created by one of Norbert Wiener's 'White Boxes' or is in the RR 'Black Box'.

Because the CEV introduced earlier in Figure I.1.2b is unknowable in itself, it is labelled '???' in Figure I.12.1a below. There is also a perception of the CEV, the output of a 'white box', a Perceptual Function. Furthermore, the perception can be made conscious, and when it is, it is experienced as being in Perceptual Reality, a 'white box' world that contains only what the Perceptual Functions can produce. It is convenient therefore to replace the '???' placeholder in Figure I.12.1a with 'Real Reality Environmental Variable' (RREV) in Figure I.12.1b, and use CEV to label the PR variable that appears to be in a complicated perceived external environment. The RREV is the output of some unknown process that occurs in RR, but the CEV in a perceived complicated environment is the output of a knowable process (the perceptual function) which is a 'white box' emulation of the black-box process in RR.

The better the 'white box' process (the perceptual function) mimics the 'black box' process that produces the sensory data that contribute to the perception, the more effective and efficient can be the control of that perception. If we accept the basic proposition that the mutual uncertainty between the reference variable and the perceptual variable is an index of control quality — the lower, the better — then we can use this as a limit on how well the perceptual input function in our 'white box' emulation models some function performed by Real Reality, insofar as the value the Perceptual Function produces is a variable that depends only on its Real Reality inputs.

The same applies to the entire ensemble of CEVs produced by a changing perceptual control hierarchy. The better the structure of interconnections among the 'white boxes' matches the interconnection structure of the 'black boxes', the more likely it is that the organism will be able to operate safely and healthily within the unknowable Real Reality in which it lives.

The structure discussed so far consists only of a one-way data flow from the sensory input surface to the input terminals of the white boxes, and from their output terminals to other white boxes. The control loop, however, has an internal counter-flow beginning in the sensory input terminals that terminates in the effectors such as muscles, which provide some of the initial inputs to the black boxes of Real Reality. This counter-flow in the white box ensemble is from the outputs of some white boxes to the input terminals of other white boxes, and finally to the actuator surface. Again, the more precisely the structure of this aspect of the white box ensemble matches that of the structure of the process interconnections in Real Reality, the better could be the control by the organism. Powers emphasised this outflow structure in his approach to reorganisation.

To create the emergent property of control, the inflow white boxes must be linked to the outflow boxes by some kind of structure of interconnection, completing feedback loops. The linkage between effectors and sensors already exists in Real Reality, though we know nothing about that linkage structure other than what can be discovered from the relationships between input and output patterns as they change over perceived time. But we can once more invoke the white-black conformance to match the *structures* of the white box interconnections with the *structures* of the black-box interconnections.

We now have three interlinked but quasi-independent sets of structures in the 'white-box world' that potentially could be matched to structures in the Real Reality 'black-box world'. The CEVs of the perceived world are one, coordinated actions are another, and the network of interconnections of the White Boxes a third. The structures of CEVs correspond to those of black-box RREVs, and the interconnection structures among the white boxes and the sensor-actuator interface match the corresponding interconnection structures in the unknown Real World. The key question is how these correspondences can be validated over evolutionary time and over individual lifetimes, and be improved in the process.

The feedback processes of reorganisation and evolution are Weiner's feedback process that co-organises a white-box collection into a structured ensemble. The structure of CEVs, and therefore of perceptual functions in ECUs, is the solution to Wiener's problem as posed in the Preface to the second edition of Cybernetics (Wiener 1948/1961).

As many philosophers at least as far back as Plato have noted, we can never know what is in Real Reality. All we can ever know about it we must base on our perceptions, including our perceptions of our outputs. And yet, we seem able competently to control many perceptions at several different levels of complexity by acting on Real Reality and basing our perceptions on what Real Reality gives us back. Figure I.12.1a is the first of a series developing the data flows.

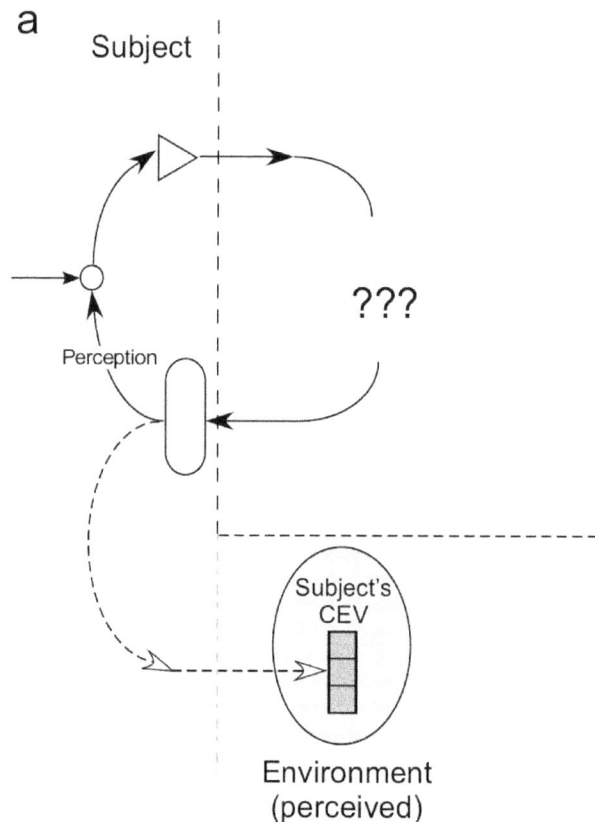

Figure I.12.1a Control loops pass through an unknown Real Reality, affecting something there and controlling a perception based on how Real Reality influences what we perceive. We can consciously experience the perception as a CEV, set in a perceived environmental context of other things perceptions produced by other perceptual functions. The CEV is not in Real Reality, but in the perceiver's mind.

How can a control loop be effective with no knowledge of what it is acting on or what it is looking at, especially when that unknown 'thing' is subject to outside influences that are equally unknown? The answer is that it is not really true that the loop has no knowledge of what happens in the passage of its output influences through Real Reality back to its perceptual function.

In Chapter I.11 we discussed 'reorganisation', which varies the configuration and parameter settings of the parts of the control loop internal to the organism, to improve not only the Quality of Control but also improves benefits to the organisms' intrinsic variables controlling *this* perception. Reorganisation functions entirely without knowledge of what is in Real Reality, but provided Real Reality is somewhat consistent over time, reorganisation, either within the individual lifetime or over the life of a species (when we call it 'evolution'), can produce useful perceptions that are well-controlled. The end result is what is shown in Figure I.12.1a.

Here is a little parable that I posted to CSGnet, lightly edited to fit the changed context and to accommodate occasional comments made on the original.[124]

> *Here's an analogy that may bring back memories to the older CSGnet readers. Once upon a time, there were lots of radio stations operating on frequencies in different 'bands'. There was a long wave band with stations such as Droitwich (UK) and Hilversum (NL). They and quite a few others were intended for broadcasts with a global range. There were medium wave stations, lots of them, intended for everyday home listening, and there were short-wave stations, many operated by amateurs called 'hams'.*
>
> *To listen to one of these stations, you selected a 'band' (expensive radios had five or six) and then turned a knob to move a pointer to a place on a dial that might be labelled with a station name and/or a number representing frequency. Turning the knob also changed the setting of a 'variable condenser' that altered the frequency to which the receiver was tuned, but the average listener knew nothing of that. All the listener knew was the choice of band and the placement of the pointer.*
>
> *Anywhere you set the pointer you would hear sounds, often fragments of speech or music, but more probably just crackly noise. If you set the pointer near to a frequency on which some station was transmitting, you would hear something coherent such as someone talking or music playing. That was what you wanted, but unless you set the pointer just right, the speech or music would be distorted. You fiddled with the pointer until what you heard was clear and true, but this was possible only for stations close to you or transmitting with high power. Usually you heard background noise along with the undistorted signal when you found the best setting for the pointer.*

---

124 To find the original message in the CSGnet archive, search for [Martin Taylor 2019.04/17.08.49].

Now think of the air full of the electromagnetic waves from the different transmitters and from elsewhere in Nature as being the part of Real Reality you could access by changing the choice of band and the position of the pointer on the dial. The tuner itself played the part of a Perceptual Function, and you were reorganising it by changing the band and the pointer position so that the sound you heard would come clearly from the station you wanted.

No matter what the band and pointer position settings might be, the resulting Perceptual Function would define a part of Real Reality and let you hear what was being transmitted in that part. With random band and pointer position parameters, the resulting Perceptual Function (tuner setting) was almost never useful in 'maintaining survival' (allowing you to hear something you enjoyed or from which you learned). But Real Reality alone determined the 'value' of the perception/CEV produced by a particular Perceptual Function.

By changing the band settings and turning the knob, you changed the parameters of the Perceptual Function (tuner), and sometimes what came out was indeed pleasant or informative. You might mark the dial so you could come back to the same setting later, thus defining a perceptual function that might on another occasion provide a useful CEV that corresponded closely to an RREV (band and frequency parameters of the radio station's influence on the flood of electromagnetic waves filling the parable's version of Real Reality).

By fiddling with the band choice and the pointer position on the dial, you might find several different parameter settings for perceptual functions that tended to provide useful CEVs, perhaps some often helped you with cooking recipes, some often let you listen to foreign propaganda, some allowed you to hear classical music most of the time, and so forth.

Always, by setting the band and pointer at random, you could create arbitrary perceptual functions and CEVs that had no relationship to the structure of Real Reality. Most of the time the resulting CEV would produce crackle and noise, but sometimes it might produce a distorted version of what was being transmitted by a station you never knew to exist. When that happened, you might want to hear this new station better, to see whether what it transmitted 'helped your survival' (gave you pleasure or was interesting to you). If it did, you might mark the setting on your dial, stabilising this new perceptual function. If it didn't you wouldn't bother and you would probably never hear that station again once you changed the settings away from those that produced a CEV that matched that RREV.

In this parable, the band and the pointer setting are two low-level variables you could change. You could freely set them independently, but only if the pair matched a pair of values used by a transmitter would you hear anything that might help your survival. It would be no good setting the dial pointer correctly for a station you wanted to hear if you had set a different band, nor to set the band correctly if the pointer was not in the correct place for the station you wanted. The RREV was a variable at a higher level of a Real Reality hierarchy, and to hear what the station transmitted, you had to define a higher-level CEV that had the same pair of values for both lower-level variables.

If you like to look at it that way, the Perceptual Function defined a CEV = F(band, pointer) that had a value near zero for most value pairs, but non-zero for occasional ones (near zero for distant low-power stations, far from zero for nearby high-power ones). The value didn't tell you anything about what was being transmitted. It just indicated how clearly what was being transmitted through that RREV could be heard. It was up to you, the listener, to determine whether more reorganisation (re-setting the tuner parameters) was required.[125]

In this parable, clarity takes the place of Quality of Control, a radio station transmitter takes the place of your own output, and the person of the radio listener takes the place of the reorganising system that acts to rearrange the hierarchy so that the CEVs of the various Perceptual Functions approach matches to RREVs (Real Reality Environmental Variables) that are influenced by particular kinds of output. Outputting something that affects odour perception is not much help in adjusting a visual property (wrong radio band).

The parable assumes that the radio listener knows nothing of electromagnetic waves and their spectra, and depends entirely on the markings on the dial to allow the listener to recover the signal from a previously detected radio station. In this, the dial serves much the same purpose as changing the direction of gaze to look at something previously seen, using markings on a compass dial. The listener controls the perception of the radio station by changing the very small piece of the electromagnetic environment being observed at any moment. To pursue our investigation of the relation between control and Real Reality we need something else, which we control directly by the effects our actions have on our perceptions. We don't need to know *what* it is, but we do need to presume *that* it is.

Based on this presumption of non-solipsism, in the first development from Figure I.12.1a we hypothesise that there exists in real reality some structure we call a Real Reality Environmental Variable (RREV). The word 'structure', like all words, relates to something we can perceive because that is all we can know.

---

125 As a personal note to this parable, in the late 1950s I shared a house in Baltimore, Maryland, USA, with three other graduate students. We rigged the partially finished basement with cheap loudspeakers of many different shapes and sizes to play music from sources in our various bedrooms. The big bass speaker was acquired as part of an old TV set with a four inch screen, equipped with Channel 1 (which had been obsolete as a TV Channel since 1948, according to Wikipedia "Channel 1 (North American TV)" retrieved 2019.04.21). The tuner did work for Channel 1, and using the technique described in the parable we were able to find a few very distant radio stations, among which I remember Radio Moscow and Radio Karachi.

This can be no more than an analogy to whatever is in Real Reality, but since the effects of our actions on the RREV include effects on our perceptions in some more or less consistent way, we can use the relationships among those effects to say something about what the RREV does. For example, if we act on what we perceive to be X, and we perceive a corresponding change in our perception of X, we can say that the RREV provides a link between action and perception through some structure that includes X.

If now we observe that when our perception of X changes for any reason, Y usually changes in some consistent way, we can say that the RREV provides a link from X through Y to our sensors. X and Y form a structure within the RREV, and consciously we may perceive a structure that would implement these linkages in a perceived external environment.

Figure I.12.1b illustrates one possible substitution for the question marks in Figure I.12.1a, but goes a stage further in the form of a dashed arrow between the RREV and the perceptual function. This arrow represents evolution or reorganisation, which over time improves the match between the structure of the RREV and the structure of the Perceptual Function. The better the match, the more closely the CEV implements the pattern of relationships created among our sensations and perceptions by the linkages among the elements of the RREV.

Of course, we know nothing of what these elements are. We know only a little something about the relationships among our sensations and our actions that are created by their interactions. That 'knowing' is implemented in the control hierarchy in the form of the perceptual functions.

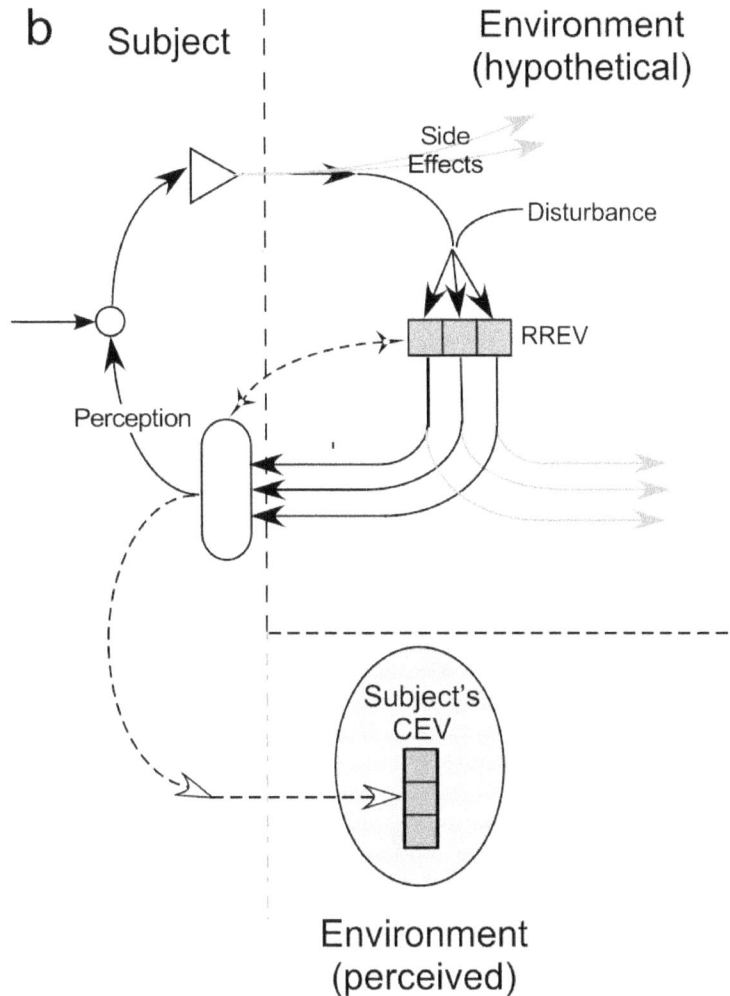

Figure I.12.1b We assume that there exists in Real Reality some structure of interacting elements (the RREV), and that these elements provide inputs to the perceptual function that generates the perception being controlled. This structure in Real Reality is unknown, but when we influence one or more of its elements directly, others of its elements are also influenced. Because we observe that the elements cohere in some way to form the perception of the RREV, we also see the CEV as being composed of interacting elements. We also observe that the perception of the structure changes when we do nothing. We hypothesise that the hypothetical RREV is being disturbed by other influences.

None of this would be of any interest if the Subject in Figure I.12.1a and I.12.1b were the only living organism in the Universe. What gives it interest is that all other organisms live in the same Real Reality. One organism that we can call an Observer could have sensors that are influenced by changes in the same RREV (Figure I.12.1c), and if the Observer has a perceptual function based on the relationships among the effects on the sensors of changes in the RREV, that perception enters the Observer's consciousness as a CEV.

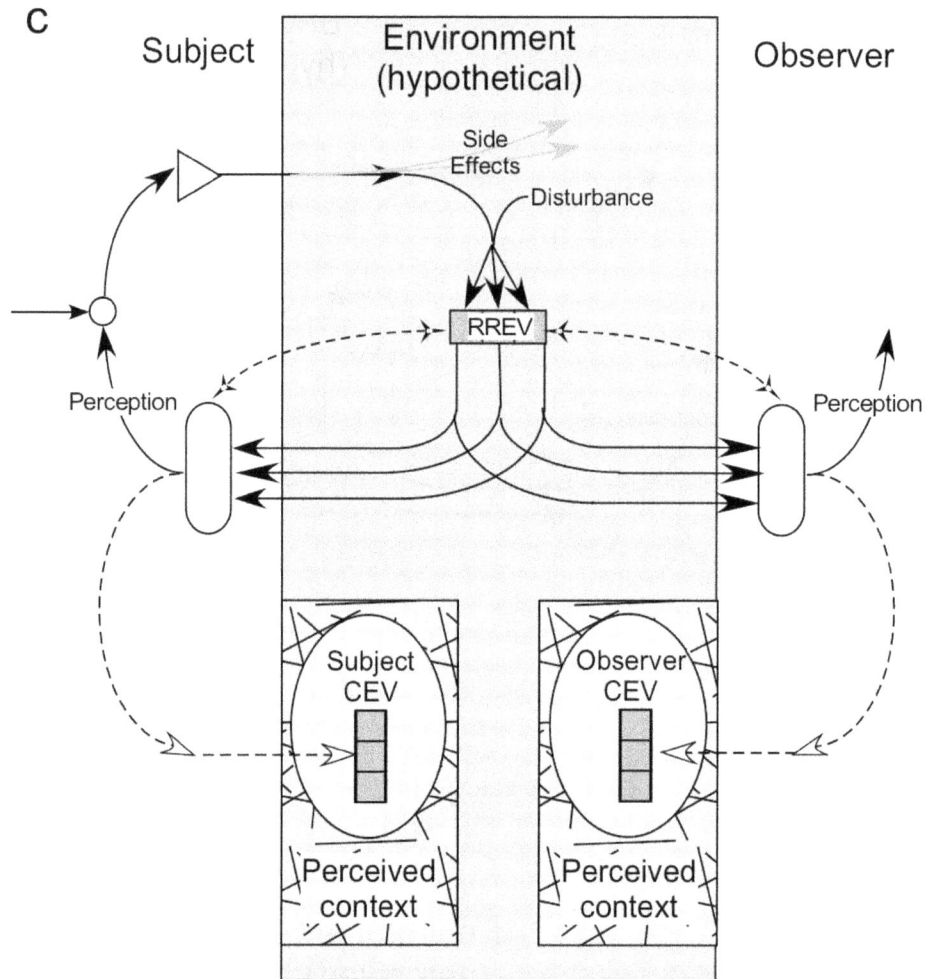

Figure I.12.1c An Observer exists in the same Real Reality as the Subject, and might observe many of the same elements of the RREV as does the Subject. The Observer creates by reorganization a perceptual function that takes advantage of much the same pattern of relationships among the elements of the RREV structure as does the Subject, but the two perceptual functions are unlikely to be identical. Through continued reorganization they both tend toward a match with the form of the RREV, but do so by different routes. Their CEVs will be related, but not identical.

Even if the Observer has exactly the same sensors as the Subject, and can observe the same changes in the sensor inter-relationships, there is no guarantee that the Observer's reorganisation processes will have created the same perceptual functions as those of the Subject. In fact, since all that can be said is that both reorganisation processes will be tending toward the same optimum match, it would be unlikely beyond reason that the perceptual functions would be identical. What the observer sees, and consciously perceives as a CEV, will probably be closely related to the CEV consciously perceived by the Subject, because both implement mimics of the relationships produced by the RREV, but they will be different.

The longer the evolution/reorganisation processes have had to work, the better will be the agreement between the Subject's and the Observer's CEVs. Within species that have much the same sensors (say, most land-living mammals), low-level perceptual functions such as those for muscle tensions, for gravitational effects, or for sensory edges between smooth periods (time) or regions (sound spectra, skin touch, visual areas) are likely to be very much the same. These perceptual structures can evolve over millions of years and are probably much alike in bats and baboons (and humans), whereas perceptual functions for animal tracks or for safety in street crossing will probably be very different (if they exist at all) in human desert dwellers, forest dwellers, and city dwellers.

We have two more developmental stages to consider in the Figure I.12.1 series. In the next (Figure I.12.1d) we recognise that so far we have explicitly limited the RREV structure to the relationships used by the Subject in creating the perceptual function, and assumed that the Observer will use those relationships and only those in creating a perceptual function. Since we presume no knowledge of Real Reality beyond the relationships that reorganization imposes on our sensory apparatus, we are not justified in assuming that the Real Reality structure is limited by what we can see of it. Figure I.12.1d shows an indefinitely extended structure, of which the Subject and the Observer might use overlapping but different sets of elements in building their perceptual functions.

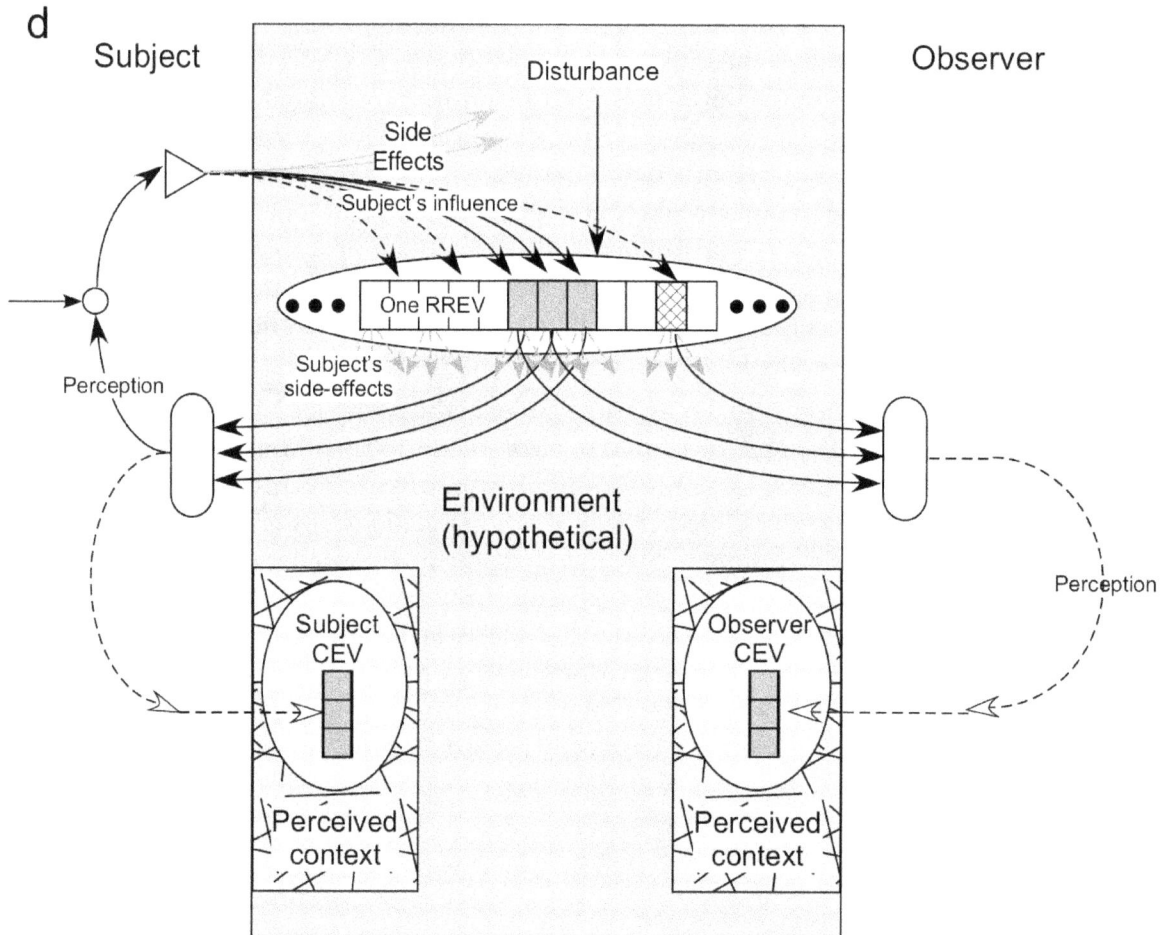

Figure I.12.1d The Real Reality structure may extend indefinitely beyond the part of it perceived by the Subject as the RREV. The Observer may create a perceptual function that omits some elements used by the Subject, and that includes others not used by the Subject. The subject's actions may directly influence elements of the structure not included in the inputs to the perceptual function. The interactions among the unperceived elements are just as much a part of the structure as are those among the perceived elements. All the influences also potentially affect aspects of Real Reality that are not parts of the coherent structure. We call those distributed influences "side-effects" of control by the Subject.

Figure I.12.1d is the same as Figure I.12.1c except that it acknowledges that the structure that produces the sensory input relationships used by the Subject to create the perception may well extend beyond the part we called the RREV. Since we do not know whether it does not, we assume the more general possibility that, as Hamlet said: "*There are more things in heaven and Earth, Horatio, / Than are dreamt of in your philosophy.*"

We may use 'RREV' also to label this extended structure, trusting that context will make clear the difference between the part and the whole. To create a perceptual function, the Observer's reorganisation may use many but not all of the elements that are used as inputs to the Subject's perceptual function and may omit others. If the common elements between the two perceptual input functions carry most of the influence between the Subject's actions and perception, the missing or added elements may add little to the difference that inevitably exists between the Subject's and the Observer's perceptions of 'the same' thing.

The final extension of Figure I.12.1 is that we allow the Observer to perform the 'Test for the Controlled Variable' (TCV). The Observer enters into conflict with the Subject by acting as a disturbance to the RREV that produces the Observer's perception of what the Subject controls, and reorganising so that the perceived effects are as though the Observer were doing the controlling. The more closely the Observer's perceptual function uses the same elements in the same way as the Subject's perceptual function, the better the Observer's apparent control of that perception (Figure I.12.1e), or the more rigid the conflict.

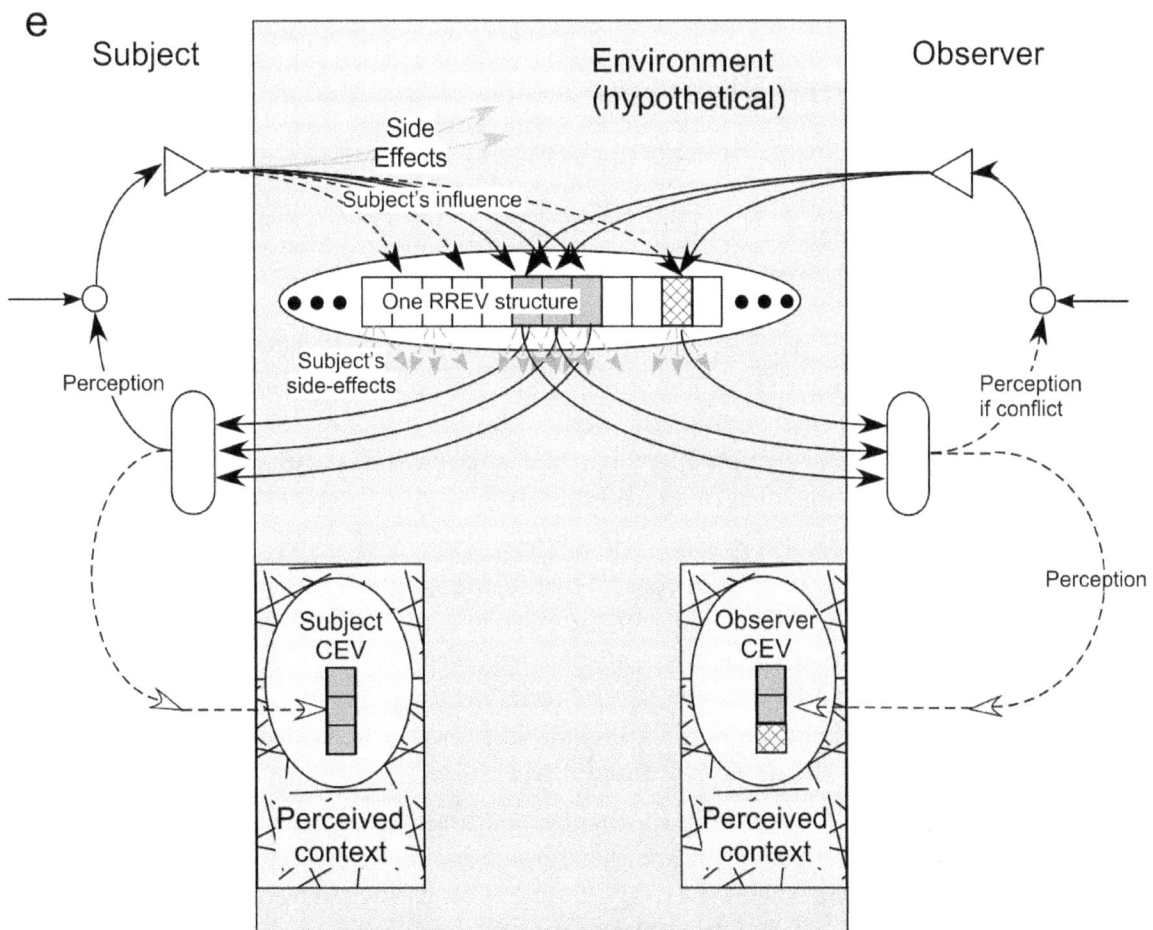

Figure I.12.1e The observer may act on the RREV to influence its own perception, either of the force applied to mimic a disturbance, as in the TCV, in which case the "perception of conflict" is not used, or of the observer's perceived state of the RREV in which case the observer's control loop is completed through that connection.

The important thing to remember is not this sequence of diagrams, which were drawn only to make the last one easier to understand, but that all of what exists in Real Reality is unknown and unknowable, while the relationships among its influences on our sensors or on any tools we might develop are discoverable (though not necessarily known). Likewise, something exists in Real Reality that processes our influences on it and produces related effects on our sensors. All else is the product of our internal processes acting on what our sensors have sensed over our individual lifetime and the lifetimes of all our ancestors. In this respect every individual is different, so one might expect everyone to perceive the same reality differently.

With that, we are able to discuss illusions, which consist of perceived relationships that are inconsistent with other evidence of relationships among our perceptions.

I.12.2 Visual Illusions

The parable of the radio dial quoted in the last section exposes the possibility of illusion, the construction of a CEV that differs appreciably from the RREV to which it seems related. For example, the Ames Room illusion depends on the viewer perceiving that a room viewed from a predetermined viewpoint is a rectangular room, whereas it has been constructed to be far from rectangular. In the radio tuner parable, it is quite possible for random variations in the waves filtered by a tuner set to a random band and pointer position to produce what sounds like music or voices. If this condition persisted for more than a short time, the listener might perceive that this particular higher-level Perceptual Function was matched to a radio station.

In Chapter I.10, we considered how extremely unlikely such conditions would be in the absence of control by outside agencies. A deceitful person might want to persuade the listener of the existence of a radio station in Real Reality, and could do so by controlling the waveform of the electromagnetic waves directly (given adequate tools). The tuner set to the frequency corresponding to a chosen band and pointer setting would provide the listener with a perception as coherent as the deceiver wanted it to be.

The same kind of thing can happen naturally under special conditions. The perception of a water surface depends in part on its reflectivity, especially at low observation angles. Low-angle refraction caused by temperature variation near hot surfaces can have the same effect, and the viewer from a distance may perceive a water surface although no water is perceived when that place is approached — a mirage. One can never find the base of a rainbow, though it looks like a material coloured arc in the air.

Illusions modify perceptions, or rather, influence the Analyst's perception of the 'truth' of a perception as a representation of the outer environment — the 'Real World'. They demonstrate that perception can never assuredly correspond to what may exist or change 'out there'. All the same, the effectiveness of an illusion as a misrepresentation of some property of the environment must depend on its use of sensory data in ways that usually do not misrepresent.

The main reason for introducing illusions here is that the processes that appear to be involved in simple visual illusions and after-effects are likely to be similar to processes operating at higher levels of perceptual control. We consider two illusions that will become metaphors or analogies for our consideration of changes in language and culture. Both illustrate the importance of 'anchoring' in perception. We then use this same construct in the analysis of figural after-effects, and later, of cultural change.

The context changes what one perceives from a given display. Figure I.12.2a shows a variant of one of the oldest and most famous illusions, the Müller-Lyer illusion, in which the angled 'arrowheads' or 'feathers' alter the perceived distances between their points. It is a variant because in the usual form of the illusion, the angles are either separated by blank space or are connected by a solid line, whereas here they are separated by a row of dots. In the standard illusion, the distance between the tips seems longer when the angles point inward than when they point outward, and this is still true in the variation with the dots. The interesting observation in the variant is what happens to the perception of the distances between corresponding pairs of dots in the two figures.

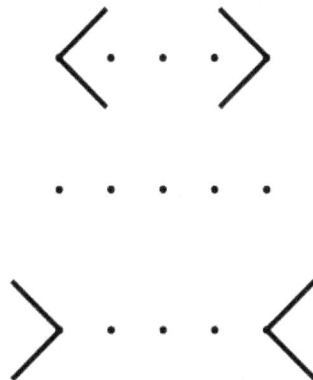

Figure I.12.2a The subdivided Mueller-Lyer illusion. The three groups of five dots (including the arrow-tips) are physically identical and have equal physical distances between consecutive dots in each of the three patterns (and the middle group is centred between the end groups). But most people do not perceive the corresponding inter-dot distances to be equal. Comparing the top and bottom groups, the end-to-end distance is usually perceived as shorter in the upper (arrowhead) figure than in the lower (feathers) figure. But the perceived inter dot distances do not follow the same pattern, the distances between dots 2-3 and 3-4 being usually perceived as longer in the arrowhead figure (dots 1 and 5 are at the points of the angles).

The variation with the dots was described to me by W. P. Tanner in a conversation in the early 1960s as something he had observed but never reported. With his permission I reproduced and reported it formally in Taylor (1965), using a six-dot row with the outermost dots at the angle points, similar to the five-dot rows of Figure I.12.2a. Subjects adjusted a visible line until it was perceived as the same length as the selected inter-dot distance, say between dots 1 and 2 or between the left arrowhead and dot 3. Figure I.12.2b shows the proportion of times that the inter-dot distance appeared longer in the (upper) arrowhead figure than in the (lower) 'feathers' figure, as a function of how many other dots intervened between the test pair of dots.

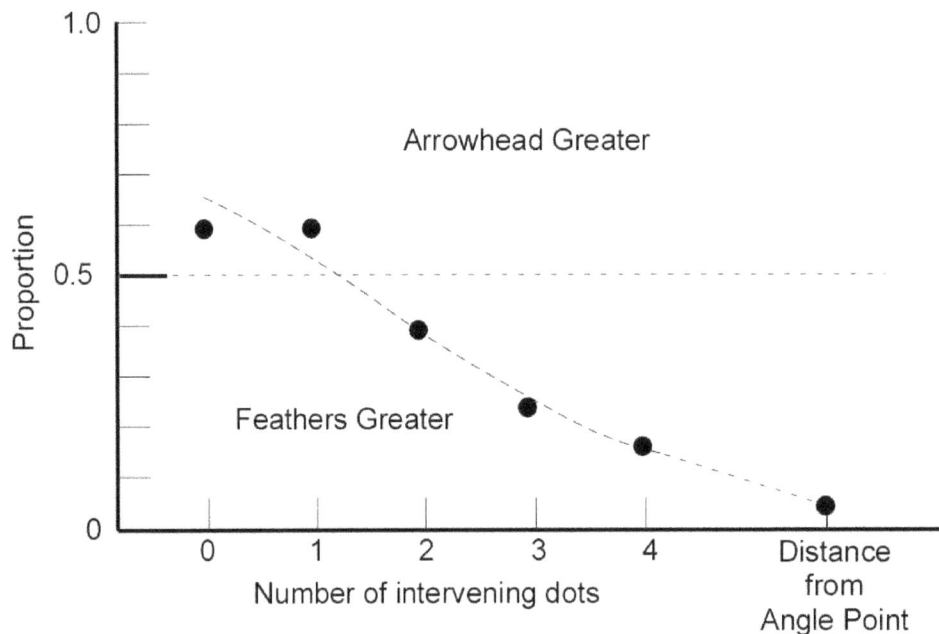

Figure I.12.2b Proportion of times the comparison line was set longer for the dot pair in the arrowhead figure than for the corresponding dot pair in the feathers figure.

Tanner's illusion demonstrates that the perception of distance is not additive, and that no consistent function can exist that allows a long perceived distance to be equated to the sum of its perceived partial distances. The separation between neighbouring or next-but-one dots is likely to be reported as longer in the arrowhead figure, whereas the distance between widely separated dots is consistently seen as longer in the feathers figure. The sum of a set of longer distances should be longer than the sum of a matched set of shorter distances, no matter whether perceived distance is a linear, logarithmic, or any other monotonic function of physical separation, but the opposite is true in this pattern. By itself, the illusion demonstrates that perception need not be true of the environment, since in the environment the sum of longer intervals along a line must be longer than the sum of matched shorter distances.

Beginning with Weber (1834), many psychologists have thought that perceived distance can be equated with the discriminability between the ends of the distance. The easier the ends can be discriminated, the greater the distance. Most have equated discriminability with the summation of just-noticeable-differences (JNDs) along the path between the end points. Tanner's Müller-Lyer effect shows that this cannot be true in any simple sense, but the central point was noted as early as the 1870s in a book of *Popular Scientific Recreations* (Anon, ca. 1882) in which it is explicitly stated that summation of JNDs is not the same as ease of discrimination.[126] This causes a significant problem with determining the discriminability of a distance without simultaneously determining its perceived length.

The problem of whether perceived distance maps onto perceived discriminability can be addressed in a more subtle way, by finding a condition that would be affected in a precisely determined way if the mapping were true. Taylor (1962b) predicted the existence and magnitude of an illusion that had apparently not previously been described in the literature, using the assumption of the 'discriminability-to-distance' mapping, the space-expanding property of anchors in the filled space illusion, and the properties of partial differential equations.

The analysis also produced the counterintuitive prediction that the subjects who were most precise in their perception would also be the ones who experienced the largest illusion. The magnitude of the illusion for each subject shown in Figure I.12.2c was found by fitting a proportionality line to their individual data, and reading off the value for a one-inch disk, not by using their one-inch result by itself.

---

126 This book was a school prize won by my grandfather. I had long searched libraries for the French original, by 'the aeronaut Gaston Tissandier' (who organised balloon spotters for the French artillery in the 1870 Franco-Prussian War), and serendipitously found it in the University of Michigan library while standing in the stacks waiting for a colleague! The English 'translation' is some three times the size of the French version, which does not contain this statement. The English version, judging from its other content, was written by one or more first-class scientists, but when I enquired of the publisher for the information, I was told that all their records had been destroyed in the London Blitz of 1941.

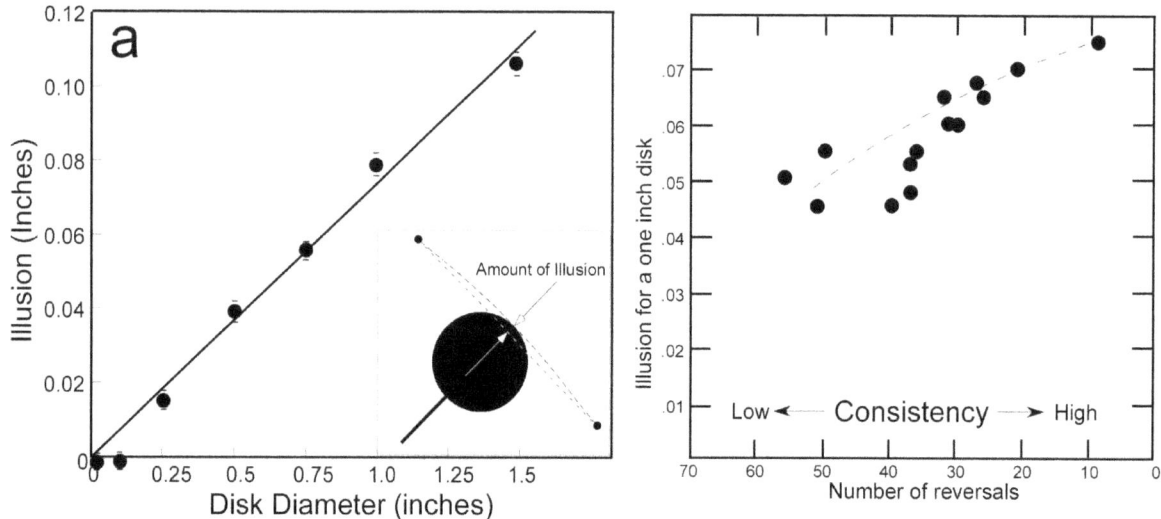

Figure I.12.2c The circle-tangent illusion. (a. left) Amount of illusion as a function of disk diameter; the line is the a priori prediction that the amount of illusion would be proportional to the disk diameter. (Inset: schematic of the display judged by the subjects; the dots were proportionately smaller than are shown here. The minimum diameter of the disk was just the width of the "handle" line, with no identifiable disk.) (b, right) The size of the illusion for a one-inch disk for the individual subjects, showing that, as predicted in advance, the most precise subjects experienced the largest illusion. (Figures based on Taylor, 1962)

A disk was placed between two dots, so that the straight path between the dots was nearly tangent to the disk. It was predicted that, for the perceived straight path between the dots to be tangent to the disk, the disk must overlap the physically straight path between the dots by an amount proportional to its radius. The predicted relation between the magnitude of the illusion and the disk diameter depends only on the following:

1. A perceived straight line is a geodesic[127] in a non-Euclidean perceptual

---

127 A geodesic is the shortest path between two points on a surface or in a space. If the surface or space is flat (often called 'Euclidean'), the geodesic is an ordinary straight line, and if one is inside the space it looks like a straight line however curved the space might be. On the Earth's surface, a geodesic is any part of what is called a 'Great Circle', of which the equator is an example. Airline routes tend to follow geodesics around the Earth. Einstein predicted that the space in which we live is locally stretched by mass, and that this changes the apparent direction of stars seen very near the disk of the sun. This prediction was proved correct at a solar eclipse, and the effect has more recently allowed the gravity of distant galaxies and galaxy clusters to be used as lenses to probe the Universe more deeply than our best telescopes could achieve unaided. The illusion of Figure I.12.2c is the perceptual equivalent. In perception, the argument is that anchoring points have a similar effect of stretching perceived space, causing ruler-straight lines to appear curved and perceptually straight lines to turn out to be curved when checked with a ruler.

space.[128]

2. Perceived length of a path is a continuous monotonic function of the discriminability along the arc between the endpoints. The minimum occurs when the path is the geodesic, the perceived straight line between the points.

3. The 'filled space illusion' applies to arcs that cross visual edges and pass near anchor points in the visual space. (The filled space illusion refers to the apparent increase in length of an interval if identifiable marker points exist inside the interval).

The twin results shown in Figure I.12.2c provide strong evidence of (2), that the perceived separation between two points is a monotonic function of the discriminability of the interval between them. The right panel illustrates the point because the subjects with finer discrimination showed the most illusion, contrary to naïve intuition.

The reason this is relevant to our enquiry into culture from a perceptual control viewpoint is that we will extrapolate this finding, which for almost two centuries has been generally assumed to be true, to other dimensions of perception. The more distinct two things are perceived to be, the more different they seem, and their distinctness when seen separately is greater if there is an anchor point in their neighbourhood than if they are seen in an otherwise empty field.

---

128 The paradoxical result of Tanner's Müller-Lyer variant might possibly be explained as a consequence of differential curvature of the perceptual space inside and outside an arrowhead.

I.12.3 Perceptual Experience

Perceptions depend on experience — in particular on experience with controlling perceptions through the external environment. One can, and necessarily does, have perceptions of aspects of the external environment that have not been tested by control, and these perceptions may differ from those in another person who has actually tested the perception of the same environmental variable.

A dramatic example was offered by Turnbull (1961) who worked with BaMbuti pygmies in the African jungle, which at the time was neither criss-crossed by logging roads nor infested by informal guerrilla armies. The inhabitants, according to Turnbull, seldom had sightlines longer than perhaps ten metres. Turnbull took one of them to a high place from which could be seen a lake on which people were fishing from canoes. The pygmy exclaimed in surprise that he could not understand how those tiny twigs could hold up real people. Apparently he did not have size constancy as such, but perceived humans as being normal size and canoes as tiny twigs that he could hold in his hand several at a time. He had much experience with controlling perceptions related to the size of people, but not with boats and other objects such as large bodies of water seen from a distance.

The relationship between perception and the external environment also changes with immediate prior experience. If you look at a white wall after having stared at a blue patch for a short while, a corresponding patch on the wall will appear yellowish or orangey. After watching something moving continuously in one direction relative to its surroundings, stationary objects in the same area of the visual field will seem to move in the opposite direction (the 'Waterfall Illusion', Addams 1834). Similar after-effects can be observed in various configurations in different sensory modalities, such as the visual tilt of a line, the temperature of a liquid, the location of a sound, and the size of an object you feel but do not see.

All the illusions show the effect of context on the way perceptions depend on what else is sensed in the local environment, or has been sensed in the recent past, and even over a lifetime — though lifetime effects are more probably due to reorganisation of the perceptual functions themselves. One class of after-effects illustrates the effect of context quite dramatically, as it combines the 'Waterfall' class with the 'colour patch' class of after-effect. The 'McCollough Effect' (McCollough 1963, McCollough-Howard and Webster 2011) is observed under a wide range of conditions (citations in McCollough-Howard and Webster, 2011). An on-line demonstration shows two patterns in slow alternation, one having red and black horizontal bars, the other having green and black vertical bars.[129] After one watches this alternation for a while, one is shown a field of black and white bars which are horizontal in some regions, vertical in others. In that field, one does not see white, but pale green (where the bars are vertical) or pale red (where they are horizontal). The effect can last for hours or even days.

---

129 See https://www.youtube.com/watch?v=nfwJxnijBno (Retrieved 2/10/2024).

The two perceptual dimensions of tilt and colour combined into a single complex that showed the after-effect, when neither tilt alone nor colour alone would have shown an after-effect because the individual contexts averaged out as neutral tilt and neutral colour.

A demonstration by Viviani and Stucchi (1992) showed a similar contextual effect in a quite different perceptual realm. When an object follows a curved path at constant velocity it seems to speed up by an amount that depends on the curvature of the path. As with the 'McCollough effect', this may be an example of the 'adaptation and contrast' after-effect at a different perceptual level involving more complex contextual variables.

Under many conditions a variety of experiments have shown that when an organism moves along a curved path, such as a finger tracing an ellipse or a fly larva seeking the source of a food odour, it slows down on curves. Typically the speed around the curve is proportional to a power of the radius of curvature between 1/3 and 1/4. That is what is measured to happen when someone or something moves purposefully along a curved path, but not necessarily when they are slowly wandering.

Over the years, one has observed many instances of this slowdown without necessarily being conscious of it, so it has presumably become a norm in the same way that a long observed shallow arc seems to become more straight, and truly straight lines seem bent in the opposite direction. Just as after exposure to the tilted line or the shallow arc, people saw a vertical line or a straight line as deviant in the opposite direction, so Viviani and Stucchi found that in order for the speed around a curve to be perceived as constant, the object should slow down. The speed that was perceived as constant was a power function of the radius of curvature, not between 1/4 and 1/3, but somewhere below 1/5, often near 1/6.

In other words, the perceived 'truth' of the environment was changed in the same way as in the static situation, not sufficiently to make the usually observed deviation from true constancy seem to be constant, but enough to bias the objectively (i.e measured) constant seem deviant in the opposite direction. These 'truths' are directly perceived. The relation between perception and environment may change, but that relation is based on the effects of the preceding and surrounding environmental context on the observer's sensory systems.

I experienced a higher-level form of the McCollough contextual illusion when typing my thesis in 1959, using a mechanical typewriter. I typed on yellow paper for several hours a day, and when after a week of this I put a sheet of white paper into the typewriter, it appeared such a strong sky-blue colour that I thought I had used the wrong stack of paper and went to put the sheet back on my stack of blue paper. But as soon as it was out of the typewriter, the paper appeared white. Putting it back in the machine turned it blue again. This effect lasted over a week after I had finished using the yellow paper. Most proposed explanations of the McCollough effect refer to retinal or early cortical processes, but the 'typewriter-colour' effect must occur at a much higher perceptual level, as must the Viviani and Stucchi curvature-speed effect.

Some of these effects, in particular the visual after-effect of staring at a coloured patch, can be readily explained as adaptation of the sensors to a new 'neutral'. Others are less easily explained as sensor adaptation. The McCollough effect and especially the typewriter effect cannot be explained that way, unless the term 'sensor' is expanded to mean the same as 'Perceptual Function' in the Powers hierarchy. They are 'contextually contingent' effects, which seem to be similar in general structure to the Tanner variant of the Müller-Lyer illusion, in which the arrowheads or feathers make the entire gap seem shorter or longer, against which the shorter inter-dot distances may be subject to a contrast effect. We will argue that similar effects are likely to occur at all perceptual levels, and to have appreciable consequences for social and political perceptions and their control.

After-effects in one dimension can be explained as being caused by changes in discrimination and hence perceived distances within the dimension in question, consequent on prior experience (Taylor 1962a). The underlying theoretical construct is that the perceived distance between two percepts on a given dimension is proportional to their discriminability, which is affected by the presence, at that moment or nearby in time, of another percept that might lie between them in that dimension.[130] (For several examples in different perceptual dimensions, see Figure I.12.2a and the references in Taylor 1962a.)

This explanation yields a 2-parameter equation for the 'distance paradox of the figural after-effect' in which displacement first increases and then decreases as a function of the experimentally manipulated environmental difference between the inspection and test presentations (Figure I.12.3a right panel shows an example). The actual equation is

$$E = \frac{hM/R}{1 + (kM/d)^3}$$

where E is the expected displacement, M is the separation in units of d, which is a measure of the perceptual precision for the variable in question, and R is the ratio between precision at the inspection and test points, usually but not always taken to be 1.0. The two parameters are h and k, which in most cases tested have been found to be 0.2 and 0.03 respectively. The discriminability measures d and R are taken from independent experiments.

By fitting the two-parameter equation to the so-called 'distance paradox' of the figural after effect in several different modalities (acoustic location, visual tilt of a straight line or a grid, the haptically felt width of a block, and the displacement of visual dots), it was found that the two parameters were the same

---

130 Notice that whereas Figure I.12.2a shows two close entities being merged into one, this is unlikely to happen if there is a time separation between them. If one disappears and the other appears after a gap, the shift of location can be seen much more readily than could the distinction between the two if they were presented together. Hence, the repulsion effect is expected in the after-effect even for small spatial separations.

for all the fits except the haptic case.[131] Taylor (1963, 1966) later used those same parameter values to fit two studies in which different variables were used, namely the effect of visual contrast in visual displacement (left panel of Figure I.12.3a) and visual curvature (right panel of Figure I.12.3a). To emphasise the point, all the parameter values entered into the equation were derived from quite different studies unrelated to figural after-effects, none from the experiments whose data were fitted by the theoretical equation.

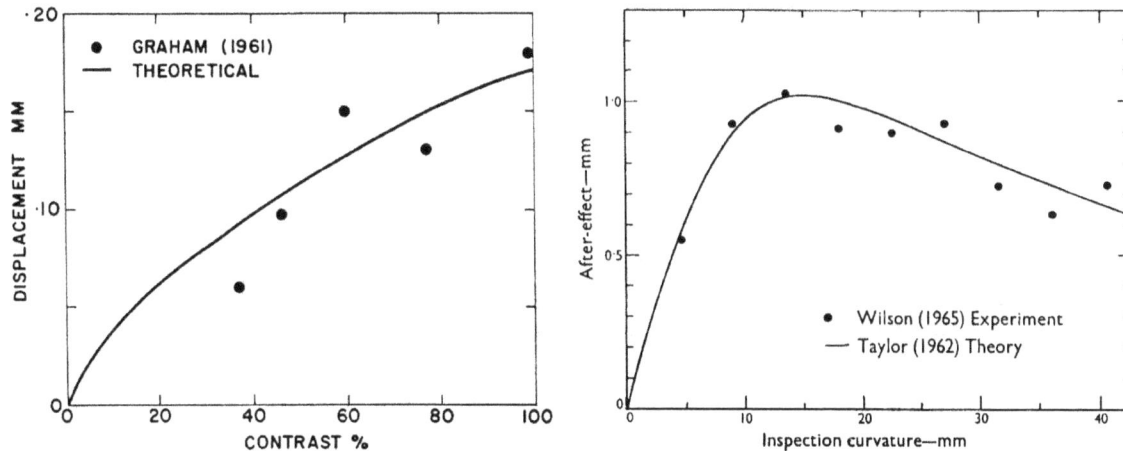

Figure I.12.3a "Zero-parameter" fits of figural after-effect theory to data that was not used in computing the theoretical functions. (Left) Displacement of a small rectangle by another as a function of visual contrast (Figure I. from Taylor. M.M., 1963). (Right) Curvature of an arc as a function of curvature of a previously inspected arc (From Taylor 1966).

The discussion of lateral inhibition before, and in particular the mutual excitation and inhibition pattern in the field of polyflops, suggests a mechanism whereby this might happen. Furthermore, the 'labelling' property of polyflop sets allows a 'truth' to be seen as the effect of a perception of one kind ('A') on the perception of another kind ("Is that 'eh' or 'aitch'?"). A 'truth' is often what something is 'seen as'; "Is that dark patch a shadow or an object?" As Taylor (1962a) said: "... the argument is intended to apply to any perceptual dimension at any level of coding." We will later apply it to social influences on people's perceptions.

Another effect that may have some bearing on the creation of perceptions is the 'reversing (or ambiguous) figure'. After looking steadily at such a figure for a while, what it represents seems to change, though the configuration of light and dark does not. Three classical examples are shown in Figure I.12.3b. The fourth is a photograph of an actual physical surface of grey plasticine dented by a table-tennis ball, which was viewed by the subjects of Taylor and Aldridge (1974). You may see it as bubbly or as dented, or you may see an alternation between the two states.

---

131 The problem with the haptic case is likely to be that appropriate measures for d and R had to be estimated from experiments that used presentation conditions considerably different from the after-effect study, in particular between two-hand comparisons and one-hand comparisons.

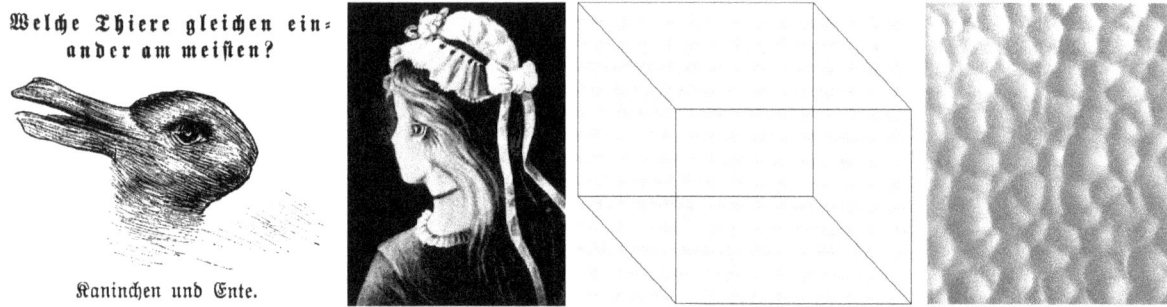

Figure I.12.3b Some ambiguous images that can be seen as different things with no change in their configurations of light and dark. (The rabbit-duck and old-young woman are in the public domain, the other two are by the author.)

Taylor and Aldridge were following up a study by Taylor and Henning (1963) which showed that, for several different kinds of static and moving visual, and verbal and non-verbal auditory presentation, subjects were likely to perceive a considerable variety of different forms. The Necker Cube (the third image in Figure I.12.3b) was seen in as many as 22 forms described by the subjects as 'different', not limited to 0the two forms that are typically said to be the only ways this figure can be seen.

The one consistency in the Taylor and Henning studies was that the cumulative number of transitions reported by a subject who had reported n different forms was very closely proportional to $n(n-1)$, as though the perception seen at any given moment roamed randomly among the forms already seen, except when a novel form appeared. This consistency over such a wide variety of display types suggested that something other than simple fatigue was dictating the perceptual changes.

To study this question, Taylor and Aldridge attempted to find a simple display (unlike the visually complex left two images in Figure I.12.3b) that was consistently perceived in only three forms, since if only three forms were available to be perceived and fatigue caused a switch to occur, the perceived forms should be more likely to cycle 'A-B-C-A-B-C-' rather than appear in random order. In this, we were unsuccessful, but we were able to use a physical surface, of which a photograph is shown as the fourth image of Figure I.12.3b. This dented grey plasticine was consistently reported as being either a dented surface (the physical 'truth') or a bubbly surface like a foam. During prolonged viewing, all subjects reported seeing the display change from bubbles to dents and back again many times, though the time to the first switch for some subjects could be many minutes. The moments of change were recorded and used in further analysis.

Although in some of the ambiguous figure illusions the viewer can choose which version to see, neither the experimenters nor any of our naïve or well-practised subjects were able to affect the bubble-dent changes intentionally, even after much effort to do so. We therefore treated the timing results as representing the behaviour of some uncontrolled bottom-up process in the perceiving system.

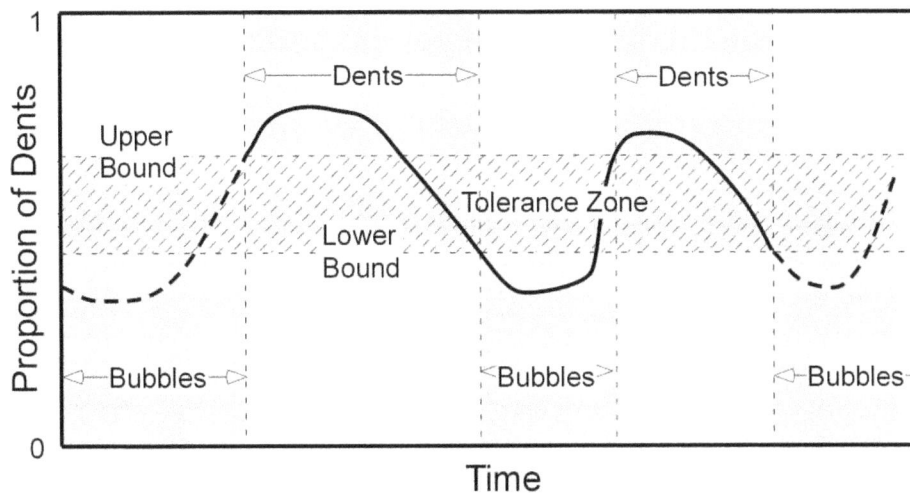

Figure I.12.3c Illustrating the effect of hysteresis. The "master" demon reports "bubbles" or "dents" according to whichever percept has the greatest support from the lower demons, but the number needs to cross a tolerance zone before it will switch from whatever it had previously been reporting.

A Pandemonium-like model (Selfridge 1959) fitted the results well. In this model a layer of low-level 'demons' individually decided whether to report 'bubbles' or 'dents' and a hysterical 'master' demon that determined the perception reported by the subject according to a vote that required a measurable excess to switch, as suggested in Figure I.12.3c and Figure I.8.5d.[132] This is the behaviour that would be expected of a flip-flop process, in which some energy is expended in performing the hysteresis loop of Figure I.8.5d.

The surprising thing about fitting the data to this model was that the number of low-level demons and the hysteresis bounds were precisely determined numerically for each of two subjects who served four nine-minute sessions on five consecutive days. For most of the trials, one subject appeared to use 33 low-level demons, and for one session used 34. The other subject used 29 low-level demons in every session. The placements of the tolerance zone bounds illustrated in Figure I.12.3c were less constant, but when they changed, the change was almost always exactly one unit, as shown in Table I.12.3.

---

132 'Hysterical' is not a comment on the demon's personality, but indicates that its behaviour shows hysteresis. The analysis suggests that hysteresis is related to perceptual tolerance over a random walk among 29 to 34 nerve fibres in a 'neural bundle' in the subjects tested.

Table I.12.3 For two subjects who did 20 9-minute sessions over 5 days, the fitted number of low-level demons and the placement of the upper and lower bounds in each session, showing the consistency of the numbers.

Session	Subject B			Subject E		
	N	Lower	Upper	N	Lower	Upper
1.1	29	13	20	33	17	23
1.2	29	15	20	33	17	23
1.3	29	13	20	33	17	23
1.4	29	13	20	33	18	24
2.1	29	15	20	33	17	23
2.2	29	13	20	33	18	24
2.3	29	13	20	33	18	24
2.4	29	13	20	33	17	24
3.1	29	13	20	33	17	24
3.2	29	13	18	33	17	24
3.3	29	13	18	33	18	25
3.4	29	13	18	33	17	24
4.1	29	13	20	33	18	24
4.2	29	13	20	33	20	25
4.3	29	13	20	33	18	27
4.4	29	11	18	34	18	28
5.1	29	11	18	33	18	25
5.2	29	11	18	33	18	25
5.3	29	11	18	33	18	27
5.4	29	11	18	33	18	27

These one-unit changes alter the timing curves quite appreciably, as shown in Figure I.12.3d, so we can argue with some force that the numbers 33 and 29 do not mean 'around 30'. They mean exactly 33 and exactly 29. The implication is that these numbers are more than just fitting parameters, but represent some property of the perceiving system. Whatever kind of perceiving unit is involved, one subject usually employed 33 of them, while the other used 29 of them. It may be stretching the bounds of speculation to breaking point, but it is tempting to suggest that these numbers represent the actual number of fibres in a neural bundle that forms one of the 'wires' carrying a 'neural current' in Powers's models.

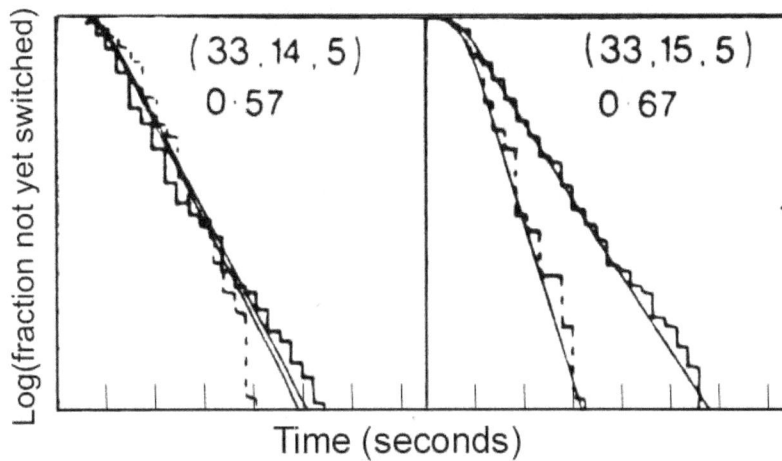

Figure I.12.3d Two representative fits to the survivorship data for subject E on consecutive sessions (3 and 4) showing the discrete nature of the fitting. The two sessions separated by a rest period differ by a shift of one unit upward of the tolerance zone. (Solid line shows switches out of "bubble" state, dashed line switches out of "dents" state). The width of each panel is 6.5 seconds. The numbers in the panel are the fitting parameters: (Number of low-level demons, Lower Bound, Zone width), and the average switching rate per low-level demon. (Extracted from Fig 12 of Taylor and Aldridge, 1974)

Today, we would probably not use a Pandemonium model, but would achieve the same effect with a two-level polyflop structure in which 29 or 33 individual low-level units independently provide input to two flip-flop-connected higher-level units. Either way, these results illustrate the probable reality and relative stability of the multi-stranded structure of the perceptual part of the control hierarchy, as well as the tenuous connection between perception and the truth of the environment in the absence of perceptual control feedback.

I.12.4 Decisions, Patterns, Habits

Whether dealing with biological or electronic systems, if there is an advantage to detecting a pattern, then a question always arises (in the Analyst) as to whether it is better to see a pattern that has actually been built from a fortuitous combination of random noise elements (a false detection) or to miss a pattern that might be vital to survival (a false rejection). To re-use an earlier example, if you are in a jungle, patterns of sunlight and shadow, which pose no threat, can sometimes create the appearance of yellow and black stripes. But what if those stripes are actually colours of the coat of a hungry tiger? Might it not be better for survival to assume it is a tiger and to take possibly unnecessary action rather than to assume it is just the play of light and be eaten by a real tiger? It might be a survival policy to try to escape if one saw even a flash of orange-yellow in the jungle.

How much like a tiger must the pattern be before evasive action becomes the best policy? Too much unnecessary action wastes energy and takes away from the time available for gathering food. Too little gets you eaten. There is a balance between risk and opportunity. A person who has usually assumed that a possible pattern represented a real danger or opportunity is more likely to have reached an age at which they are capable of passing their genes to a new generation than is one who waits for certainty or one who runs away from every possible suggestion of danger. If a tendency for 'jumping the gun' in seeing patterns, or its opposite, is heritable, then a population should be expected to contain a few extremely nervous people who see patterns based on very little evidence, and a few who resist seeing patterns even when the evidence is strong, but many who see patterns when the evidence is technically only suggestive.

The risk-opportunity balance is evident even in reading. Rausch (1981, reanalyzed by Taylor and Taylor 1983:249-250) asked normals and people with right- or left-hemisphere (RH or LH) temporal lobe brain damage to read a series of words in a list, and say whether they had seen the word before. Later in the list some words were repeated, but in addition 'foils' were included that were not the same as an earlier word but that might sound the same as, were in the same category as, or were frequently associated with, an earlier word.

People with RH damage almost never wrongly said that they had seen a word before, but missed many they actually had seen. Their results were in strong contrast to those of people with LH damage, who identified 20% to 40% of the foils as having been seen before, particularly if they were in the same category or sounded the same, while missing few of the ones they had actually seen before. Normals fell between the two kinds of brain-damaged people, accepting some foils as having been seen before, especially if they sounded the same, making a few mistakes with the other foils, and missing a few words that they actually had seen before. The people that had to rely more on their left hemisphere tended to be overcautious compared to normals, whereas the people who had to rely more on their right hemisphere tended to be more risky. One may similarly assume that in the population as a whole, there will be a range of left-right balance differences, resulting in some risk-preferring people and some safety-minded people, with most people lying between the extremes.

When two control units must use the same resource to control their different perceptions, they are in conflict. Either one of them prevails, or both perceptions remain in error as the conflict persists. But what about differing perceptions? When the same data might be interpreted by two different perceptual functions as representing incompatible states of the external environment, such as a pattern of light and shadow or a fierce tiger, what happens to control? If the actions required to control one perception do not interfere with the actions to control the other possibility, is there a conflict, or is the perception seen as some intermediate state of the world?

Figure I.9.4c and Figure I.12.8 illustrate for a simple flip-flop the nature of the problem. As these figures show, whether a flip-flop arrangement has a hard, clear, output or a soft intermediate output for a given ambiguous input depends on the gain of the lateral interaction loop and on the task or contextual stress to make a decision. At low gain, the output of the circuit may tend toward one or other pole, but at high gain it is fully one or fully the other, and is locked in so that it takes a lot of counter-evidence to change the output. This lateral loop gain determines the difference between jumping too quickly (perhaps into another tiger's mouth) and hesitating long enough for the first tiger to catch you.

There probably is no interaction between 'tiger' and 'shadow pattern', but there is a very strong interaction between 'dangerous' and 'safe'. The perception of 'danger' will become a label associated with 'tiger' in the same way that the letter-string 't-i-g-e-r' is associated with the sight of one; so also will 'safe' be associated with 'trick-of-the-light'. This implies that there is likely to be a flip-flop or a labelling polyflop interaction between the two possible (danger and safe) perceptions of the scene, so that only one is actually perceived. However, when the perception of, say, the level of safety is controlled and 'tiger' is the perception that results from the polyflop interaction, the change of scene that results from evasive action might alter the inputs to change the relative likelihoods of the two labels, so that eventually 'safe' is perceived, and no 'tiger'.

The same argument applies equally to the output side. As Powers suggested in B:CP, the Reference input function, at least at some levels of the hierarchy, is likely to take the form of an associative memory. By the argument used for the perceptual side, the associative memory may well have been created through lateral inhibition, and be subject to the same flip-flop and polyflop selection process in cases where possible output mechanisms might conflict. This mechanism avoids actual conflict, at the cost of sometimes failing to use the best available mechanism for control of a perception.

Just as in the case of the definitive resolution of an ambiguous perception, the definitive resolution of an optional output has benefits and costs. Taking too long to decide you should avoid the tiger can get you killed, and your offspring never are born. Acting too quickly and decisively might result in jumping into the mouth of the tiger you did not see.

The associative memory structure of the output side of the hierarchy has another consequence, habit. A habit is a coordinated pattern of output that is frequently used whether or not it is appropriate in some particular circumstance. One may, for example, make a familiar turn on the way home from work even though on that day one had intended to take the other direction to visit a friend. Why? An associative memory produces its output when only a part of its input 'address' is present. If indeed the profile of reference values at any moment is actually the output of a set of associative memories 'addressed' from the next higher level, it is to be expected that when much of the context is consistent with the associative memories producing a certain pattern of reference values, the polyflop structure of the memories imply that the same pattern is quite likely to be produced even in the presence of some contrary data.

The more often the 'habitual' context occurs with the same profile of reference value outputs from associative memory, the stronger the positive feedback loop, and the harder it is for contrary data to flip the output to the other state. Indeed, another separate feedback loop is involved, since perceptual controls act to restore perceptions to their reference state, and those controlled perceptions form part of the habitual context. Depending on the strength of this external feedback loop, a 'habit' may turn into an obsessive behaviour.

J. G. Taylor (1963) recognised this exterior positive feedback loop and used it in therapy for obsessive-compulsive disorder, by opposing the action component of the loop through what, in PCT, would be called conflict. This conflict is logically similar to the 'Bomb' (Section I.6.5) but instead of creating a positive feedback loop when a negative feedback path is blocked, Taylor's therapeutic procedure, when successful, blocked a positive feedback path, allowing negative feedback to stabilise the situation to an extent that the positive path could no longer dominate once the therapy was complete. This technique is by no means a precursor of MOL, but uses the same principle of seeking a blocking effect and adjusting the circuitry to avoid it.

I.12.5 Trust

'Trust' is a strange concept in PCT. If you can correct any error by varying your output, why do you need to trust anything? 'Trust' seems to suggest that you know in advance the result of an action because you are using a modelling approach to control, and adjusting output based on your prediction of what will be needed, as opposed to controlling input. This sounds like 'Predictive Coding Theory', which I called 'slow control' because of the extreme computational requirements imposed by the unpredictable nature of disturbances. But in everyday life, we trust a lot of things and act upon that trust because usually to do so has worked well. Trust is closely related to prediction in the statistical sense.

As we walk, we put a foot down without looking to see that the ground still exists where the foot will fall, because in some sense we predict that it does. That is trust. But we don't do that if we are walking on slippery seaweed-covered rocks or on a broken sidewalk. In such a place we watch where we intend to place each step. The ground is not trustworthy. In other words, the quality of the place where the foot will next fall is not predictable from our observations of nearby places. Here is another example.

> *I am walking through the field: controlling proximity to the opposite side of the field. At the same time I know or believe that there can be old mine shafts where I can fall. And I fear falling. But I don't perceive any shaft - I don't even know where they are - so I can't control the distance to a shaft. The shaft is in my imagination and I believe I recognise it if I see it in the field.*
>
> *What do I do? I walk along but at the same time I keep watching and seeking for some signs of a shaft. If I see something which can be a shaft only then I can start to control the distance to it. Before that I control in imagination.*[133]

In this example, 'I' believe that I will both see and correctly recognise a mineshaft if I am close to it. I trust that the shaft will not be covered over by a thin layer of overgrown turf through which I might fall. When I do not perceive a mineshaft in my intended path, or if I do perceive one, I believe and trust that my perception is a true mapping of that aspect of my real world environment. In both examples, we predict that if we control for not stumbling over the apparently broken sidewalk, or for not falling down the mineshaft, we will be safe from these unwelcome possibilities.

Our real-world security in these situations depends on the trust we may (or may not) have in our perceptions. All we can know of the outer world is what we perceive, but if what we perceive misrepresents the outer world, our perceptual control either will not work well or will work but not serve our intrinsic variables

---

133 Quoted from Eetu Pikkarainen (personal communication 2017.12.01).

well. Maybe what we do will have side effects that have no influence on our intrinsic variables, but this will be rare. More likely is a situation in which the actions do influence them, but the perception being controlled does or does not correspond to anything that really exists in the real world.

Suppose, for example, that being hungry we perceive a piece of lettuce to be nice and fresh and act by eating it. If it is good, we assuage our hunger, but if it is tainted with e-coli we may also get very sick. Less dramatically, imagine we want to put a glass down on a table that we perceive to exist, but the table is actually a hologram. When we let go of the glass, it drops to the floor, which does not indicate good control of the position of the glass in three dimensions. Maybe that drop has no remarkable effects on our intrinsic variables, but maybe the glass falls onto our foot and creates a bruise. If, while out for a walk, we mistake a mine shaft opening for a patch of burnt black earth, we may die from the fall. We will not go further into the area of perceptual trust at this point, since the role of feedback in creating our perceptual functions and verifying our perceptions is covered in several other parts of this book.

When we talk to a friend, or even a stranger playing a well known role, we trust that the sounds we make will be understood as our intention to make words, and that those words will be understood as being connected to the pragmatic situation known to both. Only when it is harder to trust what behavioural effect our words will have on the other person do we usually observe their individual effects carefully, and only when the results are not what we expect will we question whether the words understood were the words we intended. Until then we trust that they were and are.

We use trust in everyday life when the environment is consistent enough to allow the influence of output on perception to be predicted sufficiently accurately for our purposes. The batter who hits a monstrous home run does not start to run to first base as quickly as does one who might have hit a double. We may swing a door as we pass through, and not watch to see whether it closes properly, because it usually does. If sometimes in the past when we heard the door hit the jamb, it closed and sometimes it did not, then if we are controlling for perceiving it to be closed we will look to see whether it did close. But if in the past it always closed properly, we may well go on our way, because we maintain a World Model that produces (perhaps wrongly) the perception that it is closed.

We discuss World Models and imagination later. It is sufficient here to comment that we perceive a lot about the world that is not immediately available to our sensors. If we are married, but cannot see our spouse at the moment, he or she does not disappear from our perceptual world, but remains part of our World Model, influencing our control of other perceptions.

McClelland devotes a considerable part of his chapter in LCS IV (McClelland 2020) to the social construction of stable feedback pathways (such as the sidewalk on which we will put our feet) in which we can place trust. In his very reasonable view, the coherence of a culture depends in large part on the stability of feedback paths which depend in part on the built environment, but which also may depend on the psychologically built environment of accepted 'protocols' and facts about the world. (Protocols are described and analysed in Chapter II.14 and in my chapter in LCS IV, Taylor 2020.)

Trust may perhaps be treated as control in imagination using a world model in which only one outcome of an action is highly probable. In that world model, 'planning' degenerates into a single path by which the desired perception *will* be produced, and because of it the desired perception *has been* produced once the action is complete, no matter what the unobserved actual effect of the output on the external environment may be.

Since the output of a higher-level ECU contributes to, or perhaps on occasion provides, the reference value for lower-level control systems that perform the required action, trust may sometimes be justified in the sense that the desired effect on the environment at the higher level happens is as intended. If something had gone wrong in the trusted lower levels, the higher level perception would probably still differ from its reference value, but if something went wrong with a trusted lower level and yet the higher-level perception came to its reference value, the lower-level failure might not even be noticed.

Sometimes this form of trust is called 'fire-and-forget' mode. It is often justified at moderately high levels of the hierarchy because the lower levels on which perceptual control is based usually control very well, and will produce the desired effect on the higher-level perception. It is justified at low levels if there is little likelihood of disturbances interfering with the effect of the output on the perception, because the effect of the disturbance will be propagated by way of the lower-level perceptions. Failure of high-level control can thereby lead to replacement of 'fire and forget' by actual control against the unexpected disturbance.

If the World Model includes much uncertainty about possible outcomes or about likely ongoing disturbances, one would be more apt to control through the environment, varying output to bring the perception near its reference value and keep it there. If one asks a person to turn the light out when she leaves a room she is about to visit and she agrees, she may not do it, but if usually in the past she has reliably done what she said she would do, you probably will not get out of your chair to check whether she did so this time. That is 'trust'.

In everyday discussions of social issues, trust and tolerance both are thought to be important, an opinion that in Volume IV we will find to be supported by PCT.

I.12.6 Belief and Uncertainty

We introduced the concept of a perceptual profile in Figure I.4.5. Now we add in some lateral inhibition and imagine how such profiles might change. Figure I.12.6a shows perceptual profiles over six possible perceptions as they might be with a given set of input values (more or less equal in Profile A, and highly biassed in Profile B) but different strengths of lateral inhibition. These profiles represent the outputs of a set of lower-level perceptual functions such as are suggested as inputs in the 'labelling' polyflop diagrams (Figure I.8.5c and Figure I.8.6).

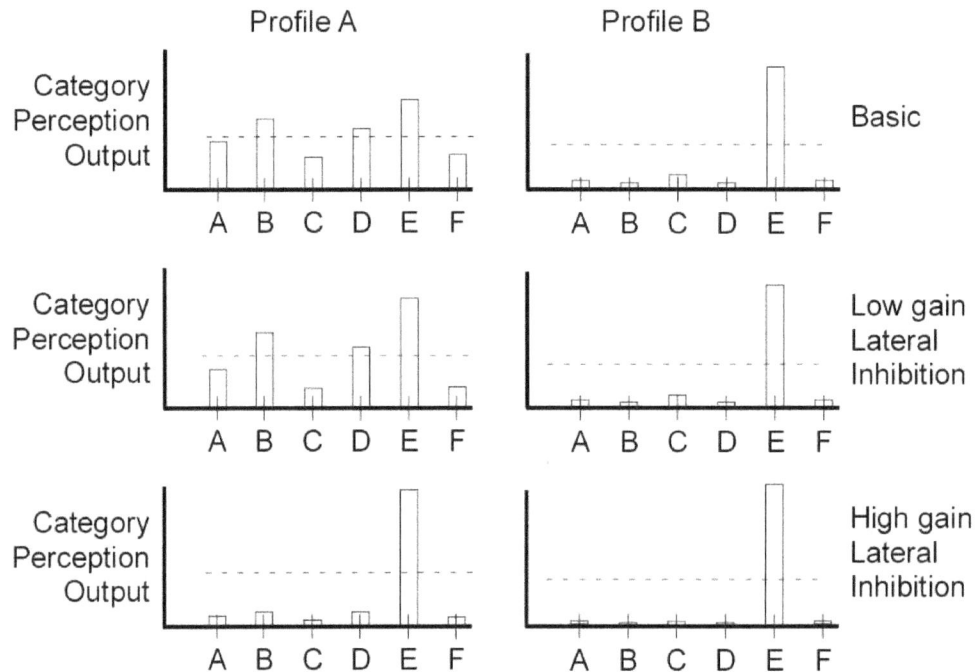

Figure I.12.6a The outputs from six "category-style" perceptual functions with different input data and different strengths of lateral inhibition. Profile A has little discrimination among the possibilities, whereas Profile B strongly agrees with item E and with none of the others. If the lateral inhibitory gain is high, only one of the items has a strong output, no matter what the input profile; a choice must be made. With low but non-zero lateral inhibitory gain, the more evenly distributed profile is sharpened, but output from all the possibilities is passed to the next level. (Dashed horizontal line is approximate average output).

Lateral inhibition operates among neighbours, whether in geometric space or in feature space. In the figure all of the perceptions labelled A, B, C, D, E, and F, are created by slightly different perceptual functions working on largely the same data inputs. If they were very different, they would not mutually inhibit one another. The question we want to answer is about the perceptual uncertainty as to which of the possibilities (A to F) represents the state of the environment, and the perceptual control implications of this uncertainty.

As we argued earlier, lateral inhibition redistributes the total output across the range of interconnected units, rather than specifically adding or subtracting from the total, though it is capable of doing either. If the lateral loop gain is low, the variation is distributed indiscriminately over the field, but if it is high, the variation consists almost entirely of the binary distinction between units that have strong output and units that have little or none, as suggested by Figure I.12.6a. The result is that even when the input variation is random 'noise', the perceptual system may well have an output profile that suggests the existence of a pattern, and this pattern is likely to serve as an input to a higher level of perceptual functions. We may hear someone talking softly when the only source of sound is the airflow through a duct; and how many mariners in the days of sail have reported seeing islands where later none was ever found (e.g., Gould 1965).

In Figure I.12.6a, items B, D, and E are all shown as having increased output in the presence of low gain lateral inhibition. Why would this be, since E would tend to suppress both B and D? The answer is that they are released from potential inhibition by A, C, and F, and if, as we argued earlier, the overall output of the set maintains the same average value, the result depends on the degree to which the reduction of suppression from A, C, and F exceeds or falls short of the suppression by inhibition from E. The same effect may occur if the input from the senses is insufficient for a clear categorization, so that all of the outputs are slightly excited, some more than others.

We all are guilty of jumping to conclusions from time to time, when mathematical analysis might suggest we should wait for more data to be sure of what we see. It is usually safer to perceive a tiger in the jungle when the senses provide a pattern of bright yellow and dark grey than it is to investigate further to be sure whether a tiger is about to pounce or the pattern is just the sun shining through foliage. Most of the time, the conclusion is more likely to be right than wrong, and our belief is then justified by success in control, but control based on such 'jumped-to' beliefs can also fail spectacularly. In Chapter III.4.3 we will start to address the interconnections between the fast control based on the hierarchy of categorical perceptions (bottom-up) and the slow control based on conscious thought. Here we deal only with the former.

The relative height of the bars in Figure I.12.6a could be seen by an analyst as representing levels of 'belief' in the items, as expressed in the influences of their outputs on higher-level perceptions. If several outputs have significant values, the input is effectively an analogue profile, many of the items being possibly what is seen, with little belief in any of them. If, however, one of the items has a high output while the others are low, either because of the pattern of input values or because of the strength of the lateral inhibition, the set is effectively categorical or digital, and the perceiver perceives only the item that has the high output.

The person in whom the set of perceptual functions in Figure I.12.6a exists may have a perception of 'belief in' a perception. What might this perception be? Where does it come from? The profiles of Figure I.12.6a are for category recognisers or other systems in which the perception is of 'what is it', but what does it mean when the question is of magnitude: "Is the sofa too big to fit in the space under the window?" "I'm not sure, let's measure it." This is the distinction between metron (what is it) and logon (how much of it is there) information or uncertainty.

From the Analyst's viewpoint, a perception is 'true' if it accurately reports a state of the outer environment *as the Analyst perceives it*. A single simple Controller, however, has no external point of view from which to determine that truth. The perception has the magnitude it has; the external world is what it is. Both are well-defined values. 'Belief' is not about either; it is about their relationship. The adage "*Measure twice, cut once*" would make no sense if one could believe that one's perception of magnitude exactly represented the real world. But how does one perceive whatever one does perceive when one says "No, it's too big" or "I think it's probably too big" or "I'm not sure"? One perceives something, but the mechanism is not immediately obvious. What is clear is that one can have a perception that is of the degree to which another perception represents the truth of the external world.

Let's think back to Oliver and his measurement of the weight of whatever is in the left pan of his scales (the 'rock pan'). When Oliver controls his perception of the direction of the scale pointer, that perception has nothing to do with his perception of the weight on the rock pan. Even when he controls his perception of the relationship between the weight in the rock pan and the weight in the scale pan by making ever smaller changes in the weight in the scale pan, that perception is only of the relationship between the two weights, and is not a perception of the weight of either.

But Oliver's perception of the weights he has placed in the pan, together with either the perception of the pointer direction or the sense of the difference between the weights in the two pans — *that* perception allows Oliver to believe that the rock weight is greater than (or less than) the weight he perceives to be in the weight pan. He can believe "The rock pan weight is greater than 11010", for example.[134] It's a perception of the relationship between another perception (the weight in the scale pan) and a state of the environment (the weight of the rock and whatever else the prankster has put in the rock pan).

How strongly Oliver believes that "The rock pan weight is greater than 11010" depends on his prior experience with *these* scales and *that* prankster, or with scales and pranksters in general. If he did a test measurement with nothing in either pan, and found that the pointer was far from vertical, he might not believe any statement about the weight very strongly. If in the past he has observed (using different sensors, of course) that the prankster sometimes put a finger on or under either pan, he might believe it even less strongly. But if his initial test showed the pointer almost vertical and he has carefully watched to be sure the prankster has not interfered with the scales, his belief might be very strong. His belief perception is *about* the truth of his perception that the weight on the scale pan is a measure of the weight on the rock pan.

---

134 11010 is a binary representation of the same number as 21+23+24 or 1+8+16 = 25
      in decimal notation.

'Belief' in the truth of a perception is the strength of a perception of the relation between a perception and its corresponding environmental state. Perceptions of perceptions are what higher-level perceptions are made of, but 'belief in X' seems to be characteristically different from 'X is a function of Y, Z, …'. For one thing, the 'belief' is *about* a perception, not a function of several. 'Belief' says things like: "I clearly see an oasis in the distance, but I don't believe it really is there".

Figure I.1.3, augmented here as Figure I.12.6b, offers another view on 'belief', as a system of possible interpretations of a given set of data. If the neural current is represented as Powers defined it, the result of averaging the firing rates over a bundle of related neural fibres, the individual fibres will not all have the same firing rates within the bundle, with zero firing rates in fibres not belonging to the bundle. Instead, some fibres will be well tuned to the incoming data, some not so well tuned, and yet others uninfluenced by the same data. Figure I.12.6b shows a one-dimensional profile of firing rates across a set of fibres sensitive and insensitive to the incoming data, including fibres whose sensitivity differs enough from the incoming data pattern that their firing rates can be treated as just noise.

Of course, since each neuron's firing is influenced by incoming excitatory and inhibitory firings at its thousands of synapses, its related neurons differ in not one dimension but in many, like the two-dimensional halos of on-centre-off surround visual regions (Hubel and Wiesel, 1962), but in more dimensions. The effect on neural currents is to segregate them by 'moats' of lateral inhibition, thereby partially justifying Powers's use of neural currents in PCT analysis, while allowing for the variation in strength of belief by way of the breadth of the firing rate profile and the distribution of firing rate peaks and valleys (as opposed to noise) associated with a particular input data set.

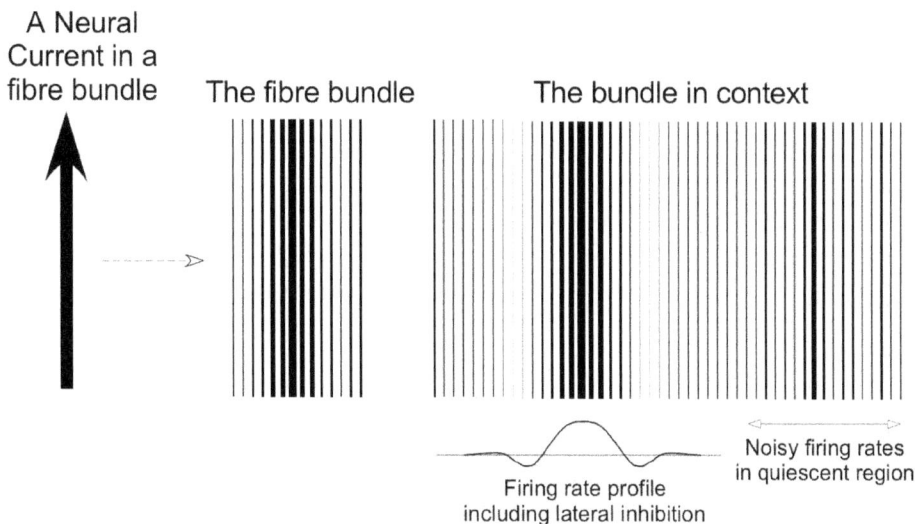

Figure I.12.6b (Figure I.1.3 augmented) Belief as a property of fibre bundles that is unavailable in a neural current. The right-hand view shows lateral inhibition of peri-neighbouring fibres by the core bundle, and the noisiness of firing in the absence of locally coherent fibre sensitivity.

If a particular incoming data pattern creates separated peaked firing bundles, subsequent processes will treat all the peaks as representing possible 'interpretations' of the data, but with different levels of belief if the interpretations conflict in other processing regions, such as action for error correction.

I.12.7 Perceiving What's Missing or Wrong

Either the polyflop mechanism or the Powers mechanism of obtaining the imaginary element from an addressed reference value in associative memory, may be the way we perceive something to be 'missing'. Figure I.12.7a shows several examples in which one might easily believe something to be 'missing'. That is to say, one perceives not the existence or magnitude of something, but the fact that the context requires the missing thing to have some magnitude — even a magnitude of zero — in order to be complete. In this section we suggest how that might happen within the PCT structure.

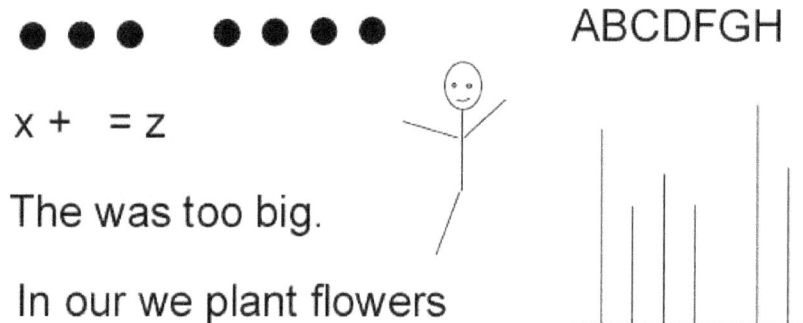

● ● ●   ● ● ● ●   ABCDFGH

x +   = z

The was too big.

In our we plant flowers

Figure I.12.7a. Eight sets of things in which something is missing. You probably can perceive exactly what is missing in each case.

In the Figure, you probably could see immediately that a dot was missing from the row, a character likely to be 'y' from the equation, a leg from the stick figure, an 'E' from the string of letters, and a line from the pattern that looks like a bar graph, and a noun from each of the two texts. In the lower text you probably perceive that noun as 'garden' or something conceptually similar, but in the other text the missing noun is impossible to perceive more precisely than as an indefinite 'thing'.

But did you see that something else is missing from the figure?

The caption to Figure I.12.7a says that the figure shows eight sets of things in which something is missing, but if you count them, there are only seven. Now you can perceive that there is an example missing, but is that perception of the same kind as the other seven? The only clue you have that an example is missing is the word 'eight' in the caption. Is that a context equivalent to the others? No, it is not. It is more akin to the context of 'surprise', in which an imagined perceptual state conflicts with the sensed value of that state. The relationship between the verbal numerosity and the visual numerosity is not that of equality, when it is expected to be. We consider 'surprise' a little later.

In six of the cases (omitting the missing example and the indefinite 'thing' that was too big), you may have perceived both the absence of the missing item and its character. You perceive that the item is missing, while at the same time you can imagine the missing dot in the row of dots, the missing 'E' in the letter string, or the missing leg on the stick figure, perhaps by the Powers 'fill-in' mechanism described above.

You might not, however, have been able to perceive all the properties of the missing item, such as the length of the missing bar in the bar graph, or whether the place we plant flowers uses the word 'garden' or 'plot' or 'yard' or 'pots' or something semantically similar. In the 'flowers' case, the conceptual and semantic space of the missing item is perceived accurately, but the plausible words that might have been missed are of very different shapes. We perceive the sense but not the shape of what is missing, which indicates that the 'fill-in' occurs at a perceptual level above the level where those property perceptions enter into the perceptual function. The missing item belongs to a category, conceptually 'the kind of place where people usually plant flowers'. If one thinks of an associative polyflop structure like that of Figure I.8.5c, the missing item would have positive feedback connections from 'plant' and 'flower' and might be perceived as a category, even though the analogue circuits did not include the appropriate data.

How can we perceive *that* something is absent? After all, at any moment almost everything that we have ever perceived over the course of our lives is at this moment out of our field of view (or sound or taste), let alone everything else that might be in the big, wide, world we could imagine to exist 'out there'. You probably do not right now perceive a fire-breathing five-toed dragon flying overhead, but you probably did not perceive it as 'missing' until you read this sentence, if then. Nor do you scatter bits of paper on the street to keep five-toed dragons away (borrow an image from an old joke).[135]

Less fancifully, with my eyes I do not see at this moment the sandy bay near which I grew up, nor do I perceive it as 'missing', as I would if I went to my old house and found that the house was now opposite a busy shipping port instead of a sandy bay. If I went to that street and found an empty lot where the house used to be, I would perceive the house to be 'missing', but at this moment I do not, despite not being able to perceive that house (other than in my imagination).

Either the Powers substitution mechanism or the polyflop structure may suggest a plausible answer. The missing value filled in from Powers's associative memory — the action context, in other words — is not available through the lower-level sensory-perceptual processes. Neither is it in the polyflop. Either way, it is produced because the context ordinarily includes it. In Section I.7.7, we said that because of the myriad individual fibres that collectively carry the perceptual signals:

---

135 The joke: A man on a bus tears little bits of paper and throws them out of the window. When asked why he does this, he answers "To keep the tigers away." When told that there are no tigers here, he says "See. It works."

In the absence of a pathological condition such as schizophrenia, if the perception derived from current sensory input is clear, fibres carrying its signal will dominate those carrying a signal derived from imagination, but if the perception is absent or unclear, the impulses from the imagination connection may substitute or support the absent or unclear current sensory input, providing an appropriate perceptual value for the next-level perceptual input functions.

Now we ask whether another consequence may be that if the signal coming from the senses differs substantially from the signal coming from the associative memory, a perception of 'wrongness' might exist. We are talking about a relationship perception, a relation between a presently perceived and a usually perceived pattern, and are not relationship perceptions controllable? If it is normally the case that we perceive the relationship comparing the value derived directly from the senses to the value obtained as input from the associative memory together with the current context, that difference is usually close to zero, its reference value.

But we do notice deviations from the normal: If we normally visit Aunt Maud for dinner on Tuesdays, but on a particular Tuesday we were unable to do so, we are likely to remember that day if we are asked about it some time later. We would not remember a normal Tuesday other than to reply that we must have gone to Aunt Maud because that's what we always do. We may well have controlled for going to Aunt Maud's even on that Tuesday. We may, however, remember that there was a Tuesday when we did not go to Aunt Maud's, since the perception of that Tuesday's events was unique and could not be subsumed into the perception of what we do on Tuesdays. At a very low perceptual level, one perceives a wide-band noise such as the hiss of escaping air as just that, a noise. But if from that sound a narrow band of frequencies is eliminated by a filter, one hears a tone at the missing frequency.

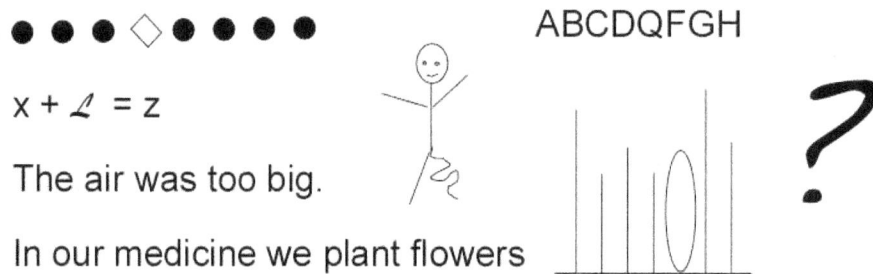

Figure I.12.7b. Surprise! The missing items from Figure I.12.7a have been filled in. Or have they?

Figure I.12.7b could illustrate 'surprise', which we discuss later, but it also illustrates the perception of 'wrongness'. It shows the same set of examples as Figure I.12.7a, but now the missing elements have been filled in. Even the eighth example is there. But do you perceive them as having been filled in? I suspect that you do not;

instead you probably perceived that in each case there is something wrong with what has been filled in. The row of dots does not need a twisted diamond in the middle (though it is possible). One does not plant flowers in medicine (though it is possible). A scrawled question mark hardly seems like an eighth example of something missing (though it is possible). And so forth.

As the case of the eighth item in Figure I.12.7a suggests, there is no perception of something missing without a context from which it is missing. A space surrounded by more space is different from the same space inside a glass bottle. The bottle is 'empty', bu0issing if the letters had been, say, MDBOESR instead of ABCDFGH. Perhaps something would have been seen as missing, however, if it had been BEDROMS, on the surface an equally 'random' ordering of the same seven letters, and something would have been seen as wrong if the result of a 'fill-in' had been BEDROVMS. Why?

The answer seems to be the same as is suggested for the detection and identification of missing items. In this case, the senses provide one value, the associative memory or the polyflop provide another, creating a relationship perception that deviates clearly from zero. If the relationship perception between current input and associative memory is controlled, a non-zero relationship should be expected to lead to action, which might be in imagination (as when one seeks the letters to fill in a crossword, possibly changing ones already filled in if they are perceived to be wrong), or might be in the real world (as when one seeks a pencil to make a note). We may perceive the 'wrongly' filled-in item as a 'surprise'.

I.12.8 Surprise and Belief Change

The same item may appear in many contexts, so that there may be several relationship perceptions with different values of the difference between sensory and associative input to a range of perceptual functions. These differences may be what is perceived as a degree of uncertainty or of belief. If we accept that input to a perceptual function ordinarily comes from both the imagination connection through the World Model and from direct sensory input, then we have a PCT explanation of 'surprise'. Surprise may be the perception we have when prior uncertainty was low, and the sense-based perception had low prior probability. "I was quite sure you would say yes. Why did you say no?"

We are surprised when we perceive something we imagined we would not perceive. What we imagine we will perceive must come from the World Model using our current outputs with our current perceptual and reference values — the 'Imagination Connection' of Figure I.7.3a or Figure I.7.3b. We are most surprised when our belief in the truth of the imagined perception is strongest. Friston's Predictive Coding approach treats surprise a little differently. There, a surprise is something that occurs when the chosen actions produce a result other than what was expected, rather than a perception having a statistically unlikely value.

What we perceive through the senses is likely to override anything imagined if the sensory input is clear enough. The imagination connection from output to perception is, however, quicker than is a connection between output and perception by way of the effects of action on the environment. So long as our ongoing sensory perception agrees with the earlier-arriving imagination-based version reasonably well, there is no surprise, but whenever the value of the disturbance changes abruptly, the two sources will conflict, causing a transient, a relatively rapid change, in the value of the perception. Soon we will identify such unexpected perceptual changes as sources of 'rattling', a measure described in Chapter II.5 of Volume II.

Whereas a controlled perception that suffers the transient merely acts to correct any resulting error, a 'surprise' transient alters the belief perception so that there is a shift in the location of the peak of the belief profile. It changes the World Model. Whether or not such a categorical change is controlled against depends on the corresponding reference profile. The belief may not even be a controlled perception. For example, we may have been watching a figure approach in a hazy distance and been perceiving it to be a friend, but when the person comes close enough, we are surprised to see that it is a stranger. We do not then act to turn this person into the friend we had imagined to be there; we just accept that we had imagined incorrectly, and incorporate the stranger-not-friend into our revised World Model of *the way the world is*.

There is no surprise if we continue to believe the same as before, whether with less or more certainty. We do consciously feel surprised if we now suddenly believe something we did not, and disbelieve what we had previously believed. The lateral inhibition of the flip-flop that is assumed to be responsible for categorical and sharpened perception allows both for a gradual shift of belief from one to the other at low lateral loop gain (which does not eliminate surprise when the balance shifts) and for an almost instantaneous shift (a bolt of insight, or even, in religious terms, a conversion) if the lateral loop gain is higher (Figure I.12.8).

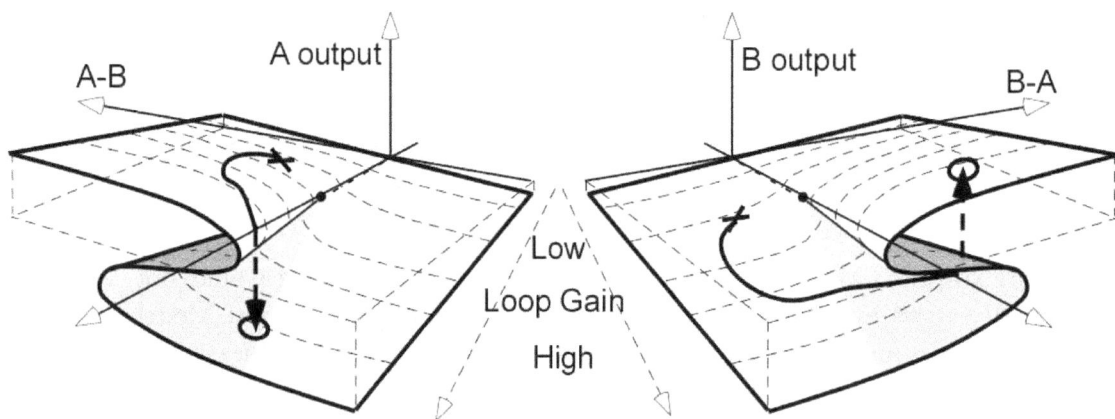

Figure I.12.8 Slowly changing data can lead to an abrupt change of perception. In this example, not only does the data change from its initial value at "X" but also the loop gain increases. At first the data increase the A-likeness of the unbiased perception, but with time, B-preferring data accumulate.

As Figure I.12.8 suggests by the dashed vertical arrows, a sudden insight could be the result of a slowly changing balance of data as the sensed data diverges from the data initially imagined through the World Model. Some religious conversions, perhaps most, occur after a period of doubt based on a succession of events or periods of contemplation. To the Analyst, such a slow divergence completed by a sudden transition suggests that the World Model had been inaccurate. The Analyst might think that the situation will result in a reorganisation that would correct the World Model, but is there any reason to think that the control system itself would reorganise as a consequence of the insight? Perhaps there is. Certainly when we start to use the rattling measure and a principle we call 'low rattling' (Chvykov et al. 2021) in Volume II, we will definitely come to expect reorganisation.

If, as was proposed in Section I.7.7, the World Model operations related to planning consist of the currently reorganised structure of the control hierarchy, that structure is the means through which all the perceptions are controlled, whether the perceptual values are created entirely from imagined input, entirely from sensory input, or from a mix of both. If control is good, then (according to Powers) reorganisation is slow, but also if control is good, the imagined data will correspond closely to the sensed data apart from the effects of ongoing changes in the disturbance. Good control systems keep their perceptual values close to changing reference values except in the moments following disturbance transients, so if the World Model is good, slowly increasing deviations between imagined and sensed data should not happen.

The 'surprise' shift of category belief, consequent on a disturbance transient that moves the flip-flop (or polyflop) to a new state, changes the current perceptual value. If the value of this perception is controlled, the system will act to eliminate the induced error, and it might succeed. If it cannot succeed with the resources available, reorganisation will speed up, and a new structure will emerge in which either the reference profile for this changed belief is altered to match the perceptual profile, or the structure can alter lower-level references in such a way that the belief switches back to its original state. Either way, the World Model changes, and if the reorganisation succeeds, the imagined perceptual values will again track the values based on sensory input. Further surprises will be avoided.

I.12.9 Shall I Compare Thee to a Summer's Day?

> *Shall I compare thee to a summer's day?*
>
> (Shakespear, Sonnet 18)

So the lovelorn swain seems to ask the object of his affections. But does he really? Is he not simply musing and asking himself the question, perhaps to judge whether to use the line when he sees her? It seems a far-fetched comparison to make, until one reads a little further into the sonnet, and remembers that this sonnet was written in England, where a summer's day may not be as predictable as are summer days in some parts of the world. He is *thinking* about it.

> *Thou art more lovely and more temperate.*
> *Rough winds do shake the darling buds of May,*
> *And summer's lease hath all too short a date.*

She is past the tempestuous teens, and into (perhaps late in) a too short-lived period of youthful beauty coupled, perhaps, with temperamental maturity. Is he evaluating for himself whether to make a play for her, whom he may know as a friend but not as a lover?

> *Sometime too hot the eye of heaven shines,*
> *And often is his gold complexion dimmed;*
> *And every fair from fair sometime declines,*
> *By chance, or nature's changing course, untrimmed;*

Ah, but she may yet sometimes get angry and lash out, so perhaps she would not be a good candidate for a long-term relationship?

> *But thy eternal summer shall not fade,*
> *Nor lose possession of that fair thou ow'st,*
> *Nor shall death brag thou wand'rest in his shade,*
> *When in eternal lines to Time thou grow'st.*
> > *So long as men can breathe, or eyes can see,*
> > *So long lives this, and this gives life to thee.*

Yes, she has qualities that outlast superficial beauty or occasional fits of anger. He will after all try to see whether she might agree to be his life-long partner.

*** *** ***

Were you surprised to see a Shakespear sonnet appear, seemingly out of nowhere? Why am I quoting and musing about this particular sonnet? In the sonnet, what is the questioner doing? The questioner and I are both thinking about a problem, but why? And what do I mean by 'thinking', after so many chapters in which the issue was never raised? Were you *conscious* of surprise or of thinking about these things? When are we conscious, and is 'thinking' necessarily conscious?

'Why' is a question that has been at the heart of all the preceding discussion of PCT, starting with some of the first words of Chapter I.1: "I want to visit Aunt Maude, but I am at home, two blocks away." Figure I.1.1 features the word 'Why' prominently at a much lower level of the perceptual hierarchy, wanting to hear the doorbell ringing and wanting the doorbell button to be pushed. We do not normally consider any of these actions to demand thought, or perhaps even for the actions to be performed consciously. We have developed a control hierarchy that 'just works'.

On the other hand, we can become conscious of them if we want. One task of a teacher of a skill such as, say, golf or piano playing is to get the student to be conscious of the muscular feelings and how they work together to produce the desired result. It is not easy for either the student or the teacher, as a rule. To experience consciously the perceptions at the lower-levels of the hierarchy is a learned skill. The higher in the hierarchy, the easier it becomes to experience them consciously, but we certainly do not consciously perceive all that is in our sensory input, even at fairly high levels.

That we do not is well attested by the so-called 'Invisible Gorilla' phenomenon in which six people pass a basketball among themselves and the viewer is asked to count the passes.[136] During this action, a person in a gorilla suit walks in from one side of the frame, faces the viewer for a moment, and strolls off the other side. Very few people report having seen the gorilla, despite it having been at least as easily seen as any of the people. Being 'conscious of' is not the same as being 'able to be conscious of'. Once the gorilla has been pointed out in the video, it becomes hard to ignore on a second viewing.

Indeed, we are typically so conscious of higher-level perceptions and not of lower-level ones that many people do not think about low-level ones until something goes wrong and gives them consciously perceived problems, such as pain. Most seem to assume that conscious perceptions are all that there can be. It is not far-fetched to imagine that Powers might even have come to his theory of hierarchical perceptual control from a position in which he had imagined his consciously experienced perceptions to be the perceptions to be controlled. Later, at least in the informal medium of the PCT mailing list called CSGnet, he took the position that we could consciously perceive only 'perceptions' in the form of the perceptual signals whose values are the controlled variables of the control hierarchy.

Consciousness has been a puzzle for millennia of philosophers, and more lately, for centuries of scientists. We make no attempt to solve it here, though Friston's 'Free Energy' formulation of Predictive Control (Friston 2010) uses it. Instead, the question asked is how consciousness and its cousin, thinking, relate to perceptual control. In so doing, we come to a view of the relationship between conscious perceptual experience and the controlled perceptions of the

---

136 For example https://www.youtube.com/watch?v=z9aUseqgCiY retrieved 2018.04.03.

reorganised hierarchy that we propose in Chapter II.8. The perception which we have a problem controlling, or which we may want to control in the future, is in consciousness, but with it are perceptions of the actions which might be used to control it. According to the normal view of PCT, the actions which result from the outputs of control units are simply not usually perceived, but when they are demanded by conscious control of perceptions (control other than in the non-conscious hierarchy), they are perceived.

Conscious experience and controllable perception are as different as the price you pay for gas when you fill the car and the gas that you pay for at that price. A conscious experience is rich, a unitary experience of many properties of many things at once; a controllable perception is a value of one of those properties, which may itself be a complex function of the vector of values of several simpler properties. The experience is not the controlled perception, but it might consist of the values of properties that form the vector — or it might not. Using Korzybski's phrase "The map is not the territory"; the vector values are the map, but the experience is the territory. The conscious experience is 'the big picture'; a controllable property is a mere detail.

To use another metaphor, a controllable perception is the magnitude of the red, green, and blue signals that define a pixel on your TV screen or smartphone, whereas the conscious experience that we call perception in everyday language is the picture on the screen. The picture on the screen depends on the red, green, and blue values of the pixels, and those pixel values depend on what picture is to be displayed. The picture and its elementary perceptions are inextricably linked in much the same way as a perceptual signal is linked to the corresponding CEV in the environment.

Can one non-consciously generate a metaphor? Introspection cannot provide evidence, but Powers's theory says that we can. When we asked a few pages ago about perceiving something that is missing or wrong, the properties that are not missing may be sufficient to excite two quite different perceptual functions that are structurally the same in those properties, but quite different in others. (We will discuss this further in Chapter II.6 where we introduce the 'crumpling' metaphor for category perception.) Both may be perceived in the hierarchy, without being consciously experienced. Each is a metaphor for the other, in the same way that a label can stand for the constellation of properties indicated by the label (Section I.9.6). Such a constellation of properties may be an abstract or concrete 'object' (Section I.11.3).

When we discuss some problem, we may say "Let's look at the Big Picture," using the metaphor of conscious experience and thinking as a simulacrum of the environment. In the 'Big Picture' the hierarchy of perceptual signals forms the pixels, and the other side of the control hierarchy is the ways we can influence them. The 'Big Picture' contains the environmental context of whatever we more narrowly want to do.

The 'Big Picture' is uncommitted, in the sense that there are no prespecified patterns in it. One sees (experiences) patterns and structures in it, some of which presumably are perceptions already wired into the control hierarchy by evolution and reorganisation because they have proved useful, but others are built by 'thinking', conscious testing of the effects of treating quasi-random patterns as though they meant something useful. Perhaps if *this* pattern is true of the world, then if we change *that*, *the other* will happen. And we may try changing *that*, observing whether *the other* does actually happen.

The implication of this suggestion is that consciousness can not only produce from the perceptions in the perceptual hierarchy a 'Big Picture' that acts in the way that the organic fascia or sheath acts as a tension component in the tensegrity structure of the biological body, it can also act on the output side of the hierarchy to complete the control tensegrity picture of the way the control hierarchy works. It allows what we often call 'problem solving', creating structured perceptions that do not (yet) exist in the control hierarchy, but that might be controllable and might be useful if they are controllable.

The world in which all this trial and re-think occurs may correspond perceptually to the real world, with the perceptions and actions forming the normal kind of negative feedback loop, but it might equally well be in imagination, using the current perceptions of *the way the world is* as variable values and using *the way the world works* to provide the necessary environmental feedback paths for conscious control by Predictive Coding. Neither of these may accurately model what would happen if a successful trial in imagination were to be attempted using the relevant parts of the control hierarchy and the external environment, but the problem being solved by conscious thought may not need them to. The problem may require only the examination of possibilities, not truths.

Some people (myself included) think preferentially in pictures, whereas others deny the possibility of thinking other than in language. The 'fascia' metaphor allows for both, since a tensegrity fascia surface fixed at discrete points to the 'rods' of the structure contains linear regions of high tension and areas of low tension. Only if the surface is 'ballooned' by internal pressure will the tension in all directions be uniform, and as yet we have no control analogue for volumetric pressure such as may exist within a cell of an organism.

The high-tension lines act as 'wires' forming a network, and where these 'wires' cross, their interactions are likely to induce stiffness (Section I.8.1) in the structure. But where are the 'fixed points' linked by the high-tension lines? Without them, the tensions in the fascia collapse. In our tensegrity picture of the hierarchy, the tensions are the 'pulls' of the perceptions towards their reference levels and the fixed points are the reference values set from a higher level control unit or limitations defined by the environment. But now we are using a metaphor of a less structured 'Big Picture' whose 'pixels' are the values of perceptions at all levels of the control hierarchy. Can we assert that there are equivalents of structured reference value sets toward which structured perceptual values are drawn in order to set those levels? Perhaps we can.

Reference values are what you want to perceive the world to be like, and problems addressed by conscious thinking are created by situations in which the world does not look and act as you would wish. You are already controlling your perceptions through the control hierarchy, and many, probably most, are currently close to their reference levels and are staying within their tolerance limits by the actions of the control hierarchy. These may be the fixed points for the control of the 'Big Picture' and its elements.

The problems are with the controlled perceptions that are outside their tolerance bounds, or with uncontrolled perceptions that contribute to their perceptual function. At any one moment, your actions, performed by changing your muscular tensions, are preferentially influencing only a small proportion of all the perceptions you might be controlling. Of the others, most do not change often and are within tolerance bounds, such as the positions of all the books on your bookshelf, or the furniture in your abode. Others change slowly and can be allowed to change for a while before you act to bring them back within their tolerance bounds. These cause no problems, and in an evolutionary sense there is no need to do anything or change anything related to them.

Some perceptions you actively control may on occasion leave their tolerance bounds, some because of their slow movement, such as changes in your blood chemistry since you last ate, some because of changing reference values, such as you now wanting to read a book that has been stably positioned on your bookshelf, some because of external disturbances, such as your front doorbell ringing. These all create a particular class of problem, which is that in order to use your muscles to control them you must stop using your muscles to control something else you are actively controlling. You have a resource limitation conflict that usually cannot be solved by simultaneously controlling both.

The departure of a perception which is currently not actively being controlled from its reference value or its expected value (surprise) creates what has been called an 'alert' (e.g. Taylor 1963b; Cunningham and Taylor 1994). An 'alert' has the effect of bringing the perception in question into consciousness. The problem is to select which, if any, of the perceptions you are actively controlling to put on the back burner in order to free resources to deal with the one that caused the alert.

Another class of problem that engages consciousness manifests itself when resources are missing that are ordinarily used to control some perceptions which are already organised into the control hierarchy. If you are nailing planks, picking up a hammer that is at hand is seldom a conscious action, but if the hammer is not where your hand has moved to pick it up, then its location becomes a problem to be solved by a 'Search'.

The main problem is not the hammer location perception. It is that the next nail is not where you want it to be, probably with its head flush with the surface of the plank. Without the hammer, you cannot control that perception of the location of the nail, so the nail problem becomes a part of conscious perception, as does the failure to perceive the current location of the hammer. A belief in the form of a non-conscious perception amenable to control that it is where it should be has changed into an uncertainty, creating the effect we have called 'surprise'. Surprise is a component of conscious experience. Whether it is a component of anything in the control hierarchy is dubious, since the control hierarchy simply acts to bring its perceptions closer to their reference values, no matter how deviant they may have suddenly become. Surprise suggests that the control problem may be better solved by conscious thinking.

A problem that engages consciousness and results in observable actions in the external environment is a dual of 'Search'. Search asks *"Where is X?"*, whereas when you are exploring you are asking *"What is in that place?"* The actions involved with either may require ceasing to control some other perceptions actively. Searching occurs because there exists a perception that is not being well controlled but that could be, whereas exploring in the external environment is more likely to occur when the organism would otherwise be relaxing. These differences are caused by the limited resources of muscle tensioning at the interface to the real world, both in the number of muscles and in the physical speed limitations caused by the forces needed to move masses.

The physical limitations of force and the movement of mass do not apply to manipulations of World Models in imagination. In principle, there is no limit to how many different perceptions can be controlled in imagination, and yet we may say that we can think of only one thing at a time, whereas through the environment we can simultaneously control several. We can search in imagination, trying to remember where we put the hammer so that it can be located in the World Model and then in the external environment. But it is hard to search simultaneously for several things at once that might be in quite different places.

Why should this be? Why does the swain in the sonnet have to think through a sequential list of possibilities in his metaphor rather than see the solution immediately, having followed all the metaphoric associations simultaneously?

I do not propose an answer to this question, but I have a suggestion, which is that everything in the world model of how the world works must be emulated. Emulating the way the world works, even for one environmental feedback path for a scalar variable, implies the use of processing power on the same order as that used in the perceptual processing and output processing parts of the hierarchy. Thinking about the Big Picture involves commensurately more, and every process involves the production of extra heat that must be dissipated. Heat dissipation is in any case a big, perhaps limiting, problem for a human brain, so evolutionarily it makes some sense that most thought processes tend to run sequentially rather than in parallel.

I.12.10 Planning and Performance

Let's use a different metaphor, a metaphor of networks. The evolved and reorganised perceptual hierarchy consists of a set of perceptual functions that are connected by directed links to comparators and then to output functions that implement actions that preferentially influence the perceptions produced by the perceptual functions. It is a very sparse network, in the sense that of all the places where connections from any point might be made to any other point in the hierarchical structure, only a very small proportion are actually used. Much of the early development of a baby's brain consists of eliminating connections, preferentially leaving and strengthening the useful ones.

On top of and mixed in with this sparse network, let us imagine another network, a randomly connected network of connections not pruned away, but potentially linking any perceptual signal to any other perceptual signal or to any output function. Maybe it consists of remnants of the baby's randomly connected dense network, or maybe it is a tentative network that is continually built and rebuilt, the way that Powers proposed for the development of the hierarchy by reorganisation, or maybe it is a bit of each, continually being built and destroyed, but made of links seldom strong enough to be considered permanent.

That 'seldom' occurs when a link allows some perception to be controlled in a new way that benefits the intrinsic variables, among which we include Quality of Control. New links that improve the Quality of Control of some perception are likely to be re-used and strengthened by Hebbian processes (Chapter I.9), which implement the 'continuing in the same direction' aspect of e-coli reorganisation. The 'randomly connected network' is the source of new perceptual functions and new possibilities for actions to control perceptions, old and new. Only the reality of the environment affects the probabilities for these new possibilities to "Live long and prosper."

Making a leap of faith, I propose that this 'network of possibilities' is where the consciously experienced 'Big Picture' has its 'pixels'. Conscious experience is the set of perceptions evoked in this as yet uncommitted set of links by the perceptual signals from the already pruned and strengthened control hierarchy, and 'thinking' is the making and remaking of links between those perceptions and the action possibilities already embodied in the perceptual control hierarchy, with results that are not directly realised in action, except in imagination. The 'external environment' of perceptual control in thinking is imagination — in models of worlds that may or may not act like the real world and may or may not contain states that match those of the real world.

When are we conscious of something? When there is something unexpected or uncertain about it. Perhaps we want to look at it more closely, or our attention has been drawn to it. A perception that we normally control non-consciously is unexpectedly not well controlled, as, for example if we are walking along not conscious of our leg movements and the feelings in the foot, but stumble over a sidewalk slab slightly raised above its neighbour. We then consciously feel the sensations in the toe, the muscle tensions changing unexpectedly in the leg, and perhaps a shock at the hip, among other low-level perceptions of which we probably could be conscious when all is going normally, but we are not.

We are conscious of perceptions that we have difficulty controlling. We may think about what might be the problem and about ways to solve it, like the person whose thoughts are examined in the Shakespear sonnet. He is uncertain whether he wants the woman to love him and whether he should act to see if she might. He is conscious of many of her properties and also of another concept (a summer's day) against which to compare some of them, knowing how the passing moods of a summer's day do not last, and likewise neither do the moods of the object of his passion.

Thinking about problems and how to solve them is an aspect of 'planning'. Planning is done in imagination, using the World Model and possible variations of the World Model. When we discussed planning in Section I.7.7, examples included World Models in which gravity was very much lower than on the surface of the Earth. We can imagine what might happen if we did X and the world responded differently than we expect it to, or if Y happened to be true. That is sometimes called 'risk management.'

Risk management happens in imagination, but when we have a plan that we want to execute, consciousness must be able to set reference values in various parts of the hierarchy at all levels, perhaps overriding reference values set level by level as perceptions are controlled. The plan must be able to stop control of some perceptions and start control of others, just as must happen when an 'alerting' situation exists. The Powers 'Imagination Loop (Figure I.7.3b) does not allow for this to happen. It is concerned only with perceiving, not with action.

The 'imagination loop' connection is part of the network that implements the 'Big Picture'. If we stick to the 'neural current' view of signals, then links that set the switches must be part of a reorganised control structure that sets the imagination loop into action. On the other hand, if we dissociate the neural currents into their component fibres with their individual firings, something else can happen. Some branchings of the nerve axon may signal the value of the imagination variable, while others serve to inhibit the active loop at the lower level.

The same may be true of executing the actions involved in an active plan; the reference value from the level above may be inhibited at the same time as the reference value from the plan is substituted. Just as a conscious perception, a 'pixel' in the 'Big Picture' can come from any part of the control hierarchy, so we might expect the action reference value from the plan to be substituted at any level of the hierarchy from which a perceptual signal can be made conscious.

In training a skill, the perception-comparison-action triad of a control unit may be emulated in the 'consciousness network', and eventually be incorporated into a growing control hierarchy by the HaH process as control through it becomes successful. It is tempting to see the same kind of process at work in the Method of Levels, which works on the assumption that conscious attention to a poorly functioning or conflicted part of the hierarchy might enhance the likelihood of reorganisation in that region.

We will discuss the interactive relationship between conscious processes and the non-conscious reorganised perceptual control hierarchy in Chapter II.10, where the concepts of conscious narrative and perceptual events are explored. To preview that chapter's main point, there is a one-to-one relationship between a perceptual event (a change in some perception) and a narrative element (a conscious, and possibly linguistic, equivalent) corresponding to the perceptual event. Both narrative elements and perceptual events follow the perceptual control hierarchy to the extent that they might almost be called synonyms of translation between the two different domains of description, the perceptual control hierarchy and a conscious narrative hierarchy.

Glossary

Some Specialised Words Used in PCT

The chapter numbers refer to where in this four Volume book they are best, not necessarily first, described.

Atenex ('ATomic Environmental NEXus', Chapter I.5) refers to an object which is used actually or potentially in service of more than one perception. Almost all, perhaps all, objects are atenexes for some perception someone is capable of wanting to control. An atenex comprises both the external object property and the skill to use it effectively. (See also **Molenex** and **Environmental Feedback Path**.)

Atenfel ('ATomic ENvironmental FEedback Link', Chapter I.2.4) refers to a link in the feedback connection of a control loop. It includes the ability to use it in controlling a particular perception. An atenfel is sometimes confused with a Gibsonian 'affordance', something the external environment offers to all comers if they want to use it, but an atenfel differs because it exists only while it is used in control of a specific perception by a control structure that incorporates the specific ability to use the atenfel in control of that particular perception. (See also **Molenfel**, **Environmental Feedback Path**, and Section I.2.4.)

Collective Control (Chapter III.1) occurs when a collection of individuals control their own individual perceptions and the effects of controlling those perceptions happen to be correlated in some dimensions they use with some of the same environmental variables in their atenfels. The effective result is the apparent existence of a controlled variable in the environment that corresponds to some virtual perception controlled by none of them to a reference value that is an averaging of the reference values of the individually controlled perceptions with a loop gain that is the sum of the projected loop gains of the individual controllers on the virtual perception.

A **Commons** (Chapter III.8) as used in this book is an extension of the millennia old idea of the village commons, a grazing and recreational resource available for use by the villagers. We extend this concept into more abstract structural areas, such as a commons of ideas or of collectively controlled stable states available to a community. Types of commons are distinguished and related to collective control.

Crumpling (Chapter II.6) is a metaphor that helps one to understand the refinement and consciously experienced discrimination of perceptual categories as a living entity matures. A 'crumpling event' can send a shock wave through a Perceptual Control structure, rattling it and perhaps leading to substantial reorganisation into a more coherent structure.

Disturbance is a name for any influence external to a control loop that in the absence of the completed loop would influence the perception. Perceptual control counters the influence of the disturbance on the perception.

Elementary Control Unit (ECU) is a name for the minimal triad of components needed for each individual control loop, at a level of perceptual complexity. It consists of (1) a perceptual input function that produces a scalar value (the controlled perception), (2) a comparator that produces some function of the difference between the perceptual value and a reference value (often a simple difference between them), and (3) an output function that provides a value that contributes to the reference values of lower-level control loops. Every control loop consists of an ECU and an Environmental Feedback Loop, which connects the output of the ECU back to the perception, usually by way of intervening layers of control loops and perceptual functions.

Environmental Feedback Function is a term in the mathematical specification of a computational model or simulation of behaviour. Unlike the environmental feedback path, it is not an object or relationship observable in the environment, it is a mathematical function expressing the net effect of links in the environmental feedback path.

Environmental Feedback Path is a name for the path by which the effects of the output of an ECU become one of the inputs to the perceptual input function of the ECU. The Environmental Feedback Path includes a junction point where the effects of the ECU output are combined with the effects of external (external to the control loop) influences collectively called 'the disturbance'. The job of control is to change the effect of the disturbance on the perception. The environmental feedback path comprises objects and relationships in the environment, called links, through which control outputs are transformed to effects upon that which is perceived as controlled input to the loop.

Giant Virtual Controller (GVC) (Chapter III.1) is a title for an array of functionally independent individual control loops that act together in any ordinary test, such as 'The Test for the Controlled Variable', as though only one variable was being controlled, even though none of the component control loops was controlling that exact variable.

HaH (Hebbian-anti-Hebbian) learning (Chapter I.9) occurs at the level of individual neurons connected through synapses. If the neuron sending the signal to the synapse fires very shortly before the receiving neuron does, the synapse strengthens, increasing the connectivity between the two neurons, whereas if the timing difference is opposite, the receiving neuron firing first, the synapse weakens, reducing their connectivity. This kind of synaptic interaction between neurons is found in many places in the brain and elsewhere.

Molenex ('MOLecular ENvironmental NEXus', on the analogy of a molecule to an atom, refers to objects which are together used actually or potentially as more than one atenex. (See also **Atenex** and **Environmental Feedback Path**.)

Molenfel ('MOLecular ENvironmental FEedback Link', Chapter I.5), on the analogy of a molecule to an atom, refers to a collection of atenfels that together form an effective link in a control loop where none of the component atenfels would serve the same function alone. A simple example is pen and paper, neither of which alone will allow one to control a perception of seeing a diagram of a circle, but in the hands of a skilled artist the two together enable the perception of that circle diagram being produced. (See also **Atenfel** and **Environmental Feedback Path**.)

A **Motif** (Chapters I.2.7 and I.8) of Perceptual Control, as I use the term in this book, is a pattern or grouping of basic perceptual control loops that is found to recur and that has some emergent property not produced by other structural arrangements of control loops. The emergent property may have its own name. The earliest and simplest I am aware of is 'conflict', an emergent property that appears when two control loops try to bring the same environmental property to different values. Motifs are to control loops as molecules are to atoms. They can be as simple as the 'conflict' motif or as complex as, for example, 'trade' which includes at least two communication motifs that each contain 19 control loops. The trade motif is like a complex chemical molecule, which has several smaller structures attached in novel ways to each other that determine the function of the molecule.

A **Narrative Fragment** (Chapter II.10) is the consciously experienced counterpart of a non-conscious perceptual event, a perceptual change in one or more instances of one or more perceptual categories. This conscious to non-conscious relationship may lead to the formation of new perceptual categories or reorganisation of the pre-existing non-conscious perceptual control structure.

Perception has a special meaning derived from the everyday idea of perception. It is a scalar value that is maintained by the control loop near some, perhaps dynamically varying, reference value.

A **Protocol** (Chapter II.14) in communication is a motif that involves at least two separate partner control structures that communicate with each other, at least some of which is conscious. The motif has 19 potentially used control loops, but in most cases very few of these are actually used. It is not necessary that both partners be aware that a communication protocol is being performed, and deceitful communication may rely on one partner not being aware of it. Protocols exist in a structure of levels, more complex protocols using simpler ones in the same way that more complex perceptual categories are built on a foundation of simpler ones.

Rattling: In the introductory paper by Chvykov et al., the individual entities were mindless physical structures called 'smarticles' whose flailing arms might hit each other (hence the name 'rattling') and affect their movements in a bounded space small enough that not all of them could entirely escape all the impacts from the arms of other smarticles. We generalise the flailing arms of 'smarticles' to the side-effects of the actions of different control systems within or over collections of individual living entities. In physical demonstrations with small numbers of smarticles, Chvykov et al. showed that the total rattling over the entire group was more likely to decrease over time as a statistical tendency than to increase or stay constant.

In simulations that used more simulated smarticles than they could use in a physical demonstration, they showed that the low-rattling minimal organisational structure contained an approximately exponential distribution of rattling across individual smarticles, most being little rattled in this structure, but a few who became highly rattled were incorporated in a long tail of the distribution. I use the tendency toward rattling reduction in collectives both when considering the 'reorganisation' of the perceptual control hierarchy within individual living beings and in analysing social processes, including revolutions and the temporary stability of autocratic polities. Rattling in this context is directly analogous to annealing of metals.

A **Role** (Chapter III.5) is analogous to a role in a stage play. A person may play many different roles, which define the details of the protocols they will use while playing that Role. A person may play the role of, say, dentist with another playing the role of patient, and on another occasion the two persons may play different roles in their interactions as members of a hockey team or a club committee. The protocols available to these different roles are quite different in detail, though identical in structure. At any point, the users of a role-pair protocol may choose to switch from role-playing to acting as persons. For example, a shopper may play a purchaser role with a person playing a cashier role, and if those persons are friends, the shopper-person may ask after the cashier-person's mother, which is definitely not part of the cashier-purchaser role-play.

Roles define a community, in that the available role-pairs for communication differ from community to community. The role of religious teacher differs among religions, being quite different for an Islamic Imam and a Buddhist monk or a Catholic priest. An isolated village of yurts in Mongolia has Roles unknown to a person who may at some time play the Role of clerk in a downtown office, and vice-versa.

References

Addams, R. 1834. "An Account of a Peculiar Optical Phenomenon Seen after Having Looked at a Moving Body." *London and Edinburgh Philosophical Magazine and Journal of Science* 5:373–374.

Ashby, W.R. 1952, 1960. *Design for a Brain: The Origin of Adaptive Behavior.* 2nd ed. London, Chapman & Hall. https://ia802605.us.archive.org/34/items/designforbrainor00ashb/designforbrainor00ashb.pdf (Retrieved 2021.09.22).

Bagno, S. 1955. "The Communication Theory Model and Economics." *IRE Convention Record Part 4, Computers, Information Theory, Automatic Control.* URL: https://www.iapct.org/publications/other/the-communication-theory-model-and-economics/.

Bar-Gad, I., and Bergman, H. 2001. "Stepping Out of the Box: Information Processing in the Neural Networks of the Basal Ganglia." *Current Opinion in Neurobiology* 11, 689-695.

Bateson, Gregory (1972). *Steps to an Ecology of Mind.* Scranton, PA: Chandler Publishing Company.

Bell, C.C., Caputi, A., Grant, K., and Serrier, J. 1993. "Storage of a Sensory Pattern by Anti-Hebbian Synaptic Plasticity in an Electric Fish." *Proceedings of the National Academy of Sciences* 90:4650-4654.

Bell, C. C., Han, V. Z., Sugawara, Y., and Grant, K. 1997. "Synaptic Plasticity in a Cerebellum-like Structure Depends on Temporal Order. *Nature* 387:278-281.

Boltzmann, L. 1877. Über die Beziehung zwischen dem zweiten Hauptsatze des mechanischen Wärmetheorie und der Wahrscheinlichkeitsrechnung, respective den Sätzen über das Wärmegleichgewicht. *Sitzungberichte der Kaiserlichen Akademie der Wissenschaften. Mathematisch-Naturwissen Classe.* Abt. II, LXXVI 1877, 373-435 (Wien. Ber. 1877, 76:373-435). [Repr. *Wiss. Abhandlungen*, Vol. II, reprint 42, 164-223, Barth, Leipzig, 1909.] [Tr. by Kim Sharp & Franz Matschinsky 2015, "On the Relationship between the Second Fundamental Theorem of the Mechanical Theory of Heat and Probability Calculations Regarding the Conditions for Thermal Equilibrium," *Entropy* 17.4:1971-2009; DOI: 10.3390/e17041971.]

Brady, T.F., & Oliva, A. 2012. "Spatial Frequency Integration During Active Perception: Perceptual Hysteresis when an Object Recedes." *Frontiers in Psychology* 3:462.

Burns, R. 1785. "To a Mouse, On Turning Her Up in Her Nest with the Plough," November, 1785. Retrieved from Wikipedia 2020.01.24.

Carey, T.A. 2006. *The Method of Levels*, Menlo Park, CA: Living Control Systems Publishing.

Carey, T.A. 2008. "Conflict, as the Achilles Heel of Perceptual Control, Offers a Unifying Approach to the Formulation of Psychological Problems." *Counselling Psychology Review* 23:5-16.

Carey, T. A., W. Mansell, W., and S. J. Tai. 2015. *Principles-Based Counselling and Psychotherapy: A Method of Levels Approach.* London: Routledge.

Carlson, A. 1990. "Anti-Hebbian Learning in a Non-linear Neural Network." *Biological Cybernetics* 64.2:171-176.

Carroll, L. 1871. *Through the Looking Glass, and What Alice Found There.* London: MacMillan.

Chvykov, P, Berrueta, T.A., Vardham, A., et al. 2021. "Low Rattling: A Predictive Principle for Self-organization in Active Collectives." *Science* 371:90-95. DOI:10.1126/science.abc6182.

Curry, Andrew (2018). "Early Mongolians Ate Dairy, but Lacked the Gene to Digest It: Ancient DNA and Protein Help Solve Evolutionary Riddle." *Science* 362.6415:626-627.

Cossell, L., Iacaruso, M.F., Muir, D.R., Houlton, R., et al. 2015. "Functional Organization of Excitatory Synaptic Strength in Primary Visual Cortex." *Nature* 318:399-403.

Craik, K.H. 1970. "Environmental Psychology." In *New Directions in Psychology, 4*, Holt Rhinehart and Winston.

Cunningham, W.B. and Taylor, M.M., 1994, "Information for Battle Command." *Military Review* 74/11:81-84.

Darwin, C. 1899. *The Expression of Emotions in Man and Animals.* New York: D. Appleton & Co. URL: https://www.gutenberg.org/files/1227/1227-h/1227-h.htm.

Dedekind, R. 1901. "Continuity and Irrational Numbers." In *Essays on the Theory of Numbers.* Translated by W.W. Berman. Chicago, The Open Court Publishing Company. (Retrieved from http://www.gutenberg.org/ebooks/21016, December 24, 2014).

Donders, F. 1862. "Die Schnelligkeit Psychischer Processe." *Archiv Anatomische Physiologie und wissenschaftliche Medizin,* Berlin, 657-681. Translated by W. G. Koster 1969. "On the Speed of Mental Processes," *Acta Psychologica,* 30:412–431.

Eliot, T. S. 1963. *Collected Poems, 1909-1962.* New York: Harcourt Brace.

Engel, F.L. and Haakma, R. 1993. "Expectations and Feedback in User-System Communication." *International Journal of Human-Computer Studies* 39.3:427-452.

Falconbridge, M.S., Stamps, R.L., & Badcock, D.R. 2006. "A Simple Hebbian/anti-Hebbian Network Learns the Sparse, Independent Components of Natural Images." *Neural Computation* 18.2:415-429.

Földiak, P. 1990. "Forming Sparse Representations by Local Anti-Hebbian Learning." *Biological Cybernetics* 64:165-170.

Fourier, J. 1822. *Théorie de la Chaleur.* Paris, Firmin Didot. [Tr. Alexander Freeman, 1878, *The Analytical Theory of Heat*, Cambridge: U. Cambridge Press.]

Friston, K. 2010. "The Free-energy Principle: A Unified Brain Theory?." *Nature Reviews: Neuroscience* 11:127-138.

Gabor, D. (1946). Theory of communication. *Journal of the Institution of Electrical Engineers - Part I: General*, 94:58-58.

Garner, W. R. 1954. Context effects and the validity of loudness scales. *Journal of Experimental Psychology*, 48.3:218–224. DOI: 10.1037/h0061514

Garner, W.R. 1962. *Uncertainty and Structure as Psychological Concepts*. New York; Wiley.

Garner, W.R., and D. E. Clement 1963. "Goodness of Pattern and Pattern Uncertainty." *Journal of Verbal Learning & Verbal Behavior* 2.5-6:446–452.

Garner, W.R., and McGill, W.J. 1956. "The Relation Between Information and Variance Analyses." *Psychometrika* 21:219-228.

Gibbs, J.W. 1902. *Elementary Principles in Statistical Mechanics*. New York: Charles Scribner's Sons.

Gibson, J.J. 1966. *The Senses Considered as Perceptual Systems*. London, Allen and Unwin.

Girolami, M., and Fyfe, C. 1997. "An Extended Exploratory Projection Pursuit Network with Linear and Nonlinear Anti-Hebbian Lateral Connections Applied to the Cocktail Party Problem." *Neural Networks* 10:1607-1619.

Gould, R.T. 1965. *Oddities*. New York: University Books.

Gouyet, Jean-François 1996. *Physics and Fractal Structures*. Paris/New York: Masson Springer.

Habets, B., Bruns, P, and Röder, B. 2017. "Experience with Crossmodal Statistics Reduces the Sensitivity for Audio-visual Temporal Asynchrony." *Scientific Reports* 7.1:1486. DOI: 10.1038/s41598-017-01252-y.

Handel, S., & W. R. Garner, 1966. "The Structure of Visual Pattern Associates and Pattern Goodness." *Perception & Psychophysics* 1:33-38.

Hardin, G. 1968. "The Tragedy of the Commons." *Science* 162:1243-1248.

Harris, Z. S. (1955). From phoneme to morpheme. *Language*, 31:190–222. DOI: 10.2307/411036

Hebb, D.C. 1949. *The Organization of Behavior: A Neuropsychological Approach*. New York: Wiley.

Heimann, M. (ed.). 2003. Regression periods in human infancy. Norwood, NJ: Ablex Publishing Corporation.

Herculano-Houzel S. 2010. "Coordinated Scaling of Cortical and Cerebellar Numbers of Neurons." *Frontiers in Neuroanatomy* 4.12. DOI: 10.3389/fnana.2010.00012.

Hock, H.S., Kelso, J.S., and Schöner, G. 1993. "Bistability and Hysteresis in the Organization of Apparent Motion Patterns." *Journal of Experimental Psychology: Human Perception and Performance* 19.1:63-80.

Hogben, L. 1964. *The Mother Tongue*. London: Secker and Warburg.

Hubel, D.H., and Wiesel, T.N. 1962. "Receptive Fields, Binocular Interaction and Functional Architecture in the Cat's Visual Cortex." *The Journal of Physiology* 160.1:106–154.2

Hyvärinen, A., and Oja, E. 1998. "Independent Component Analysis by General Nonlinear Hebbian-like Learning Rules." *Signal Processing* 64.3:301-313.

Kauffman S. 1995. *At Home in the Universe: The Search for the Laws of Self-Organization and Complexity*. [Paperback repr. 1996.] New York & Oxford: Oxford University Press.

Kennaway, R. 2008. "Appendix.," In W. T. Powers, *Living Control Systems III: The Fact of Control*, 175-193. Bloomfield NJ.: Benchmark Publications.

Koch, G., Ponzo, V., Di Lorenzo, F., Caltagirone, C., and Veniero, D. 2013. "Hebbian and anti-Hebbian Spike-Timing-Dependent Plasticity of Human Cortico-Cortical Connections." *The Journal of Neuroscience* 33:9725-9733.

Kolmogorov, A.N. 1965. "Three Approaches to the Quantitative Definition of Information." *Problems of Information Transmission* 1.1:1–7.

Korzybski, A. 1933. *Science and Sanity. An Introduction to Non-Aristotelian Systems and General Semantics*. The International Non-Aristotelian Library Pub. Co.

Levin, Stephen 2015. "16. Tensegrity, The New Biomechanics." In Hutson, Michael; Ward, Adam (eds.). *Oxford Textbook of Musculoskeletal Medicine*. Oxford University Press, 155–56, 158–60.

Lieb, E.H. and Yngvason, J. 2000. "A Fresh Look at Entropy and the Second Law of Thermodynamics." *Physics Today* 53:32-37. DOI: 10.1063/1.883034

Mackay, Donald M. 1950. "XXIV. Quantal Aspects of Scientific Information." *The London, Edinburgh, and Dublin Philosophical Magazine and Journal of Science* 41.314:289-313.

Mackay, Donald M. 1953. "Quantal Aspects of Scientific Information." *Transactions of the IRE Professional Group on Information Theory*, 1:60-80.

Mackay, Donald M. (1969). Information, Mechanism and Meaning. Cambridge, MA: MIT Press. DOI: 10.7551/mitpress/3769.001.0001

Mansell, W. ed. 2020. *The Interdisciplinary Handbook of Perceptual Control Theory: Living Control Systems IV*. London: Academic Press.

Mansell, W, Carey, T.A., and Tai, S.J., 2013. *A Transdiagnostic Approach to CBT Using Method of Levels Therapy*. New York: Routledge.

Marken, R.S. 1986. "Perceptual Organization of Behavior: A Hierarchical Control Model of Coordinated Action." *Journal of Experimental Psychology: Human Perception and Performance* 12:267-276. [Repr. in Marken (1995).]

Marken, R.S. 1999. "PERCOLATe: Perceptual Control Analysis of Tasks." *International Journal of Human-Computer Studies* 50:481-487.

Marken, R.S. 2002. *More Mind Readings: Methods and Models in the Study of Purpose*. St. Louis, MO: newview.

Marken, R.S. 2014. *Doing Research on Purpose*, St. Louis, MO: New View (report available from http://www.mindreadings.com/BehavioralIllusion.pdf)

Marken, R.S., and Powers, W. T. 1989a. "Levels of Intention in Behavior." In *Volitional Action: Conation and Control*, ed. by W. A. Hershberger, 409-430. Amsterdam: Elsevier Science Publishers.

Marken, R.S., and Powers, W. T. 1989b. "Random-Walk Chemotaxis: Trial and error as a Control Process." *Behavioral Neuroscience* 103:1348-1355.

Marken, R.S., Mansell, W., and Khatib, Z. 2013. "Motor Control as the Control of Perception." *Perceptual and Motor Skills* 117:236-247. DOI: 10.2466/24.24.PMS.11715z2

Markram, H., Lübke, J., Frotscher, M., & Sakmann, B. 1997. "Regulation of Synaptic Efficacy by Coincidence of Postsynaptic APs and EPSPs." *Science* 275.5297:213-211.

Martinez-Condé, S., Macknik, S.L., and Hubel, D.H. 2004. "The Role of Fixational Eye-Movements in Visual Perception." *Nature Reviews Neuroscience* 5:229-240.

McClelland, K. 1993. *Conflictive Control*. Presentation to the Control Systems Group meeting, Durango, CO, USA. Video: https://www.youtube.com/watch?v=lrHgrMNMqYM.

McClelland, K. 1994. "Perceptual Control and Social Power." *Sociological Perspectives* 37:461-496.

McCelland, K. 2020. "Social Structure and Control: Perceptual Control Theory and the Science of Sociology." In *The Interdisciplinary Handbook of Perceptual Control Theory: Living Control Systems IV*, ed. by Warren Mansell, 229-297. London: Academic Press.

McCollough, C. 1965. "Color Adaptation of Edge Detectors in the Human Visual System." *Science* 149 (3688):1115-1116. DOI: 10.1126/science.149.3688.1115

McCollough, C. 2000. Do McCollough effects provide evidence for global pattern processing?. *Perception & Psychophysics* 62:350–362. DOI: 10.3758/BF03205555

McCollough Howard, C. and Webster M. 2011. "McCollough Effect." *Scholarpedia* 6.2:8175. DOI: 10.4249/scholarpedia.8175.

Miller, G.A. 1956. "The Magic Number Seven Plus or Minus Two: Some Limits on Our Capacity for Processing Information." *Psychological Review* 63:91-97.

Miller, G. A., E. Galanter, and K. H. Pribram 1960. *Plans and the Structure of Behavior*. New York: Henry Holt.

Nevin, B. 2020. "Language and Thought as Control of Perception." In *The Interdisciplinary Handbook of Perceptual Control Theory: Living Control Systems IV*, ed. by Warren Mansell, 351-459. London: Academic Press.

Ohala, J.J. 1992. "What's Cognitive, What's Not, in Sound Change." In *Diachrony within Synchrony: Language History and Cognition*, ed. by G. Kellerman and M. D. Morrisey. Frankfurt am Main, Peter Lang Verlag.

Pigeau, R.P., Naitoh, P., Buguet, A., McCann, C., Baranski, J., Taylor, M., Thomson, M., and Mack, I. 1995. "Modafinil, D-amphetamine and Placebo During 64 Hours of Sustained Mental Work. I. Effects on Mood, Fatigue, Cognitive Performance and Body Temperature." *Journal of Sleep Research* 4:212-228.

Plooij, F.X. & Rijt-Plooij, H.H.C. van de. 1990. "Developmental Transitions as Successive Reorganizations of a Control Hierarchy." In R.S. Marken (ed.), *Purposeful Behavior: The Control Theory Approach* [Special Issue]. *American Behavioral Scientist*, 34:67-80.

Plumbley, M.D. 1993a. Efficient information transfer and anti-Hebbian neural networks. *Neural Networks* 6:823-833.

Plumbley, M.D. 1993b. "A Hebbian/anti-Hebbian Network which Optimizes Information Capacity by Orthonormalizing the Principal Subspace." In *1993 Third International Conference on Artificial Neural Networks* (86-90). Brighton, UK. IET.

Popa, L.S., Hewitt, A.L., and Ebner, T.J. 2014. "The Cerebellum for Jocks and Nerds Alike." *Frontiers in Neuroscience* 8, Article 113: 1-13.

Powers, W.T. 1973/2005. *Behavior: The Control of Perception*. New York: Aldine. [Rev. ed. New Canaan, NJ.: Benchmark Publications.]

Powers, W. T. 1978. "Quantitative Analysis of Purposive Systems: Some Spadework at the Foundations of Scientific Psychology." *Psychological Review* 85.5:417-435. [Repr. Powers (1989:129-235).]

Powers, W. T. 1979a. "The Nature of Robots Part 1: Defining Behavior." *Byte* 4.6:1-9.

Powers, W.T. 1979b. "The Nature of Robots Part 2: Simulated Control System." *Byte* 4.7:1-14.

Powers, W.T. 1979c; "The Nature of Robots Part 3: A Closer Look at Human Behavior." *Byte* 4.8:1-16.

Powers, W.T. 1979d. "The Nature of Robots Part 4: Looking for Controlled Variables." *Byte* 4.9:1-15.

Powers, W.T. 1989. *Living Control Systems: Selected Papers of William. T. Powers*. Gravel Switch, KY: The Control Systems Group.

Powers, W.T. 1992. *Living Control Systems II: Selected Papers of William. T. Powers*, Gravel Switch, KY: The Control Systems Group.

Powers, W.T. 1994. "An 'Artificial Cerebellum' Adaptive Stabilization of a Control System." In *Perceptual Control Theory: Proceedings of the 1st European Workshop on Perceptual Control Theory Gregynog, The University of Wales 23rd - 27th June 1994*, ed. by M. A. Rodrigues and M. H. Lee, 41-49. Aberystwyth: U. of Wales.

Powers, W.T. 2008. *Living Control Systems III: The Fact of Control.* Bloomfield NJ: Benchmark Publications.

Powers, W.T., Clark, R.K., and McFarland, R.I. 1960a. "A General Feedback Theory of Human Behavior: Part I." *Perceptual and Motor Skills* 11:71-88.

Powers, W.T., Clark, R.K., and McFarland, R.L. 1960b. "A General Feedback Theory of Human Behavior: Part II," *Perceptual and Motor Skills* 11:309-323.

Powers, W.T., McFarland, R.L., and Clark, R.K. 1957. "A General Feedback Theory of Human Behavior: A Prospectus." *American Psychologist* 12:462.

Prigogine, I., Nicolis, G., and Babloyantz, A. 1972a. "The Thermodynamics of Evolution." *Physics Today* 25.11:23–28. DOI: 10.1063/1.3071090.

Prigogine, I., Nicolis, G., and Babloyantz, A. 1972b. "The Thermodynamics of Evolution." *Physics Today* 25.12:38–44. DOI: 10.1063/1.3071140.

Pulfrich, C. 1922. "Die Stereoskopie im Dienste der isochromen und heterochromen Photometrie." *Naturwissenschaften* 10.35:751-761.

Röder et al. (2016) Röder, P.V., Wu, B., Liu, Y., and Han, W. "Pancreatic Regulation of Glucose Homeostasis." *Experimental & Molecular Medicine.* 48.3:e219. DOI: 10.1038/emm.2016.6. PMID: 26964835; PMCID: PMC4892884.

Rausch, R. 1981. "Lateralization of Temporal Lobe Dysfunction and Verbal Encoding." *Brain and Language*, 12.1:92-100.

Rijt-Plooij, H.H.C. van de, and Plooij, F. 1992/2019. *The Wonder Weeks*, New York: W.W. Norton. [Eng. tr. by S. Sonderegger and G. Kidder of *Oei, ik groei!* 2012, Ede and Antwerp: Zomer & Keuning Boeken BV.]

Roberts, P.D., and Bell, C.C. 2002. "Spike Timing Dependent Synaptic Plasticity in Biological Systems." *Biological Cybernetics* 87:392-403.

Roberts, P.D., and Leen, T.K. 2010. "Anti-Hebbian Spike-Timing-Dependent Plasticity and Adaptive Sensory Processing." *Frontiers in Computational Neuroscience* 4:156.

Runkel, P. 2003. *People as Living Things: The Psychology of Perceptual Control.* Menlo Park, CA: Living Control Systems Publishing.

Scarr, G. 2014. *Biotensegrity: The Structural Basis of Life*, Pencaitland, UK: Handspring Publishing.

Schouten, J.F, and Bekker, J.A.M. 1967. "Reaction Time and Accuracy." *Acta Psychologica* 27:143-153.

Selfridge, O.G. 1959. "Pandemonium: A Paradigm for Learning." In *Proceedings of the Symposium on Mechanisation of Thought Processes*, ed. by D. V. Blake and A. M. Uttley, 511–529. London: HM Stationary Office.

Seth A.K., and Friston K.J. 2016. "Active Interoceptive Inference and the Emotional Brain." *Philosophical Transactions of the Royal Society B* 371:20160007. DOI: 10.1098/rstb.2016.0007.

Shannon, C.E. and W. Weaver, W. 1949. *The Mathematical Theory of Communication*, Urbana and Chicago: University of Illinois Press.

Snelson, Kenneth and Heartney, Eleanor. (n.d.) *Kenneth Snelson, Art and Ideas.* [Essay by Eleanor Heartney Additional text by Kenneth Snelson. (Kendall School of Art and Design of Ferris State University. Kenneth Snelson in Association with Marlborough Gallery, NY, NY.] URL: https://kcad.ferris.edu/uploads/docs/readings_tensegrity.pdf

Standing, L.G., Dodwell, P.C., and Lang, D. 1968. "Dark Adaptation and the Pulfrich Effect." *Perception & Psychophysics* 4:118-120.

Stevens, S. S. 1957. "On the Psychophysical Law." *Psychological Review* 64.3:153–181.

Stevens, S.S. 1966. "Matching Functions Between Loudness and Ten Other Continua." *Perception & Psychophysics* 1.1:5-8.

Stroop, J.R. 1935. "Studies of Interference in Serial Verbal Reactions." *Journal of Experimental Psychology* 18:643–662.

Sun Tzu. ca. 600 B.C.E. *The Art of War*. Translated by L. Giles, London: Luzac, 1910.

Swanson, RL (2013). "Biotensegrity: a unifying theory of biological architecture with applications to osteopathic practice, education, and research-a review and analysis." *The Journal of the American Osteopathic Association*. 113.1:34–52. DOI:10.7556/jaoa.2013.113.1.34. PMID 23329804

Swift, J. 1726. *Travels into Several Remote Nations of the World. In Four Parts. By Lemuel Gulliver, First a Surgeon, and then a Captain of Several Ships*, London: Benjamin Motte.

Taylor, I.K. 1963. "Phonetic Symbolism Re-examined." *Psychological Bulletin* 60:200-209.

Taylor, I.K. and Taylor, M.M. 1962. "Phonetic Symbolism in Four Unrelated Languages." *Canadian Journal of Psychology* 16:344-356.

Taylor, I.K. and Taylor M.M. 1983. *The Psychology of Reading*. New York, Academic Press.

Taylor, J.G. 1963. "A Behavioural Interpretation of Obsessive-Compulsive Neurosis." *Behaviour Research and Therapy* 1.2-4:237-244.

Taylor, J.G. 1966. "Perception Generated by Training Echolocation." *Canadian Journal of Psychology* 20.1:65-81.

Taylor, M.M. 1962a. "Figural After-Effects: A Psychophysical Theory of the Displacement Effect." *Canadian Journal of Psychology* 16.3:247-277.

Taylor, M.M. 1962b. "Geometry of a Visual Illusion." *Journal of the Optical Society of America* 52.5:565-569.

Taylor, M.M. 1963. "Tracking the Decay of the After-effect of Seen Rotary Movement." *Perceptual and Motor Skills* 16:119-129.

Taylor, M.M. 1965. "Non-additivity of Perceived Distance with the Mueller-Lyer Figure." *Perceptual and Motor Skills* 20:1064.

Taylor, M.M. 1966. "Adaptation and Repulsion in the Figural After-effect" and the Psychophysical Theory." *Quarterly Journal of Experimental Psychology* 18.2:175-7. DOI: 10.1080/14640746608400025

Taylor, M.M. 1973a. "The Problem of Stimulus Structure in the Behavioural Theory of Perception." *South African Journal of Psychology* 3:23-45. URL: https://www.iapct.org/publications/other/the-problem-of-stimulus-structure-in-the-behavioural-theory-of-perception/

Taylor, M.M. 1973b. "Principal Components Display of ERTS Imagery." *Proceedings of the Third ERTS Symposium*, Washington, 1973, 1877-1898.

Taylor, M. M. 1984 "The Bilateral Cooperative Model of Reading: A Human Paradigm for Artificial Intelligence." *Proceedings of the International NATO Symposium on Artificial and Human Intelligence*, ed. by A. Elithorn and R. Banerji, 239-250. New York: Elsiever North-Holland.

Taylor, M.M. 1988a. "Layered Protocols for Computer-Human Dialogue. I: Principles." *International Journal of Man-Machine Studies* 28:175-219.

Taylor, M.M. 1988b. "Layered Protocols for Computer-Human Dialogue. II: Some Practical Issues." *International Journal of Man-Machine Studies* 28:219-257.

Taylor, M.M. 1989. "Response Timing in Layered Protocols: A Cybernetic View of Natural Dialogue." In *The Structure of Multimodal Dialogue*, ed. by M. M. Taylor, F. Neel, and D. G. Bouwhuis, Chapter 19. Amsterdam: Elsevier Science Publishers.

Taylor, M.M. 1992. "Principles for Intelligent Human-Computer Interaction: A Tutorial on Layered Protocol Theory." *DCIEM Report 93-32*, Dept. Nat. Defence Canada, Paris
September 7 1992. Presentation material. URL: https://www.iapct.org/publications/other/principles-for-intelligent-human-computer-interaction/.

Taylor, M.M. 1993. "Perceptual Control and Human Data Fusion." *Presentation to Joint Directors of Laboratories Data Fusion Seminar* (June 14 1993). URL: https://www.iapct.org/publications/other/perceptual-control-and-human-data-fusion/.

Taylor, M.M. 1995. "Fitting Data to Models." Presentation to the 1995 Annual Meeting of the Control Systems Group. URL: https://www.iapct.org/publications/other/fitting-data-to-models/.

Taylor, M.M. 1996. "Effects of Modafinil and Amphetamine on Tracking Performance During Sleep Deprivation." *Proceedings of the 37th Annual Conference of the Military Testing Association,* Canadian Forces Personnel Applied Research Unit, Toronto.

Taylor, M. M. 2020. "Perceptual Control in Cooperative Action." In *The Interdisciplinary Handbook of Perceptual Control Theory: Living Control Systems IV*, ed. by Warren Mansell, 299-249. London: Academic Press.

Taylor, M.M., and Aldridge, K.D. 1974. "Stochastic Processes in Reversing Figure Perception." *Perception & Psychophysics* 16:9-27.

Taylor, M.M., Farrell, P.S.E., and Hollands, J.G. 1999b. "Perceptual Control and Layered Protocols in Interface Design: II. The General Protocol Grammar." *International Journal of Human-Computer Studies* 50:521-555.

Taylor, M.M., Forbes, S.M., and Creelman, C.D. 1983. "PEST Reduces Bias in Forced Choice Psychophysics." *Journal of the Acoustical Society of America* 74:1367-1374.

Taylor, M.M., and Henning, G.B. 1963. "Transformations of Perception with Prolonged Observation." *Canadian Journal of Psychology* 17.4:349-360.

Taylor, M.M., Lederman, S.J., and Gibson, R.H. 1973. "Tactual Perception of Texture." In *Handbook of Perception III: Biology of Perceptual Systems*, ed. by E.C. Carterette and M.P. Friedman, 251-272. New York: Academic Press.

Taylor, M.M., Lindsay, P.H., and Forbes, S.M. 1967. "Quantification of Shared Capacity Processing in Auditory and Visual Discrimination." *Acta Psychologica* 27:223-229. [Volume title: *Attention and Performance*.]

Taylor, M.M. and Taylor, I.K. 1965. "Another Look at Phonetic Symbolism." *Psychological Bulletin* 64:413-427.

Taylor, M.M. and Taylor, I.K. 1995/2014. *Writing and Literacy in Chinese, Korean and Japanese*. Amsterdam: John Benjamins.

Taylor, M.M., and Waugh, D.A. 1992. "Principles for Integrating Voice I/O in a Complex Interface." In *Advanced Aircraft Interfaces: The Machine Side of the Man-Machine Interface.* AGARD Conference Proceedings 521, Madrid.

Taylor, M.M., and Waugh, D. 2000. "Dialogue Analysis Using Layered Protocols." In *Abduction, Belief and Context in Dialogue*, ed. by H. Bunt, and W. Black. Amsterdam: John Benjamin.

Tigaret, C.M., Olivio, V., Sadowski, J.H.L.P., Ashby, M.C., and Mellor, J.R. 2016. "Coordinated Activation of Distinct Ca2+ Sources and Metabotropic Glutamate Receptors Encodes Hebbian Synaptic Plasticity." *Nature Communications* 7:10289. DOI:10.1038/ncomms10289.

Tribus, M.; McIrvine, E. C. (1971). "Energy and information." *Scientific American*. 224:178–184. URL: https://www.jstor.org/stable/10.2307/24923125

Tuchman, B.W. 1984. *The March of Folly: From Troy to Vietnam*, New York: Random House.

Turchin, P., Brennan, R., Currie, T.E., Feeney, K.C., Francois, P., Hoyer, D., ... and Peregrine, P. 2015. "Seshat: The Global History Databank." *Cliodynamics: The Journal of Quantitative History and Cultural Evolution* 6:77-107. URL: https://escholarship.org/uc/item/9qx38718.

Turchin, P and fifty others. 2018a. "An Introduction to Seshat: Global History Databank." *Journal of Cognitive Historiography* 5.1-2. DOI: 10.1558/jch.39395.

Turchin, P., Currie, T.E., Whitehouse, H., François, P., Feeney, K., Mullins, D., … and Mendel-Gleason, G. 2018b. "Quantitative Historical Analysis Uncovers a Single Dimension of Complexity That Structures Global Variation in Human Social Organization." *Proceedings of the National Academy of Sciences* 115.2:E144-E151.

Turnbull, C.M. 1961. "Some Observations Regarding the Experiences and Behavior of the BaMbuti Pygmies." *American Journal of Psychology* 74:304-308.

Tzounopoulos, T., Rubio, M.E., Keen, J.E., & Trussell, L.C. 2007. "Coactivation of Pre-and Postsynaptic Signaling Mechanisms Determines Cell-Specific Spike-Timing-Dependent Plasticity." *Neuron* 54:291-301.

Tzounopoulos, T., and Kraus, N. 2009. "Learning to Encode Timing: Mechanisms of Plasticity in the Auditory Brainstem." *Neuron* 62:463-469.

Vaidiyanathan, G. 2021. "The World's Species are Playing Musical Chairs: How Will It End?" *Nature.* 596.7870:22-25. DOI: 10.1038/d41586-021-02088-3. PMID: 34349287.

van Schaik, C.P., Damerius, L., and Isler, K. 2013. "Wild Orangutan Males Plan and Communicate Their Travel Direction One Day in Advance." *PLoS ONE* 8.9:e74896. DOI: 10.1371/journal.pone.0074896.

Velasco, C., Woods, A.T., Deroy, O., and Spence, C. 2015. "Hedonic Mediation of the Crossmodal Correspondence Between Taste and Shape." *Food Quality and Preference* 21:151-18. DOI: 10.1016/j.foodqual.2014.11.010.

Velasco, C., Woods, A.T, Pe, O., Cheok, A.D., and Spence, C. 2016. "Crossmodal Correspondences Between Taste and Shape, and Their Implications for Product Packaging: A Review." *Food Quality and Preference* 52:17-26. DOI: 10.1016/j.foodqual.2016.03.005.

Vibrans, F. C. 1931. "Anti-oxidants in Edible Oil Preservation." *Journal of the American Oil Chemists' Society* 8: 223-227. DOI: 10.1007/BF02574658. [Presented at Fall Meeting, American Oil Chemists' Society, Chicago, November 14, 1930.].

Viviani, P. and N. Stucchi. 1992. "Motor-Perceptual Interactions." In *Tutorials in Motor Behavior II*, ed. by G. E. Stelmach & J. Requin, 229–248. Amsterdam: North-Holland.

Volta, Alessandro.1800. "On the Electricity Excited by the Mere Contact of Conducting Substances of Different Kinds." *Philosophical Transactions of the Royal Society of London* [in French] 90:403–431. DOI: 10.1098/rstl.1800.0018.

von Karman, T. 1931. "Mechanical Similitude and Turbulence." *National Advisory Committee for Aeronautics, Tech. Memorandum 611*. URL: https://ntrs.nasa.gov/api/citations/19930094805/downloads/19930094805.pdf. [Tr. of Karman, Th. 1930, Mechanische Ahnlichkeit und Turblenz, *Nachrichten von der Gesellschaft der Wissenschaften zu Gottingen, Fachgruppe I (Mathematik)* 5:58-76.]

von Neumann, J., and Morgenstern, C. 1947. *Theory of Games and Economic Behavior*, Princeton: Princeton University Press.

Vuong, H.E. & Hsiao, E.Y. 2017. "Emerging Roles for the Gut Microbiome in Autism Spectrum Disorder." *Biological Psychiatry* 81:411–423.

Wagemans, J., Elder, J.H., Kubovy, M., Palmer, S.E., Peterson, M.A., Singh, M., and von der Heydt, R. 2012a. "A Century of Gestalt Psychology in Visual Perception I. Perceptual Grouping and Figure-Ground Organization." *Psychological Bulletin* 138:1172–1217. DOI: 10.1037/a0029333.

Wagemans, J., Feldman, J., Gepshtein, S., Kimchi, R., Pomerantz, J.R., van der Helm, P.A., and van Leeuwen, C. 2012b. "A Century of Gestalt Psychology in Visual Perception II: Conceptual and Theoretical Foundations." *Psychological Bulletin* 138:1172–1217. DOI: 10.1037/a0029333.

Warren, R. M. 1970. "Perceptual Restoration of Missing Speech Sounds," *Science* 167:392-393.

Watanabe, S. 1969. *Knowing and Guessing: A Quantitative Study of Inference and Information.* New York: Wiley.

Watkins, C. 2000. *The American Heritage Dictionary of Indo-European Roots,* 2nd ed. Boston and New York: Houghton Mifflin Harcourt.

Watson, S.K., Townsend, S.W., Schel, A.M., [& five others]. 2015. "Vocal Learning in the Functionally Referential Food Grunts of Chimpanzees." *Current Biology* 25:495–499, DOI: 10.1016/j.cub.2015.12.032.

Weber, E.H. 1834/1996. *De tactu. Annotationes anatomicae et physiologicae.* Leipzig: Koehler. Translated in Ross, H.E. and Murray, D.J. (eds.) *E.H. Weber on the Tactile Senses,* 2nd edition. Hove: Erlbaum (UK) Taylor & Francis, 1996.

Wiener, N. 1948/1961. *Cybernetics: Or Control and Communication in the Animal and the Machine.* Paris: Hermann & Cie. [2nd rev. ed. 1961 Cambridge, Mass: MIT Press.]

Wiener, N. 1950. *The Human Use of Human Beings.* Cambridge, MA: The Riverside Press.

Williams, W.D. 1989. "The Giffen effect: A Control Theory Resolution of an Economic Paradox." *Advances in Psychology* 62:531-547.

Williams, W.D. 1990. "The Giffen Effect: A Note on Economic Purposes." *American Behavioral Scientist* 34.1:106-109.

Wise, J. A. 1988. "Control is Beautiful: Measuring Facility Performance as if People (and Buildings) Really Mattered." *ASTM Symposium on Overall Facility Performance: Use, Operation and Cost,* October 13-14, 1988; also in STP 1029, *Performance of Buildings and Serviceability of Buildings* [STP=ASTM Special Technical Publication], ed. by G. Davis and F. Ventre.   Philadelphia. PA: ASTM, 121-143.

Whorf, B.L. 1944. "The Relation of Habitual Thought and Behavior to Language." *ETC: A Review of General Semantics* 1.4:197-211. URL: https://www.time-binding.org/resources/documents/1-4-whorf.pdf

Whorf, B.L. 1950. "An American Indian Model of the Universe." *International Journal of American Linguistics* 16: 67-72. URL: https://www.time-binding.org/resources/documents/1-4-whorf.pdf

Xu Jiang et al. 2020. "Brain Control of Humoral Immune Responses Amenable to Behavioural Modulation." *Nature* 581:204-208. DOI: 101038/s41586-020-2235-7.

Yelen, D.R. 1980. "A Catastrophe Model for the Effects of a Response Set on a Discrimination Task." *Perception & Psychophysics* 28:177-178.

Youn, H., Bettencourt, L.M.A., Strumsky, D., and Lobo, J. 2014. "Invention as a Combinatorial Process: Evidence from US Patents." *Santa Fe Working Paper 2014-06-020.* DOI: 10.1098/rsif.2015.0272.

Yuen, D.M.C. 2008. "Deciphering Sun Tzu." *Comparative Strategy* 27.2:183-200. DOI: 10.1080/01495930801944727.

Yuen, D. M. C. 2014. *Deciphering Sun Tzu: How to Read 'The Art of* War'. Oxford: Oxford University Press.

Zadeh, L.A. 1965. "Fuzzy Sets." *Information and Control* 8:338-353. DOI: 10.1016/S0019-9958(65)90241-X.

Index

www.ingramcontent.com/pod-product-compliance
Lightning Source LLC
Chambersburg PA
CBHW061739210326
41599CB00034B/6730